International Conference on the Physics of Low Temperature Plasma (PLTP-2017)

Journal of Physics: Conference Series Volume 927

Kazan, Russia
5 – 9 June 2017

ISBN: 978-1-5108-5311-9
ISSN: 1742-6588

Printed from e-media with permission by:

Curran Associates, Inc.
57 Morehouse Lane
Red Hook, NY 12571

proceedings.com

Some format issues inherent in the e-media version may also appear in this print version.

Printed by Curran Associates, Inc. (2018)

For permission requests, please contact the Institute of Physics
at the address below.

Institute of Physics
Dirac House, Temple Back
Bristol BS1 6BE UK

Phone: 44 1 17 929 7481
Fax: 44 1 17 920 0979

techtracking@iop.org

Additional copies of this publication are available from:

Curran Associates, Inc.
57 Morehouse Lane
Red Hook, NY 12571 USA
Phone: 845-758-0400
Fax: 845-758-2633
Email: curran@proceedings.com
Web: www.proceedings.com

Table of contents

Volume 927

The International Conference "The Physics of Low Temperature Plasma" (PLTP-2017)

5–9 June 2017, Kazan University, Kazan, Russian Federation

Accepted papers received: 6 November 2017
Published online: 23 November 2017

Preface

Papers

The All-Russian (with international participation) conference "Physics of low-temperature plasma" (PLTP-2017) was held in 5-9 of June in the capital of the Republic of Tatarstan in Kazan on the basis of the Kazan Federal University.

The conference was the 16th in a row and continued the glorious tradition of holding such forums started more than half a century ago back in 1966. For a long time the conference was held on the basis of Petrozavodsk University, and since 2014 the tradition to hold this large-scale event has moved to Kazan. The organizing committee of the conference was presented by leading world scientists and specialists in the field of physics and applications of low-temperature plasma.

Registered participants, and there were more than 500 of them, represented institutes of RAS, research institutes, Russian and foreign universities and companies producing advanced plasma equipment that is used in many industries.

The conference discussed issues related to fundamental research in the physics of low-temperature plasma, as well as a wide range of its application. The work conducted in 8 areas - from "Physics of gas discharges" to "Plasma in the synthesis of nanomaterials." Particular attention was paid to the plenary reports of leading foreign and Russian researchers.

Speakers were academicians and corresponding members of the RAS, professors and doctors of advanced universities in Russia, Europe and the United States.

From the Joint Institute for High Temperatures of the Russian Academy of Sciences, with plenary lectures, Academicians Eduard Son and Oleg Petrov spoke. One of the leading universities in the US and the world, Princeton University, was presented by two scientists: Mikhail Schneider and Andrei Starikovsky. The report of M. Schneider was devoted to the behavior of nanoparticles in an arc discharge plasma, and the speech of Professor A. Starikovski - plasma aerodynamics. Plasma science of France was also represented quite widely: the conference was attended by professors Jacques Amouroux, Jean-Pierre Boeuf, Leanne Pitchford and Ana Lacoste. A large delegation came from Sofia University (Bulgaria). Professors M. Benilov and V. Guerra presented the University of Madeira and Lisbon University (Portugal).

As for Russian scientists, many of them are employees of the leading institutions of the Russian Academy of Sciences, such as the JIHT, TIPS, ITAM SB, JINR and others. Russian university science was represented by participants from the

IOP Conf. Series: Journal of Physics: Conf. Series **927** (2017) 011001 doi:10.1088/1742-6596/927/1/011001

M.V. Lomonosov Moscow State University, St. Petersburg State University, MIPT, MEPI, Peter the Great St. Petersburg Polytechnic University, TSU, TSPU, etc. Tatarstan universities represented by scientists of KFU, KNITU-KAI and KNITU-KHTI did not stay aside. Also scientists of institutes subordinated to state corporations ROSKOSMOS and ROSATOM took part in the conference.

The welcoming and warm atmosphere at the conference made it an excellent platform for exchanging experience and generating new ideas in the field of low-temperature plasma physics.

By the decision of the organizing committee, Kazan will be the place of the next conference - PLTP-2020.

IOP Conf. Series: Journal of Physics: Conf. Series **927** (2017) 011002 doi:10.1088/1742-6596/927/1/011002

Peer review statement

All papers published in this volume of *Journal of Physics: Conference Series* have been peer reviewed through processes administered by the proceedings Editors. Reviews were conducted by expert referees to the professional and scientific standards expected of a proceedings journal published by IOP Publishing.

Effect of anode material on the breakdown in low-pressure helium gas

S F Adams[1], V I Demidov[1,2], A A Kudryavtsev[3], I P Kurlyandskaya[4], J A Miles[1] and B A Tolson[5]

[1] Air Force Research Laboratory, WPAFB, OH 45433, USA
[2] Department of Physics, West Virginia University, Morgantown, WV 26506, USA
[3] Department of Optics, St Petersburg State University, St Petersburg 199034, Russia
[4] INTEPH Technologies LLC, Springboro, OH 45066, USA
[5] UES, Inc., Dayton, OH 45432, USA

E-mail: vladimir.demidov@mail.wvu.edu

Abstract. An experimental study of the electric breakdown in helium gas for the plane–parallel electrode configuration has been conducted using a copper cathode and a variety of anode materials: copper, aluminum, stainless steel, graphite, platinum-plated aluminum and gold-plated aluminum. According to the Paschen law for studied electrode configuration, the breakdown voltage is a function of the product of gas pressure and inter-electrode gap. The breakdown processes on the left, lower pressure side of the Paschen curve have been the subject of this investigation. For those pressures, the Paschen curve may become multi-valued, where any given pressure corresponds to three breakdown voltage values. It was experimentally demonstrated that the form of the Paschen curve might strongly depend on the material of the anode and the cleanness of the anode surface. A possible explanation for this phenomenon is that electrons streaming from the cathode are reflected by the surface of the anode.

1. Introduction

The electric breakdown of gases is one of the fundamental phenomena of gas discharge physics. It has been studied for a long time but still attracts incessant interest of researchers [1–4]. Besides the interesting physics, breakdown is important for many applications including development of reliable electric insulation in electric grids and the study of different aspects of gas discharge physics [5-7].

According to the Paschen law for plane–parallel electrode configurations, the breakdown voltage, V_B, is a function of the product of gas pressure, p, and inter-electrode gap, d, that is $V_B = f(pd)$ [8]. The Paschen curve has the lowest breakdown voltage for a certain value of pd, which depends on the gas medium and the material of the cathode. V_B increases as pd increases on the right side of the Paschen curve and increases with respect to the lowest breakdown value on the left side of the Paschen curve. For most gases, V_B is a single-valued function of pd, but in Helium gas, for certain pd below about 1.7 Torr·cm, breakdown may take place for three different values of the breakdown voltage [1]. While there is no complete quantitative explanation of the multi-valued behavior of the Paschen curve, it is shown that in large extent this behavior is connected to the energy dependence of the electron emission yield for ion impact onto the cathode surface and the ion impact ionization of buffer gas atoms [1,8,9]. Modelling has shown that the following processes are the main contributors of the particular shape of the Paschen curve in helium at low pressures [3]: the dependence of the (ion-induced) secondary electron emission yield on the ion energy, the appearance of ion impact ionization of the gas at high electric fields, and the secondary electron emission from the cathode due to fast neutral atoms. To test these modelling results, previous experiments have been conducted with a copper cathode and anode. Modelling also demonstrated that backscattering of electrons from the anode can influence the breakdown voltage at low pressures [3]. However, there have been no systematic experimental studies of those effects.

In this work, systematic experiments were conducted using the same copper cathode and a variety of materials for the anode. This was done to demonstrate the effect of the anode material on the breakdown in low-pressure helium gas. This technique helped avoid variation of the breakdown conditions due to processes not connected to the influence of the anode and to study only the effect of the anode's material on voltage breakdown in low-pressure helium gas. In the experiments, the following materials were used for the anode: copper, aluminum, stainless steel, graphite, platinum-plated aluminum, and gold-plated aluminum. In the last two cases, 300 nm platinum and gold layers were used.

2. Experimental setup

In the experiments, an automated system for measurements of Paschen curves, shown in figure 1, was used [10]. A wineglass discharge tube is shown in figure 2. The discharge tube had two flat, disk-shaped electrodes facing each other at a variable distance from 2 to 20 mm. Most experiments were conducted at distance of 5 mm between electrodes. The cathode was made from copper and was 50 mm in diameter. The anode has the same diameter, but was made from different materials. In this work, experiments were conducted with copper, aluminum, stainless steel, graphite, platinum-plated aluminum, and gold-plated aluminum anodes.

Figure 1. An automated system for production of reliable Paschen curves over extended range of *pd*.

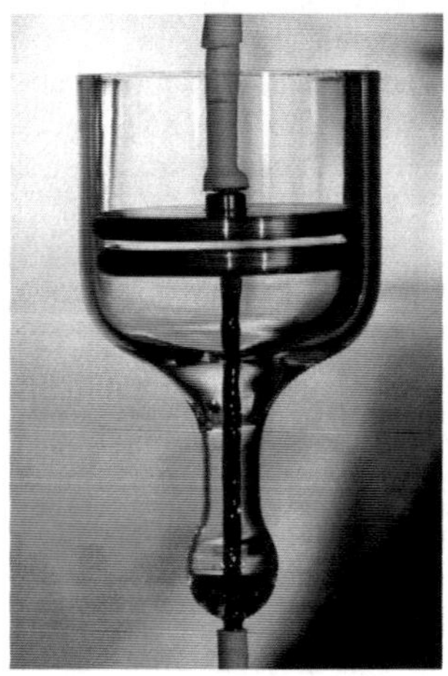

Figure 2. A wineglass discharge tube. Distance between cathode and anode is 2 mm (as shown).

The experiments were conducted in two regimes. In the first regime, for a certain *pd*, voltage between the cathode and anode was slowly increased from zero to breakdown voltage. In the second regime, for a certain voltage between the cathode and anode, gas pressure was slowly increased from a value without breakdown up to breakdown value. The first regime was suitable for investigation of single-valued parts of the Paschen curves, while the second regime allowed for studies of the multi-valued Paschen curves, and was most suitable for the research conducted in this work. The Paschen curve for any particular anode material was measured ten times and all curves show below were each averaged over ten curves.

3. Results of the experiments and discussion

Results of measurements of the left sides (lower *pd*) of the Paschen curves for the copper cathode and the different (copper, aluminum, stainless steel, graphite, platinum-plated aluminum and gold-plated

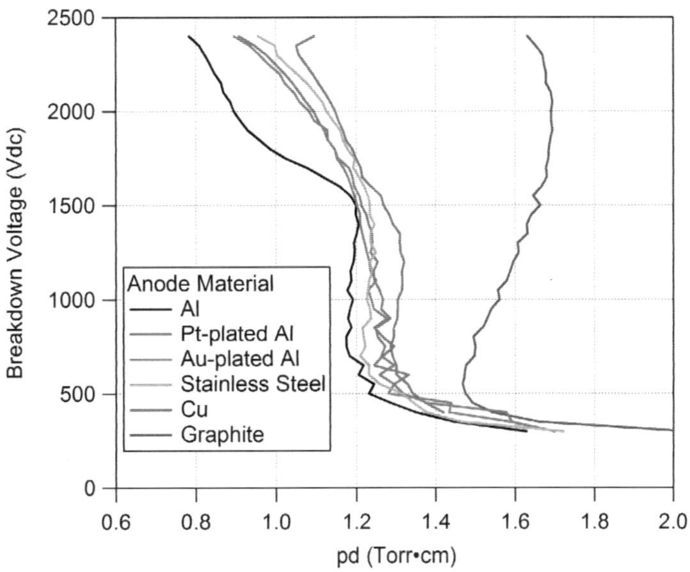

Figure 3. Paschen curves with different anode materials: aluminum, platinum-plated aluminum, gold-plated aluminum, stainless steel, copper, and graphite.

aluminum) anodes are shown in figure 3. It can be seen that all curves, with the exception of graphite, generally had similar behaviour. The curve for graphite is substantially shifted to the right (a much higher *pd*) and has a more pronounced multi-valued section. This general behaviour may be explained as follows. Anode surfaces may have substantial reflection of electrons with energies of several hundreds (thousands) of electron-volts. Those electrons for the lower *pd* range of our data arrive at the anode from the cathode without a substantial energy loss in contrast to similar electrons for the far right side of the Paschen curve (not shown in figure 3), which have more collisions with atoms in the discharge gap. As a result, the far right side of these Paschen curves would only weakly depend on the anode material. Electrons reflected from the anode participate in the ionization processes in the discharge volume, making the breakdown simpler. As the reflection is somewhat different for different materials, the Paschen curves are not exactly the same for different anodes. The graphite surface has a porous structure. This leads to the suppression of the reflection of electrons from the graphite surface and, as a result, to more difficult ionization of helium. This explains to the breakdown at a higher *pd* and the pronouncement of the multi-valued section of the curve. Experiments also demonstrate that an insufficient cleanliness of the anode can lead to substantial modification of the Paschen curve.

A higher manifestation of the multi-valued part of the Paschen curve allows, for example, one to demonstrate experimentally the various breakdown voltages at that part of the curve. In figure 4, the *pd* value was first held at 1.6 Torr·cm, while the voltage between the cathode and anode was slowly increased from zero to about 400 V, where breakdown occurred. In a second experiment, for *pd* of near 1 Torr·cm, voltage between the cathode and anode is increased from zero to 1.5 kV. Then, the *pd* value was increased without breakdown from 1 Torr·cm to 1.6 Torr·cm. Afterwards, *decreasing* the voltage eventually led to breakdown at about 1.2 kV. This somewhat paradoxical behavior was a direct manifestation of the multi-valued Paschen curve. Note that reflection of electrons from the plasma volume walls is important in gas discharge physics. For example, this effect may have profound influence on probe measurements [11].

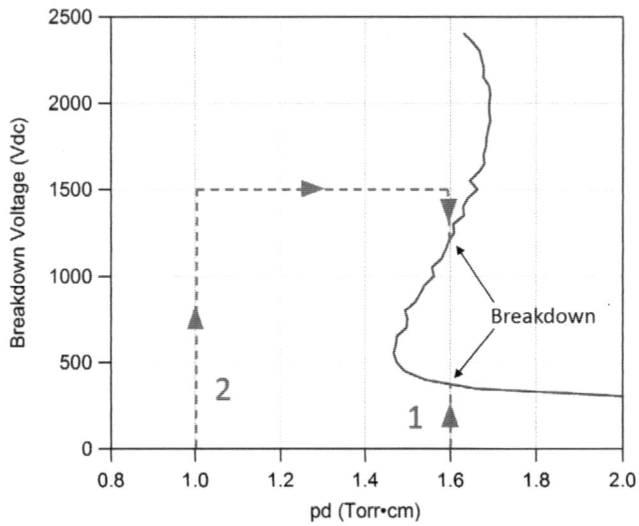

Figure 4. Plot of 2 experiments demonstrating multi-value breakdown points for a graphite anode accessed by (1) increasing and (2) decreasing the applied voltage.

4. Conclusions

An experimental study of the electric breakdown for the plane–parallel electrode configuration using a variety of anode materials demonstrates that material of the anode can substantially affect the measured Paschen curve. This effect is especially pronounced at the left part of the Paschen curve in helium gas, causing the curve to be multi-valued. This indicates that for a certain pressure and a specific gap between the cathode and anode there are three distinct breakdown values of voltage. It was experimentally demonstrated that the form of the Paschen curve might strongly depend on the material of the anode and the cleanness of the anode surface. This may be explained by the reflection of electrons at the anode surface.

References

[1] Penning F M 1931 *Proc. R. Acad. Amst.* **34** 1305
[2] Ward A L and Jones E 1961 *Phys. Rev.* **122** 376
[3] Hartmann P, Donkó Z, Bano G, Szalai L and Rozsa K 2000 *Plasma Sources Sci. Technol.* **9** 183
[4] Korolov I and Donkó Z 2015 *Physics of Plasmas* **22** 093501
[5] Raizer Y P 1991 *Gas discharge physics* (Berlin: Springer)
[6] Demidov V I, Kudryavtsev A A, Kurlyandskaya I P and Stepanova O M 2014 *Phys. Plasmas* **21** 094501
[7] Adams S F, Demidov V I, Bogdanov E, Koepke M E, Kudryavtsev A A and Kurlyandskaya I P 2016 *Phys. Plasmas* **23** 024501
[8] Francis G 1956 The glow discharge at low pressure *Encyclopedia of Physics* ed S Flugge (Berlin: Springer)
[9] Hillmann H, Muller F and Wenz H 1994 *Plasma Sources Sci. Technol.* **3** 496
[10] Adams S F, Huang J, Howe K C, Demidov V I, Tolson B and Hensley A 2015 *Bull. American Phys. Soc.* **60**(19) 205
[11] Demidov V I, Adams S F, Kaganovich I D, Koepke M E and Kurlyandskaya I P 2015 *Phys. Plasmas* **22** 104501

Influence of composition of electrolyte on electric discharge in liquid

M F Akhatov, F M Gaisin, F R Iskhakov, R R Kayumov, A I Kuputdinova, R A Mukhametov, A I Shpilev

Faculty of Physics and Mathematics, Kazan National Research Technical University named after A.N.Tupolev - KAI, Kazan, 420111, Russia

E-mail: almaz87@mail.ru

Abstract. Within this work results of researches of electric discharge with liquid nonmetallic electrodes are presented. Features of formation of electric discharge depending on structure and concentration of electrolyte, electrophysical processes are described, the volt-ampere characteristic of the category is constructed. Distribution of electric field strength along an electrolyte stream is presented.

1. Introduction

One of the new directions in researches is ignition of the gas discharge in a steam-gas cover. Unlike classical types of the discharge generated between firm electrodes [1-3], features of this discharge is that one or both electrodes are liquid nonmetallic electrodes. The main attention of experts is directed to a research of composition of plasma, distribution of concentration and energy of components of plasma, a research of thermal and hydrogasdynamic processes on limit of the section of environments in a zone of formation of the discharge [4 - 8]. Along with basic researches, this type of the discharge finds broad practical application in various branches of industrial production, such as modification and cleaning of a surface of materials and products from various type of metals and alloys [9]. In the field of mechanical engineering there are relevant tasks of increase in a class of roughness, superficial hardness of details and knots of cars. However, despite a large number of publications on this type of the discharge, he remains investigated insufficiently. There are questions on influence of concentration and composition of electrolyte on physics of formation of the discharge in a steam-gas cover. Generally in the existing researches as electrolytes water solutions of salts are applied. However there is a large number of kinds of electrolytes which differ from each other in the chemical compositions, so both physical and chemical properties. Also the results of a research which are provided in electrolytic categories namely conditions of ignition of the discharge, a form of electric discharge, volt-ampere characteristics of the discharge will depend on composition of electrolyte. Results of the volt-ampere characteristic of the discharge between the jet electrolytic cathode and the firm anode for various composition of electrolyte are given in this work.

2. Experiment

The gas discharge between the jet and dropping liquid cathode and metal anode was studied at the pilot plant with the direct current source and controlled DC voltage up to 4000 V at the rated current of up to 10A.

The following research methods were applied to address the tasks specified:

1. Current and tension of the category were measured by multimeters of various class of accuracy.
2. The process of gas discharge combustion was recorded by the digital photo- and video devices SONY FDR-AX33.

Combustion of the gas discharge within the electrode spacing was identified at the voltage of U=100-1500B, ambient pressure p=〚10〛^5 Pa, a copper plate d_a=1 mm was taken as the anode.

In the figure 1 photos of formation of the discharge between the jet electrolytic cathode and the metal anode are submitted. After application of potential on electrodes on border of interaction of a stream of liquid and a metal electrode there is a breakdown. The discharge is formed in the form of a cone on limit of the section of environments in a zone of formation of steam-gas bubbles. After a discharge influence on a surface of a metal electrode his cleaning of organic and inorganic pollution is observed.

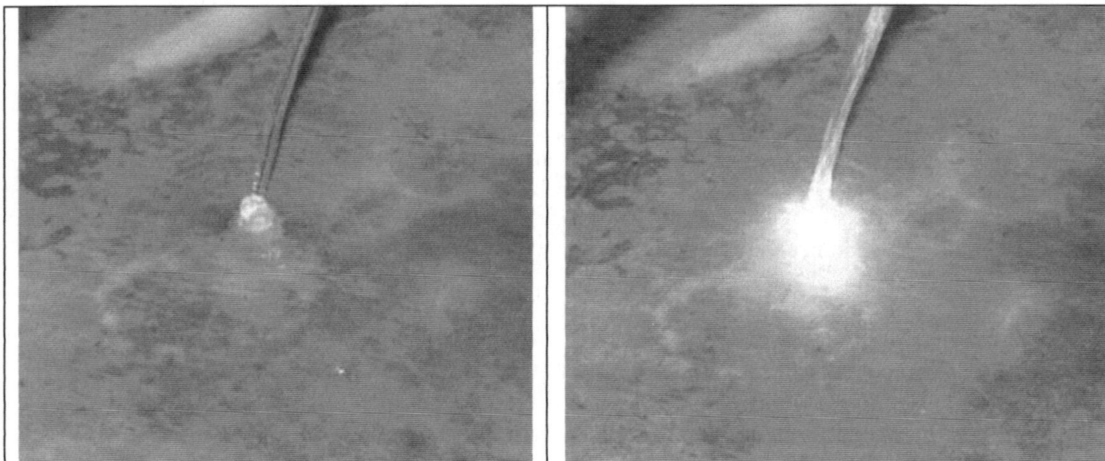

1 2

Figure 1. The photo of the discharge between the string cathode and the metal anode 1 - before breakdown, 2 - after breakdown

The analysis of the volt-ampere characteristic of the discharge between the jet electrolytic cathode and the firm anode for NaCl, KCl and CuSO4 solutions in technical water in fig. 1 has shown that in case of saturated CuSO4 solution (a curve 1) the category burns at small currents of I = 0,03 A and the big tension of U > 1200 V in the form of a ring near the metal anode. In a case use of saturated NaCl solution (a curve 2) in technical water leads to the fact that in the range of I = 250 ÷ 450 мA size U increases almost linearly. In case of use as the jet electrolytic cathode saturated KCl solution (a curve 3) size U in the range from 200 to 250 мA increases in technical water slowly, and in the range of I = 250 ÷ 380 мA tension of the discharge increases sharply.

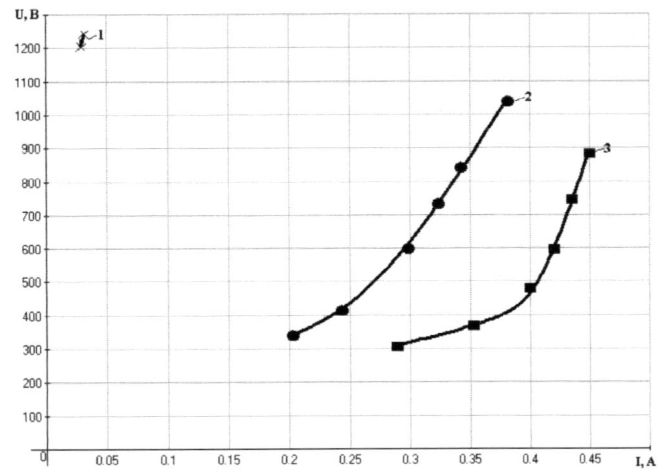

Figure 2. The volt-ampere characteristic of the discharge between the string electrolytic cathode and the metal anode

Distribution of floating potential φ on an axis of discharge it was measured by means of the tungsten probe with a diameter of 0.7 mm and a static voltmeter of C50 of a class of accuracy 1.0. All measurements were taken with the help a coordinate space. On measured φ calculations a distribution of electric field intensity on a formula $\bar{E} = -grad\varphi$ with accuracy of ±5% are performed. Distribution of potential was measured in the bit camera. Results are provided in a figure 3.

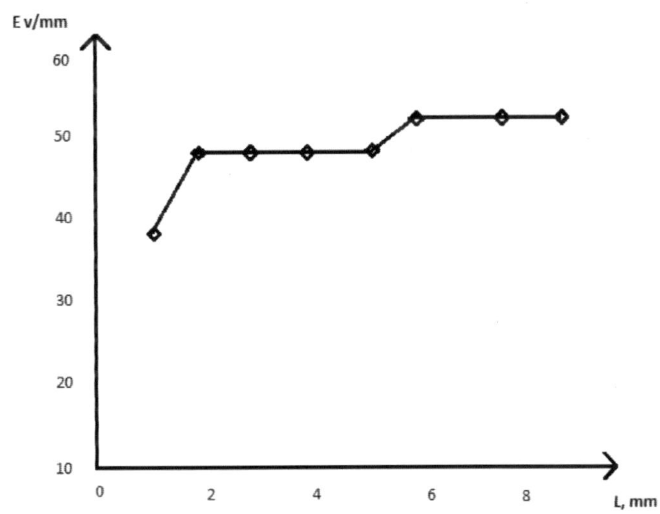

Figure 3. Strength of electric field of the discharge along the jet electrolytic cathode at U = 400 - 750 V and I = 0.1 - 0.6 A with an atmospheric pressure.

3. Results

Formation of the discharge in the form of a cone on limit of the section of environments between streams of electrolyte and metal electrodes is established. It is established formation of the discharge in the interval of voltage of U = 300 – 1250 V, discharge current I = 0.03 – 0.45 A. It is established distribution ща electric field strengths along a stream in the range of E = 38 – 52 V/mm.

Acknowledgments
This work was supported by state contract regarding resolution 220 of the Ministry of Education and Science of the Russian Federation, project no. 14.Z.50.31.0023.

References

[1] Reiser Yu.P., *Physics of Gas Discharge (Edition 3, updated and revised). Dolgoprudny, 2009. - P. 734.*

[2] Fortov V.E., Son E.E., Gaysin F.M., Bromberg L., Son K.E., O John Khe, I Khe Iong. *Plasma Technology (in Korean) Moscow Institute of Physics and Technology, KOFST, 2006. –P. 135.*

[3] Dautov G.Y., Dautov I.G, Fayrushin I.I, Kashapov N.F. 2013 *Journal of Physics: Conference Series* 1 012014

[4] Gaisin A.F. 2013 *High temperature* **6** 863 – 866

[5] Gaisin A.F., Sarimov L.R. 2011 *Plasma Physics Reports* **6** 535 – 540

[6] Gaisin Al. F., Son E. E., Petryakov S. Yu. *Plasma Physics Reports* 7 741 -748

[7] Gaisin Al.F., Son E. E., Efimov A. V., Gilmutdinov A. Kh., Kashapov N. F. *High temperature* **3** 457 – 460

[8] Gaisin Al.F. *High temperature* **1** 145–146

[9] Gaisin Al.F. *Inorganic Materials: Applied Research* **3** 392–395

Combined method of stabilization of the glow discharge in gas flow

T Ya Asadullin, I G Galeev

Kazan National Research Technical University named after A.N. Tupolev, Department of General Physics, Russia, Kazan

Abstract. Many applications of the glow discharge need the discharge power and gas pressure grow up. As usual this cause the appearance of discharge instabilities which violate the normal glow mode. Here the combination of two methods of stabilization of the glow discharge with gas flow is considered.

1. Introduction

Volumetric glow discharge can be effectively used for pumping active medium of the optical lasers with higher power, but to achieve it, it is necessary to create the conditions for homogeneous volume pumping of the active medium. To increase the limiting characteristics of the stable volumetric discharge gas pumping is used, including the supersonic one.

In most cases in practical implementation of glow discharge to make an influence on the surface the glow discharge with a transverse direction of the gas flow is often used. In this case it is possible to use camera with the variety of geometry, to imply different ways of preionisation, etc. When arranging the glow discharge in a transverse direction of gas flow sectionalized cooled cathode blocks are widely employed. Cathode sections of cathode blocks are usually separated from each other with dielectric layers.

In most cases in practical implementation of glow discharge it is necessary to provide the stable homogeneous volumetric burning mode of the discharge at high current values and specific energy input. The increase of the characteristics of the discharge is limited because of the discharge instability development, firstly – ionization-overheating instability which causes discharge contraction.

To achieve stable values of electric and thermal characteristics of a glow discharge [1-3], effective methods are applied to the cathode region of a glow discharge. This is due to the fact that the transition of the discharge to the contracted state is accompanied by compression of the cathode spot.

For glow discharge it is normal that at low pressures cathode spot occupies much more place on the cathode, what corresponds to the Gehl law for the normal cathode current density $j/N^2 = \text{const}$, where j – current density on the cathode, N – concentration of neutral gas particles. With the increase of gas pressure cathode spot decreases (inverse square law) quite strongly. However glow discharge positive column occupies rather big space of discharge chamber. As it is known cathode heat dissipation is rather big, so that it might be the reason for emergence and development of the thermal instability on the cathode. Consequently there are limitations for maximum possible currents of stable volumetric glow discharge, at the same time homogeneous pumping of active laser medium decreases significantly.

There are a number of methods of glow discharge stabilizing, efficiency of which depends on several factors. The general principle for all these methods of stabilization is not to allow conditions for instabilities. To achieve it swamping resistors, gas flow swirl, different ways of preionisation, cathode sectioning, effects of spinning magnetic field etc. are used.

The ways to chip away at instability development can be divided into two types:

- effects, leveling discharge characteristics according to the chamber capacity, for example preionisation and gas flow swirl
- effects of speeding the processes of instability development, for example effect of spinning magnetic field.

2. Passive stabilization

In this research we consider the way combining the two types of stabilizing effect. As it was mentioned above, the glow discharge in transverse gas flow is considered. Here we use a cathode block, consisting of not very big copper sections, separated from each other dielectrically with cooling distilled water. On the upper surface of the cathode plates the layer of intermissive dielectric enamel coating was put [4]. The basic rectangular cell of cathode uninsulated section was the size of 3x4 мм^{2}. Anode is a copper plate.

At rather low values of pressure, gas flow and electric current, the discharge glows in a diffused mode and fills in with itself the whole surface of the cathode sections. While increasing the voltage on the discharge or increasing the pressure, the discharge contracts to a spot on the cathode, usually it happens downstream, and then at a further increasing of discharge current, it continues contracting to a point on the cathode, where there are preferential conditions for discharge burning.

In most cases the conditions for the development of ionized-overheated instability are realized nearby the cathode, where the processes of intensive ionization and significant heat release occur. Thus, the stabilizing effect in the examined approach will be caused by currents control, flowing through the cathode sections.

On the one hand there is an efficient stretching of a cathode spot by putting on it an intermissive dielectric coating. It makes the cathode spot to spread along the cathode section.

Consequently the cathode heat dissipation is distributed along the cathode section more uniformly. Therefore the water cooling can effectively draw off the heat dissipating in the near cathode space. Thus, the conditions for the glow discharge instability growth near cathode surface become less realizable. So the higher upper limit values of the discharge current without contraction can be reached.

On the other hand an advanced control action is taken on the electrical potentials of the cathode sections to prevent the compression of the cathode spot, followed by contraction of the discharge.

3. Active stabilization

Stretching of the cathode spot along the surface of the cathode section prevents compression of the cathode spot within one section. And the control of the potentials of the sections makes it possible to further increase the limiting values of currents, voltages and pressures while maintaining a stable diffuse discharge regime.

The characteristic time for the development of the ionization-overheating instability is of the order of 10-3 s. Consequently, the time for which the stabilizing effect must eliminate the conditions for the development of the ionization-overheating instability should have a significantly lower value.

The essence of the method is to continuously monitor the dynamics of changes in the values of the currents through individual cathode sections. With a stable diffuse burning mode of a glow discharge, the law of the constancy of the current density at the cathode is usually satisfied. If the stable burning regime is violated, and a transition to the contracted combustion form of the discharge occurs, this will be accompanied by the appearance of an increasing uneven distribution of the current among the cathode sections. Through some sections the current will increase, and through the remaining sections the current values will decrease. Timely determining those sections through which large currents begin to flow, it is necessary to increase their ballast resistances. At the same time, it is necessary to reduce the resistance of neighboring sections, the currents through which begin to decrease. Thus, the cathode spot will be forced to move, stretching across the cathode, and the development of instability will be prevented [5].

Obviously, because of the rather small characteristic time of development of the ionization-overheating instability (10-3 s), the stabilization system must be automated. It should implement the next functions:

- measurement of discharge parameters,
- calculation of the parameters change rate,
- comparison of the parameters with previous values for a given cathode section,
- comparison of the parameters with the values of adjacent cathode sections,
- selection of the control algorithm,
- formation of the control action,
- evaluation of the result of the control action,
- protection from approaching critical conditions of the discharge,
- protective discharge shutdown.

The modern development of digital information and measuring tools makes it possible to implement such a system of dynamic stabilization of a glow discharge. In general, the stabilization system consists of a block of measuring resistors, a matching circuit, a commutator with a multiplexer, an analog-to-digital converter (ADC), a computer or microcontroller, a cathode section potential control unit.

The current through each section is determined by the voltage drop across the measuring resistors with low resistance. After passing the matching circuit, the converted signal from each section is serially connected via a switch to the ADC. After that, the voltage (current) data of each section is fed to the microcomputer for processing. The data is processed in the microcomputer. The values of the currents of adjacent sections and the dynamics of the change in the currents in all sections are compared. An analysis of these changes is performed depending on the change in the discharge burning conditions: the values of the pressure distribution in the discharge chamber, the gas flow, voltages and currents. After processing the data, a control signal is generated, which is fed to the cathode section potential management system. The cathode section potential control circuit consists of a digital-to-analog converter, a demultiplexer and switching elements whose resistance can vary depending on the control signal.

The hardware of the stabilization system is supplemented with appropriate software. The program manages the processes of measuring the parameters of the experimental setup, transferring measurement results to a microcomputer, analyzing the flow of experimental data, generating control commands, monitoring execution, and so on. A special role of the software is to provide a dynamically maintained unstable discharge equilibrium near the critical (maximal) current values through the cathode sections.

On the one hand, the stabilization system should ensure a stable diffuse discharge burning regime. On the other hand, it should not be possible to reduce the currents to the minimum values due to excessive regulation. Depending on the operator's settings, the system can support discharge burning in different modes in relation to proximity to the critical mode. The use of the preionization system built into the experimental setup greatly extends the range of regulation of the limiting currents, the magnitude of which is limited by the efficiency of the electrode cooling system.

4. Conclusion.

By influencing the cathode discharge region, it is possible to increase the homogeneity of the distribution of the average discharge characteristics over the entire volume of the discharge region and to increase the discharge stability limits.

References

[1] Galeev I G Tikhomirov I A and Asadullin T Ya. 1995, Rus. Phys. J. 38, Is.10, 1036-39.
[2] Galeev I G Timerkaev B A. 1986, J. of Eng. Phys. 51, 1112-16.
[3] Asadullin T Ya Galeyev I G and Timerkayev B A. Journal of Physics: Conference Series (2016) 669 012001.
[4] Galeev I G Asadullin T Ya IOP Journal of Physics: Conference Series (2016) 669 012015.
[5] Galeev I G Asadullin T Ya IOP Journal of Physics: Conference Series (2017) 789 012012.

Acknowledgments

The work is supported by grant of Ministry of Education and Science of Russian Federation 3.6564.2017/8.9

Using two-chamber photoplasma for creating photovoltaic converter

S A Astashkevich[1], E A Bogdanov[1], G V Kirsanov[1] and A A Kudryavtsev[2,1]*

[1]Faculty of Physics, St. Petersburg State University, 3 Ulianovskaya Street, Petrodvorets, St. Petersburg, 198504, Russia
[2]Department of Physics, Harbin Instute of Technology, Harbin 150001, China
[*]E-mail: akud53@gmail.com

Abstract. Two-chamber photoplasma is developed for study photovoltaic converters. 2D simulation of conical geometry (similar to one of a solar collector) photoplasma parameters for Na-Ar gas mixture was fulfilled. The present study shows strong spatial non-uniformity of electron density and the electric potential distributions in the two-chamber cell that provides notable value of photo-emf.

1. Introduction

The importance of creating photovoltaic plasma converters is obvious from the point of view of diversification of energy, environment and responds to the global trend of increasing the relative share of renewable energy in the modern energy industry [1]. Such sources may have certain advantages compared to commonly used semiconductor solar cells [2, 3]. There are some approaches to convert solar energy into electrical energy based on photovoltaic effect in photoplasma [4, 5] containing a gas of an alkali metal (mainly Na and Cs): MHD generators [2], thermionic converters [6], and non-homogenous plasma [3]. In our recent papers [7, 8] it is shown that two-chamber 2D plasma cell allows obtaining large spatial gradients of electron density and temperature and thus create appreciable electric potential between chambers. So far the investigations of photovoltaic plasma converters (see [3] and bibliography in this paper) limited by one dimension case.

The aims of the present work are developing two-chamber plasma model for study photovoltaic plasma converters and 2D simulation of parameters photoplasma for Na–Ar gas mixture on the example two-chamber cell of conical geometry which conforms to one of a solar collector.

2. Model

2.1. Conical cell as an example of the photoplasma geometry

Let us consider conical two-chamber plasma geometry which conforms to one of a solar collector (see figure 1) and is similar the geometry of the experimental work [2].

IOP Conf. Series: Journal of Physics: Conf. Series **927** (2017) 012004 doi:10.1088/1742-6596/927/1/012004

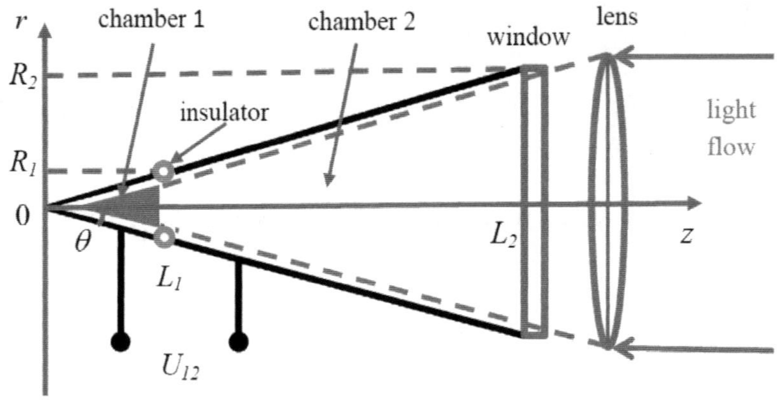

Figure 1. Cross section of a conical photoplasma two-chamber cell.

2.2. Elementary photoplasma reactions

In the present work we study two-chamber resonance photoplasma for the Na–Ar mixture. The Ar as buffer gas serves to provide diffusive regime of destruction of the charged particles in both chambers. The list of main accounted reactions is given in table 1.

Table 1. Reactions taken into account in the present work.

No	Reactions	Constants	Comments
1	$Ar+e \rightarrow Ar+e$	Cross section [9]	Momentum transfer
2	$Na(3s)+h\nu \rightarrow Na(3p)$	…	Resonance photoexcitation
3	$Na(3p)+Na(3p) \rightarrow Na^{**}+Na(3s)$	Reaction constant [10]	Excitation transfer
4	$Na(3p)+Na^{**} \rightarrow Na^{+}+Na(3s)+e$	Reaction constant [10]	Penning ionization

2.3. Theoretical model

Two-chamber cell (α is a chamber number: α=1,2) with open joint border (see figure 1 for the case studied in the present work) may basically have different cross profile both in shape and size. The continuity equation with the drift-diffusion approximation for each chamber α (see e. g. [5]) is

$$\frac{\partial n_j}{\partial t} - \nabla\left(D_{j\alpha}\nabla n_j\right) = I_{j\alpha} - R_{j\alpha}, \tag{1}$$

where subscript j indicates jth specie, n is the density, I and R is creation and destruction rates, and D is the diffusion coefficient.

It was assumed that the spatial distribution of rate of resonance photoexcitation (reaction (2) in table 1) and electronic temperature has the following (r, z)-dependence:

$$s(r,z) = \begin{cases} S_1 e^{-(z-z_0)^2/z_1^2} \cos\left(\dfrac{\pi r}{2R(z)}\right), & \text{if } s(r,z) \geq S_0; \\ S_0, & \text{otherwise.} \end{cases} \tag{2}$$

Here $R(z)$ is the radial coordinate at the cell board for a given value of z.

The electric potential is calculated using the Poisson's equation.

3. Calculation and analysis

The conditions of 2D simulation are follows: the Ar pressure p_{Ar} =1 torr; the ratio of Ar and Na pressures p_{Ar}/p_{Na}=100; the geometric sizes of the first and second chambers are (L_1=10^{-2} m; R_1=5×10^{-3} m) and (L_2=6×10^{-2} m; R_2=3×10^{-2} m) respectively. The spatial profiles (Eq. 2) of the rate of resonance photoexcitation (for S_0=0; S_1=1.6×10^{24} m^{-3}sec^{-1}) and electronic temperature (for S_0=0.026 eV; S_1=0.3 eV) for z_0=5×10^{-3} m and z_1=10^{-2} m are shown in figures 2 and 3. The area of photoexcitation is mainly in the first chamber that corresponds concentrating of radiation in this chamber by means optical lens. 2D simulation was fulfilled in the COMSOL Plasma Module accounting equations (1)–(2) and reactions listed in table 1.

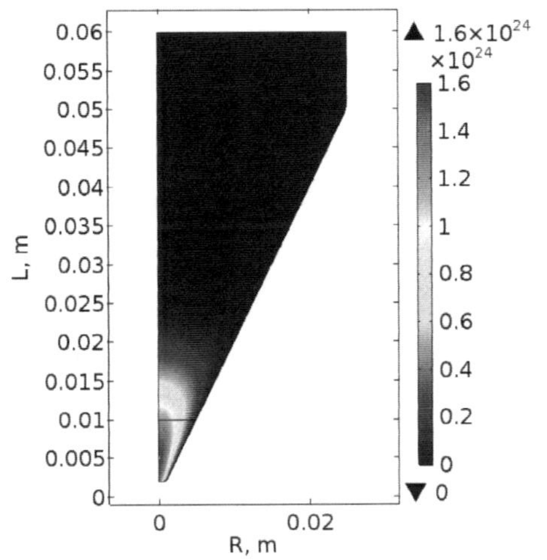

Figure 2. 2D profile of the rate of photoexcitation reaction [m^{-3}sec^{-1}].

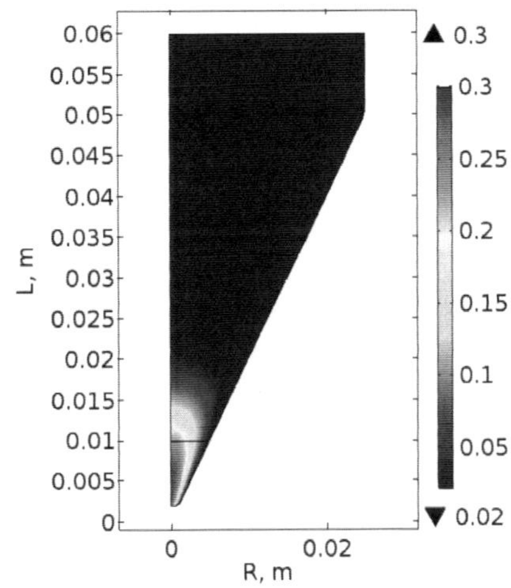

Figure 3. 2D profile of electronic temperature [eV].

For the stationary case of Eq. (1) calculated 2D profile of the electron density is shown in figure 4 and one at the z-axis is given in figure 5. It is seen strong non-uniformity of electron density distribution through two chambers with the maximum in the second chamber.

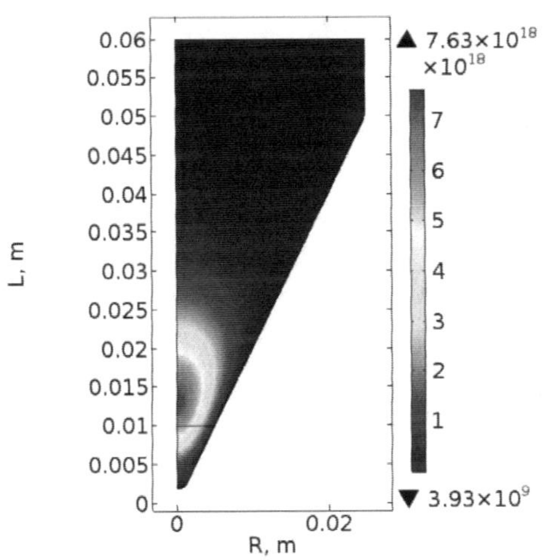

Figure 4. 2D profile of the electron density [m^{-3}].

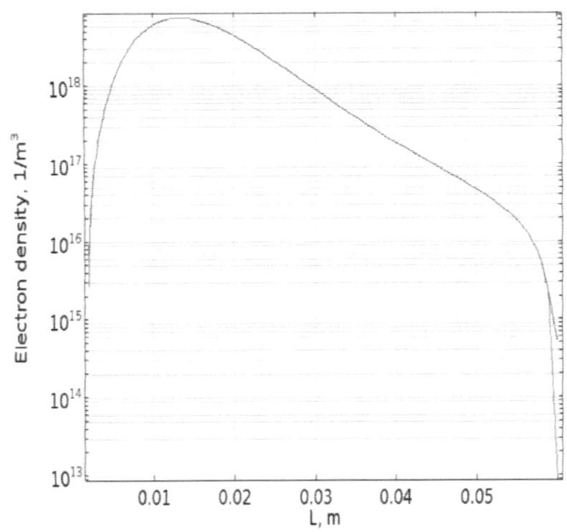

Figure 5. The electron density at the z-axis [m^{-3}].

The spatial distribution of the electric potential is shown in figure 6 and the electric potential at the cell border is given in figure 7. Calculated photo-emf for considered conditions U_{12} is 1.28 V.

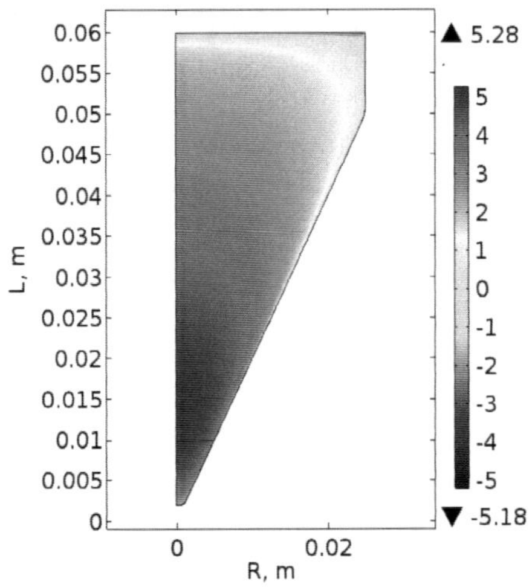

Figure 6. 2D profile of the electric potential [V].

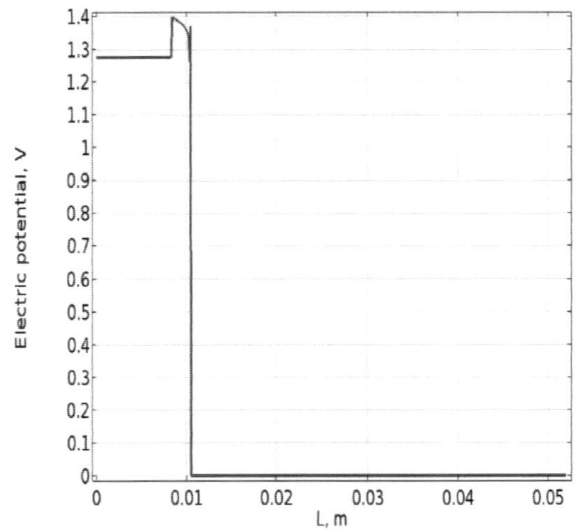

Figure 7. The electric potential [V] at the cell border at points (r=$R(L)$, z=L).

4. Conclusions

The present study shows strong spatial non-uniformity of electron density and the electric potential distributions in the photoplasma two-chamber cell that provides notable value of photo-emf. The further research suggests more detailed accounting of photoplasma reactions and optimization of plasma geometry and conditions.

Acknowledgements
This work was supported by grant SPbGU 11.37.212.2016.

References

[1] *World Energy Resources: Full Report* 2016 (London: World Energy Council) https://www.worldenergy.org/wp-content/uploads/2016/10/World-Energy-Resources-Full-report-2016.10.03.pdf
[2] Dunning G J and Palmer A J 1981 *J. Appl. Phys.* **52** 7086
[3] Gorbunov N A and Flamant G 2015 *Plasma Chem. Plasma Process.* **35** 799
[4] Beterov I M, Eletskii A V and Smirnov B M *Sov. Phys. Usp.* 1988 **31** 535
[5] Smirnov B M 2007 *Plasma Processes and Plasma Kinetics* (Weinheim: WILEY-VCH)
[6] Khalid K A A, Leong T J and Mohamed K 2016 *IEEE Trans. El. Dev.* **63** 2231
[7] Kudryavtsev A A and Serditov K Yu 2012 *Phys. Plasmas.* **19** 073504
[8] Bogdanov E A, Kudryavtsev A A, Ochikova Z S and Chirtsov A S 2015 *Tech. Phys.* **60** 1570
[9] Phelps A V 1985 *JILA Information Center Report* **28** (University of Colorado)
[10] Mahmoud M A and Gamal Y E E 2012 *Indian J. Phys.* **86** 659

The electron beam diagnostic of the clustered supersonic nitrogen jets

S V Avtaeva, T S Yakovleva, V V Kalyada and A E Zarvin
Novosibirsk State University, 2 Pirogova Street, Novosibirsk, 630090, Russia
E-mail: s_avtaeva@srd.nsu.ru

Abstract. Axial and radial distributions of the rotational temperature and density of N_2 molecules in supersonic nitrogen jets formed with conic nozzles (critical diameters d_{cr} of 0.17 and 0.21 mm) were studied using the electron beam fluorescence technique at stagnation pressures P_0 of 0.1-0.6 MPa. A rotational temperature T_r, equaling a gas temperature T_g owing to fast RT relaxation, was obtained using the rotational line relative intensity distribution in (0-1) vibrational band of the N_2 first negative system. Gas density profiles in the jets were obtained using the integral intensity of the band. It is found, T_r at the nozzle outlet is of the order of a few tens of Kelvin and at further expansion T_r drops up to 15-20K at distance of (100-200) d_{cr}. The gas temperature and density distributions in the studied supersonic nitrogen jets are not similar to the isentropic distributions. It is shown that the lower is the stagnation pressure the faster the gas density and temperature decrease with distance from the nozzle. Increase in P_0 leads to elevating T_g in the jets. A reason for this effect may be cluster formation in the jets. Estimations of cluster mean sizes in the jets using Hagena's parameter show presence of large clusters ($M \geq 200$) at P_0=0.4-0.6 MPa.

1. Introduction

For the first time cluster beams have been received in the middle of a past century in Germany at evaporation of a vapor, formed in a source, through a small nozzle into empty space [1]. Now the homogeneous condensation of gas at its expansion at outflow through a nozzle is the most widespread method of cluster production. Cluster beams are usually used for preparation of thin films, fine polish of surfaces and developing new materials [2, 3].

Cluster size depends on stagnation pressure and temperature, nozzle orifice and gas nature [4]. On the other hand, clusters continue to form and grow in expanding free jet until gas density is high enough and gas temperature is low enough. The cluster growth influences on gas density and temperature in the jet, as gas density decreases and some amount of heat is released at gas condensation. Here we study gas density and temperature distributions in supersonic nitrogen free jets using the electron beam fluorescence (EBF) method developed by Muntz [5]. The EBF method was frequently utilized to measure rotational temperature distributions in supersonic nitrogen free jets [6, 7]. However, Marrone pointed that the measurement by the EBF method might be affected by secondary electrons. Some deviations from the Boltzmann distribution (BD) in population of the rotational states were observed in [8, 9]. Coe *et al.*[7] proposed an analytical model including a quadrupole interaction with an ejected electron as well as a dipole interaction with a primary electron and reduced the spectroscopic data to the BD, concluding that there were no deviations from the BD in the rotational states in a wider range of $P_0 \cdot d$ (P_0 is the source pressure and d is the orifice diameter) from 16 to 1016 Torr·mm. Sharafutdinov *et al.* compared the experimental data obtained by EBF with coherent anti-Stokes Raman spectroscopy (CARS) in the range of $P_0 \cdot d$ from 60 to 1500 Torr mm, the good agreement was obtained when the density in the jet was high enough to keep the equilibrium conditions [10]. They mentioned that the contribution of secondary electrons is small in low-density conditions. In our experiments $P_0{\cdot}d$ was varied in a range of 128-767 Torr· mm and we suppose the BD in the rotational states. Gas density and temperature distributions at different source pressures were obtained and an attempt to observe some effects of clusters formation in the jets on the distributions was performed.

2. Experimental setup

The jets were obtained in the vacuum chamber of the experimental setup LEMPUS-2 [11]. The expansion chamber is horizontal cylinder with a diameter of 70 cm and a length of 120 cm. The vacuum pumping system provided a total pumping rate of up to 3500 L/s in N_2. Stagnation chamber was mounted inside the expansion chamber on the four-directional positioning device possessing three translational and the rotational degrees of freedom. The chamber has laser positioning unit which provide the necessary alignment of the jet axis and the axis of the chamber. Two supersonic nozzles with stationary modes of expansion were used in the experiments: 1) the conic nozzle #1 with inlet diameter d_{cr} of 0.17 mm, outlet diameter d_{out} of 2.4 mm and length L of 8.2 mm, and 2) the conic nozzle #2 with d_{cr} =0.21 mm, d_{out} =3.5 mm and L= 17.5 mm. The stagnation pressure P_0 was varied in the range of 0.1-0.6 MPa, gas pressure in the expansion chamber P_h was changed in the range of 0.35-3.6 Pa.

The electron beam (e-beam) was generated by an electron source installed in the upper part of the chamber. The electron source based on a hollow cathode discharge has a ballast volume evacuated by turbomolecular pump and generates a well focused e-beam with electron energies up to 10 keV and the beam current up to 100 mA. The e-beam was crossed the gas jet in perpendicular direction; a position of the jet relatively to the e-beam was assigned by the positioning device. The EBF of the jet was observed through the optical window of 160 mm diameter in the flange on the expansion chamber sidewall. The radiation was focused on one of optical fiber ends while other end of the fiber was installed in front of the spectrometer entrance slit. The spectrometer had a diffraction grating with 1200 grooves/mm and with 30 μm entrance slit it had a resolution of about 0.03 nm. The EBF spectra were recorded by a CCD array connected with a computer via an ADC.

3. The electron beam diagnostics

The electron beam fluorescence (EBF) technique is particularly well suited to low density hypersonic and supersonic flows (< 10^{16} cm^{-3}) because high fluorescence yields are obtained from excitation of the molecules or atoms of gas flows with high energy electrons. Since its first application in 1953, it has been used in numerous facilities to perform local and non-intrusive measurements of density, vibrational and rotational temperatures and velocity of different species such as N_2, NO, CO, CO_2 and He [12]. Hear we use EBF technique to study rotational temperature and gas density profiles in the supersonic nitrogen jets.

The rotational temperature T_R of N_2 molecules was obtained using the relative intensity distribution of rotational lines in (0-1) band of the N_2 first negative system (FNS). The theoretical prediction of the relative intensities of the rotational lines in the e-beam excited FNS was developed by Muntz [5]. In case of direct excitation of the $N_2^+(B^2\Sigma_u^+)$ state by e-beam at electron collisions with nitrogen molecules and the subsequent re-emission unaffected by gas kinetic collisions as well as provided that N_2 molecule distribution over rotational levels is a Boltzmann distribution, the T_R is obtained by measuring the relative intensities of the rotational lines in a vibrational band of the FNS and plotting $-\ln[(I_{K',K''})_{v'v''}/(K'+K''+1)/\nu^4 G(K',T_R)]$ versus $K'(K'+1)$. Here $(I_{K',K''})_{v'v''}$ is intensity of FNS rotational lines in the (v', v'') vibrational band, K' and K'' are rotational quantum numbers (Hund's case (b)), v' and v'' are vibrational quantum numbers of $N_2^+(B^2\Sigma_u^+)$ and $N_2^+(X^2\Sigma_g^+)$ electronic states, $G(K',T_R)$ is a function of K' and T_R, ν is the wave number of the emission. Since

$G(K',T_R)$ depends on the rotational temperature T_R and on K', it is necessary to assume a temperature and then obtain the appropriate value $G(K',T_R)$ from a previously calculated look up table. One-two iteration is usually sufficient to obtain a satisfactory temperature measurement.

The total intensity of the vibrational molecular band does not depend on gas or rotational temperature and in case of low gas temperatures (<800K) when practically all molecules are in zero vibrational level (v''=0) is proportional to density of the molecules in the ground state $I_{n''v''}^{n'v'} \propto n_0$ (n' and n'' are electronic quantum numbers) [5]. We carried out measurements of the total intensity of (0-1) band of the N_2 FNS along and across the jet to obtain gas density profile in the jet recording the spectrum with 300 μm width of the spectrometer entrance slit. A typical band shapes with resolved and unresolved rotational structure are presented in Fig. 1.

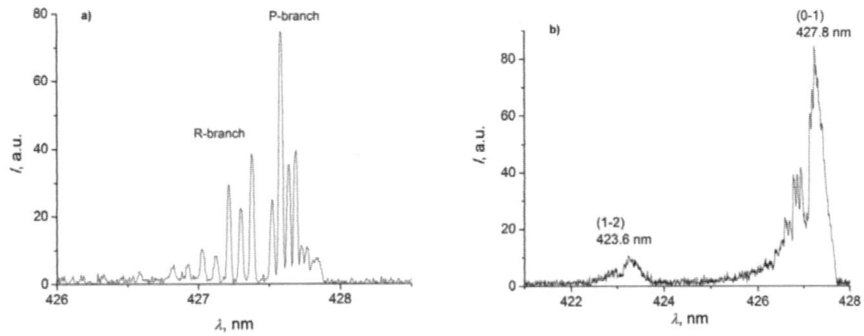

Fig. 1. Typical views of the (0-1) vibrational band of the nitrogen FNS in e-beam fluorescence spectra of the supersonic nitrogen jets with a) resolved and b) unresolved rotational structure; distance from the nozzle is 5 mm; a) nozzle #1, b) nozzle #2.

Several experiments were performed to make sure that the effect of secondary electrons and gas heating are small in our conditions. We observe EBF originated from stagnant room temperature nitrogen at the background gas pressure, varying e-beam current in a range of 10-100 mA, and studied dependences of intensity of the (0-1) band of FNS and T_r on e-beam current. The intensity was proportional to e-beam current i_{eb} in a range of 10-50 mA and then declined from the linear dependence showing tendency to saturation. In this range of i_{eb} the T_r was equal to room-temperature within the experimental error (≤5%) and at i_{eb} of 80-100 mA the T_r was higher on 30-50 K. In addition, T_r was measured in X-area of the supersonic jet (nozzle#2, P_0=0.4 MPa) at different i_{eb}, the measurements also display rise in T_r at i_{eb} of 80-100 mA compared with T_r obtained at i_{eb} of 20-40 mA, in the range of e-beam currents of 20-40 mA T_r was unchanged within the experimental error. Therefore, measurements were carried out at e-beam currents of 20-40 mA.

4. Results and discussion

Photos of the EBF of nitrogen supersonic jets expanding in the vacuum chamber are shown in Fig. 2.

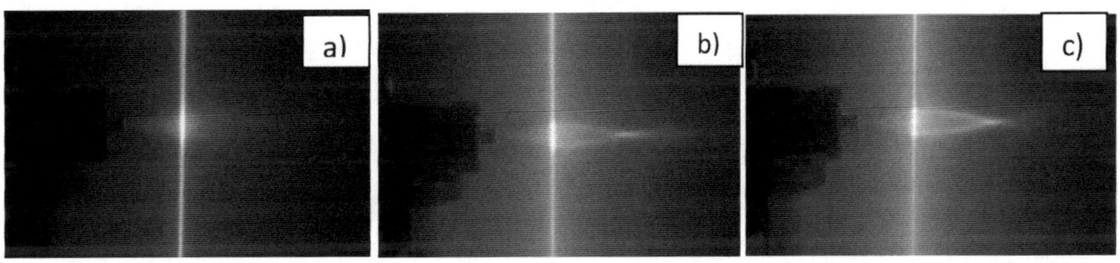

Fig. 2. Photos of the EBF of the expanding nitrogen supersonic jets (nozzle #1) at stagnation pressure of a) 0.2, b) 0.4 and c) 0.6 MPa.

Fig. 3 show axial profiles of radiation intensity of the (0-1) vibrational band of the FNS and rotational temperature of N_2 molecules for the supersonic nitrogen jets. Provided that the intensity is proportional to N_2 density n_0 the intensity profile displays n_0 distribution along the jet centerline. The T_r profile is like the gas temperature distribution along jet centerline because of $T_r=T_g$ due fast rotational-translational relaxation. As it is seen in Fig. 2a at $P_0 \leq 0.2$ MPa there is not formed a typical shape of supersonic jet with one or more barrels [14] so gas density and temperature in the jet decreases at mooving away from the nozzle (Fig. 3, curve *1*). At $P_0 \geq 0.4$ MPa the EBF photographs clearly show outlines of the first barrel and the shok wave intersection region (X-region). T_g at the nozzle outlet is of the order of a few tens of Kelvin. At mooving the jet away from a nozzle the gas density on the jet centerline first decreases at expansion then begins grow due to lateral shock waves influence up to X-region and after then decreases as the jet starts expand again (Fig. 3a, curves 2, 3); at jet expantion the T_g elevates and at compression it falls up to very low values of 15-20 K (Fig. 3b). The T_g and n_0 distributions in the nitrogen jets are unlike to the isentropic distributions, for isentropic approximation it is typical decreasing gas density and temperature at jet expansion away from a nozzle.

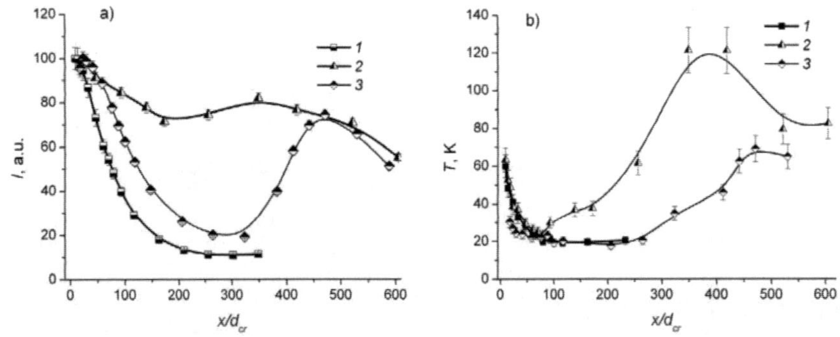

Fig. 3. Axial profiles of a) radiation intensity of the (0-1) band of the N_2 FNS and b) rotational temperature of N_2 molecules for the supersonic nitrogen jets; *1* – nozzle #2, P_0= 0.2 MPa, *2* – nozzle #2, P_0= 0.4 MPa, *3* – nozzle #1, P_0= 0.6 MPa.

Radial profiles of the (0-1) band intensity and T_r show that gas density decreases at mooving away from jet axis in radial direction and T_r remains almost unchanged in some area around axis and then increases; the last implies on penetration of background gas in to the jet.

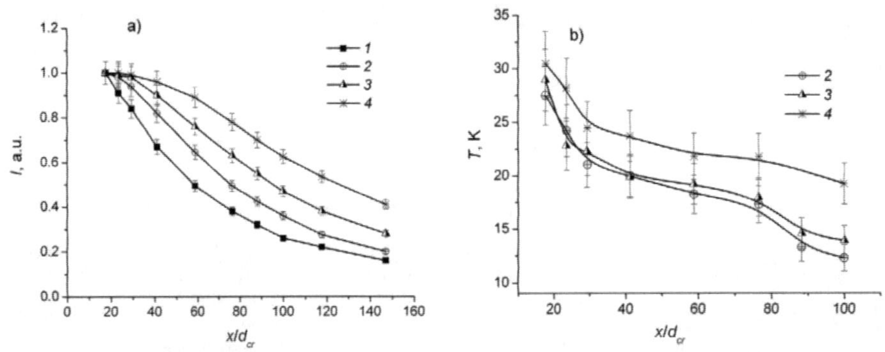

Fig. 4. Axial profiles of a) radiation intensity of the (0-1) band of the N_2 FNS and b) rotational temperature in expanding supersonic nitrogen jets (nozzle #1) as functions of stagnation pressure; *1* – P_0=0.1 MPa, *2* – 0.2 MPa, *3* – 0.4 MPa and *4* – 0.6 MPa.

Fig. 4 shows axial profiles of radiation intensity of the (0-1) band of the N_2 FNS and T_r in expanding nitrogen jets (nozzle #1) at different P_0. One can see the lower is P_0 the faster the gas density and T_r on the jet centerline decrease with distance from the nozzle. It is worth nothing that the larger is P_0 the higher is T_r. The reason for this effect may be formation of clusters in the jets.

Cluster mean size in the supersonic nitrogen jets was estimated with Hagena's parameter Γ^* [4] $\Gamma^* = K_{ch} P_0 d_{eq}^{0.85} / T_0^{2.29}$, where P_0 is stagnation pressure in mbar, K_{ch} is a coefficient depending on gas nature (528 for nitrogen [8]), T_0 is stagnation temperature in K; $d_{eq} = 0.74 d_{cr} / tg\alpha$ is equivalent nozzle diameter in μm, where d_{cr} is critical nozzle diameter and tgα is tangent of a nozzle half-angle. In accordance with [13] the cluster mean size is related to the Hagena's parameter Γ^* for large clusters as $\overline{N} = 2.94 \cdot 10^{-6} (\Gamma^*)^{2.35}$ and for small clusters as $\overline{N} = 4.62 \cdot 10^{-4} (\Gamma^*)^{1.64}$. The calculated values of Γ^* and $\overline{N}$ for the studied jets are presented in the Table. At cluster formation and growth a heat is released so the larger is the cluster mean size the higher is the gas temperature in supersonic jet. In jet forming with the nozzle #2 at stagnation pressure 0.4 MPa large clusters with mean size of 257 molecules are formed, this is greater than in the jet of nozzle #1 at stagnation pressure of 0.6 MPa. As it is seen in Fig. 3 T_g in jet of nozzle #2 at P_0 =0.4 MPa is higher compared with T_g in jet of nozzle #1 at P_0 =0.6 MPa. This agrees with the hypothesis of the cluster formation effect on gas temperature in the jet.

Table. The supersonic nitrogen jet parameters.

Nozzle #	P_0, MPa	P_0/P_h	Γ^*	$\overline{N}$	Nozzle #	P_0, MPa	P_0/P_h	Γ^*	$\overline{N}$
1	0.1	285714	355	7	1	0.4	285714	1421	68
1	0.2	303030	710	22	2	0.4	103359	2397	257
2	0.2	215054	1199	52	1	0.6	167131	2131	195

5. Conclusion

Axial and radial distributions of the gas temperature and density in supersonic nitrogen jets were studied using the EBF technique at stagnation pressures of 0.1-0.6 MPa. At the outlet of the nozzle, the gas temperature is of the order of a few tens of Kelvin and at distance of 100-200 d_{cr} the temperature drops up to 15-20K. The gas temperature and density distributions in the studied clustered nitrogen jets are unlike to the isentropic distributions. At pressures ≥0.4 MPa the EBF clearly shows outlines of the first barrel and the shok wave intersection region. It is shown that the lower is P_0 the faster decrease the gas density and T_r on the jet centerline with distance from the nozzle and the larger is P_0 the higher is the temperature. The reason for this effect may be gas condensation in to clusters in the jets. Mean size of clusters forming in the supersonic nitrogen jets was estimated with Hagena's parameter. It was found that at P_0 ≥0.4 MPa clusters with mean size of ≥ 200 M are formed in the jets. It is shown that there can be interdependence between cluster formation and gas temperature in the jets.

References

[1] Becker E W, Bier K und, Henke W 1956 *Zeitschrift fur Physik.* **146** S333
[2] Yamada I 2014 *Appl. Surf. Sci.* **310** 77
[3] Smirnov B M 1994 *Physics-Uspekhi* **164** (7) 665
[4] Hagena O F 1981 *Surface Science* **106** 101
[5] Muntz E P 1962 *Phys. Fluids* **5** 80
[6] Marrone P V 1967 *Phys. Fluids* **10** 521
[7] Coe D, Robben F, Talbot L, and Cattolica R 1980 Phys. Fluids **23** 706
[8] Belikov A E, Zarvin A E, Karelov N V, Sukhinin G I, Sharafutdinov R G 1984 *J. Appl. Mech. Techn. Phys.* **25** 331
[9] Mori H, Niimi T, Akiyama I, and Tsuzuki T 2005 Phys. Fluids **17** 117103
[10] Sharafutdinov R G, Ilyukhin A A, Smirnov V V, Belikov A E, Sukhinin G I, and Pykhov R L *1998 Chem. Phys.* **233** 127

[11] Zarvin A E, Kalyada V V, Madirbaev V Zh, Korobeishchikov N G, Khodakov M D, Yaskin A S, Khudozhitkov V E, and Gimelshein S F 2017 *IEEE Transactions on Plasma Science* **45** 819
[12] Mohamed A K, Bonnet J, Larigaldie S, Pot T, Soutadé J 2009 *Aerospace Lab* **01**-08
[13] Korobeishchikov N G, Zarvin A E, Kalyada V V, Khodakov M D 2012 *Vestnik of Novosibirsk State University* **7** 85
[14] Zischang J and SuhmM A 2013 *J. Chem. Phys.* **139** 024201

On electron attachment effect on characteristics of the DBD in chlorine and its mixtures with xenon

S V Avtaeva

Novosibirsk State University, 2 Pirogova Street, Novosibirsk, 630090, Russia

E-mail: s_avtaeva@srd.nsu.ru

Abstract. The electron attachment effect on DBD characteristics in chlorine and its mixtures with xenon has been studied. Characteristics of the DBDs in pure chlorine and in xenon-chlorine mixtures with a chlorine fraction of 0.1-5% were modeled using the fluid model. It is shown that the electron attachment limits a magnitude of the DBD current, contributes to formation of multiple current spikes, appearance of a double layer near the dielectric surface and formation of $XeCl^*$ excimer molecules, and leads to a redistribution of the power deposited into the discharge: more power is deposited into ions and less power is deposited into electrons.

1. Introduction

The electron attachment in plasma of electronegative gases and their mixtures with electropositive gases made the plasma properties very different from that of electropositive plasma [1]. The properties of the electronegative plasma depend on electronegativity n_n/n_e and a whole series of other parameters. The theoretical and experimental studies of dc discharges and slightly less rf discharges show that the discharges in electronegative gases are structured with an ion-ion core and an outer electron-positive ion plasma [2, 3]. The electron attachment effects on property of the transient discharges quickly changing their characteristics in space and time as, for example, the dielectric barrier discharges (DBD) are much less studied. DBDs in inert gas-halide mixtures are widely used in excilamps to generate VUV radiation [4, 5] and can be applied for diverse surface treatment. The future development of the DBD plasma sources will depend on a better understanding the underlying physics. There we present some finding on the electron attachment effect on properties the DBD in $Xe\text{-}Cl_2$ mixtures and in pure chlorine, which are revealed at fluid modeling of the discharges.

2. Model

Characteristics of the DBD in $Xe\text{-}Cl_2$ mixtures were simulated using 1D fluid model described in detail elsewhere [6]. To decrease a time of simulation the simplified kinetic scheme of reactions was used [7, 8]. Characteristics of the DBD in pure chlorine were simulated using similar 1D fluid model of the DBD described in detail in [9]. We analyzed characteristics of a DBD between plane-parallel dielectric-covered electrodes to which a harmonic voltage with amplitude U_0 and frequency f is applied. The one-dimensional model of a homogeneous DBD is based on the continuity equations for charged components (electrons, negative and positive ions), balance equations for neutral components coupled via a set of reactions under the assumption of a local electric field, Poisson's equation, and equation for the external electric circuit.

Set of partial differential equations with allowance for initial conditions, boundary conditions, and equation for the external circuit was solved by the method of lines. In this method, the spatial derivatives are discretized by the finite-difference method. As a result, the nodal values of each dependent variable become unknown functions of time. Discretization was performed by the control volume method on a quasi-uniform mesh. The electron and ion fluxes were approximated by means of the Sharfetter–Gummel exponential scheme. The resulting set of coupled differential-algebraic equations was solved by the multistep backward differentiation method (the Gear method). The numerical algorithm was described in detail in [6, 10].

3. Results and discussion

Characteristics of the DBDs in pure chlorine and in xenon-chlorine mixtures (0.1-5% Cl_2) were modeled using the fluid model. The discharge gap length d_g is varied in a range 4-8 mm, electrodes are covered with dielectric layers. Two dielectric layers have an identical thickness d_b of 1- 2 mm and a relative permittivity of 4 (glass). The harmonic voltage $U_S=U_0 sin(2\pi ft)$ with amplitude of 4.25-8 kV and frequency of 25-100 kHz was applied to electrodes. The gas pressure P was varied in a range of 15-300 Torr and gas temperature is assumed to be equal 300-400 K.

3.1. Xe-Cl_2 mixtures

The calculations show that, as the chlorine concentration in the mixture increases, the voltage drop across the discharge gap lifts, while the amplitude of the current spike decreases. The lower the chlorine concentration is, the greater the voltage drop across the discharge gap after the current spike is. The discharge accumulated at the surfaces of the dielectric layers decreases with increasing chlorine concentration in the mixture. These effects are explained by electron attachment to chlorine atoms, e + Cl2--> Cl + Cl^-.

It was obtained both experimentally and theoretically that as the chlorine concentration n_{Cl2} in Xe-Cl_2 mixtures increases, the radiation intensity of the second continuum of Xe_2* decreases substantially, while the intensity of the $XeCl^*$ (B→X) band increases. Already at n_{Cl2} =0.2%, the radiation of the $XeCl^*$ (B→X) band with a head at 308 nm becomes dominant. The increase in the intensity of the $XeCl^*$ (B→X) band with increasing chlorine concentration is observed only at n_{Cl2} ≤1%. The further increase in the chlorine concentration leads to the rapid decrease in both the intensity of the $XeCl^*$ (B→X) band and the total intensity of discharge radiation. According to our experimental examination and fluid modeling, the chlorine concentration optimal for formation of the $XeCl^*$ (B→X) band is 0.4-0.5% [11]. At this chlorine concentration about 81% of the discharge radiation falls on the $XeCl^*$ radiation, 18% - on Xe_2^* radiation and only 1% - on Xe_r^* radiation.

Consideration of the main processes influencing the formation and quenching of working XeCl* molecules during a various phases of voltage cycle shows that during short current spikes the harpoon reaction $Xe^*+Cl_2 \rightarrow XeCl^*+Cl$ provides the greatest contribution into generation of $XeCl^*$ exciplex molecules while the ion-ion recombination $Xe_2^++Cl^- \rightarrow XeCl^*+Xe$ provides the greatest contribution during afterglow phase [12]. Quenching of $XeCl^*$ molecules is a result of the radiative decay $XeCl^* \rightarrow Xe+Cl+hv$ (308 nm). However during current spikes the dissociative ionization $e+XeCl^* \rightarrow Xe^++Cl+2e$ also plays a dominant role in the quenching of XeCl* molecules.

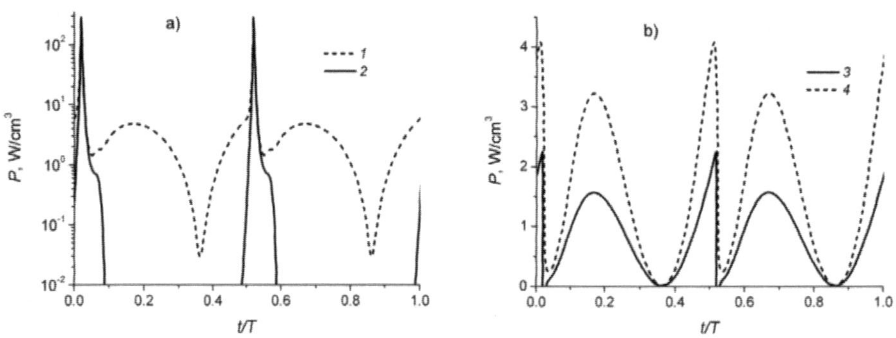

Fig. 1. Gap-averaged instantaneous discharge power *1* and its distribution between electrons *2* and positive *3* and negative *4* ions during a cycle; U_0=5 kV, f=100 kHz, and P=200 Torr [12].

The discharge radiation power strongly changes during a cycle and correlate with the power deposited into electrons. It was observed that electrons adsorb discharge power in each period during small intervals of a time correlating with current spikes (Fig. 1). Between current spikes, discharge energy is mainly absorbed by ions. It is because of during periods between current spikes densities of electrons and Xe^+ ions strongly decrease because of attaching of electrons to chlorine atoms $e+Cl_2 \rightarrow Cl+Cl^-$ and conversion of Xe^+ ions into Xe_2^+ ions $Xe^+ + 2Xe \rightarrow Xe_2^+ + Xe$. However during current spikes the electron density strongly increases owing to ionization of gas and in the sheath near dielectric layers it exceeds densities of Cl^- and Xe_2^+ ions.

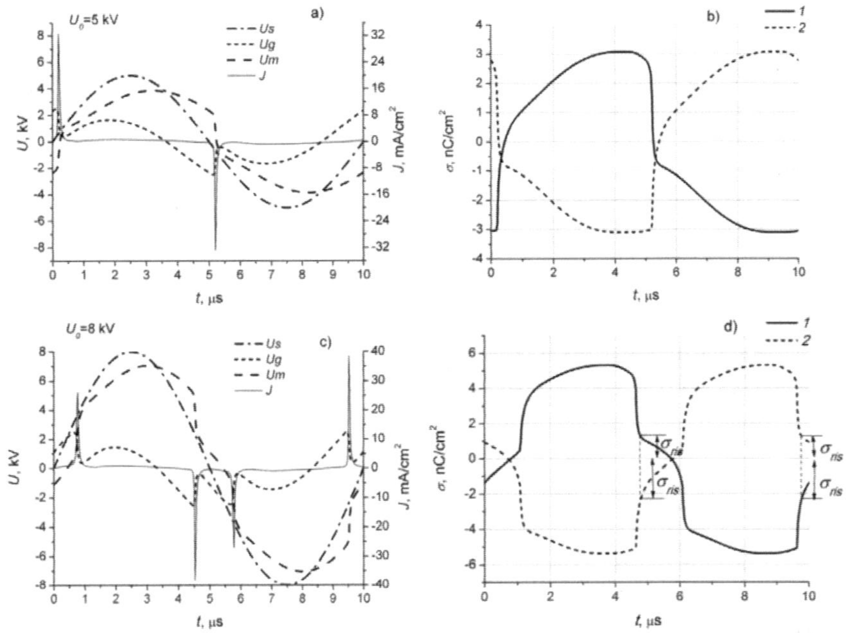

Fig. 2. Time behavior (a), (c) of the source voltage U_s, voltage drops across the discharge gap U_g and dielectric barriers U_m and the total current density J, and (b), (d) the surface charge density on the dielectric layer adjacent to (*1*) the left (grounded) electrode and (*2*) the right (loaded) electrode at the voltage with f= 100 kHz and U_0= 5 (a, b) and 8 kV (c, d) in the DBD in Xe-Cl_2 (0.5% Cl_2), P=200 Torr; d_g= 4 mm, d_b= 2 mm [7].

Fig. 2 shows time variations of the source voltage U_s, the voltage drops across the discharge gap U_G and the dielectric barriers U_m , the total current density J and surface charge densities at the dielectric surfaces adjoining to the left (grounded) and right (loaded) electrodes in the DBD in Xe-Cl_2 mixture with chlorine concentration 0.5%.

In case of moderate U_0 (a low over-voltage) the charge deposited on dielectric surfaces during the current spike quickly decreases the dielectric surface charge up to zero, so U_G and U_m fall up to small values (Fig. 3 a, b) and it leads to a local collapse of the electric field in the area defined by the surface charge. After the current spike the surface charge again gradually increases owing to charge transferred by the plasma. Ionized gas in the gap is almost in quasi-neutral condition. The most part of a source voltage falls across the dielectric layers and smaller part falls across the gas gap. As a result, after the spike a voltage drop on a gap decreases with time, not having reached the breakdown voltage during the voltage half-cycle once again [7].

In case of high U_0 (a sizeable over-voltage) the charge deposited on dielectric surfaces during current spike is not sufficient for neutralization of the surface charge accumulated on the dielectric surfaces before the spike (Fig.3 c, d). Thus, after the current spike a residual charge with density σ_{ris} $\approx$1–2 nC/cm^2 remains on dielectric surfaces (Fig. 3 d). As a result, an electric field near to the dielectric surface (the cathode fall) decreases not completely. After 'primary' spike across dielectrics and across the gas gap there are voltage drops having opposite signs. The great part of the source voltage U_S falls on the gap. The gap voltage accelerates seed electrons to dielectric to neutralize the residual surface charge. Leaving of electrons from the plasma to the dielectric surface increases electric field in the sheath near the dielectric. The voltage drop across the gap promptly reaches the breakdown value once again. In the strong electric field a 'second' ionization wave is developed to create a 'secondary' current spike [7].

Thus, the reason of developing the 'secondary' current spike is that in case of the great over-voltage there is strong electric field across the gap and a great charge deposited on dielectric surfaces before the 'primary' current spike. At the same time, magnitude and time of the current spike depend on U_0 a little and are determined by the electron mean energy and density culminated at breakdown voltage, and by rates of ionization, recombination and electron attachment during the current spike. As a result, the charge, transferred during a current spike, depends on U_0 a little.

3.2. Pure chlorine

Fig. 3 shows the time dependences of the power supply voltage, the voltage drops across the discharge gap and the dielectric barriers, and the total current density of the DBD in chlorine over 1.5 voltage cycles. It can be seen that at a chlorine pressure of 15 Torr a breakdown occurs when the voltage drop across the discharge gap reaches a value of ~0.5 kV. Several weak current spikes, a number of which increases with decreasing pressure, are observed after the first spike. The peak current density in the spikes at the chlorine pressure of 15-100 Torr is about 4 mA/cm^2.

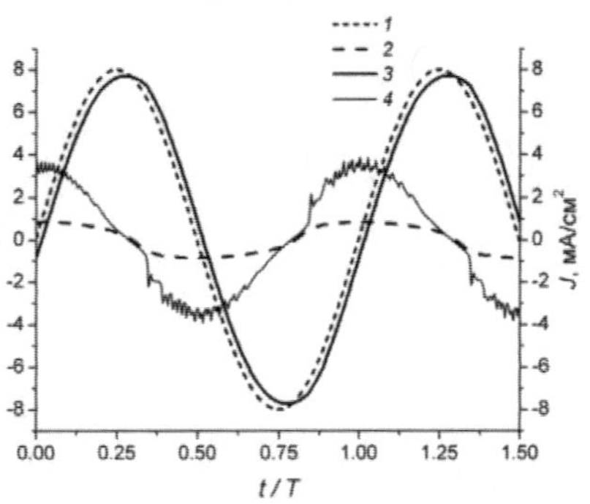

Fig. 3. Time dependences of the (*1*) electrode voltage U_S, (*2*) voltage drop U_g across the discharge gap, (*3*) voltage drop U_m across the dielectric layers, and (*4*) total current density J in the DBD in Cl_2 with d_g =8 mm, d_b=2 mm at $f = 100$ kHz, $U_0 = 8$ kV and P= 15 Torr [9].

Two phases can be distinguished in the discharge dynamics: the active phase with multiple current spikes, which follows the breakdown of the discharge gap, and the passive phase. The active phase is characterized by the current spikes, a high current amplitude compared to that in the passive phase, small surface charge density on the dielectrics, and high voltage drop across the discharge gap. The passive phase (with no current spikes) is characterized by a low current amplitude, large surface charge density on the dielectrics, and small voltage drop across the discharge gap [9].

The chlorine plasma is electronegative, $n_n >> n_e$, n_n and n_e are density of negative ions and electrons, correspondingly. During current spikes the electron density in both space charge sheaths (~0.5 mm) increases and double layer appears near the electrode to which an ionization wave moves (the left electrode in Fig. 4): the positive space charge sheath adjoining the dielectric is followed by the negative space charge sheath extending up to the boundary of the quasi-neutral plasma. The appearance of a double layer is related to the fast increase in the electric field near the dielectric as it is approached by the ionization wave. The mean electron energy at the front of the ionization wave increases substantially, reaching 7 eV near the dielectric, and the dissociation and ionization rates of chlorine molecules, the rate constants of which increase with increasing E/n, also grow. At the same time, the rate constant of electron attachment to chlorine atoms depends weakly on E/n, and the rate of electron attachment to Cl atoms decreases after the mean electron energy reaches 5 eV. As a result, ionization of chlorine molecules near the dielectrics dominates over electron attachment; the density of

Cl_2^+ ions near the dielectric increases abruptly, forming a positive space charge; and the electrons escape onto the dielectric surface, thereby neutralizing the surface charge. Electron attachment dominates over ionization behind the front of the ionization wave, where the electric field is weak. In this case, the field of the positive space charge near the dielectric surface facilitates charge separation behind the front of the ionization wave by entraining negative ions and repelling positive ones. As a result, a double layer forms near the dielectric toward which the ionization wave propagates [9].

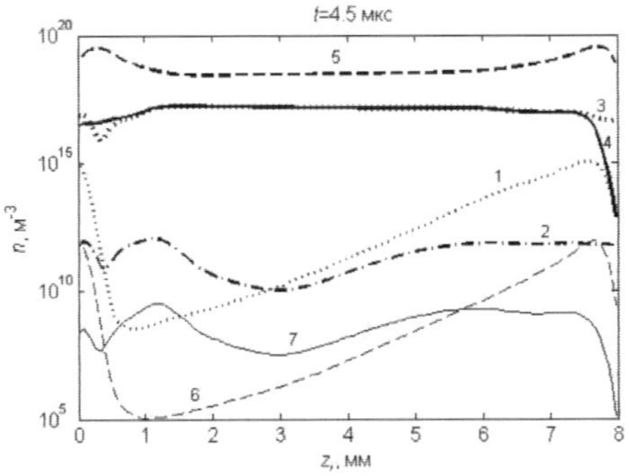

Fig. 4. Distributions of the densities of the plasma species along the discharge gap 4.5 μs after the beginning of the voltage period at a chlorine pressure of 15 Torr: (*1*) electrons, (*2*) Cl^+, (*3*) Cl_2^+, (*4*) Cl^-, (*5*) Cl, (*6*) Cl*, and (*7*) Cl_2^* [9].

After first current spike, the surface charge on the dielectrics decreases. However, the electron density remains low, although it increases by two to three orders of magnitude. Since the mobility of Cl^- ions is much lower than the electron mobility, the current spike is weak and rapidly terminates, because the electric field decreases due to a decrease in the surface charge density and mean electron energy in the course of ionization, dissociation, and excitation. As a result, the surface charge density decreases a little, and a large surface charge still remains on the dielectrics after termination of the current spike. This leads to a rapid growth of the electric field near the dielectric toward which the ionization wave propagates, thereby giving rise to a new current spike. This spike is also weak for the same reason as the preceding spike, and the process repeats again. Thus, there appear several successive current spikes, the amplitudes of which initially grow from spike to spike. When the voltage across the discharge gap drops to value that is lower the breakdown voltage the spikes disappear and the current gradually decreases as the surface charge accumulates on the dielectrics [9].

4. Conclusions

The electron attachment determines the DBD characteristics in chlorine and has the great effect on the DBD behaviors in Xe-Cl_2 mixtures. The modeling shows that the electron attachment limits a magnitude of the DBD current, contributes to formation of the multiple current spikes, appearance of a double layer near the dielectrics surface and formation of $XeCl^*$ excimer molecules, and leads to a redistribution of the power deposited into the discharge: more power is deposited into ions and less power is deposited into electrons. The reason of developing the multiple current spikes is that a charge, transferred by a current spike, weakly depends on a voltage, so at big voltage the charge is insufficient for neutralization of the dielectric surface charge. As a result, voltage drop across the gap promptly reaches the breakdown value once again and the 'next' current spike develops. At moving an ionization wave to the electrode covered by dielectric layer the electron attachment leads to appearance of a double layer near the dielectrics surface because of sharp changing an electric field, ionization and attachment rates and separated effect of the field on charged species.

References

[1] Franklin R N 2002 *Plasma Sources Sci. Technol.* **11** A31.
[2] Lichtenberg A J, Lichtenberg M A, Kouznetsov I G and Chung T H 2000 *Plasma Sources Sci. Technol.* **9** 45.
[3] Ivanov V V, Klopovsky K S, Lopaev D V, Rakhimov A T and Rakhimova T V 1999 *IEEE Trans. Plasma Sci.* **27** 1279.
[4] Kogelschatz U 2012 *J. Opt. Technol.* **79** 484.
[5] Lomaev M I, Skakun V S, Sosnin E A, Tarasenko V F, Shitts D V, Erofeev M V 2003 *Phys.Usp.* **46** 193.

[6] Avtaeva S V, Saghi B and Rahmani B 2011 *IEEE Trans. Plasma Sci.* **39** 1814.
[7] Avtaeva S V 2014 *IEEE Trans. Plasma Sci.* **42** 229.
[8] Saghi B, Avtaeva S V, Rahmani B, and Zissis G 2016 *IEEE Access* **4** 792.
[9] Avtaeva S V 2017 *Plasma Phys. Rep.* **43** 876.
[10] Avtaeva S V and Kulumbaev E B 2008 *Plasma Phys. Rep.* **34** 452.
[11] Avtaeva S V, Sosnin E A, Saghi B, Panarin V A, and Rahmani.2013 *Plasma Phys. Rep.* **39** 768.
[12] Avtaeva S 2014 *Eur. Phys. J. D.* **68** 90.

IOP Conf. Series: Journal of Physics: Conf. Series **927** (2017) 012007 doi:10.1088/1742-6596/927/1/012007

Interaction of low-temperature plasma with knitted fabric based on natural cellulose fibers

A A Azanova, I A Borodaev, A A Shakhirov, V A Sysoev and V S Zheltukhin

Kazan National Research Technological University, 68 K. Marks str., Kazan, 420015, Russia

E-mail: vzheltukhin@gmail.com

Abstract. The influence of low-temperature plasma on a knitted fabric of natural cellulose fibers is considered. It is shown that the plasma ion energy is sufficient for the waxy layer destruction. It was determined that the ion bombardment and the paraffin vaporization lead to disintegration of the natural waxy fiber layer and cause the defects in paraffin layer. The local ruptures of carbon chains of alkane molecules of the paraffin layer are caused by the low-energy plasma ions bombardment and the ions recombination. The alkyl radicals which react with the plasma gas particles and with each other were identified. As a result of reactions of radicals, the hydrophilic functional groups was evaluated by IR spectroscopy and secondary ion mass spectrometry. All these changes facilitate the penetration of the working solutions in the fiber. Plasma effect leads to an increase in the hydrophilic properties of the material and can replace the liquid processes of preparing knitted fabrics for dyeing.

1. Introduction

Plasma technologies are widely used for the processing of textile materials of different nature. The advantage of plasma treatment is the versatility. Plasma allows changing the properties of the textile material in the desired direction. These changes depend on the type of plasma-supporting gas and processing parameters, such as discharge wattage – *W*, processing time – *t*, etc. Plasma processing can be used at any stage of fiber finishing. The main advantages of plasma treatment are the use of the gas medium and the minimum expenses for recycling. The economic efficiency of plasma and plasma chemical technologies mostly depends on the level of the equipment use and the duration of the production cycle [1]. Reduction of the latter can be achieved by the intensification of traditional processes or by the exclusion of some liquid-phase stages from the technological chain.

All types of plasma used for the modification of textile materials can be divided into two main groups - atmospheric pressure plasma and low pressure (vacuum) plasma [2]. The main advantages of the atmospheric pressure plasma (APP) are higher processing speed and lower cost of the equipment. On the other hand, the low-pressure plasma (LPP) provides the flexibility and easiness of process control and the reproducibility of effects. The LPP facility can be simply adapted to various materials treatment [2]. So, it is possible to regulate the obtained effects widely, using different plasma-supporting gases [3]. There are many other factors that prove the advantages of LPP: utilization of ozone and nitrogen oxides is not necessary; lower applied voltages can be used; plasma effects over time are more stable.

According to some researchers [4, 5] it is shown, that the intensity of textile materials processing is higher in the LPP than in the APP, on equal watts input. It is obvious that the lifetime of active particles is higher in the LPP than in the APP. In vacuum, active particles (e.g. atomic oxygen) can diffuse inside the structure of the textile material deeper. Conversely in the APP a very thin surface layer is mainly processed.

Among other things, the APP processing effects are sensitive to atmospheric humidity and moreover to the material moisture content. The latest can vary (sometimes quite significantly) from the edge to the center, across the fabric. Alternately, the material processed in the LPP is degassed, what results in the higher treatment uniformity.

The influence of the LPP produced by a radio-frequency capacitive coupled discharge (RFCCD) on knitted materials of cellulose fibers is under consideration in the given article.

2. Materials and methods

Knitted materials are obtained by knitting of yarns from cellulose fibers. The paraffin is applied to the yarns to facilitate knitting and reduce yarn breakage, but it should be removed before dyeing. Traditionally, the paraffin and other non-cellulosic contaminants are removed by boiling in an alkaline medium at high temperatures (95-105 °C). As a result, the cotton fiber loses its strength and the fabric hand properties become worse. Plasma processing is offered by the authors to prepare knitted fabrics for dyeing, as an alternative to the traditional liquid processes.

The object of the research is unbleached cotton knitted fabric with a surface weight - 165 g/m^2. Refined oil paraffin was used as a model of traditional wax coating. The paraffin was applied on a glass substrate and then was treated with plasma. The paraffin layer thickness was from 0.005 mm to 0.01 mm.

The plasma processing was carried out on the facility with the following characteristics of RFCCD: the charged particle density - n = 10^{16}- 10^{19} m^3, the ions energy in the ionic layer of positive charge - W_i = 70-100 eV and the ion current density on the sample surface - j_{is} = 0.3-1.0 A/m^2 [6].

3. Results and discussion

The original unbleached knitted fabric is hydrophobic and is not wetted by water. So, there is no capillary rise of water through the fabric. But, the fabric becomes hydrophilic and gets perfect wetting ability after plasma exposure. The height of the capillary rise of water is 180-220 mm depending on the processing conditions. It was also determined, that the hydrophobic surface of model paraffin becomes hydrophilic after plasma treatment (Figure 1). Consequently, the hydrophilization of the unbleached cotton knitted fabric depends on the hydrophilization of paraffin, covering the fibers and filaments.

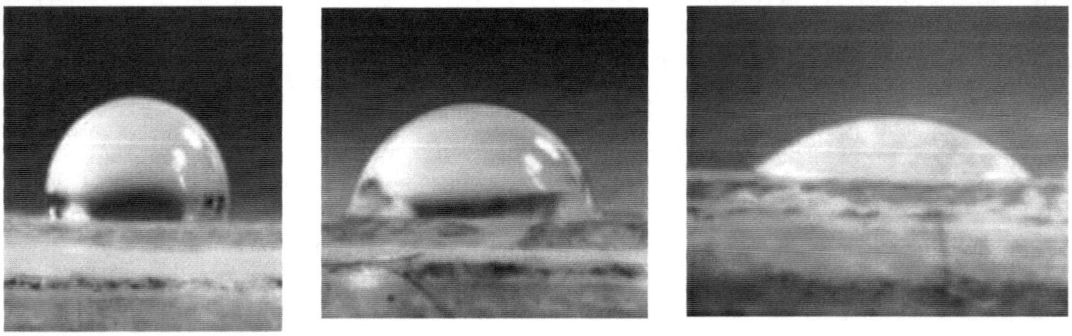

a b c

Figure 1. View of the distilled water drop on the paraffin: a – an initial sample (θ = 107^0), b – the sample treated by argon plasma (θ = 88^0), c – the sample treated by air plasma (θ = 41^0).

The content of paraffin on the fabric decreases on average by 18-25% in air plasma and by 5-8% in argon plasma after plasma treatment during 300-500 sec (Figure 2).

Plasma treatment changes the surface structure of the cotton fibers (Figure 3). The surface becomes bulkier after treatment in the argon plasma, but no noticeable damage is observed. The discharges on fibers and an increase in the fiber roughness are fixed after its' processing in air plasma. The size of etched areas is 10-20 nm [7].

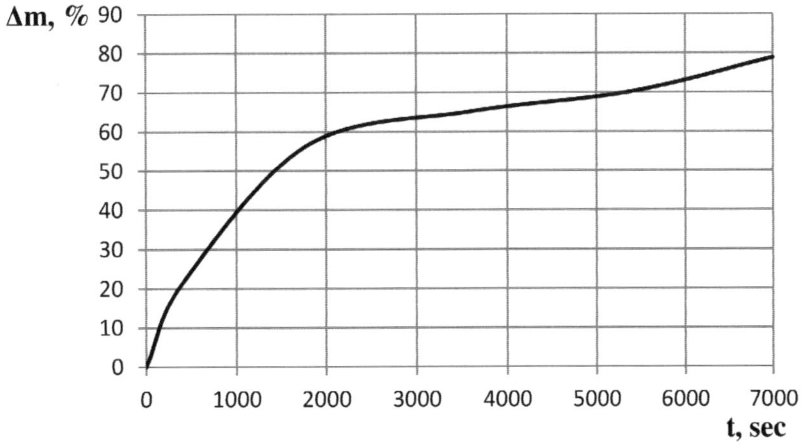

Figure 2. The removal rate of paraffin (determined by extraction with CCl_4) from the unbleached cotton fabric as a function of the processing time in air plasma.

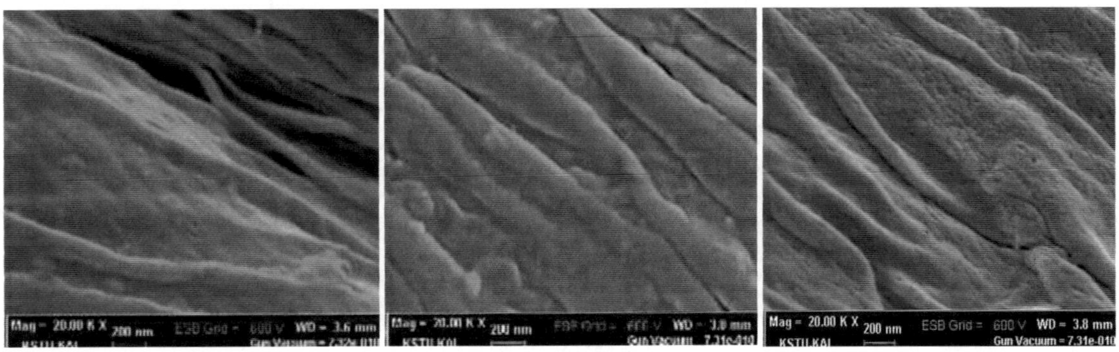

a b c

Figure 3. Microphotographs of cotton fibers (× 20000): a - check sample, b – the sample treated in argon plasma, c - the sample treated in air plasma.

Thus, according to the results of experimental studies, the unbleached fabric properties are similar either after processing in air plasma or after alkaline boiling. So, the paraffin is partially removed, the knitted fabric is hydrophilic and the wholeness of the outer layer of cotton fiber is damaged in places.

It is necessary to evaluate the plasma effect on cotton fiber and mostly on the upper paraffin layer. For this reason, the modeling by the Monte Carlo method was performed and the changes in the paraffin surface during the bombardment by low-energy plasma ions were studied.

The initial data for the model were the following: the expected composition of argon and air plasma [8] and the molecular weight distribution of the paraffin [9]. The breaking process of carbon chains of paraffin molecules was simulated because of impact of plasma-supporting gas ion. Since the body placed in the plasma is charged negatively [10], then positively charged ions were considered. Only the singly charged ions were included in the simulation. So, the wattage W_i of all ions was chosen equal. The model of the paraffin layer was simplified and was based on a set of molecules of alkane hydrocarbons of different lengths. It was supposed that molecules are arranged relative to each other in a random manner.

The unit cell area of 100 nm^2 was studied within the simulation. The plasma processing of paraffin layer for 420 secs with the gas ions energy W_i = 50-100 eV was simulated. Such ions energy corresponds to the discharge wattage – W_d = 0.8-2.0 kW. The processing of the fibers surface in the interfibre space was simulated in breakdown conditions (assisted discharges). Those discharges are formed due to the energy - W_i = 18 eV, which released by the ions recombination [6]. The processing intensity was set equal to 7 ions / (nm$^2\cdot\times$ sec).

The Figures 4-6 depict the results of numerical simulation of argon and air plasma processing with the ions energy W_i = 70 eV. The fractional composition of paraffin and generated alkyl radicals are shown in the given figures.

According to the calculation results, the impact of plasma ions with the energy W_i = 50-100 eV causes some changes in the fraction composition of the paraffin. These changes are more obvious after air than after argon plasma processing. Paraffin is also processed due ions recombination in the interfibre space, with the low energy - 18 eV [6], so the changes in the molecular weight distribution are not so significant in this case.

Among the alkyl radicals that are formed during the plasma treatment over than 70% correspond to the gaseous fraction (Figure 6). Since, with a decrease in pressure, the boiling point of paraffin is reduced under vacuum and gas purge, fractions with a carbon number of 16 and less will be removed. The removal of the gaseous fraction results in a change in the molecular weight distribution of paraffin. In this case, according to calculations, the average molecular weight M_{av} of paraffin increases from 331.4 to 334.8 in air plasma and to 334.7 in argon plasma.

Therewith the calculated weight loss of paraffin Δm (when W_i = 50-100 eV) is (0.69 – 1.09) × 10^{-8} mg/m^2 in argon plasma, and (0.96-1.72) × 10^{-8} mg/m^2 in the air plasma (Figure 7). The calculated minimum depth of penetration of the plasma-supporting gas ions is 1.5-8 nm.

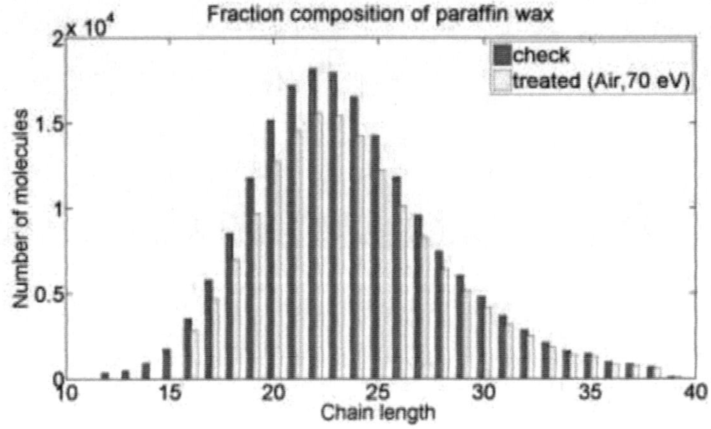

Figure 4. Fractional composition of paraffin before and after the treatment in air plasma, W_i = 70 eV.

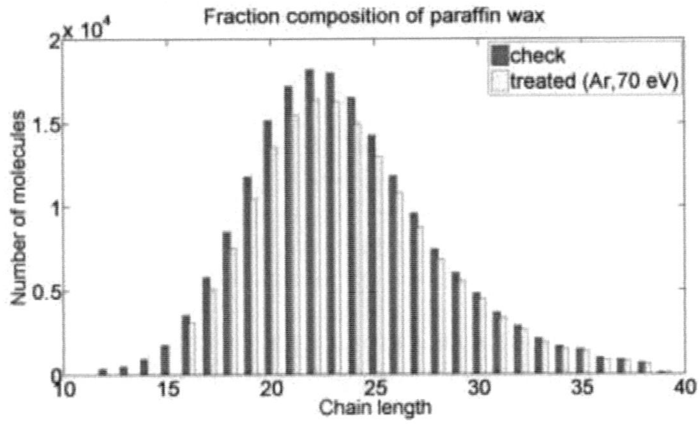

Figure 5. Fractional composition of paraffin before and after the treatment in argon plasma, W_i = 70 eV.

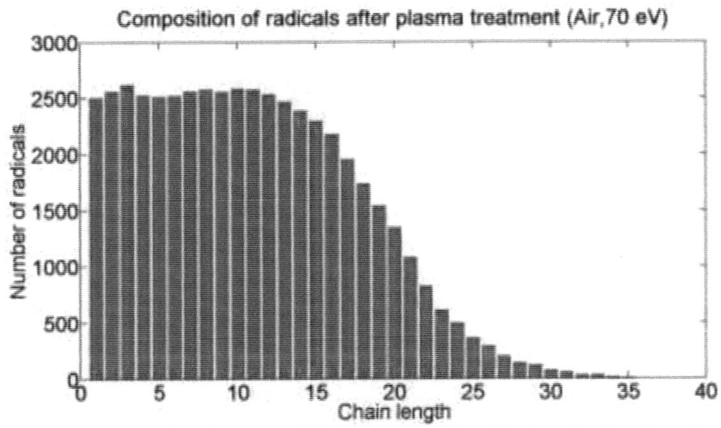

Figure 6. Distribution of fractional composition of paraffin alkyl radicals after the treatment in air plasma, W_i = 70 eV.

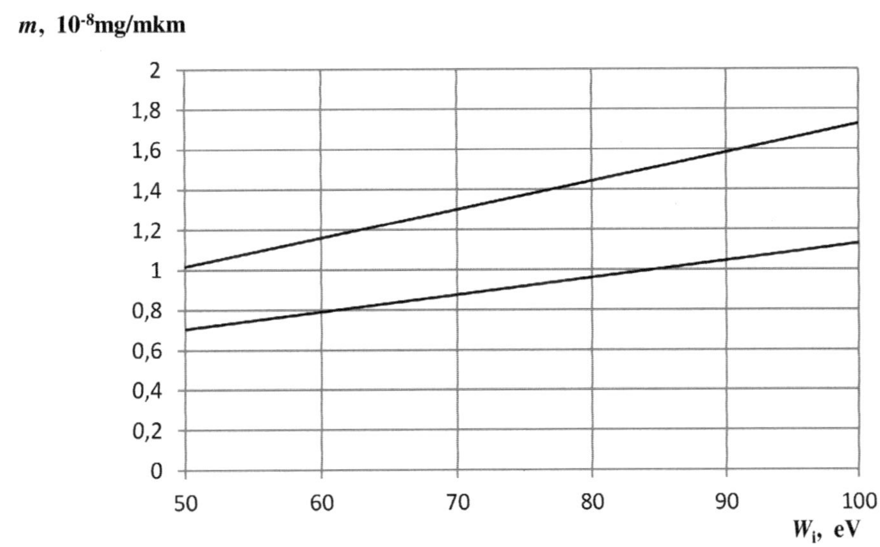

Figure 7. Mass *m* of removed paraffin fractions, as a function of the ion energy W_i: 1 - air, 2 – argon.

According to the results of simulation and experimental studies it is shown that the physical effect of plasma is insignificant. Such effect doesn't lead to increase of the parameters of hygroscopic properties (wetting ability, capillarity) of unbleached knitted fabrics. Quite the opposite, the chemical effect of the plasma has the greatest contribution, while the fabric is processing in air plasma.

The developed model is approximate, because it doesn't consider the chemical interaction between the radical residues and their reactions with the particles of the plasma-supporting gas. However, the obtained simulation results are sufficient to explain and evaluate the achieved effects. To determine the adequacy of the proposed model, additional experimental studies were carried out.

The average molecular weight of the paraffin M_{av} correlates with the self-diffusion coefficient D, which was determined by NMR method. It is known that an increase of polymers' M_{av} complicates the molecules diffusion within the system. According to NMR results, plasma treatment leads to a change in the diffusion attenuation of the sample signal. Along with this, the number of the components with the lower self-diffusion coefficient is directly proportional to the plasma treatment time. According to calculations, there is a reduction of the average self-diffusion coefficient of paraffin in 1.7 times: $D_{check} = 4\times10^{-9}$ m^2/sec, $D_{treated} = 2.3\times10^{-9}$ m^2/sec. The given results indicate an increase in the paraffin M_{av} after plasma treatment.

According to the proposed model, ion bombardment of the paraffin surface leads to the formation of alkyl radicals. These short-lived radicals can react with plasma particles; thus the products of such reactions were found. The oxygen-containing functional groups in the treated paraffin were registered by different methods. So, acid groups are determined in the amount of 0.95 mg KOH/g after treatment in air plasma. There are no acid groups in the check and in the argon treated samples and acid number is 0.00. In addition to it, the IR spectrums in ATR mode were studied. The occurrence of signal at 1720 cm^{-1} is recorded, which is relevant to the asymmetric stretching vibrations of the C=O groups. IR spectrums of the initial paraffin sample and after the treatment in argon plasma are identical.

It is known [11] that for paraffin oxidation reactions rather rigid conditions are required: elevated temperature, the action of ultraviolet or radioactive radiation and/or the presence of initiators as sources of active free radicals. In our case plasma is the source of active free radicals and ultraviolet radiation. Except but, the oxidation reactions occur locally and only at the interface, causing changes in the thin surface layer of the paraffin.

The oxygen is in the active form in the gas medium of air plasma. It can be in the form of ions, excited molecules, atoms, ozone, etc. So, all these oxygen forms can cause the complete oxidation of paraffin and organic substances in the contact layer. It is known that such reactions proceed by the release of carbon dioxide CO_2. When the amount of oxygen is insufficient for complete paraffin oxidation, then partial oxidation products are formed: carbon monoxide CO and carbon. Mass-spectrometry was used to reveal these oxidation products in a plasma medium.

There are the changes in the gases ratio within the processing medium during the plasma treatment. A noticeable increase in N_2, CO_2 and CO and reduction of O_2 and N_2 are determined in the air plasma treatment process. There are no significant changes in the gases ratio after the paraffin processing in argon plasma. Only the increase of H_2 quantity is recorded. It is known [12] that the release of hydrogen into the gas phase is usual for the plasma treatment of polymeric materials. It occurs due to radical reactions because of transfer of excitation energy from plasma particles to macromolecules [5]. The expenditure of oxygen in oxidation reactions can be indicated by the increased CO and CO_2 and reduced O_2 quantity. The alkyl radicals are also involved in oxidation reactions. Radicals appear in gaseous medium due to local rupture of paraffin carbon chains and have the C_{16} or less carbon number in the chain.

The balance of atomic molecular composition of the gaseous medium before and after plasma treatment was determined. It was found that the O_2 consumption is higher than it is required for reactions of CO and CO_2 formation. On the contrary, the number of carbon and hydrogen atoms increases. These changes indicate the oxygen consumption in the reactions with non-volatile components formation.

The cellulose fiber surface was studied by secondary ion mass spectrometry (SIMS). The appearance of the products of the cellulose interaction with air plasma was detected. SIMS method provides the surface composition analysis to a depth of 0.3-10 nm. There is a sharp increase of ions corresponding to oxidized hydrocarbons (up to 20-50 times): $C_{17}H_{34}O_2^+$, $C_{18}H_{39}O_2^+$, $C_{19}H_{38}O_2^+$, after RFCCD air plasma treatment. Along with this, the quantity of nitrogen-containing ions increases. The SIMS results indicate reactions of oxygen and nitrogen of the plasma-supporting medium with some substances located on the fiber surface.

According to the above-mentioned simulation results, the alkyl radicals are formed in the process of paraffin plasma treatment, where the largest proportion (40-50%) take the radicals with the $C_{16\text{-}19}$ amount of carbon atoms in the chain (Figure 6). They react with each other and with plasma particles including oxygen and nitrogen, what results in their contents increase as recorded by SIMS. This position is approved by the increase of oxygenates ions content with the number of $C_{17\text{-}19}$ carbon atoms on the fiber surface.

As the waxy substances on cellulosic fibers contain saturated hydrocarbons, so such processes are likely to occur in the fibers wax layer.

4. Conclusion

LPP changes on the thin surface layer of cotton fiber including the paraffin layer, which is used to cover yarns for facilitating the knitting process during knitted fabrics manufacture. The local ruptures of carbon chains of alkane molecules of the paraffin layer are caused by the low-energy plasma ions bombardment and the ions recombination. Further, the alkyl radicals emerge and react with the plasma gas particles and with each other. As a result, the volatiles form and then they are removed by the gas purge under reduced pressure. This procedure helps to clean knitted fabric. The radicals react with oxygen on the paraffin layer surface and then the hydrophilic functional groups appear. These changes lead to the increase of wetting ability and capillarity characteristics.

The paraffin heats and spreads over the fibers and filaments surface during the plasma treatment. At low pressure, the boiling point of the paraffin is reduced and the evaporation of its low molecular weight fractions starts. The defects of paraffin layer are formed in the evaporation field. Since paraffin composition on the fibers and filaments surface is uneven, so in the areas with the smallest paraffin layer thickness and with the absence of paraffin the ion bombardment affects waxy substance. The ion energy is sufficient for the waxy layer destruction. Therefore, the areas with "naked" cellulose appear. The ion bombardment and the paraffin vaporization lead to disintegration of the natural waxy fiber layer and cause the defects in paraffin layer. All these changes facilitate the penetration of the working solutions in the fiber and increase cellulose accessibility for dyes.

By this means, the plasma processing is more gentle method as compared with traditional boiling processes, because the plasma treatment does not lead to degradation of the cellulose fibers and has an influence only on a thin surface layer.

Acknowledgments

This work was supported by Russian Foundation of Basic Research, project № 15-41-02672.

References

[1] Shishoo R 2007 *Plasma technologies for textiles* (England) p 322
[2] 2010 *Plasma-Technologie für Funktionstextilien* vol 4-5 (Melliand Textilber) p 171
[3] Oulet R and Barbier M B 1983 *Technological application of NTP* (Moscow: Energolitizdat) p 144
[4] Gorberg B L 2003 *Textile chemicals* **1(21)** 59-68
[5] Kutepov A M, Zaharov A G and Maksimov A I 2004 *Vacuum plasma and plasma-solution modification of polymeric materials* (Moscow: Nauka) p 370
[6] Abdullin I Sh, Zheltukhin V S and Kashapov N F 2000 *High-frequency plasma jet processing of materials at reduced pressures: Theory and Practice* (Kazan: Kazan tehnol. univ.)

[7] Azanova A A, Abutaupova L N, Kulevtsov G N, Tikhonova N V and Ivshin Ya V 2016 *Izvestiya Vysshikh Uchebnykh Zavedenii.* Teknologiya Tekstil'noi Promyshlennosti. **5** 80-83

[8] Sakiyama Y, Graves D B, Chang H W, Shimizu T and Morfill G E 2012 *J. Appl. Phys.* D **45** 425201

[9] Skirda V D, Shkalikov N V 2011 *A method for determining the molecular weight distribution of the paraffins in the hydrocarbon mixture by the method of nuclear magnetic resonance* Patent RF 2423686 Published 10 07 2011

[10] Mitchner M and Kruger Ch 1976 *Partially ionized gases* (Moscow: Mir) p 496

[11] Sunyaev Z I 1980 *Oil carbon* (Moscow: Chemistry) p 272

[12] Parkhomenko V D, Soroka P I and Krasnokutsky Yu I 1991 *Plasma-chemical technology* (Novosibirsk: Science) p 392

Numerical solution of the initial boundary value problems of radio-frequency capacitive coupled discharge.

I B Badriev[1], V S Zheltukhin[2] and V Ju Chebakova[1]

[1]Kazan Federal University, 18 Kremlevskaya Street, Kazan, 420008, Russia
[2]Kazan State Technological University, 68 Karl Marx Street, Kazan, 420015, Russia

E-mail: vchebakova@mail.ru

Abstract. We present the numerical method, which allow to calculation of the charged particles flow density in the areas with heavy gradients of charged particles density and changing of convection-diffusion non-steady equation coefficients for charged particles concentration finding by the solution of electrons balance initial boundary value problem.

1. Introduction

Comparison of the dissipation time and relaxation length of the electron energy with period field and computation domain dimensions shows that different (local and nonlocal) approximations should be used on modeling of RF capacitive discharge with high and low pressures [1-3]. A peculiarity of the RF discharge models is the appearance of their implementation a large gradients of charged particle density and electric field strength, electron temperature in the near-electrode layers at the boundaries of the computational domain. Thus in a quasi-neutral domain takes place a process with diffusion domination and the considerable convection domination in the near-electrode domains is observed. As a result, there are domains of considerable changing of the initial boundary values problems solutions and equation coefficients that are nonlinearly dependant on those problems solution. During the calculation of the flow of charged particles by means of the usual numerical differentiation of the solution, significant errors arise in the mentioned domains. Therefore, the previously developed methods for solving non-linear problems (see, for example, [4-9]) can not be used for the problem under consideration. The flow-through method, which makes it possible to avoid these difficulties, in the presence of strong convection reduces to approximating the convective term using central differences to obtain a band type matrix. This leads to the need to introduce a strict restriction on the relationship between the spatial and time steps to ensure the stability of the calculations [10]. In this work numerical solution is offered through the example of solution of the initial boundary value problem of the charged particles balance. Proposed method is included in the class of implicit methods for solving the non-steady problems. It is based on a modification of the explicit two-step Hummel's method. In the proposed modification, the order of calculations is changed. At the first step of the method, the concentration of charged particles is calculated using implicit difference approximation, constructed by the integro-interpolation method using oriented differences. The second step calculates the flux density of charged particles. For this aim, the first step of the method proposed in [11] is used, but it varies due to the inconstancy of the mobility of charged particles.

2. Statement of the problem

We consider the RF capacitive coupled discharge between two plane-parallel electrodes, one of which is grounded and another one is connected with RF capacitive generator. We assume that distance between electrodes is less than electrode size. Mathematical model of a RF capacitive discharge in the non-local approximation at low pressures in argon contains electron balance equation

$$\partial n_e / \partial t + \partial G_e / \partial x = R_1 n_e N + R_2 n_m^2 + R_3 n_m n_e - R_4 n_e n_+ - R_5 n_e^2 n_+ + R_{10} n_m^2 - \\ - R_{13} n_e n_{2+} + R_{16} n_{2*}^2 + R_{17} n_m n_{2*}, \quad 0 < x < b, t > 0 \tag{1}$$

$$G_e = -D_e \partial n_e / \partial x - \mu_e n_e E, \tag{2}$$

where $n_e, n_+, n_m, n_{2*}, n_{2+}$ are concentrations of electrons, atomic positive ions, metastable argon atoms, dimmers and molecular positive ions of argon, G_e is density of electron flow, μ_e is electron mobility, D_e is diffusion coefficient. Here and below R_i, i=1,...,20, are speed coefficients of the corresponding plasma-chemical reactions $Ar^*+Ar^* \to Ar_2^+ + e$; $2Ar+Ar^+ \to Ar_2^+ + Ar$; $Ar_2^+ + e \to Ar + Ar^+ + e$; $Ar_2^+ + e \to Ar + Ar$; $Ar_2^* \to 2Ar + h_\nu$; $Ar^* + 2Ar \to Ar_2^* + Ar$; $Ar_2^* + Ar_2^* \to e + 2Ar + Ar_2^+$; $Ar_2^* + Ar^* \to e + Ar + Ar_2^+$; $e + Ar_2^* \to 2Ar^* + e$; $Ar^* + 2Ar \to 3Ar + h_\nu$; $Ar_2^+ + Ar \to 2Ar + Ar^+$. Boundary conditions are the following

$$G_e(0,t) = \begin{cases} (-n_e \upsilon_e - \mu_e n_e E)|_{(0+,t)}, & E \ge 0 \\ (-n_e \upsilon_e - \delta)|_{(0+,t)}, & E < 0 \end{cases}, \quad G_e(b,t) = \begin{cases} (n_e \upsilon_e - \mu_e n_e E)|_{(b-,t)}, & E < 0 \\ (n_e \upsilon_e - \delta)|_{(b-,t)}, & E \ge 0 \end{cases}. \tag{3}$$

Here $\upsilon_e = \sqrt{8kT_e / \pi m_e} / 4$ is average thermal velocity of electron, T_e is electron temperature, m_e is electron mass, γ is secondary coefficient of electronic emission from electrodes, μ_+ is atomic ion mobility, μ_{2+} is molecular ion mobility, $\delta = \gamma(\mu_+ n_+ E + \mu_{2+} n_{2+} E)$.

3. Approximate method

Difference approximation for electron convection-diffusion equation (1,2) is constructed by integro-interpolation method using method of oriented differences as well. On the interval [0, b] we construct the uniform space mesh $\omega_h = \{x_l = lh, l = 0,1,\dots, M\}$, $h = (b-0)/M$, and the time mesh $\omega_\tau = \{t_s = s\tau, s = 0,1,\dots\}$, were τ is a time step. Mesh functions are defined on the mesh $\omega_h \times \omega_t$ (symbols are the same as for the differential function). Values of the mesh function φ in the points $x_l^s = (x_l, t_s) \in \omega_h \times \omega_t$ are noted by φ_l^s. During difference scheme construction values of the differential equation coefficients descend to the low time layer t_{s-1}. Equation (1,2) is integrated on the interval $[x_{l-1/2}, x_{l+1/2}]$ for $t = t_s$. Integral $\int_{x_{l-1/2}}^{x_{l+1/2}} \partial n_e / \partial t \, dx$ is calculated using the central rectangles quadrature formula, the time derivative is approximated by the backward difference relation. We have

$$J_2 = \int_{x_{l-1/2}}^{x_{l+1/2}} \partial(-D_e \partial n_e / \partial x + \mu_e n_e(-E)) / \partial x \, dx = (D_e \partial n_e / \partial x)|_{(x_{l-1/2}, t_s)} - (D_e \partial n_e / \partial x)|_{(x_l + h/2, t_s)} + J_{21}, \tag{4}$$

where $J_{21} = \int_{x_{i-1/2}}^{x_{l+1/2}} \partial(\mu_e n_e(-E)) / \partial x dx$. The first two terms in the right-hand side of (4) are calculated by formulas $D_e \partial n_e / \partial x|_{(x_l \pm h/2, t_s)} \approx D_{e\, l \pm h/2}^{s-1} n_{e\bar{x}}(x_{l+1/2 \pm 1/2}, t_s)$. To calculate J_{21} we transform its integrand decomposing function E into positive and negative parts, $E = E^+ + E^-$, $E^\pm = (E \pm |E|)/2$. Hence,

$\partial(\mu_e n_e(-E)) / \partial x = \mu_e n_e \partial(-E^+) / \partial x + (-E)^+ \partial(\mu_e n_e) / \partial x + \mu_e n_e \partial(-E^-) / \partial x + (-E)^- \partial(\mu_e n_e) / \partial x$. Thus,

$J_{21}^{\pm}=J_{210}^{\pm}+J_{211}^{\pm}$, $J_{210}^{\pm}=\int_{x_{l-1/2}}^{x_{l+1/2}}\mu_e n_e\partial(-E)^{\pm}/\partial x dx \approx \mu_e n_e\big|_{x_l}\int_{x_{l-1/2}}^{x_{l+1/2}}\partial(-E)^{\pm}/\partial x dx=$

$=\mu_e n_e\big|_{x_l}((-E)^{\pm}(x_{l+1/2},t_{s-1})-(-E)^{\pm}(x_{l-1/2},t_{s-1}))$. Integrals $J_{211}^{\pm}=\int_{x_{l-1/2}}^{x_{l+1/2}}(-E)^{\pm}\partial(\mu_e n_e)/\partial x dx$ are approximated by usage of upwind differences, at the same time, electron flow streamline is considered to be in the backward direction in relation to E field. Positive direction of E field is a positive direction of X-axis. In such a way integral $J_{211}^{\pm}$ is replaced on the left rectangles quadrature formula, and integral J_{211}^{-} by the right rectangles quadrature formula. Thus, we have

$$J_{211}^{+}\approx(-E)_{x_{l-1/2}}^{+}(\mu_e(x_l,t_{s-1})n_e(x_l,t_s)-\mu_e(x_{l-1},t_{s-1})n_e(x_{l-1},t_s)),$$

$$J_{211}^{-}\approx(-E)_{x_{l+1/2}}^{-}(\mu_e(x_{l+1},t_{s-1})n_e(x_{l+1},t_s)-\mu_e(x_l,t_{s-1})n_e(x_l,t_s)).$$

At approximation of a quadratic nonlinearity in the right-hand side of the equation we use a Newton's linearization $n_e^2(x_l,t_s)=n_e^2(x_l,t_{s-1})+2n_e(x_l,t_{s-1})\times(n_e(x_l,t_s)-n_e(x_l,t_{s-1}))$. As a result, a finite difference scheme for convection-diffusion equation of the electron gas is written in the form

$$h(n_{e_l}^{s}-n_{e_l}^{s-1})/\tau-(D_{e_{l+1/2}}^{s-1}(n_{e_{l+1}}^{s}-n_{e_l}^{s})/h-D_{+_{l-1/2}}^{s-1}(n_{e_l}^{s}-n_{e_{l-1}}^{s})/h)-(-E)_{l-1/2}^{+s-1}\mu_{e_{l-1}}^{s-1}n_{e_{l-1}}^{s}+$$

$$+(-E)_{l+1/2}^{-s-1}\mu_{e_{l+1}}^{s-1}n_{e_{l+1}}^{s}+(-E)_{l+1/2}^{+s-1}\mu_{e_l}^{s-1}n_{e_l}^{s}-(-E)_{l-1/2}^{-s-1}\mu_{e_l}^{s-1}n_{e_l}^{s}=h(R_{1_l}^{s-1}n_{e_l}^{s}N_l^{s-1}+R_{3_l}^{s-1}n_{m_l}^{s-1}n_{e_l}^{s}+$$

$$+R_{10_l}^{s-1}(n_m^2)_l^{s-1}-R_{4_l}^{s-1}n_{e_l}^{s}n_{+_l}^{s}-R_{13_l}^{s-1}n_{e_l}^{s}n_{2+_l}^{s}+R_{16_l}^{s-1}(n_{2*}^2)_l^{s-1}+$$

$$-R_{5_l}^{s-1}n_{+_l}^{s}((n_e^2)_l^{s-1}+2n_{e_l}^{s-1}(n_{e_l}^{s}-n_{e_l}^{s-1}))+R_{17_l}^{s-1}n_{m_l}^{s-1}n_{2*_l}^{s-1}+R_{2_l}^{s-1}(n_m^2)_l^{s-1});.$$

The second equation in (3) we integrate on the interval [$h/2$, $x_l+h/2$] to approximate boundary conditions at x=0 in the case of symmetric electrodes, the following result is obtained:

$$h(n_{e_1}^{s}-n_{e_1}^{s-1})/\tau+(G_{e_{3/2}}-G_{e1/2})=h(R_{1_1}^{s-1}n_{e_1}^{s}N_1^{s-1}-R_{5_1}^{s-1}n_{+1}^{s}((n_e^2)_1^{s-1}+2n_{e1}^{s-1}\times(n_{e_1}^{s}-n_{e1}^{s-1}))-$$

$$-R_{13_1}^{s-1}n_{e1}^{s}n_{2+1}^{s}+R_{161}^{s-1}(n_{2*}^2)_1^{s-1}+R_{17_1}^{s-1}n_{m1}^{s-1}n_{2*_1}^{s-1}+(n_m^2)_1^{s-1}(R_{10_1}^{s-1}+R_{21}^{s-1})-R_{41}^{s-1}n_{e_1}^{s}n_{+1}^{s}+R_{3_1}^{s-1}n_{m1}^{s-1}n_{e1}^{s})$$

. Then to find difference approximation for $G_{e3/2}$ we integrate formula for the flow (2) on the interval [x_1,x_2]. We obtain $\int_{x_1}^{x_2}G_e dx=\int_{x_1}^{x_2}(-D_e\partial n_e/\partial x-\mu_e n_e E)dx$. The left-hand side of equation is approximated by the central rectangles quadrature formula $\int_{x_1}^{x_2}G_e dx \approx G_{e_{3/2}}h$. The right-hand side of the equation we decompose into the sum of two integrals. Integral $\int_{x_1}^{x_2}(-D_e\partial n_e/\partial x)dx$ is approximated by the central rectangles quadrature formula, $\int_{x_1}^{x_2}(-D_e\partial n_e/\partial x)dx\approx -D_{e_{x_1+h/2}}^{s-1}((n_e(x_2,t_s)-n_e(x_1,t_s)))$. Integral $J_3=\int_{x_1}^{x_2}(-\mu_e n_e E)dx$ we also decompose into sum of two integrals $\int_{x_1}^{x_2}(-E)^{+}(\mu_e n_e)dx+$ $+\int_{x_1}^{x_2}(-E)^{-}(\mu_e n_e)dx$ by separating the positive and negative parts of E. The first integral is replaced by the left rectangles quadrature formula, the second one by right rectangles quadrature formula.

Hence, $J_3 \approx h(-E)^{+\,s-1}_{1}\mu_e n_e(x_1,t_s) + h(-E)^{-\,s-1}_{2}\mu_e n_e(x_2,t_s)$. Then according to the oriented differences method the flow G_+ can be approximated in the point x=0 by the same formula (independently on sign of E) $G_{e_{1/2}} = -n_{e_1}\sqrt{8kT_{e_1}/\pi m_e}/4 + E_1^-\gamma(\mu_{+_1}n_{+_1}E + \mu_{2+_1}n_{2+_1}E)$. Indeed, if field is directed backwards to electrode, i.e., $E > 0$, then $E_1^- = 0$, and $G_{e1/2} = -n_{e_1}\sqrt{8kT_{e_1}/\pi m_e}/4$. Thus, the boundary condition at x=0 is replaced by the difference one

$$h(n_{e_1}^{s} - n_{e_1}^{s-1})/\tau - D_{e_{3/2}}^{s-1}(n_{e_2}^{s} - n_{e_1}^{s})/h + (-E_{1/2} - E_{3/2})^{+\,s-1} n_{e_1}^{s}\mu_{e_1}^{s-1}/2 + n_{e_1}^{s}\sqrt{8kT_{e_1}/(\pi m_e)}/4 +$$
$$+(-E_{5/2} - E_{3/2})^{-\,s-1}\mu_{e_2}^{s-1}n_{e_2}^{s}/2 - \mu_{e_1}^{s-1}n_{e_1}^{s}(-E_{1/2} - E_{3/2})^{-\,s-1}/2 = h(R_{1_1}^{s-1}n_{e_1}^{s}N_1^{s-1} +$$
$$+R_{3_1}^{s-1}n_{m_1}^{s-1}n_{e1}^{s} - R_{4_1}^{s-1}n_{e_1}^{s}n_{+_1}^{s} - R_{5_1}^{s-1}n_{+_1}^{s}((n_e^2)_1^{s-1} + 2n_{e_{l1}}^{s-1}(n_{e1}^{s} - n_{e1}^{s-1})) - R_{13_1}^{s-1}n_{e1}^{s}n_{2+_1}^{s} +$$
$$+R_{16_1}^{s-1}(n_{2*}^2)_1^{s-1} + R_{10_1}^{s-1}(n_m^2)_1^{s-1} + R_{17_1}^{s-1}n_{m_1}^{s-1}n_{2*_1}^{s-1} + R_{2_1}^{s-1}(n_m^2)_1^{s-1}) -$$
$$-\gamma(\mu_{+_1}^{s-1}n_{+_1}^{s}(E_{1/2} + E_{3/2})^{-\,s-1})/2 - \gamma(\mu_{2+_1}^{s-1}n_{2+_1}^{s}(E_{1/2} + E_{3/2})^{-\,s-1})/2.$$

Upon loaded electrode (x=b) differential boundary conditions are developed by the integration the first equation in (3) on the interval [b-3h/2, b-h/2] as in a case when x=0.

Flow calculation is conducted by modification of Hummel's method offered in the work [11]. Using such modification values of electrons concentration are taken from the high time layer. Besides dependence of mobility coefficients on electrical field is taken into account. This method is based on mesh approximation of analytical solution in regard to n_e linear non-uniform equation (2). That concentration is known we receive the following result:

$$G_{l+1/2}^{s} = D_{l+1/2}^{s-1}(n_l^s - \exp(\alpha)n_l^s)/(hG_0) \quad (5)$$

where, $\alpha = \mu_l hE_{l+1/2}^{s-1}/D_{l+1/2}^{s-1}$, $G_0 = (\exp(\alpha) - 1)/\alpha$. This method of flow calculation has following restriction for a space step $h \ll 2D_{l+1/2}^{s-1}/(\mu_l^{s-1}/|\Delta E_l^{s-1}|)$. If this stability condition doesn't satisfy, then the calculations are performed by the following way. An interpolative step is introduced by formula $h_\nu = \sqrt{\varepsilon 2D_{l+1/2}^{s-1}h/|W_{l+1} - W_l|}$, where $0 < \varepsilon < 1$, $W_{l+1/2} = -\mu_{el+1/2}^{s-1}E_{l+1/2}^{s-1}$, $W_l = (W_{l+1/2} + W_{l-1/2})/2$. Taking into account flow direction at $h_\nu < h$ we have

$$G_{l+1/2}^{s} = \begin{cases} D_{l+1/2}^{s-1}(n_L - \exp(\alpha_\nu)(1 + \alpha_\nu\beta_\nu)n_R)/(h_\nu G), |W_L| \ge |W_R| \\ D_{l+1/2}^{s-1}(\exp(-\alpha'_\nu)(1 + \alpha'_\nu\beta'_\nu)^{-1}n_L - n_R)/(h_\nu G'), |W_L| < |W_R| \end{cases},$$

Here $W_{L,R} = W_{l+1/2} \mp \Delta W_\nu/2$, $\Delta W_\nu = h_\nu(W_{l+1} - W_l)/h$. Terms $n_{L,R}$ are calculated by the formula $n_{L,R} = (n_l^s + 1)\exp\{a(x_{l+1/2} \mp h_\nu/2 - x_l\}$, $x_{eL,R} = (x_l + x_{l+1} \mp h_\nu)/2$, $a = h^{-1}\log((n_{l+1}^s + 1)/n_l^s)$. Therefore, $G' = (\alpha'^{\,2}_\nu - 2\alpha'_\nu\beta'_\nu)/\alpha'^{\,3}_\nu - \exp(-\alpha'_\nu)(\alpha'^{\,3}_\nu\beta'_\nu + \alpha'^{\,2}_\nu + 2\alpha'_\nu\beta'_\nu + 2\alpha'^{\,2}_\nu\beta'_\nu)/\alpha'^{\,3}_\nu$, $G = (\alpha_\nu^2 + 2\alpha_\nu\beta_\nu)/\alpha_\nu^3 - \exp(\alpha_\nu)(\alpha_\nu^3\beta_\nu + \alpha_\nu^2 + 2\alpha_\nu\beta_\nu - 2\alpha_\nu^2\beta_\nu)/\alpha_\nu^3$, $\alpha_\nu = -h_\nu W_L/D_{e_{l+1/2}}$, $\beta_\nu = (W_R - W_L)/(2W_L)$, $\alpha'_\nu = -h_\nu W_R/D_{e_{l+1/2}}$, $\beta'_\nu = (W_R - W_L)/(2W_R)$. If $h_\nu > h$ and $E_l = 0$, there is uncertainty in the calculation of the flow according to (5), so flow is calculated as $G_{l+1/2}^s = (\exp(-\alpha')(1 + \alpha'\beta')^{-1} - 1)D_{l+1/2}^{s-1}n_l^s/hG''$, $\beta' = \Delta E_l^{s-1}/(2E_{l+1}^{s-1})$, $\alpha' = h\mu_{l+1}^{s-1}E_{l+1}^{s-1}/D_{l+1/2}^{s-1}$, $E_l^{s-1} = E_{l+1}^{s-1}(1 + 2\beta'\eta)$, $G'' = (\alpha'^2 - 2\alpha'\beta')/\alpha'^3 - \exp(-\alpha')(\alpha'^3\beta' + \alpha'^2 + 2\alpha'\beta' + 2\alpha'^2\beta')/\alpha'^3$, $\eta = (x_l - x_{l+1})/h$. When $|W_L| = |W_R| = 0$, flow is approximated as $G_{l+1/2}^s = D_{l+1/2}^{s-1}(n_l^s - n_{l+1}^s)/h$. Charged particles flows in the boundary points are approximated in the same way as during development of difference scheme for charged particles balance according to implicit scheme. This method was used to make calculations for RF capacitive discharge models with different approaches of modeling. Results of numerical experiments showed good qualitative and quantitative coincident with experimental data [12-15].

Acknowledgments
The work is performed according to the Russian Government Program of Competitive Growth of Kazan Federal University and this work was supported by Russian Science Foundation, project no. 16-11-10299.

References

[1] Kudryavtsev A A, Smirnov A S and Tsendin L D 2010 Physics of glow-discharge (Saint Petersburg: Publishing house "Lan'")
[2] Chebakova V J, Zheltukhin V S and Dubrovin V T 2016 Problem of radio-frequency discharge at atmospheric pressure in local statement *Applied Mathematical Sciences* **10** (21-24) 1013-22
[3] Zheltukhin V S, Chebakova V J, Abaidullin R N and Kadyrov R F 2015 On solving of some nonlinear problems of plasma physics *Applied Mathematical Sciences* **9** (85-88) 4351-59
[4] Badriev I B and Karchevskii M M 1989 Convergence of the iterative Uzawa method for the solution of the stationary problem of seepage theory with a limit gradient *Journal of Soviet Mathematics* **45** (4) 1302-9
[5] Badriev I B and Nechaeva L A 2013 Mathematical simulation of steady filtration with multivalued law *PNRPU Mechanics Bulletin* (3) 37-65
[6] Badriev I B, Makarov M V and Paimushin V N 2016 Numerical Investigation of Physically Nonlinear Problem of Sandwich Plate Bending *Procedia Engineering* **150** 1050-5
[7] Badriev I B, Garipova G Z, Makarov M V and Paymushin V N 2015 Numerical Solution of the Issue about Geometrically Nonlinear Behavior of Sandwich Plate with Transversal Soft Filler *Research Journal of Applied Sciences* **10** (8) 428-35.
[8] Badriev I B, Makarov M V and Paimushin V N 2016 Mathematical Simulation of Nonlinear Problem of Three-point Composite Sample Bending Test *Procedia Engineering* **150** 1056-62
[9] Badriev I B and Zadvornov O A 2003 A decomposition method for variational inequalities of the second kind with strongly inverse-monotone operators *Differential Equations* **39** (7) 936-44
[10] Samarskii A A and Vabishevich P N 1999 *Numerical methods of solution of convection-diffusion problems*(Moscow, Editorial URSS)
[11] Kulikovsky A A 1995 A More Accurate Scharfetter-Gummel Algorithm of Electron Transport for Semiconductor and Gas Discharge Simulation *Journal of computational physic* **119** 149–55.
[12] Dubrovin V T, Chebakova V Ju and Zheltukhin V S 2016 Radio-Frequency Discharge at Low Pressure: a Non-Local Problem Statement Approach *Procedia Engineering* **150** 1041–5
[13] Askhatov R. M., Chebakova V. Yu., Zheltukhin V. S. 2016 Capacitive coupled RF discharge: modelling at the local and not local statement of the problem *IOP Conf. Ser.: Mater. Sci. Eng.* **158** 012009
[14] Badriev I B, Chebakova V Y and Zheltukhin V S 2017 Capacitive coupled RF discharge: Modelling at the local statement of the problem *J. Phys.: Conf. Ser.* **789** 012004
[15] Chebakova V J, Gerasimov A V and Kirpichnikov A P 2016 On the solving of one type of problems of mathematical physics *IOP Conf. Ser.: Mater. Sci. Eng.* **158** 012023

Pulse-periodic HF induction-type discharge for pumping lasers on self-terminating transitions

V M Batenin[1], V T Karpukhin[1], M M Malikov[1], M A Kazaryan[2]

[1] Joint Institute for High Temperatures, Russian Academy of Sciences, 13 Bd. 2 Izhorskaya str. 125412 Moscow, Russia

[2] P N Lebedev Physical Institute, Russian Academy of Sciences, 53 Leninskiy Prospekt, 119991 Moscow, Russia

e-mail: mmalikov@oivtran.ru

Abstract. We present the results of numerical investigation regarding the physical processes in the working medium of a copper vapour laser pumped by trains of high-frequency (f_{tr}≈30 MHz) current oscillations with f≈10 kHz repetition rates. An electrodeless discharge achieved by employment of an induction-type process, which is new for this type of lasers, was considered. Elimination of the electrodes in the design of such inductive copper vapour laser will, in principle, provide a means to increase the lifetime and reliability of sealed discharge tubes and reduce their costs.

1. Introduction

The possibility to efficiently pump the working medium (Ne+Cu) of an inductive copper vapour laser (ICVL) using high-frequency discharge was first demonstrated in [1–4]. In the present work we used the physical model and computer codes developed in [2, 3]. These results provide a means to obtain self-consistent values of all plasma, electric, kinetic and heat parameters in the stationary operation mode. The goal of the present study is to reveal physical conditions and technical specifications that would provide the improvement of the ultimate output parameters of gas-discharge lasers based on self-terminating atomic transitions of copper. The basic approach involves numerical modeling of kinetic and thermal processes in discharge plasma with simultaneous solution of the differential equation system describing the electric circuit of induction pumping.

2. Results and discussion

We consider the design of the ICVL (figure 1a) with a one-turn inductor (6 cm in diameter) which is solid along the tube length. The working medium is disposed in an annular volume limited by two cylinders (5 and 7 cm in diameter and 90 cm in length). Numerical experiments were conducted with predefined values of the storage capacitance C=1.5 nm, initial voltage at this capacitance $U(0)$=35 kV, neon pressure P_{Ne}=250 Torr, temperature of the inner wall T_{in}=1823 K (concentration of copper atoms was $1.5\cdot10^{15}$ cm^{-3}). The calculated value of the gas temperature in the middle of the working gap did not exceed 2140 K, which is allowable for laser operation.

The calculated ring current pulse J_2 and active voltage pulse (over one azimuthal turn) $U_{ac}=J_2 \cdot R_{pl}$ are presented in figure 1b. The dynamics of the plasma resistance $R_{pl}(t)$, electron temperature $T_e(t)$ and

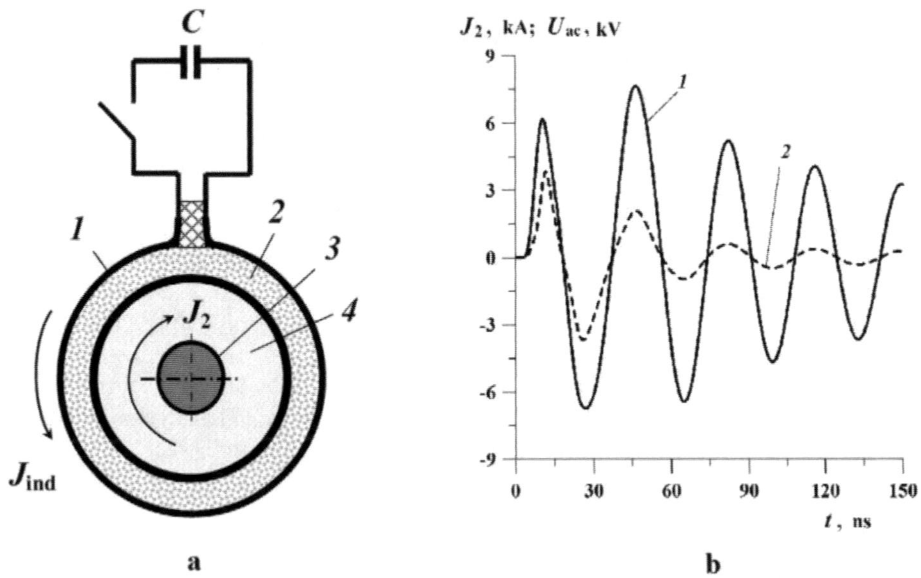

Figure 1. (a): Schematic representation of ICVL: 1– inductor, 2 – heat insulation, 3 – ceramic insert, 4 – plasma; (b): Current J_2 – 1 and voltage U_{ac} – 2 pulses.

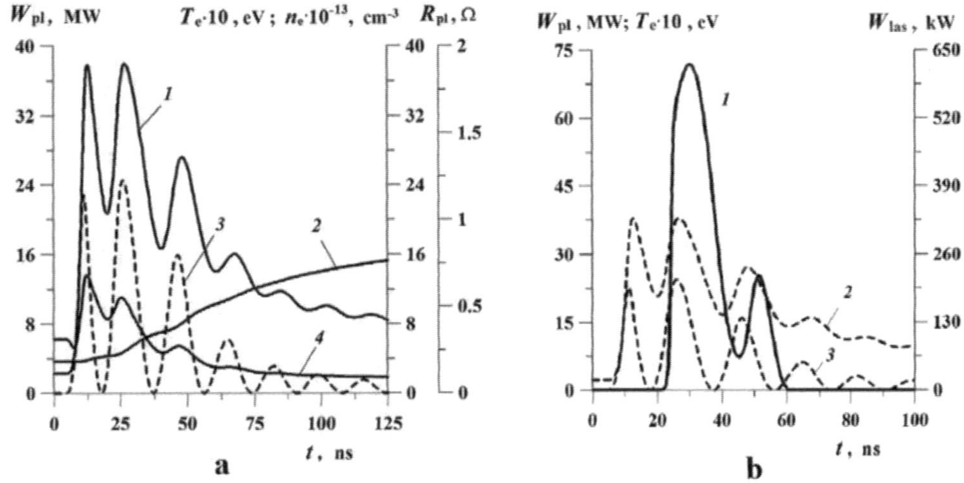

Figure 2. Plasma parameters and laser emission pulse at wavelength of 0.51 μm. (a): 1 – T_e, 2 – n_e, 3 – W_{pl}, 4 – R_{pl}; (b): 1 – W_{las}, 2 – T_e, 3 – W_{pl}.

electron concentration $n_e(t)$ as well as the Joule heat power $W_{pl}(t)=J_2^2(t) \cdot R_{pl}(t)$ released in the plasma is shown in figure 2a. It can be seen from the figure that J_2 and U_{ac} pulses appear as trains of decaying HF oscillations. This behaviour is due to the fact that the inductance of the inductor, inductance of the plasma turn and its resistance are much lower than the inductance and resistance of the discharge circuit involving electrodes in a conventional copper vapour laser (CVL) operating with the same storage capacity.

Consequently, the power W_{pl} reveals pulsed character with a doubled pulse frequency $2f_{tr}$ (figures 2a and 2b). In this case, higher rates of injection of energy into the plasma can be achieved (in every pulse) as compared to the conventional electrode-based CVL with a long (~250 ns) nonperiodic pump pulses. It should be noted that at the initial stage of ~0 – 50 ns the value of R_{pl} is increased by a factor 2 – 3 (figure 2a) due to the growth of the frequency of elastic collisions between electrons and atoms resulting from a rapid increase of T_e. At this time interval the growth of n_e is slow and cannot inhibit the growth of R_{pl}. The latter is associated with the fact that the characteristic ionization time of the medium is shorter than the period of HF oscillations of the pump current. Due to this behaviour of the plasma up to ~70% of the electric power pumped into the plasma is released at the considered time interval. All of this gives rise to inverse population and increases the laser efficiency.

On the other hand, we observe pronounced pulsation of the electron temperature T_e which is in time to W_{pl}; alongside with that, pulsations of the lasing power W_{las} may appear (see figures 2a and 2b). According to our calculations such oscillations of T_e do not affect the rise and maintenance of inverse population for self-terminating transitions between laser levels. This is due to the fact that on the drop of $T_e(t)$ in the oscillations the population rate for the upper laser level n_r still exceeds the population rate for the lower laser level n_m as for the upper level the rate of its electron impact-induced depopulation to the above levels and to the continuum is decreased. This dynamics of n_r and n_m is clearly seen in figure 3a at the time interval from approximately 7.5 ns to 23 ns.

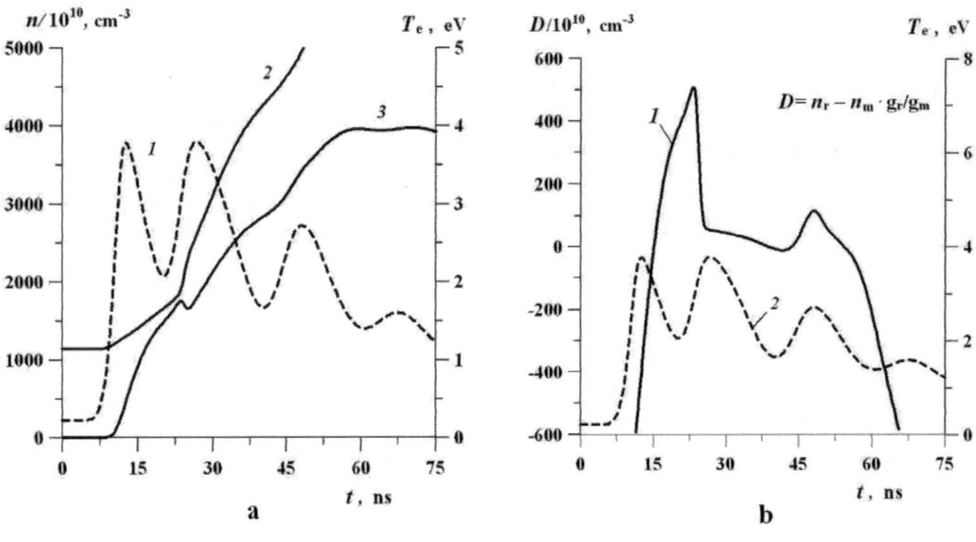

Figure 3. Dynamic behaviour of operating level populations and inverse population obtained for 0.51 μm laser emission. (a): 1 – T_e, 2 – n_m, 3 – n_r; (b): 1 – D, 2 – T_e.

At the same time interval we observe a rapid growth of inverse population $D(t)=(n_r\text{-}n_m{\cdot}g_r/g_m)$ (see figure 3b), which reaches its maximum at $t\approx 23$ ns (here g_r and g_m are statistical weights of the levels). It should be noted that the gain constant in the working medium of the laser is proportional to the value of D (in $D>0$ region). For $t\gtrsim 23$ ns a small rapid drop of the population of the upper laser level n_r is observed (figure 3a) alongside with a drastic (almost to zero) fall of the D value (figure 3b). Therefore, laser gain constant drops as well. This is due to the rapid evolution of the lasing pulse at the considered time instants (see figure 2b) and, consequently, to intensive depopulation of the upper laser level to the lower one by virtue of induced transitions. However, the following pulsation of T_e (at t~50 ns time scales) again results in a substantial growth (peak) of the D value (figure 3b) and, consequently, it results in the appearance of a second peak in the laser pulse (figure 2b). The subsequent (at $t\gtrsim 60$ ns) pulsations of T_e do not give rise to D and W_{las} peaks as the temperature drops below 2 eV and the rate constant for the electron impact-induced excitation of the upper laser level of copper atoms becomes smaller than the corresponding constant for the lower laser level.

In our ICVL optimization calculations [5] for the specified physical and geometrical parameters ($U(0)$, C, f_{tr}, f, T_{in}, P_{Ne} and others) we managed to find the regions that provided from five to six peaks of the inverse population D and emission W_{las} that followed with a frequency of $2f_{tr}$ within the pump train. In these conditions the total duration of the lasing pulse grew up to ~80 ns. Such laser generation was obtained in the numerical experiments when low values of the pre-pulse concentration of electrons $n_e(0)$ were involved. In other regions of the specified parameters for high values of $n_e(0)$ the inverse population D and the emission W_{las} contained a single peak. The width of the emission peak (at half-maximum) was ~5 − 8 ns and its amplitude exceeded 1 MW.

3. Conclusion

In general, the dynamic behaviour of the kinetic parameters and laser generation of the ICVL pumped by HF current pulse trains reveal a more complicated and diverse character as compared to a conventional copper vapour laser. The calculations predict a number of new effects and features of laser generation. It should be noted that in the present work the output power was as high as 170 W for the discharge chamber volume of 1.7 l with physical and technical efficiencies being equal to 3.7% and 1.9%, respectively. Certainly, all the conclusions drawn here require experimental verification. However, even now it is possible to argue that the obtained results may serve as the ideological basis for future applications in the development of analogous lasers based on self-terminating transitions of other metal atoms as well as for designing various state-of-the-art devices involving HF discharge, such as plasmotrons, low-temperature plasma, etc.

The results of the numerical experiments demonstrate the potential for the development of efficient ICVLs and their practical applications.

Acknowledgments
This work was supported by the Russian Foundation for Basic Research, project no. 17-08-00410.

References

[1] Malikov M M, Kazaryan M A and Karpukhin V T 2015 *Bull. Lebedev Phys. Inst.* **42** 138
[2] Batenin V M, Kazaryan M A, Karpukhin V T, Lyabin N A and Malikov M M 2016 *Plasma Phys. Rep.* **42** 1057
[3] Direktor L B, Karpukhin V T and Malikov M M 2014 *High Temp.* **52** 428
[4] Kazaryan M A, Malikov M M, Karpukhin V T and Lyabin N A 2017 *J. Phys. Conf. Ser.* **826** 012003
[5] Batenin V M, Kazaryan M A, Karpukhin V T and Malikov M M 2017 *High Temp.* **55**

IOP Conf. Series: Journal of Physics: Conf. Series **927** (2017) 012010 doi:10.1088/1742-6596/927/1/012010

ICCD-imaging of a plasma glow during the prebreakdown stage of nanosecond discharges at both polarities in nitrogen, air, and argon

D V Beloplotov[1,2], **M I Lomaev**[1,2], **V F Tarasenko**[1,2], **D A Sorokin**[1,2]
[1]Institute of High Current Electronics, 2/3 Akademicheskii Ave., Tomsk, 634055, Russia.
[2]National Research Tomsk State University, 36 Lenin Ave., Tomsk, 634050, Russia.

E-mail: rff.qep.bdim@gmail.com

Abstract. Dynamics of a plasma glow during the prebreakdown stage of a nanosecond discharge in the «cone-to-plane» gap with length of $d = 3$ mm was investigated with a HSFC PRO 12-bit four-channel ICCD camera. The gap was filled with nitrogen, air, argon. Gas pressure was ranged from 12,5 to 400 kPa. Voltage pulses of negative ($U = 25$ kV, $\tau_{0,5} = 3$ ns, $\tau_{0,1\text{–}0,9} = 0{,}7$ ns) and positive ($U = 25$ kV, $\tau_{0,5} = 10$ ns, $\tau_{0,1\text{–}0,9} = 3$ ns) polarities were applied across the gap. Images of the plasma glow at different stages of streamer formation are presented. It was established that a diffuse discharge is formed due to formation of a large streamer. It was found that plasma appears at a certain distance from the conical electrode at both polarities. These and other features of streamer formation are discussed.

1. Introduction

Great attention has been given to study of nanosecond discharges in different gases. This is due to wide possibilities of applications of nanosecond discharges to create sources of spontaneous and stimulated radiation [1,2], as well as sources of low-temperature non-equilibrium plasma [3]. Such plasma is applied in airflow control systems, ignition fuel systems, as well as for a treatment of liquid and solid materials [4–6].

Currently, the problems of formation and propagation of streamers are being studied. As is known, parameters of a streamer depend on the value of the product of gas pressure by gap length, as well as on the gap geometry, kind of gas and the overvoltage rate. The last one is determined by the rate of voltage rise (dU/dt). For relatively long gaps (~10 cm) filled with nitrogen at atmospheric pressure and dU/dt of $\sim 10^{12}$ V/s, a streamer has a branching structure [7]. Moreover, there is significant effect of the polarity of voltage pulse. As compare to a cathode-directed streamer, an anode-directed one can disappear after crossing a half of the gap at the same amplitude of voltage pulse. Oxygen admixtures lead to a decrease in the branching frequency of the streamer and to the breakdown of the gap at negative polarity.

A large number of experimental data shows that in gaps having the length of ~1 cm and dU/dt of $\sim 10^{13}\text{–}10^{14}$ V/s a diffuse discharge is formed [8–10]. From some theoretical studies it follows that the diffuse discharge under such conditions is the result of the formation of a streamer having a transverse size comparable with the gap length [11,12]. However, there is no exhaustive experimental data on the structure of the streamer under such relatively extreme conditions.

The objective of present work is to study the structure of the streamer during the phase of the diffuse discharge formation at positive and negative polarities in nitrogen, air, and argon at in the pressure range of 12.5–400 kPa.

2. Experimental setup

The block-scheme of the setup is presented in figure 1. Two FID pulsers of negative (GIN-100-1, $U \approx 25$ kV, $\tau_{0.5} \approx 3$ ns, $\tau_{0.1-0.9} \approx 0.7$ ns) and positive (GIN-50-1, $U \approx 25$ kV, $\tau_{0.5} \approx 10$ ns, $\tau_{0.1-0.9} \approx 2$ ns) polarities were used. Voltage pulses were applied across the «cone-to-plane» gap placed into a vacuum chamber. The solution angle at the apex of the cone-shaped electrode was 30°. The gap length d was 3 mm. The chamber was equipped with a capacitive voltage divider (CVD), a current-viewing resistor (CVR) based on chip-resistors as well as a collector allowing to register runaway electron beams (RE beams). Waveforms from CVD, CVR, and collector were recorded with a LeCroy WaveMaster 830Zi-A oscilloscope (30 GHz, 80 GS/s). Design of the discharge chamber allowed to register simultaneously voltage and discharge current or the voltage and current of a RE beam.

The images of plasma glow were registered with a HSFC PRO 12-bit four-channel ICCD camera. Its minimum exposure time is 3 ns. However, discharge processes shorter than 1 ns could be registered due to the precise synchronization and running the ICCD camera 2–3 ns before the moment of arrival of a voltage pulse to the high-voltage electrode. Both a jitter of the synchronization system and a jitter of the breakdown allowed to observe a streamer on different stages of its formation from pulse to pulse. As a result, dynamics of streamer development was reconstructed from the instant of its appearance to the instant of bridging the gap. Note that in this case there is no time reference between separate images.

Additionally, to estimate a streamer velocity in nitrogen, a Hamamatsu C10910-05 streak camera equipped with an Acton SpectraPro SP-2300 monochromator (Princeton Instruments) was used. The streak camera allowed to register the time dependence of the discharge plasma radiation intensity from different zones along the gap with high temporal resolution (time-base sweep was 5 ns per 1024 pixels). Time-resolved radiation intensity of 0-0 band of the second positive system of a nitrogen molecule (peak wavelength 337.1nm) was registered. It is well known that the second positive system of nitrogen is effectively excited during the prebreakdown stage of nanosecond discharges [13]. About 300 pulses were registered from each zone. The streamer velocity was estimated as a propagation velocity of the light front along the gap.

The chamber was pumped out with a forevacuum pump and then filled with nitrogen, air, and argon. Pressure was ranged from 12.5 to 400 kPa.

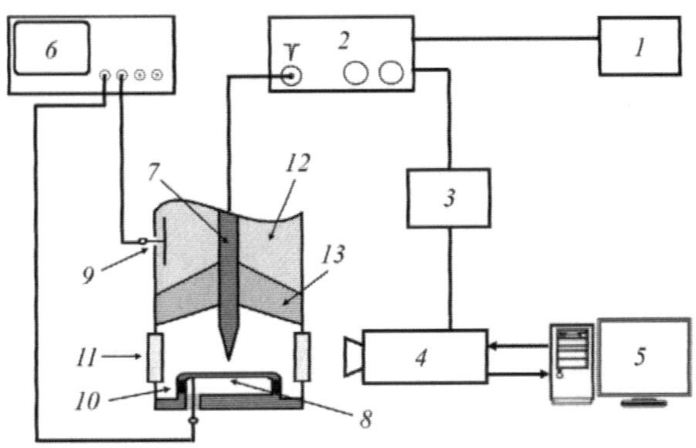

Figure 1. Block-scheme of the experimental setup. *1* – triggering pulser, *2* – high-voltage pulser, *3* – delay pulser, *4* – ICCD camera or streak camera, *5* – computer, *6* – oscilloscope, *7* – high-voltage electrode, *8* – grounded electrode, *9* – capacitive voltage divider, *10* – current viewing resistor, *11* – window, *12* – transmission line, *13* – insulator.

3. Results

3.1. Waveforms

Waveforms of voltage, discharge current, as well as RE beam current pulses are presented in figure 2. Note that a RE beam was registered in individual series of experiments as described in Section 2. It was found that a RE beam is registered at the maximum voltage (before the voltage drop) when lengths of registration paths were equal. At conditions under the study a duration of the RE beam current was ≈70 ps that corresponds to the pulse response of the collector. The increase in pressure led to a decrease in the RE beam current amplitude.

Figures 2a,b show how unstable the breakdown is. The duration of the prebreakdown stage varies from 1 to 2 ns. A RE beam is also unstable relative to the voltage pulse onset. Instability of the breakdown can be caused by the appearance of initial electrons (statistical lag). In such conditions (nanosecond voltage pulses, negative polarity, «cone-to-plane» gap), the field emission should provide initial electrons ensuring breakdown stability. However, conical electrode surface condition, dielectric inclusions, relatively low electric field (dU/dt ~10^{13} V/s) affect the stability of the field emission. Stability of the discharge increases when values of dU/dt reach 10^{14}–10^{15} V/s providing significantly higher values of the electric field [8]. In addition, it can be seen, the earlier start of ionizations processes leads to a slower voltage drop (figure 2a).

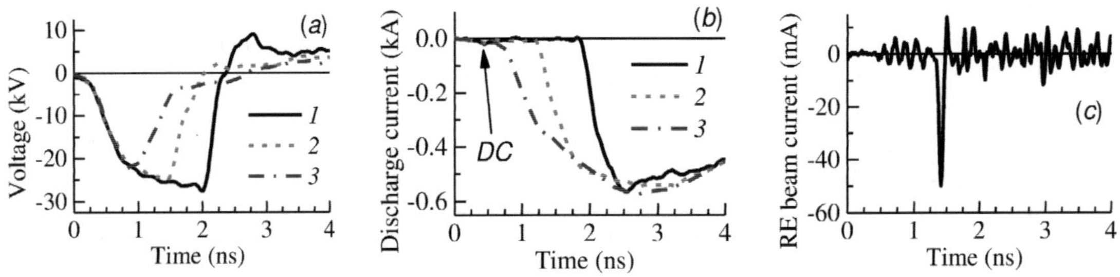

Figure 2. Waveforms of voltage (a), discharge current (b), as well as RE beam current pulses (c). A RE beam was registered behind the 10-μm-Al anode. Nitrogen pressure is 100 kPa. *d* is 3 mm. *1*, *2*, *3* – different discharge implementations. Positive surges on the voltage pulses are caused by the insufficient value of *RC*-constant of CVD. *DC* – displacement current.

3.2. ICCD-imaging

At conditions under the study, a diffuse discharge was formed in the gas pressure range of 12,5–400 kPa. The formation of a large streamer was observed with the ICCD camera (figure 3). The transverse size of the streamer was equal to the longitudinal one at pressure up to 50 kPa. The increase in pressure led to the expected reduction in its transverse size. Sometimes formation of two streamers was observed at pressure of 400 kPa. Formation of the explosive center on the cathode was observed after bridging the gap by the streamer. Rarely (~1/1000), the formation of the explosive center was observed before.

It was found that the plasma appears at some distance from the cathode tip at the initial stage (figure 3). The distance depends on the gas pressure. The higher pressure, the shorter the distance. The plasma appears on the cathode surface at a pressure of 400 kPa. On the one hand, the reason for the observed dynamics is the optimal value of the reduced electric field E/p for the impact ionization coefficient α. Due to enhancement of electric field strength near the cathode tip, E/p takes very high values at low pressure and α reaches maximum value at some distance from the cathode. The increase in pressure leads to a decrease in value of E/p and the zone with the largest α is displaced towards cathode tip. On the other hand, an electron avalanche should pass a certain distance until it transforms to a streamer and creates enough (threshold) number of excited nitrogen molecules whose radiation intensity will be enough to register by the ICCD camera. However, at positive polarity electron avalanches move toward the conical anode and we should have observed the plasma on the anode surface, but experimental data show that the same dynamics of the streamer formation is observed at positive polarity of the voltage pulse, as well as with other gases.

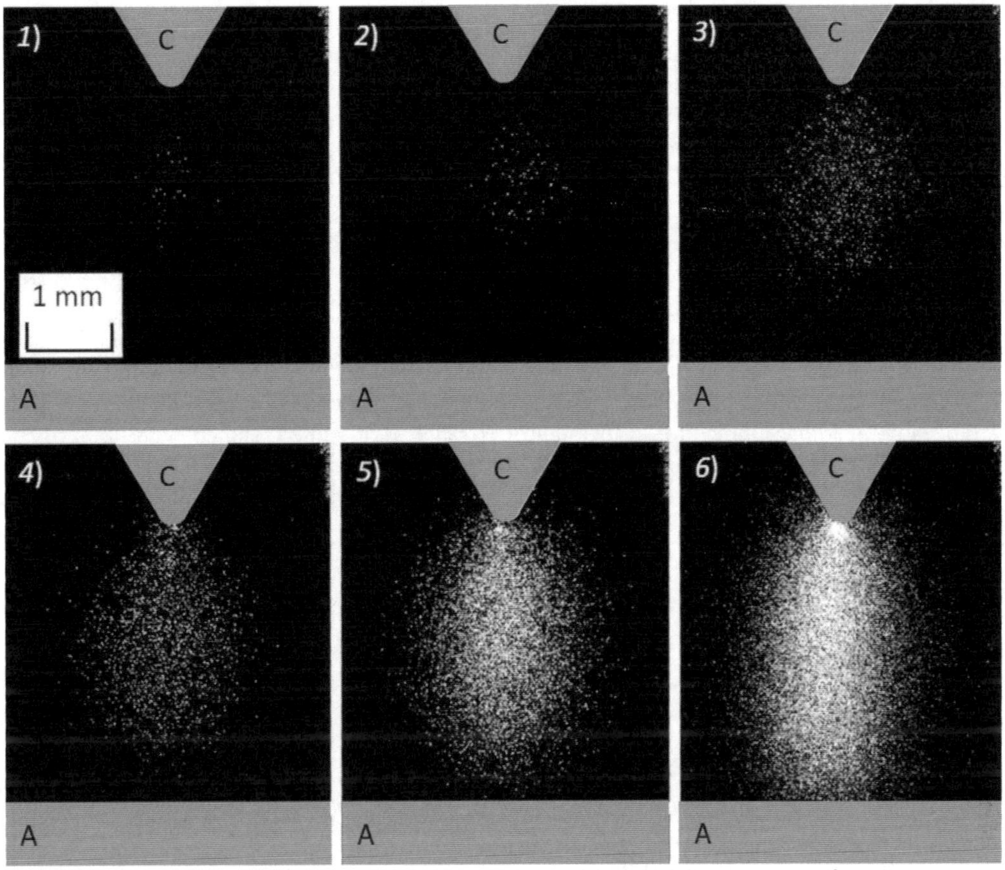

Figure 3. Images of the discharge plasma glow taken with the ICCD camera. Nitrogen pressure is 100 kPa. GIN-100-1, negative polarity. C – cathode, A – anode.

3.3. Streamer velocity

Experimental data obtained with Hamamatsu C10910-05 streak camera show that an average velocity of the anode-directed streamer is $(2\div4)\cdot10^9$ cm/s at atmospheric pressure of nitrogen. This value increases up to $6\cdot10^9$ cm/s at nitrogen pressure of 50 kPa. Unfortunately, there is no data at pressure above 100 kPa.

4. Discussion

From the experimental results it follows that at applying of nanosecond voltage pulses of both polarities across a gap with a non-uniform electric field strength distribution the diffuse discharge is formed in different gases in a wide range of pressures. ICCD-imaging of a glow of the discharge plasma during the prebeakdown stage showed that a streamer with large (2–3 mm at $d = 3$ mm) transverse sizes develops. It is well known that at a breakdown of a gap in a quasi-static electric field, the diameter of a streamer is ~100 µm. At high overvoltage an electric field strength on the streamer head is high enough and can be higher than the critical value for generation of runaway electrons. Runaway electrons moving ahead an anode-directed streamer pre-ionize a gas. Electron avalanches forming from seeded electrons overlap and provide moving of the plasma front. In the case of a cathode-directed streamer, runaway electrons move towards the streamer head. In this instance, a characteristic radiation generated as a result of the interaction of runaway electrons with molecules of a gas pre-ionizes the gas [14]. Note that bremsstrahlung can be generated at the initial stage and participate in a preliminary gas ionization. Moreover, the radial component of the electric field at the streamer front can become significant. Together, this creates conditions for the formation of a streamer having large transverssize.

5. Conclusion

The obtained experimental data show that the diffuse discharge occurs at applying nanosecond voltage pulses of both polarities across the «cone-to-plane» gap due to formation a large streamer. At conditions under the study, there was no significant influence of the polarity of voltage pulses and kind of gas on streamer dynamics.

We assume that a large streamer is formed due to generation of runaway electrons on the streamer front, where electric field is enhanced. Runaway electrons and characteristic radiation pre-ionize a gas and create conditions for the formation of a large-size streamer.

In the very beginning, the plasma appears at a distance from the conical electrode. Increase in pressure leads to displacement of the zone in which the plasma appears towards the conical electrode. It was assumed that a reduced electric field strength near the conical electrode is very high and maximum value of impact ionization coefficient α is reached at some distance. The higher pressure, the shorter the distance from the conical electrode, where the plasma appears.

Acknowledgments

The work is performed in the framework of the Russian Science Foundation (the project #17-72-20072).

References

[1] Baksht E Kh, Burachenko A G, Lomaev M I, Panchenko A N and Tarasenko V F 2015 *Quantum Electronics* **45(4)** 366–370
[2] Panchenko A N, Panchenko N A 2016 *Atmospheric and Ocean Optics* **29(4)** 390–394
[3] Lu P, Cullen P J, Ostrikov K 2016 Atmospheric Pressure Nonthermal Plasma Sources *Cold Plasma in Food and Agriculture: Fundamentals and Application* ed N N Misra, O K Schüter, P J Cullen (Elsevier) chapter 4 pp. 83–116
[4] Adamovich I V, Choi I, Jiang N, Kim J.-H, Keshav S, Lempert W R, Mintusov E, Nishihara M, Samimy M and Uddi M 2009 *Plasma Sources Science and Technology* **18(3)** 034018
[5] Bruggeman P J et al. 2016 *Plasma Sources Science and Technology* **25(5)** 053002
[6] Shulepov M A, Erofev M V, Ripenko V S, Tarasenko V F 2017 *Journal of Physics: Conference Series* **830(1)** 012090
[7] Won J Yi, Williams P F 2002 *Journal of Physics D: Applied Physics* **35(3)** 205–218
[8] Tarasenko V F et al. 2014 *Runaway Electrons Preionizaed Diffuse Discharges* ed V F Tarasenko (New York: Nova Science Pub Inc) p 598
[9] Yatom S, Gleizer J Z, Levko D, Vekselman V, Gurovich V, Hupf E, Hadas Y and Krasik Ya E 2011 *Europhysics Letters* **96(6)** 65001
[10] Tardiveau P, Magne L, Marode L, Ouaras E, Jeanney P and Bournonville 2016 *Plasma Sources Science and Technology* **25(5)** 054005
[11] Naidis G V 2009 Physical Review E **79(5)** 57401
[12] Pechereau F, Delliou P L, Jansky J, Tardiveau P, Pasquiers S and Bourdon A 2014 *IEEE Transaction on Plasma Science* **42(10)** 2346–2347
[13] Hoder T, Bonaventura Z, Bourdon A and Simek M 2015 *Journal of Applied Physics* **117(7)** 073302
[14] Tarasenko V F, Baksht E Kh, Burachenko A G and Lomaev M I 2017 *Plasma Physics Reports* **43(7)** 792–795

IOP Conf. Series: Journal of Physics: Conf. Series **927** (2017) 012011 doi:10.1088/1742-6596/927/1/012011

The gas heterogeneous flows cleaning technology from corona discharge field

A Bogdanov[1], A Tokarev[1], V Judanov[1], V Vinogradov[2]

[1]Kyrgyz-Russian Slavic University /KRSU/ named after Boris Yeltsin; 720000, Bishkek; 44 Kiev str.; Kyrgyzstan
[2]Chemistry and Chemical Technology Institute from the Kyrgyz Republic National Academy of Sciences; 720071, Bishkek; 265 Chui ave., Kyrgyzstan

E-mail: dedushka79@mail.ru

Abstract: A nanogold capture and extraction from combustion products of Kara-Keche coal, description the process: a coal preparation to experiments, nanogold introducing in its composition, temperature and time performance of combustion, device and function of experimental apparatus, gas-purification of the gas flow process and receiving combustion products (condensate, coke, ash, rags) is offerred.

1. Introduction

Coal so far remains the most affordable and acceptable source of thermal energy. 51% of the world's energy is generated from coal combustion. Recently, coal is considered as a carrier of useful elements: rare earths and platinoids [1,2]. Modern gas cleaning and electrofilter technologies are already able not only to reduce harmful emissions from exhaust industrial gases, but also to catch large particles of useful elements from combustion products along the way [3]. In this case, there is a general improvement in the processes of electrofiltration, with the possibility of trapping submicron and nanoscale particles of useful elements [4,5]. Coal reserves of Kyrgyzstan are more than 2.2 billion tons, but their gold content has not yet been studied [6,7]. However, given the relative tectonic proximity of coal mines and various gold ore deposits, the assumptions about the gold content and platinoids in the Kara-Keche, Min-Kush, Tash-Komur and other coal mines are not unfounded and are waiting for their scientific confirmation. The total consumption of coal for heating needs in the municipal and household sectors is 2 million tons / year. When burning such volumes of coal, a large number of combustion products are formed: smoke, gases, as well as slag carbonaceous residues (ash), which contain useful elements. The foregoing raises the question of the relevance of research, both on the determination of the gold content of the coals of Kyrgyzstan, and on the development and industrial implementation of the technology of fine purification of heterogeneous gas streams, with the possibility of capturing and chemical recovery of gold from the products of the burning of brown coals of our Republic.

The description of the achieved results on the development of technology for the associated capture, deposition and chemical recovery of gold from the combustion products of Kara-Keche coal is the goal of this work

IOP Conf. Series: Journal of Physics: Conf. Series **927** (2017) 012011 doi:10.1088/1742-6596/927/1/012011

2. Experiment

Pre-crushed and screened coal 2-4 mm in size and 1 kg in weight, in which, by optimal sorption as described in [8], 0.02 g of nanogold was injected into the GDE facility (1) through the upper Hole with the cap removed (6) (Fig.1, Fig.2.). Inside the housing (1) from the oxygen cylinder
(12) through the reducer (13), through the oxygen pipeline (14), the rotameter (15) and the union (16), oxygen was supplied at a flow rate of 5 liters per minute. After ignition and the beginning of smoke formation, a high voltage power supply (18) was switched on and a high voltage (18-8 kV) was applied axially inside the lower heat exchanger (8), which was reduced in proportion to the volume of smoke formation and the temperature of the gas Environment in the chamber of GDE, to prevent breakdowns of the crown. Further, a corona discharge is ignited from the corona wire to the inner surface of the lower heat exchanger. Smoke gases and pyrolysis products of coal (5) begin to move towards the upper heat exchanger (7), where they cool and enter the middle of the GDE, which is manufactured in the form of an inverted T (Fig. 2). Under the influence of corona discharge in GDE, coal gasification products are charged with a positive potential and under the influence of an electric field they settle on the inner surface of the lower heat exchanger (8). The resulting condensate of coal pyrolysis (21) flows down the bottom surface Heat exchanger outside it, and collected in a special vessel (22).
It is assumed that virtually all the nanogold will be in the condensate of coal pyrolysis, so its unimpeded drain and careful assembly is of particular importance.

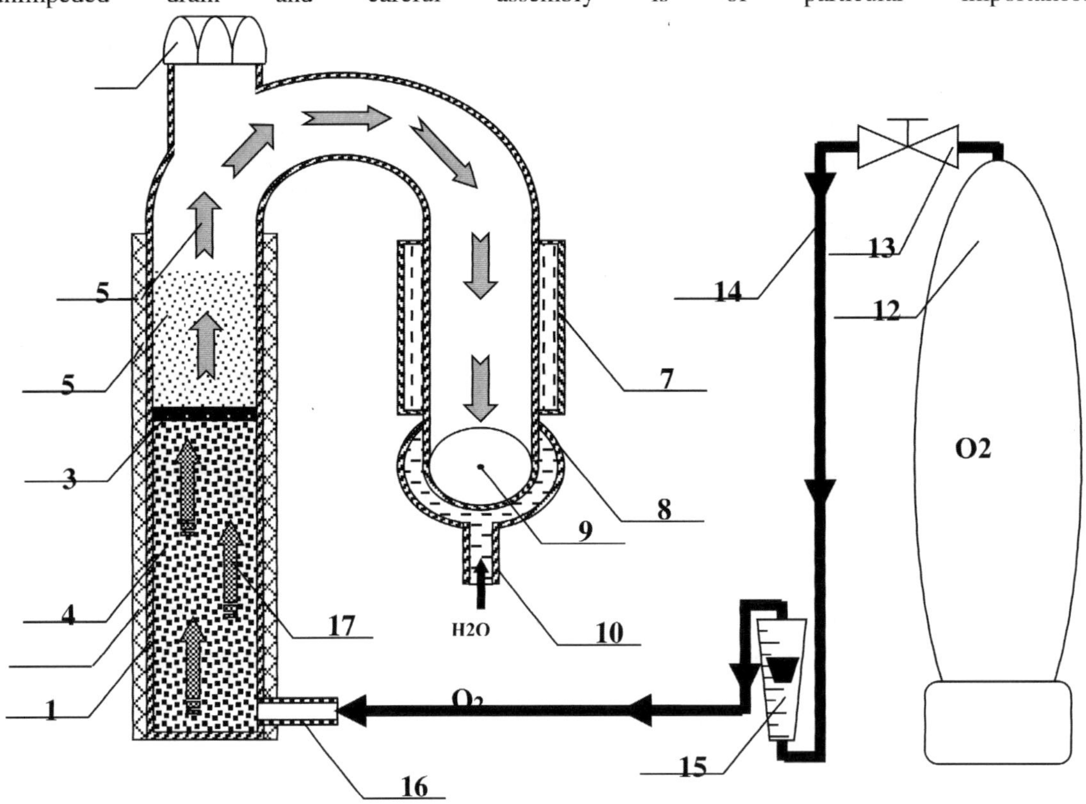

Fig.1. Schematic diagram of GDE (Side view). 1. Corpus of GDE. 2.Ugol, 3.FGi gasification of coal, spreading towards the flow of oxygen, 4. Coke residue, 5. Exhaust of products of combustion and gasification of coal, 6.Shutting, 7.The top heat exchanger. 8. Lower heat exchanger, 9. Corona wire, 10. Coolant (water) supply nozzle, 12. Oxygen cylinder, 13. Reducer. 14. Oxygen pipeline, 15.Rotameter, 16. Oxygen supply connector. 17.Potok oxygen in the coal. . The outlet of warm water from the lower heat exchanger and the inlet and outlet connection of the upper heat exchanger can be seen in Fig.6

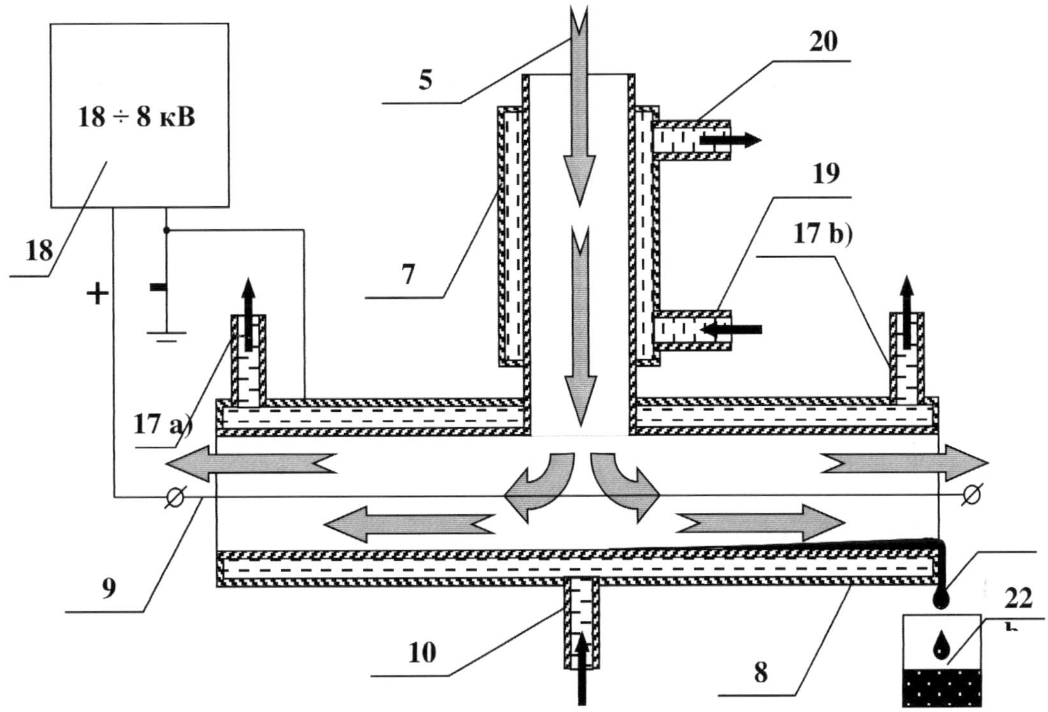

Fig.2. Schematic diagram of the device GDE (Front view). 7. Lower heat exchanger, 10.Input of cooling liquid (water) to the lower heat exchanger, 9. Corona wire, 17a) and b). Water outlet from the lower heat exchanger, 18. High voltage power supply, direct polarity (0 ÷ 20kV), 6. Upper heat exchanger, 5. Direction of movement of gaseous products of burning and pyrolysis of coal. 19. Water input into the upper heat exchanger, 20. Water outlet from the upper heat exchanger, 21. Condensate of coal pyrolysis products. 22 - Vessel for collecting condensate of coal pyrolysis.

3. Discussions

The series of experiments carried out showed that the maximum temperature of the coal combustion front is 990-1010 ° C, the maximum temperature in the combustion chamber, above the surface of the coal is 370 ° C, And the temperature of the gases at the entrance to the DGE is 75 ° C, with an average rate of passage of the front of combustion of coal in the case of oxygen combustion - 2.14 cm / min. The values changed in the direction of decrease with the duration of the experiment and the combustion of coal [9, 10]. The combustion and electrorefining experiment continued linearly, without any features for 1 hour and 40 minutes, with constant observation and fixation of attendant events, with the available corona breakdowns. As a result of the experiment, the following products were obtained: coke residue - 414.7 gr., Condensate - 138 mg. And a cloth rag (napkins, cotton wool, rags), which was used for wet cleaning of the internal surfaces of the experimental setup [11]. The samples obtained are subjected to afterburning and subsequent chemical analysis for the quantitative determination of trapped and recovered gold. The results will be presented later.

4. Conclusion

Assumptions about the gold content of the coals of Kyrgyzstan, the complex measures carried out-the preparation of Kara-Keche coal: crushing to the optimum fraction (2-4 mm), introduction with its composition of calloid gold (taking into account objectivity and profitability - 0.02 g / kg) drying; The developed experimental installation: a feature of the applied structural and technological solutions of GDE, on the basis of which the method and device for continuous coking of coal has been declared to the patent [12] - allowed to study the physical processes of coal combustion, calculate the optimal mode of oxidizer consumption (5 l / h), calculate the optimal worker Corona discharge mode (10-11kV). On the basis of the results obtained, it is possible to develop the newest technology for fine purification of heterogeneous gas flows formed during the combustion of coal with the practical extraction of the desired particles of useful elements (gold) from all the combustion products.

References

[1] Gamov M.I., Granovskaya NV, Levchenko S.V. "Metals in Coals" Southern Federal University, Rostov on Don - 2013, page 65
[2] Arbuzov S.I., Rikhvanov L.P., Maslov SG, Arkhipov V.S.,Pavlov Z.I. "Anomalous gold concentrations in brown coals and peats of the southeastern part of the West Siberian plate" / Izvestiya TPU №7, vol.307 / Engineering georesources, Tomsk - 2007, p. 78-84
[3] "Ecotechnics". Edited by L.V. Chekalov. Yaroslavl, "Rus", 2004.
[4] Tokarev A.V."Corona discharge and its application",Izd. KRSU,Bishkek-2009, pp.42-45
[5] Bogdanov A.S. Tokarev A.V., Lelevkin V.M., Yudanov V.A. / Features of deposition of submicron particles in a coaxial electrostatic precipitator // Vestnik KNU, special issue 2015, p.21
[6] Nazarova N. Qualitative Characteristics the Coals of Kirghizia/Frunze1970;Pages 31-37
[7] A.C. Jamanbaev Coals of the Kyrgyzstan and their rational use / Frunze 1980 p. 35-50
[8] Bogdanov A.S., Vinogradov V.V. / The nanogold introduction method for Kara-Keche coal// (in the edition)
[9] Bogdanov A.S, Tokarev A.V., Yudanov V.A, / Specificity of Kara-Keche coal burning // Izvestia KGTU 2016, №4 (40). Page 114
[10] Bogdanov A.S., Petroshenko T.V. /Gas stream in oxygen coals combustion cleaning// Vestnik KRSU 2016, volume 16 №9, p.151.
[11] Bogdanov A.S., Tokarev A.V., Yudanov V.A. / Continuous and economic coal coking in a shaft furnace // Izvestia KGTU 2016, №3 (39), p. 1, p. 233

Processes of electron attachment and detachment in near breakdown conditions in air plasma

N Ardelyan[1], V Bychkov[2], A Belousov[3] and K Kosmachevskii[3]

[1] Lead researcher, the Department of Computational Mathematics and Cybernetics, M.V.Lomonosov Moscow State University, Moscow, Russia
[2] Lead researcher, the Physical department M.V. Lomonosov, Moscow State University, Moscow, Russia
[3] Graduate student, the Physical department M.V. Lomonosov, Moscow State University, Moscow, Russia
[4] Senior researcher, the Department of Computational Mathematics and Cybernetics, M.V. Lomonosov Moscow State University, Moscow, Russia

E-mail: bychvl@gmail.com

Abstract. Results of theoretical research on ionization of dry air at different low and moderate values of constant electric field are described. Special attention to the processes of electron attachment to oxygen molecules and electron detachment from negative ions has been paid. New data on the detachment of electrons from oxygen molecular negative ions are taken into account.

1. Introduction

Ionization processes in air, resulting in formation of plasma, are of practical interest, related to the numerous use of gas discharges [1], to understanding of low and moderate values of electric fields in discharge devices and plasma at a surface and inside aircrafts in thunderstorm conditions [2,3]. For analysis of issues related to processes in dry air plasma, in this paper we model the discharge plasma at low and moderate (relative to the threshold) electric fields. The study is based on detailed plasma chemical model for the discharge between large electrodes at constant over the volume electric field. We model the situation close to those of [4,5] for which there are measurements for subthreshold and nearthreshold values of reduced electric field; they correspond to self-maintained glow discharge.

2. Plasma chemical model

As in our paper [1], at numerical simulation of plasma chemistry of dry air we use a system of plasma-chemical and chemical reactions corresponding to works [1,5-7]. This model includes 25 component (neutrals, positive and negative ions, electrons and excited particles) and more than 180 plasma-chemical reactions with their participation. Relatively complete description of plasma-chemical model is presented in [1].

Rate constants of electron-molecule reactions correspond to the works [1,4], where they were calculated on a basis of the Boltzmann equation. The function of electron energy distribution was considered corresponding to glow discharge in air. Costs of excitation, dissociation and ionization of molecules by fast particles are taken from [8].Attachment and detachment rate constants were taken from [1,9-11].

Into the equation for the gas temperature we have included terms corresponding to heating as a result of relaxation of vibrational and electronic degrees of freedom of molecules, the heating and cooling of the gas in chemical and plasma-chemical reactions, heating of electrons in collisions – the Joule heating. In the simulation were considered reactions with plasma and fast electrons e, e'; positive ions O^+, O_2^+, O_4^+, O_6^+, O_8^+, N^+, NO^+, N_2^+, $O_2^+N_2$; negative ions O^-, O_2^-, O_3^-, atoms N, O, molecules O_2, O_3 , N_2, N_2O, NO; nitrogen excited vibrational states $N_2(v)$, excited electronic states of nitrogen $N_2(B^3\Pi_g)$, $N_2(X^1\Sigma_g^+)$, oxygen excited vibrational states $O_2(v)$, excited electronic states of oxygen and nitrogen. For ion-molecule reactions we used experimental and theoretical dependences via *E/N*, gas and ion temperatures, all of them are presented in [1]. They were mainly taken from [5-7] and references from them.

3. Basic elementary processes involving negative ions in air plasma

We single out reliable reactions [1,5-6,9-11] taking place at ionization and attachment-detachment processes in the analysis of simple situations in dry air in pressure range of 100-760 Torr. They are:

- Atmospheric background ionization by fast electrons

  $e'+O_2 \rightarrow O_2^+ + e' + e$, $e'+N_2 \rightarrow N_2^+ + e' + e$. (1)

- 2. Direct ionization of molecules of O_2 and N_2 by slow electrons e

  $e+O_2 \rightarrow O_2^+ + 2e$, (2a)

  $e+N_2 \rightarrow N_2^+ + 2e$, (2b)

accompanied by the rapid charge-exchange of ions N_2^+ with oxygen molecules.

- Dissociative attachment of electrons to molecules O_2, with the creation of the negative ion O^- and oxygen atom O : $e+O_2 \rightarrow O^- + O$. (3)

Processes (2) and (3) depend on electrical field value, have thresholds, and play a crucial role at the values of *E/N*> 30-60 Td [1,2].

- Three-body attachment of slow electrons to molecules plays an important role in the balance of ions in air [12]: $e+O_2+O_2 \rightarrow O_2^- + O_2$. (4)

- We take into account the process of electron detachment from molecular oxygen ion in collisions with neutral molecules in air mixture

$O_2^- + M \rightarrow e + O_2 + M$, (5)

and processes detachment of electrons from O^- and O_3^- ions in air

$O^- + M \rightarrow e + O + M$, (6) $O^- + N_2 \rightarrow e + N_2O$, (7)

- Plasma ions O_2^- and O_3^- are produced in reactions

  $O^- + O_2 \rightarrow O_2^- + O$, (8) $O^- + O_2 + O_2 \rightarrow O_3^- + O_2$. (9)

- Balance of positively charged particles at initial stage of the background plasma is determined by reactions of recombination:

  $e + M_2^+ \rightarrow M + M$, (10) $O_2^- + M_2^+ \rightarrow O_2 + M_2$, (11) where M_2^+ is a positive ion.

Reactions (6) - (10) are effective when dissociative attachment of electrons to the oxygen molecules with production of O^- has already occurred, i.e. at the stage close to the avalanche ionization in the cold gas. But namely reaction (5) is a source of background electrons at the stage when the electron temperature is close to room temperature.

In [5] it was shown using numerical analysis of experiments that there could be an effective process $O_3^- + M \rightarrow e + \text{Products}$, (12)

which was not accounted for in [9]. The rate constant of this process in the range 50-150 Td (13.5-40.5 kV/cm) ranges (2.2 - 6.5)$\cdot$ 10^{-14} cm^3/s [5] and it improved an agreement between the theory and the experiment [5]. Estimates made on the basis of [9] show that the rate constant of (12) is smaller than 10^{-15} cm^3/s below *E/N*= 50 Td and we do not consider it there. In Fig.1 we present results of charged particles balance calculations at *E/N*= 50 Td.

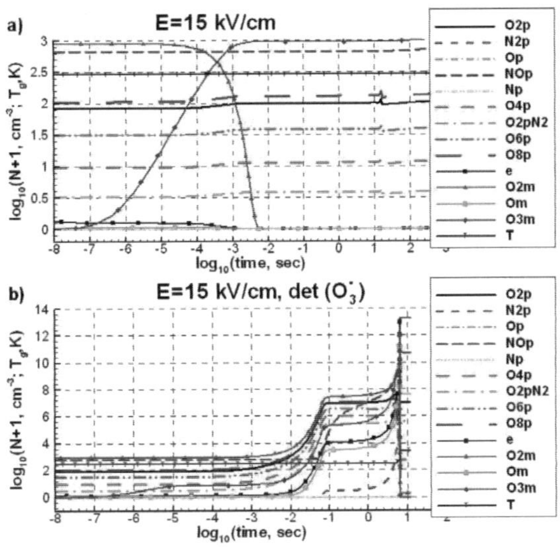

Fig.1. Balance of charged particles a) without accounting for the process (12) and b) with it.

One can see that accounting for the process (12) at 50 Td leads to sharp rise of electron concentrations at long times and cannot be experimentally detected without special measures [5]. However it can explain results of [4] with decrease of the threshold of breakdown electric field in humid air. At smaller *E/N* values results are similar to Fig.1 (a) and no breakdown takes place at them.

4. Role of the process $O_2^- + M \rightarrow e + O_2 + M$

Consider a role of the process (5) at the background and pre-threshold stages. First consider a simplified balance of charged particles at low external electric fields, when a background ionization of molecules occurs. For this, we write the balance equations for electrons, positively and negatively charged ions in the processes (1)-(2), (4)-(5), (10)-(11):

$$\frac{dN_e}{dt} = Q - \nu_{tr} \cdot N_e + k_{\text{det}} \cdot N_{O_2^-} \cdot N_{O_2} - \alpha_{ei} N_e \cdot N_{O_2^+}, \quad (13)$$

$$\frac{dN_{O_2^-}}{dt} = \nu_{tr} \cdot N_e - \alpha_{ii} \cdot N_{O_2^-} \cdot N_{M_2^+} - k_{\text{det}} \cdot N_{O_2^-} \cdot N_{O_2}, \quad (14)$$

$$\frac{dN_{M_2^+}}{dt} = Q - \alpha_{ii} \cdot N_{O_2^-} \cdot N_{M_2^+} - \alpha_{ei} \cdot N_{O_2^-} \cdot N_{M_2^+}, \quad (15)$$

here $\nu_{tr} = k_{tr} \cdot N_{O_2}^2$ is a frequency of the process (4), α_{ei}, α_{ii} - coefficients of electron-ion and ion-ion recombination (11)-(12), at that α_{ii}/α_{ei}>>1 [7], k_{det} - rate constant of the process (5), N_{O_2} is the concentration oxygen molecules, $N_{O_2^-}$ is the concentration of O_2^- ions, and $N_{M_2^+}$ - is the concentration of ions M_2^+ (at initial stage we consider that a concentration of positive molecular ions is close to those of the positive molecular ions of oxygen $N_{M_2^+} \approx N_{O_2^+}$). Usually at atmospheric pressure at low ion concentration $\nu_{tr} >> \alpha_{ei} \cdot N_{O_2^+}$. For a system of equations (13)-(15) is valid [1] a relation

$$Q = \alpha_{ii} \cdot N_{O_2^-} \cdot N_{M_2^+} ,$$

in stationary background conditions or accounting the plasma quasi-neutrality $N_{O_2^-} \approx N_{O_2^+}$ one has

$$N_{O_2^-} = \sqrt{Q / \alpha_{ii}} . \quad (16)$$

At atmospheric pressure $Q = 4$ cm^{-3}·s^{-1} [1,7], and $\alpha_{ii} \approx 2 \cdot 10^{-6}$, cm^{-3}/s [6,7], the concentration of ions is about experimentally observed $N_{O_2^-} \approx 1.4 \cdot 10^3$, cm^{-3}.

Electron concentration in this case is defined by the equation

$$N_{e0} = (Q + k_{\det} \cdot N_{O_2^-} \cdot N_{O_2}) / \nu_{tr} . \quad (17)$$

For example in the reference [9] at low electric field in semi-empirical calculation was obtained a value $k_{\det} \approx 3.7 \cdot 10^{-19}$ cm^{-3}/s, and in the work [11] in modeling by Monte-Carlo of attachment of electrons to oxygen molecules was shown that at the same conditions this value is $k_{\det} \approx 10^{-15}$ см$^{-3}$/c. Using (17), the rate constant from [9] (and $k_{tr} \approx 10^{-30}$ см$^{-6}$/c [10]) one has $N_e \approx 1.0 \cdot 10^{-4}$ cm^{-3} and from [11] $N_e \approx 0.2$ cm^{-3}. So the data from [11] demonstrate better conditions for the background ionization.

Let us consider the simplified balance of electrons at near threshold electric fields, when the ionization frequency (2) by electrons in external electric field ν_i is close to the frequency of dissociative attachment ν_a (3) or $\nu_i \approx \nu_a$, and atomic negative ions are not yet generated then

$$\frac{d N_e}{d t} = Q - \nu_{tr} \cdot N_e + k_{\det} \cdot N_{O_2^-} \cdot N ,$$

and the electron concentration in near threshold conditions is

$$N_e = \frac{Q + k_{\det} \cdot N_{O_2^-} \cdot N_{O_2} - \left(Q - \nu_{tr} \cdot N_{e0} + k_{\det} \cdot N_{O_2^-} \cdot N_{O_2}\right)}{\nu_{tr}} \exp(-\nu_{tr} t) ,$$

at small time, $\nu_{tr} t << 1$, it transforms to

$$N_e = N_{e0} + k_{\det} \cdot N_{O_2^-} \cdot N_{O_2} \cdot t = N_{e0} + g t .$$

Let us estimate a value of g which shows a velocity of electron concentration rise at the threshold. The corresponding *E/N* threshold value was estimated in [1,5] as $E/N \approx 110$ Td (the computed breakdown value [1,5]). Inserting $k_{\det} = 3 \cdot 10^{-12}$ cm^3/s [11] at this value one has $g = 2.5 \cdot 10^{10}$ cm^{-3}s^{-1} what is much greater than the background ionization (1). We see that the process (5) plays essential role in near threshold field since it facilitates the start of the ionization.

5. Conclusions

We have conducted a quantitative analysis of ionization phenomena at low and subthreshold electric fields in air at atmospheric pressure. The detailed analysis revealed the necessary features of the complex processes leading to the breakdown phenomena in air. The important source of background electrons is a reaction of $O_2^- + M \rightarrow e + O_2 + M$ which rate constant, is large enough to explain the initial prebreakdown stage in different discharges. At the near-threshold electric field all the considered processes of detachment play an important role. At higher fields, a main role is played by processes of ionization of molecules and dissociative attachment of electrons to oxygen molecules. At accounting for the processes (12) at long time can be realized a sharp rise of plasma concentration and development of thermal instability. Avalanche ionization in discharges near aircrafts [4] most likely takes place in the cathode region of the device at electric field higher than indicated in this work. Air discharge at E/N= 40 Td theoretically proposed in [3] using a simple plasma model does not occur.

Acknowledgments

Authors thank professors Y. S. Akishev, G.V. Naidis and Dr. I.V. Kochetov for fruitful discussions.

References

[1] Ardelyan N, Bychkov V, Bychkov D and Kosmachevskii K 2013 *Plasma assisted combustion, gasification and pollution control* vol 1, Ed I Matveev (Denver, Colorado: Outskirts press) pp 183-372
[2] Gunn R 1948 *J. Appl.Phys.* **19** pp 481-484
[3] Lowke J, Smith D, Nelson K, Crompton R and Murphy A 2012 *J. Geophys. Res.* **117** D19107
[4] Raether H 1964 *Electron avalanche and breakdown in gases* (London: Butterworths)
[5] Akishev Y, Deryugin A, Karalnik V, Kochetov I, Napartovich A and Trushkin N 1994 *Plasma Phys. Rep.* **20**, pp 511-524
[6] Kossyi I, Kostinsky A, Matveyev A, and Silakov V 1992 *Plasma Sources Sci.Technol.* **1** pp 207-220
[7] Smirnov B 2008 *Reference Data on Atomic Physics and Atomic Processes.* (Berlin: Springer-Verlag)
[8] Konovalov V and Son E 1987 *Plasma chemistry.* vol 14, Ed B Smirnov (Moscow: Energoatomizdat) pp 194-227
[9] Mnatsakanyan A and Naidis G 1991 *Reviews of Plasma Chemistry,* Ed. B Smirnov (New York: Consultants Bureau) pp 259–285
[10] Aleksandrov N 1981 *Plasma chemistry.* vol 8, Ed B Smirnov (Moscow: Energoizdat) pp 90-122
[11] Ponomarev A and Aleksandrov N 2015 *Plasma Sources Sci.Technol.* 2**4** 03501

IOP Conf. Series: Journal of Physics: Conf. Series **927** (2017) 012013 doi:10.1088/1742-6596/927/1/012013

Argon gas atoms trapping by carbon dendrites

M P Danilaev[1], **E A Bogoslov**[1], **Y E Polskii**[1]
[1] Kazan National Research Technical University named after A.N.Tupolev - KAI, 10, Karl Marx st., Kazan, 420111, Russia.

Abstract. The conditions of argon gas atoms trapping by carbon dendrites, which growing in atmospheric pressure gas-discharge plasma, are considered in that paper. It's showing that the argons atoms trapping by the carbon diamond-like cell can occur in arc gas discharge with current density more than $j_0 \sim 45 \ {}^{\text{mA}}\!/_{\text{sm}^2}$ and provided that the eximer molecules of noble gas and carbon atoms (e.g., ArC) can be formed.

During the last years, investigations of carbon nanoparticle properties have attracted considerable attention of many researchers. Some properties of these particles depend on the way of their fabrication. The formation of carbon nanoparticles in gas discharge is one of the most perspective ways of their fabrication. [1-3] Many carbon structures (e.g., scaly [4], vertical[5], dendrite[6] species) with reproducible physico-chemical properties were formed in gas-discharge plasma at atmospheric pressure. The gas discharge is usually switched on in the inert gas atmosphere in the presence of initial products of carbon nanoparticle synthesis, e.g., monomer vapor[7]. The formation of structurally ordered carbon nanoparticles, such as graphene sheets, carbon diamond-like cell and their conglomerates, is possible in the temperature range from 800 °C to 3000 °C[8]. Such temperatures are provided by arc discharges. The electron energy in such discharges[9] is sufficient to initiate the excimer molecule formation, e.g., ArC. These molecules can appear as growth centers of the graphene sheet conglomerates in the discharge volume. In this case, the argon atom can be trapped by carbon diamond-like cell if the growth characteristic time of the cell is less than the excimer molecule characteristic life time. It should be noted that the characteristic carbon diamond-like cell size (0,142 nm) is of the same order as the argon atom characteristic size (0,142 nm)[10]. Therefore, the argon atom does not get out of the cell without cell destroying.

In the present paper, the growth conditions of the carbon dendrites with argon atoms trapped by carbon diamond-like cell are considered.

The investigation of carbon dendrite growth conditions was performed experimentally. The scheme of experimental device is shown in Fig. 1.

Carbon dendrites were obtained in alternating current arc-discharge plasma (frequency ~40 kHz) from various monomers: styrene, acrylonitrile, methyl methacrylate, benzene and propylene. The argon flow was 200 ml/min and was regulated with an uncertainty less than 10%. The electrodes have a "needle-needle" configuration: steel needle electrodes with curvature radius of 0.5 mm. The condition $\left({}^{T_e}\!/_{T}\right) \sim 10$ was hold in the experiments, where T_e – electron temperature in arc discharge, T – thermodynamic temperature near the needle electrode[3]. The carbon dendrite growth stopped when the relation T_e/T was reduced due to the decrease of T_e or the increase of T. The

IOP Conf. Series: Journal of Physics: Conf. Series **927** (2017) 012013 doi:10.1088/1742-6596/927/1/012013

thermodynamic temperature was 800÷1200 °C depending on monomer type. It was measured by a pyrometer with measurement error no more than 5%. The carbon dendrite growth occurred from the needle surface in opposite direction until the short circuit of the discharge gap.

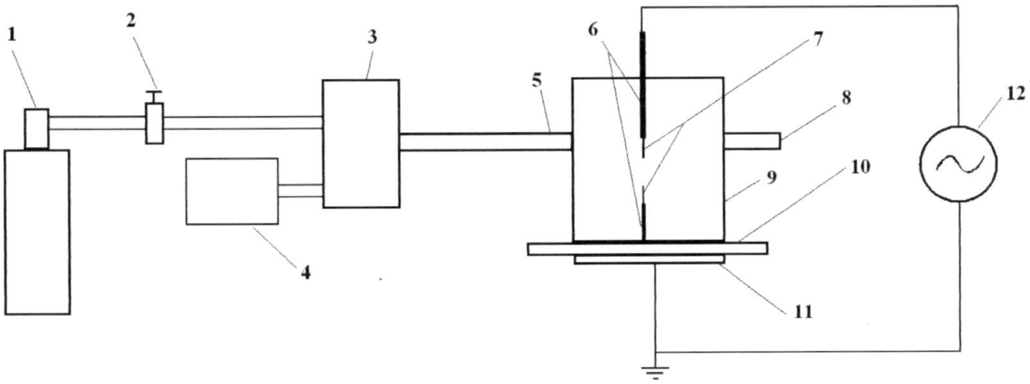

Figure 1. The scheme of experimental device: 1 – cylinder with argon, 2 – valve, 3 – multi-phase gas flow formation device, 4 – monomer ejector, 5 – inlet connection pipe, 6 – needle electrodes, 7 – carbon dendrite, 8 – outlet connection pipe, 9 – reactor, 10 – dielectric barrier, 11 – electrode, 12 – high voltage pump source.

The carbon dendrite growth started from the formation of one or more channels on the top of needle electrode. The number of branches depends on the field intensity and increases with increasing field intensity. So, the cauliflower-like structures were formed (Fig. 2).

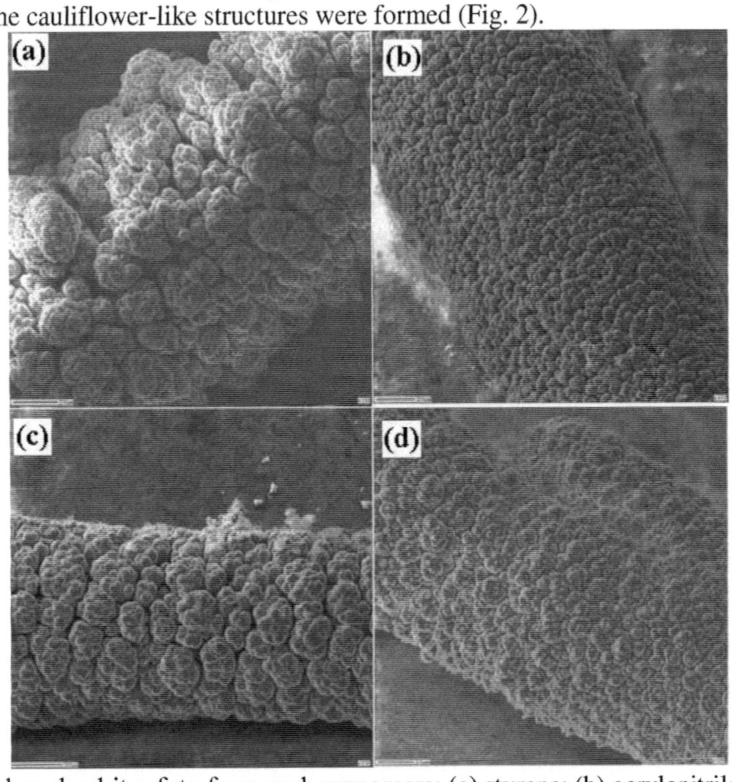

Figure 2. Carbon dendrites foto from such monomers: (a) styrene; (b) acrylonitrile; (c) benzene; (d) methyl methacrylate.

The minimum discharge power necessary for the formation of carbon dendrites was ~200 W. The carbon dendrite growth rate increases with increasing discharge power[3]. When the discharge power became more than ~300 W, the carbon dendrite growth stopped because of the strong electrode heating.

While the field intensity changed within the range 0.6 – 3 kV/sm (the precision of measurements was ~10%), the average carbon dendrite growth rate was ~3.5 mm/min for dendrite from benzene, ~1 mm/min for dendrite from styrene, ~0.7 mm/min for dendrite from propylene, ~0.5 mm/min for dendrite from methyl methacrylate and acrylonitrile. In case of benzene, the highest carbon dendrite growth rate is due to the fact that benzene ring does not require additional reactions for the formation of carbon unit cell hexagonal structure.

The analysis of carbon dendrite structure was carried out on the auger electron spectrometer JAMP 9510F ("JEOL", Japan). Experimental conditions: ultra-high vacuum better than $1 \cdot 10^{-9}$ mm Hg; temperature +23 °C. The micrographs and auger spectra are presented in Fig. 3 for carbon dendrite from propylene.

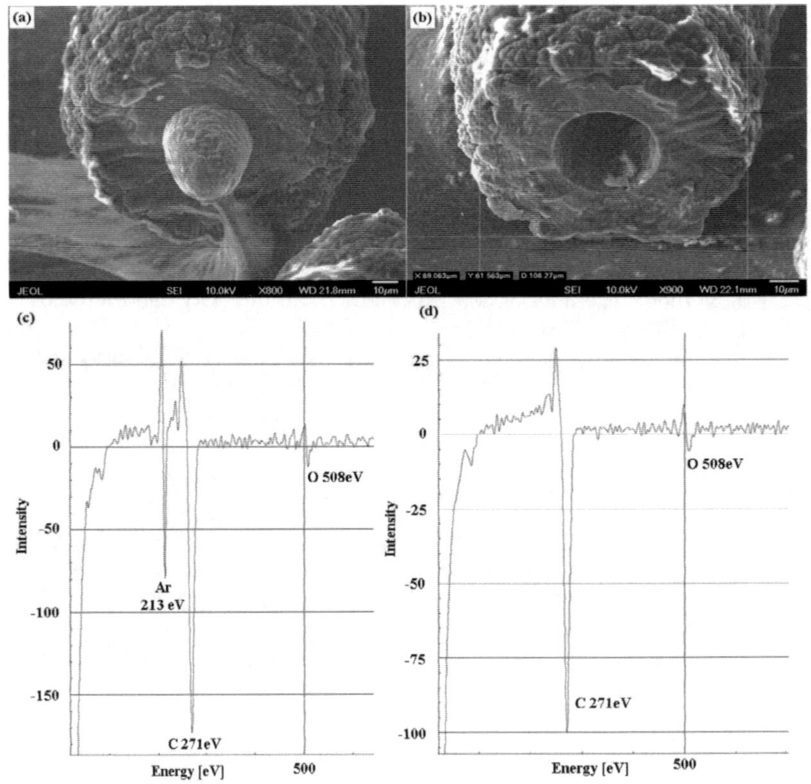

Figure 3. Micrographs and auger spectra for carbon dendrite from propylene.

The auger spectra (Fig. 3c) obtained from the central part of dendrite contain line corresponding to argon. The amplitude of this line did not change after keeping the sample at $1 \cdot 10^{-9}$ mm Hg for three days. It indicates that the central part of dendrite has crystal structure containing diamond-like cells. This fact is also confirmed by the electron beam diffraction (Fig. 4). The latter was obtained at the transmission electron microscope Zeiss Libra 120 with integrated OMEGA filter. Experimental conditions: accelerating voltage of –120 kV; electron source – LaB_6.

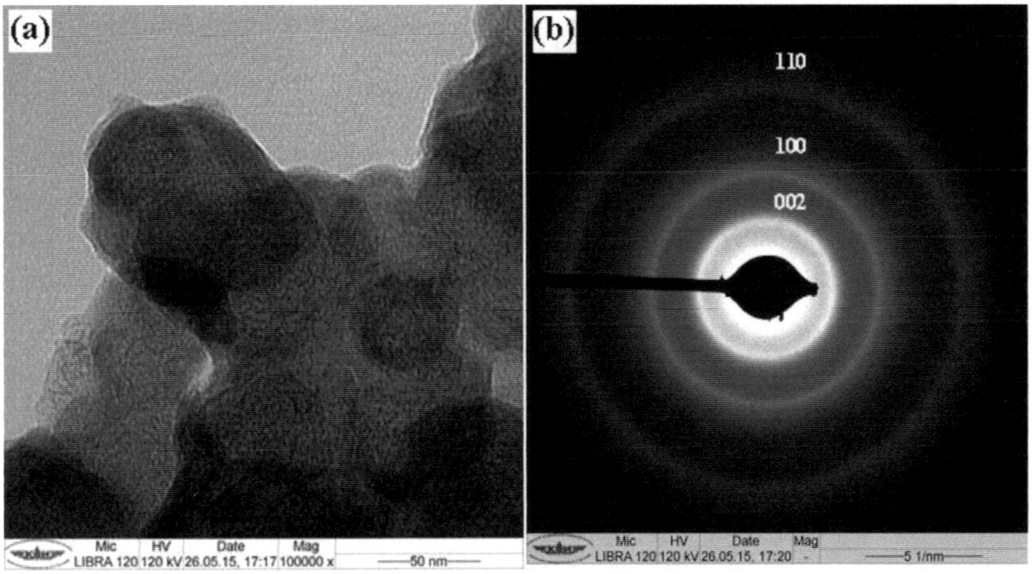

Figure 4. The carbon dendrite from propylene: micrograph (*a*) and electron beam diffraction (*b*).

The samples contain regions with ordered structure, which produces pronounced diffraction peaks, as well as regions with an amorphous structure producing poorly-defined diffraction rings. The carbon amorphous phase prevails on the periphery. This phase cannot trap argon atoms. The presence of argon atoms in the central part was observed only in case of dendrite formed from propylene. It is because of higher current density in gas discharge ($j_0 \sim 45\ \mathrm{mA/sm^2}$) compared to current density for dendrites formed from other monomers: styrene ($j_0 \sim 25\ \mathrm{mA/sm^2}$), acrylonitrile ($j_0 \sim 28\ \mathrm{mA/sm^2}$), methyl methacrylate ($j_0 \sim 35\ \mathrm{mA/sm^2}$), benzene ($j_0 \sim 30\ \mathrm{mA/sm^2}$).

The characteristic time of carbon diamond-like cell formation for carbon dendrite from propylene is ~1 ns. This time is less than the excimer molecule characteristic life time (~10 ns)[11]. Thus, the argons atoms trapping by the carbon diamond-like cell can occur in arc gas discharge with current density more than $j_0 \sim 45\ \mathrm{mA/sm^2}$ and provided that the eximer molecules of noble gas and carbon atoms (e.g., ArC) can be formed.

References

[1] G. Lupina, C. Strobel, J. Dabrowski, G. Lippert, J. Kitzmann, H. M. Krause Ch. Wenger, M. Lukosius1, A. Wolff, M. Albert, J. W. Bartha 2016 *Appl. Phys. Lett.* **108** 193105
[2] T. Labbaye, A. Canizarès, M. Gaillard, T. Lecas, E. Kovacevic, Ch. Boulmer-Leborgne, T. Strunskus, N. Raimboux, P. Simon, G. Guimbretière, M. R. Ammar 2014 *Appl. Phys. Let.* **105** 213109
[3] M.P. Danilaev, E.A. Bogoslov, Yu.E. Polsky 2014 Technical Physics Letters. **40.** No.10. pp. 857-859
[4] Hunga Nsung-Chi, Chenb Chua-Fu, Chenc Chien-Chung, Whang Wha-Tzong 2009 *Appl. Surf. Sci.* **255** 3676
[5] Bo Zheng, Kehan Yu, Ganhua Lu, Wang Pengxiang, Mao Shun, Junhong Chen 2011 *Carbon.* **49** 1849

[6] Dmytro Kozak, Etsuro Shibata, Atsushi Iizuka, Takashi Nakamura 2014 *Carbon.* **70** 87
[7] Kadek W. Hemawan, Huiyang Gou, Russell J. Hemley 2015 *Appl. Phys. Lett.* **107** 181901
[8] Michon, S. Vézian, E. Roudon, D. Lefebvre, M. Zielinski, T. Chassagne, M. Portail, J. 2013 *Appl. Phys.* **113** 203501
[9] Toshiki Sugai, Hideki Omote, Shunji Bandow, Nobuo Tanaka, Hisanori Shinohara 2000 *J. Chem. Phys.* **112** 6000
[10] F. A. Cotton, G. Wilkinson 1988 *Advanced Inorganic Chemistry (5th ed.)* (New York: Wiley–Interscience)
[11] J.B. Birks 1975 *Rep. Prog. Phys.* **38**, 903

Ion current density distribution in a pulsed non-self-sustained glow discharge with a large hollow cathode

V V Denisov, S S Kovalsky, N N Koval, I V Lopatin, E V Ostroverkhov, P M Shchanin

Institute of High Current Electronics SB RAS, 634055, Tomsk, Russia

E-mail: volodyadenisov@yandex.ru

Abstract. The paper reports on azimuthal and radial distributions of the ion saturation current density measured with a single cylindrical and a plane probe in the plasma of a low-pressure (≈1 Pa) pulsed non-self-sustained glow discharge with a large (0.21 m^3) hollow cathode at discharge currents higher than 100 A. It is shown that increasing the discharge operating voltage and decreasing the operating pressure in the range 0.4–1 Pa improves the uniformity of the ion current density distribution.

1. Introduction

Anomalous glow discharges are widely used for thermochemical surface treatment of metals and alloys [1], including nitriding, cementing, diffusion surface saturation with boron or other elements [2]. The volume of working chambers can reach several cubic meters, and the charge weight more than ten tons [3]. However, there are series shortcomings that affect the treatment time. Among them are relatively high pressures ranging from several tens to several hundreds of pascals. This factor does not allow efficient surface cleaning from oxides, which can lengthen the treatment time in an anomalous glow discharge to 50 h. In the last two decades, it has been shown that low-pressure (≈1 Pa) discharges with surface cleaning by high-energy ions can provide a several-fold decrease in nitriding time and higher nitriding efficiency [4–6]. Today, however, such discharges are hardly used in industry because it is difficult to provide plasma uniformity in large (≥0.5 m^3) vacuum volumes. A way to increase the plasma uniformity is to use several plasma sources around the periphery of a chamber whose walls will serve as an anode, but this greatly increases the cost of equipment and decreases the efficiency and reliability of the entire system. One of the most promising methods for plasma generation at low pressure is to use a hollow-cathode glow discharge. In this type of discharge, the azimuthal distribution of ion current density is symmetric about the center [7] and allows one to obtain currents of 30 A in a volume of 0.2 m^3 at a voltage of ~600 V. In the plasma of this discharge, the ion current density reaches ~2–3 mA/cm^2. Recent research demonstrates that steels and titanium are most efficiently nitrided at an ion current density of about 4 mA/cm^2 and higher [8]. To attain the so high ion current density in a large volume, it is required that the discharge current in this volume be greater than 100 A. The discharge current can be increased through injection of additional electrons which, when arrived in the plasma of a glow discharge, are accelerated in the region of potential fall near the hollow cathode, oscillate inside the hollow, and efficiently ionize the gas [9]. This allows one to obtain discharge currents of about 100 A in continuous modes and up to several hundred amperes in pulsed modes at a discharge operating voltage of ≈200 V [10]. In a non-self-sustained glow discharge, electrons are normally injected from a small area of an emission electrode separating the regions of main and auxiliary discharges [9, 11]. As a result, the current density of these electrons is two or three orders of magnitude higher than that of γ-electrons formed at the hollow cathode walls via secondary ion-electron emission. Therefore, the plasma density and hence the ion saturation current density to a probe locally increases in the direction of electron emission from the auxiliary discharge at about the free path from the face of the grid electrode. The aim of our study was to determine how the operating conditions and parameters of a low-pressure pulsed non-self-sustained glow discharge with a large (≈ 0.2 m^3) hollow cathode influence the azimuthal and radial ion current density distributions.

2. Material and research techniques

The azimuthal and radial distributions of ion saturation current densities in a hollow-cathode pulsed non-self-sustained glow discharge were studied on an experimental test bench shown schematically in Fig. 1. The inner walls of a vacuum chamber of dimensions 600×600×600 mm formed a hollow cathode of volume 0.21 m^3 for the main glow discharge. The chamber was pumped with a TMN-500 turbomolecular pump to an ultimate pressure of $5 \cdot 10^{-3}$ Pa. The operating pressure was controlled in the range 0.4–1 Pa through gas supply (nitrogen).

The main glow discharge was ignited between hollow cathode *1* (chamber wall) with a surface area S_c = $2.3 \cdot 10^4$ cm^2 and uncooled plane anode *2* with a surface area S_a = 310 cm^2. The chamber material was stainless steel (12Cr18Ni10Ti). The anode was arranged through the side flange of the chamber. The anode to cathode area ratio was S_a/S_c = 1/74. The glow discharge was powered from an inverter dc/pulse voltage source with the following parameters: voltage amplitude U_d = 30–300 V, pulse current I_d = 0–550 A at maximum average output current I_{d_av} = 120 A, maximum average power P = 30 kW, pulse repetition frequency f = 1–1000 Hz, and pulse duty factor γ_i = 1–100 %. For decreasing the drop in voltage amplitude during a discharge pulse, a capacitor bank with C = 7.2 mF was installed at the output of the glow discharge power supply. The power supply prevented from microarcs at the surface of the hollow cathode. The discharge current $I_d(t)$ was measured using a Honeywell CSNJ481 Hall probe with data transmission to a Tektronix TDS2014C oscilloscope. The discharge voltage $U_d(t)$ was measured with an oscillographic probe (1:100) between the glow discharge anode and cathode.

For stable operation of a steady glow discharge at voltages of several tens of volts and ignition of a pulsed discharge, an arc electron source with an integrally cold hollow cathode was used [12]. The arc was ignited via a dielectric flashover when nitrogen was suppled through gas inlet *4* and a high-voltage pulse was applied between trigger electrode *5* and cylindrical hollow cathode *6*. The auxiliary arc discharge operated between cylindrical hollow cathode *7* and conical anode *8* covered with a fine grid of geometric transparency 45% (mesh size 0.4×0.4 mm). Such a conical shape deflects electrons emitted in the main discharge from the axis of the plasma source, thus improving the plasma uniformity in the hollow cathode. Conical grid anode *8* with its concave central part was at the potential of the glow discharge hollow cathode and was an emission electrode through which electrons from the auxiliary arc discharge were injected to the main glow discharge. The cathode spot moved over the inner surface of the cylindrical hollow cathode at the maximum tangential component of an axial magnetic field induced by coil *3*. The auxiliary arc steadily operated through a hole in arc arrester *7* which, being at floating potential, prevented the cathode spot switching to the face of the plasma generator and the arc passing from diffuse operation in the anode region to constriction with the formation of an anode spot at the nearest site of conical grid anode *8*. The arc source was powered from an ARC150 stabilized current source [13] which provided an arc current of up to I_a = 150 A at a voltage of up to U_a = 60 V. In all experiments, the magnetic induction on the axis of the electron source was B = 3.8 mT.

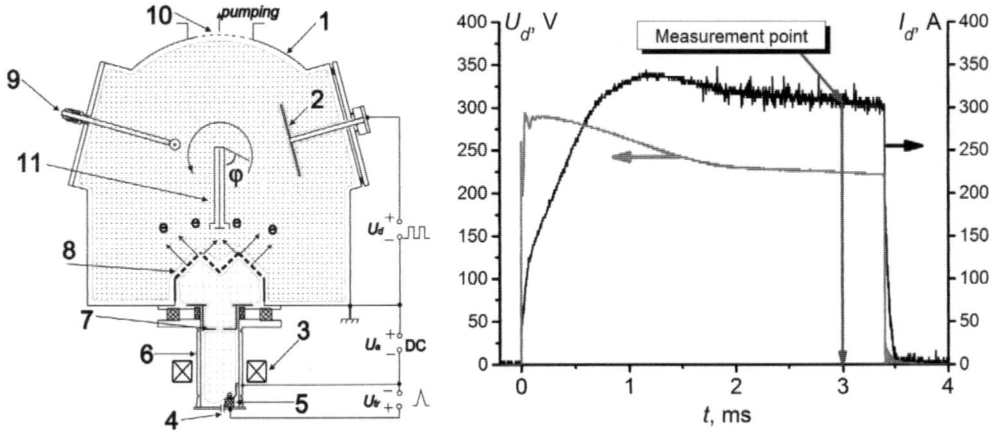

Figure 1. Experimental test bench (a) and current and voltage of non-self-sustained glow discharge at I_a = 150 A, p (N_2) = 1 Pa (b). In test bench: *1* –cathode of glow discharge, *2* – anode of glow discharge, *3* – magnetic coil, *4* – gas inlet, *5* – trigger electrode, *6* – hollow cathode of auxiliary arc, *7* – arc arrester, *8* – conical grid anode of auxiliary arc, *9* – cylindrical Langmuir probe, *10* – grid, *11* – plane probe with guard ring.

The azimuthal distribution of ion current densities was measured using plane probe *11* of diameter 5 mm with a guard ring; the probe had the potential of the glow discharge hollow cathode. The radial distribution was measured using cylindrical Langmuir probe *9*. For measuring the azimuthal distribution, the probe was rotated about the center of the chamber at a distance of 18 cm from its axis, being positioned at the height of the axis of the auxiliary plasma source from which electrons were emitted. The probe was fixed in a flange with a rotary feedthrough at the chamber bottom. The radial and azimuthal distributions of ion current densities were analyzed at different pressures and different discharge voltages and currents. The parameter characterizing the azimuthal nonuniformity of ion current density was the ratio of maximum density deviation from average to average ion current density:

$$k = \frac{|j_n - j_{av}|_{\max}}{j_{av}} \times 100\% \quad , \qquad (1)$$

where j_n is the ion saturation current density for rotation through a certain angle n = 1–18, and j_{av} is the ion current density averaged over all measurements.

3. Results and discussion

The radial distribution of ion saturation current densities is presented in Fig. 2; plane *A–A* in which the probe moved is shown in Fig. 3. Almost in all cases, the maximum ion current density falls on the chamber center, i.e., j_i = 11 mA/m^2 at I_d = 90 A, and at the edges, it decreases. This is because the probability of nitrogen ionization is higher near the chamber center than at the periphery. The plasma density calculated by Bohm's formula is ≈10^{12} cm^3 at T_e = 1 eV. If we neglect the extreme points of radial distributions near the chamber walls, the distribution nonuniformity is up to 20 %. Varying the operating pressure in the range 0.4–1 Pa can cause plasma density redistribution in the chamber: local maxima appear in the radial distribution (Fig. 2a, curve *2*), which can be due a change in the free path such that regions with a higher ionization probability, compared to others, arise at this distance from the chamber walls.

IOP Conf. Series: Journal of Physics: Conf. Series **927** (2017) 012014 doi:10.1088/1742-6596/927/1/012014

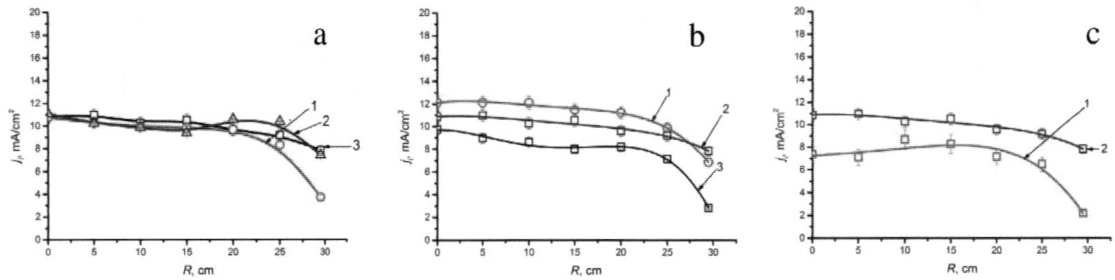

Figure 2. Radial distributions of ion saturation current density to probe, chamber center 0 cm: (**a**) for $U_d = 182$ V, $I_d = 90$ A at $p(N_2)$ equal to 1 Pa (*1*), 0.65 Pa (*2*), and 0.4 Pa (*3*); (**b**) for $U_d = 182$ V, $p(N_2) = 0.65$ Pa at I_d equal to 125 A (*1*), 90 A (*2*), and 60 A (*3*); and (**c**) for $p(N_2) = 0.65$ Pa, $I_d = 90$ A at U_d equal to 130 V (*1*) and 182 V (*2*).

As Fig.3a suggests, decreasing the operating pressure in the range from 1 Pa to 0.4 Pa improves the uniformity of azimuthal current density distributions by more than 50 %, which is also confirmed by calculations of *k* in Table 3. Increasing the glow discharge current from 60 A to 125 A increases the ion current density but the nonuniformity parameter *k*, in this case, remains almost unchanged. Increasing the discharge operating voltage from 130 V to 235 V decreases the nonuniformity by 20 %, all other things being equal.

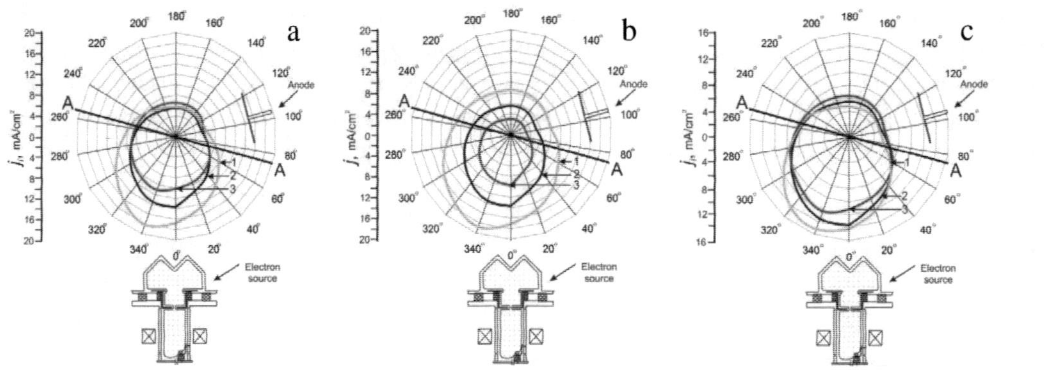

Figure 3. Azimuthal distributions of ion saturation current density to probe, chamber center 0 cm: (**a**) for $U_d = 182$ V, $I_d = 90$ A at $p(N_2)$ equal to 1 Pa (*1*), 0.65 Pa (*2*), and 0.4 Pa (*3*); (**b**) for $U_d = 182$ V, $p(N_2) = 0.65$ Pa at I_d equal to 125 A (*1*), 90 A (*2*), and 60 A (*3*); and (**c**) for $p(N_2) = 0.65$ Pa, $I_d = 90$ A at U_d equal to 130 V (*1*), 182 V (*2*), and 235 V (*3*).

Varying the anode area S_a from 300 cm^2 to 600 cm^2 does not provide any noticeable change in the azimuthal ion current density distribution and its parameter *k*.

Table 1. Nonuniformity *k* for azimuthal distribution of ion saturation current density in low-pressure non-self-sustained discharge with hollow cathode

Mode No	Mode	Key parameter	Nonuniformity *k*, %
1	$U_d = 182$ V, $I_d = 90$ A	$p(N_2) = 1$ Pa	89

2	U_d = 182 V, I_d = 90 A	$p(N_2)$ = 0.65 Pa	54
3	U_d = 182 V, I_d = 90 A	$p(N_2)$ = 0.4 Pa	34
4	$p(N_2)$ = 0.65 Pa, U_d= 182 V	I_d = 125 A	45
5	$p(N_2)$ = 0.65 Pa, U_d= 182 V	I_d = 90 A	54
6	$p(N_2)$ = 0.65 Pa, U_d= 182 V	I_d = 60 A	48
7	$p(N_2)$ = 0.65 Pa, I_d = 90 A	U_d = 130 V	50
8	$p(N_2)$ = 0.65 Pa, I_d = 90 A	U_d = 182 V	54
9	$p(N_2)$ = 0.65 Pa, I_d = 90 A	U_d = 235 V	31

4. Conclusion

Thus, almost in all modes studied, the radial plasma density peaks at the chamber center and decreases at the chamber walls. The key factors that influence the uniformity of ion saturation current densities in a non-self-sustained glow discharge are the operating pressure and the discharge operating voltage. Decreasing the operating pressure from 1 Pa to 0.4 Pa decreases the azimuthal nonuniformity of ion current density by 50 %, and increasing the discharge operating voltage from 130 V to 235 V decreases the factor k by 20 %. Increasing the glow discharge current in in the range 60–125 A and increasing in the anode area from 300 cm^2 to 600 cm^2 fails to provide any considerable change in the azimuthal distribution and its nonuniformity factor. By varying the gas pressure and the discharge operating voltage, it is possible to provide ion saturation current densities nonuniform to about ≈30% even with a single electron source and to form nitrided layers of acceptable uniformity in bulky products.

Acknowledgments
The work was supported by the Russian Science Foundation (project No. 14-29-00091).

References

[1] Berlin E V, Koval N N, Seidman L A 2012 *Plasmochemical Processing of Steel Surfaces* (Tekhnosfera, Moscow) 464 [in Russian].
[2] Rie K T 1999 *Surf. Coat. Technol.* **112** 56–62.
[3] Budilov V V, Koval N N, Kireev R M, Ramazanov K N 2013 *Intergrated Methods of Processing of Constructional and Instrumental Materials by Glow and Vacuum-Arc Discharges* (Mashinostroenie, Moscow) 320 [in Russian].
[4] Borisov D P, Koval N N, Schanin P M 1994 *Izv. Vyssh. Uchebn. Zaved., Fiz.* **3** 115–120 [in Russian]
[5] Gavrilov N V, Menshakov A I 2011 *Prib. Tekh. Eksp.*, **5** 140–148 [in Russian].
[6] Andreev A A, Sablev L P, Grigoriev S N 2005 Vacuum Arc Coatings (Kharkov: NSC KIPT Press) 236 [in Russian].
[7] Lopatin I V, Schanin P M, Akhmadeev Yu H, Kovalsky S S, Koval N N 2012 *Plasma Phys. Rep.* **38** 7 585–589.
[8] Gavrilov N V, Mamaev A S, Chukin A V 2016 *Pis'ma Zh. Tekh. Fiz.*, **42** 9 97–104 [in Russian]
[9] Vizir A V, Oks E M, Schanin P M, Yushkov G Yu 1997 *Tech. Phys.*, **42** 611–614
[10] Akhmadeev Yu H et. al. 2017 *Plasma Phys. Rep.* **43** 1 67–74
[11] Lopatin I V et. al. 2011 *Prib. Tekh. Eksp.* **1** 151–156 [in Russian]
[12] Schanin P M et. al. *Zh. Tekh. Fiz.* **74** 5 24–29 [in Russian]
[13] Yakovlev V V et. al. 2015 *Izv. Vyssh. Uchebn. Zaved. Fiz.* **58** 9/2 288–292 [in Russian]

Thrust and thrust-to-power ratio in electrohydrodynamic propulsion electrode systems

D V Dremin, V Yu Khomich, I E Rebrov
Institute for Electrophysics and Electric Power RAS, Dvorzovaya Naberezhnaya 18, 191186, Saint- Petersburg, Russia

E-mail: rbrv.igor@gmail.com

Abstract. The influence on thrust parameter and efficiency from geometric parameters and design features of the ionocraft electrode is considered, dependences were obtained. The most effective mode of work lifters of the considered structures is achieved at low current. The increase of distance between the corona electrode and the collector gives the growth of thrust and thrust-to-power ratio up to some distance. It is advisable to use emitters with smaller diameter to reduce losses to the formation of the corona.

1. Introduction

The ionocraft is a device that provides a lifting force due to the formation of the electrohydrodynamic (EHD) flow. EHD flow is a name for a physical phenomenon of collective motion of gas caused by a drift of charged particles in strong electric field and by transfer of their momentum to neutral environment. Generation of ions occurs in field of corona discharge, which is characterized by high heterogeneity of electric field in space. A typical scheme of this device consists of two conductive parts. The first one is represented by a thin wire or a tip, called a plasma emitter (PE) or corona electrode. The second one (the collector) is designed as a wire grid, a tube or a frame covered with foil. High voltage is applied to the corona electrode; which results ionization and subsequent transfer of electric field energy to ions occur that leads to the appearance of thrust. There are no moving parts in the ionocraft and surrounding atmosphere is working agent, thus only electric energy is converted. It would find application in flying micro-robots and autonomous aircrafts on solar batteries. Now there are not devices able to perform lifting of its own weight, including ionocraft power supply system. A design of the power supply implies low weight and high power-to-weight ratio for lifting of the device with the supply system and additional payload. Thus, optimization and simulation of each of such devices' parts together is required for obtaining their high efficiency [1-3].

2. Numerical simulation

In this paper, we consider the influence on thrust parameter and efficiency from geometric parameters and design features of the ionocraft electrode systems (IES) in air at atmospheric pressure and natural humidity. Designs of IES models shown in Figure 1. They include the diameter of the corona electrode, the distance between it and the collector, the number of emitters in usage and the framework of the collector. Thrust parameter is represented by magnitude of a generated thrust F measured in grams, and the efficiency is evaluated by magnitude of thrust-to-power ratio, which is equal to ratio of thrust to power consumption F/P [g/W].

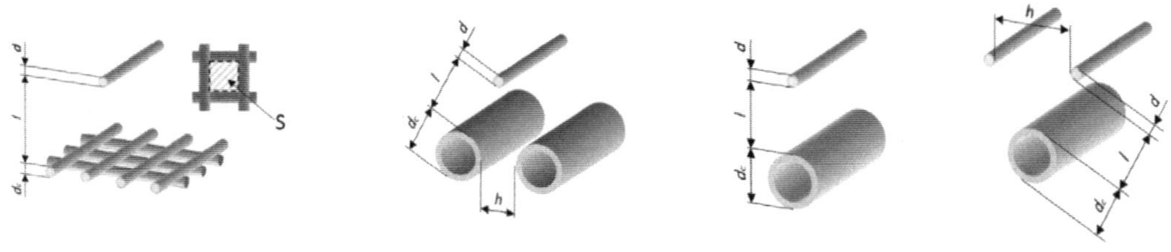

Figure 1. Designs of IES models: *1* – single wire emitter, grid collector with mesh size of $S = 2x2$ mm^2 and diameter $d_c = 0.42$ mm; *2* – single wire emitter, double tube collector with distance $h = 10$ mm; *3* – single wire emitter, the single tube collector; *4,5* – double wire emitter with distance $h = 10$ mm and $h = 30$ mm, single tube collector.

Dependences for the efficiency of the system and the magnitude of the thrust were obtained. During the research several systems have been examined, including one or two PE – nichrome wires with a diameter $d = 0.08$ mm and $d = 0.02$ mm and a series of collectors: one or two parallel aluminum tubes with a diameter $d_c = 10$ mm, a steel grid with a mesh size of $S = 2x2$ mm^2 and a wire diameter $d_c = 0.42$ mm. According to [4] formula for estimating thrust for ionocraft is as follows:

$$F = \frac{I * l}{\mu} \quad (1)$$

where I is intensity of current; l is distance between PE and the collector; μ is ion mobility in air. Considering (1), it's clear that thrust F will increase in proportion to increasing distance l if other parameters stay the same. In practice, this dependence cannot be fully performed (see Figure 2a, b).

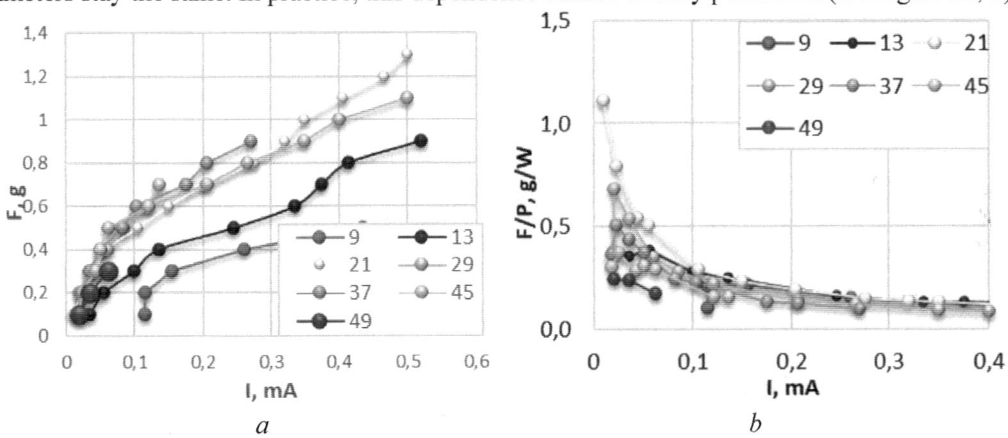

Figure 2. Thrust (*a*) and thrust-to-power ratio (*b*) depending on current for different distances between electrodes l, mm for model 3 (emitter - wire $d = 0.08$ mm, collector – tube $d_c = 10$ mm, positive corona)

Increasing of the zone of acceleration of charged particles leads to an increase of the thrust up to the distance of the collector-emitter 21 mm. As l increases, the growth of thrust F is slowing down.

Using (1) and expressions for power consumption and I–V curve of corona discharge [5], thrust-to-power ratio for ionocraft is obtained:

$$\frac{F}{P} = 2l(\mu V_k\left(1 + \sqrt{1 + \frac{4I}{CV_k^2}}\right))^{-1} \quad (2)$$

where V_k is the critical corona voltage and C is a constant value defined for a particular geometry of the system. Hyperbolic dependence for thrust-to-power ratio from current shows necessity of work of ionocraft in modes small values of current I as the most effective.

Dependences of the thrust and the efficiency of the system for different polarities of the corona and the distance between the collector and emitter is shown in Figure 3a, b.

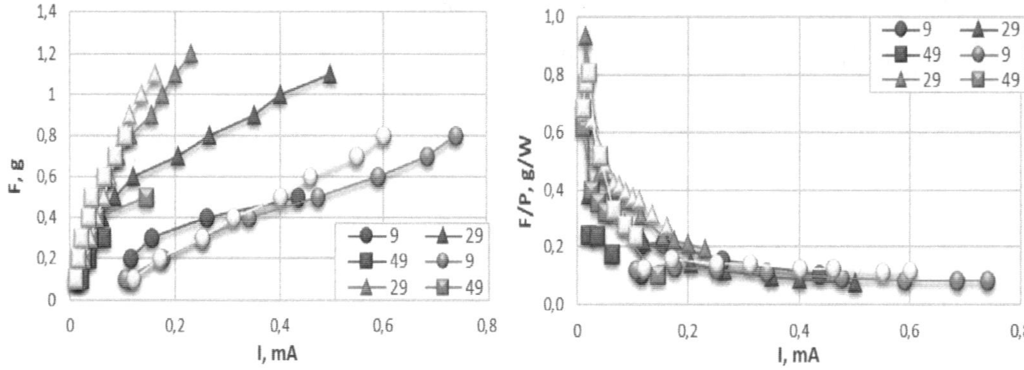

a *b*

a

Figure 3. Thrust (*a*) and thrust-to-power ratio (*b*) depending on current for different distances between electrodes *l*, mm for model 3: *red* – wire $d = 0.08$ mm, tube $d_c = 10$ mm, positive corona; *blue* – wire $d = 0.08$ mm, tube $d_c = 10$ mm, negative corona; *yellow* – wire $d = 0.02$ mm, tube $d_c = 10$ mm, negative corona

At $l = 9$ mm, all types of devices showed approximately equal performance. With increasing l, the system with a negative corona becomes more effective in comparison with the device with the positive corona with the similar configuration. This is due to the difference in composition of ions in the air for different polarities.

To assess power losses on the corona the ratio can be used [6]:

$$P_{losses} = E_0 r_w I \delta \left(1 + \frac{0.03}{\sqrt{\delta r_w}}\right) \ln\left(\frac{l}{r_w}\right) \quad (3)$$

$$\delta = \frac{0.392p}{T} \quad (4)$$

where P_{losses} is the power losses on the corona, W; E_0 is the intensity of the electric field at breakdown in the air, $E_0 = 3*10^6$ V/m; r_w is the radius of emitter wire, m; I is the intensity of current, A; l is the distance between the corona electrode and the collector, m; p is the pressure, mm Hg and T is the temperature, K. So at $l = 29$ mm, $I = 0.1$ mA, the share of losses is: for emitter wire diameter $d = 0.02$ mm: $\frac{P_{losses}*100\%}{P} \approx 12\%$; for emitter wire diameter $d = 0.08$ mm: $\frac{P_{losses}*100\%}{P} \approx 21\%$.
Influence of configuration of IES on the efficiency of the system is shown on Figure 4a, b. Five models of IES have been compared. In all five models emitter wire with diameter $d = 0.02$ mm is under negative potential. The tube collector has diameter $d_c = 10$ mm. Results of the experiment show that thrust-to-power ratios of system with the grid collector and system with double tube collector are comparable, although for the last one a greater thrust value is observed at the same discharge current. Compared to the single tube collector such configurations give an increase of value of F/P to 1.5 – 2 times. Using of double PE leads to a drop of output parameters, and with decreasing distance between wires characteristics fall also. This is due to the mutual shielding of PE, when the ratio h/l is lower value of 1.15 [4].

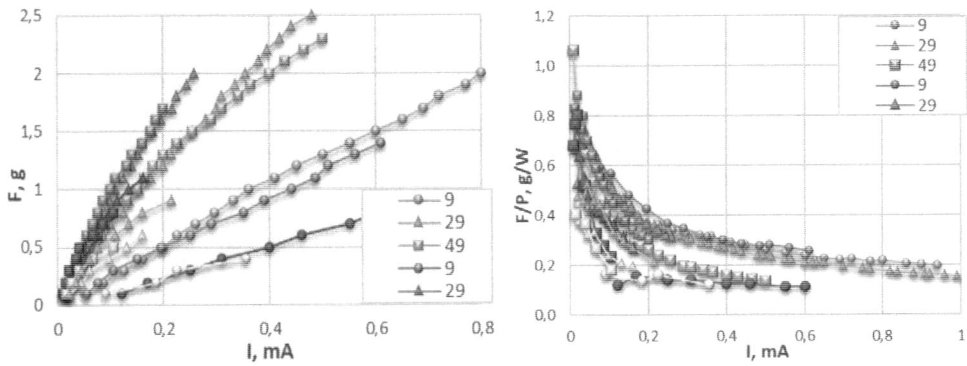

a *b*

Figure 4. Thrust (*a*) and thrust-to-power ratio (*b*) depending on current for different distances between electrodes *l*, mm for models: *1* – green; *2* – blue; *3* – red; *4* – yellow; *5* – orange.

Dependences of the thrust and the thrust-to-power ratio from power consumption is shown on Figure 5a, b. Note the ability to work in a wide range of currents and capacities in the system with a grid collector. The IES of this type has reached the maximum for this series of experiments of thrust to about 2.8 g.

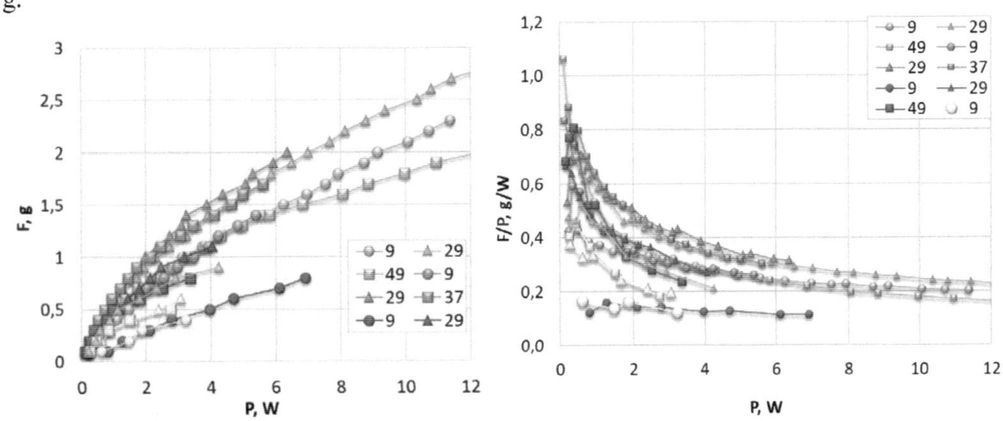

a *b*

Figure 5. Thrust (*a*) and thrust-to-power ratio (*b*) depending on power consumption for different distances between electrodes *l*, mm for models: *1* – green; *2* – blue; *3* – red; *4* – yellow; *5* – orange.

3. Conclusion

The thrust and efficiency of ionocraft electrode systems is investigated for different working polarity of the corona electrode. The most effective mode of work lifters of the considered structures is achieved at low current. The thrust-to-power ratio reaches a value of 1.1 g/W. The increase of current, according to (2), leads to a decrease efficiency up to a value of 0.2 g/W. The increase of distance between the corona electrode and the collector gives the growth of thrust and thrust-to-power ratio up to 21 mm. In the subsequent distancing statistically significant increase of characteristics is not observed. It is advisable to use emitters with smaller diameter to reduce losses to the formation of the corona. The using of emitter with wire diameter $d = 0.02$ mm increase thrust-to-power ratio by 10% compared with wire diameter $d = 0.08$ mm. Collector in form of a grid or two parallel tubes gives a significant increase of thrust and thrust-to-power ratio of EHD propulsion system compared with the single tube collector.

This work was supported by RFBR 17-08-01409 A

4. References

[1] Guo S 2013 *An Investigation of Dielectric Barrier Discharge Based Plasma Actuator Designs With Enhanced Performance In Active Flow Control* (Minnesota: University of Minnesota)
[2] Go B D, Garimella S V_and Fisher T S 2007 *J. of Appl. Phys.* ***102*** 053302
[3] Kuryachii A P, Rusyanov D A, Skvortsov V V 2011 *TsAGI Sci. Journal* **42** 227
[4] Khomich V Yu, Moshkunov S I, Nebogatkin S V et al. V A 2011 *Quant. Elect.* **41** 1093
[5] Cooperman P, 1960 *Trans. of the Commun. and Electr.* **79** 47
[6] Wolf R A 2012 *Atmospheric Pressure Plasma for Surface Modification* (New York: Wiley–Scrivener)
[7] Ong A O 2013 *Electrohydrodynamic Microfabricated Ionic Wind Pumps for Electronics Cooling Applications* (Cleveland: Western Reserve University)
[8] Khomich V Yu, Malashin M V, Moshkunov, Rebrov I E 2013 *EPE Journal* **23** 51
[9] Moshkunov S I, Rebrov I E, Khomich V Yu 2013 *Advances in Appl. Phys.* **5** 630
[10] Karpov S V, Krichtafovitch I A. 2005 *COMSOL Multiphysics User's Conference*
[11] Honer K A, Jewell-Larsen N E, Ran H, Zhang Y 2008 *ESA Annual Meeting on Electrostatics*
[12] Mahmoudi S R, Adamiak K, Cstle G S P 2011 *ESA Annual Meeting on Electrostatics*
[13] Kogelschatz U 2003 *Plasma chemistry and plasma processing* **23** 46
[14] Patel V 2012 *Chemical Kinetics* (Rijeka: InTech)

IOP Conf. Series: Journal of Physics: Conf. Series **927** (2017) 012016 doi:10.1088/1742-6596/927/1/012016

The experimental-theoretical model of the jet HF induction discharge of atmospheric pressure

R Gainullin and A Kirpichnikov
Kazan National Research Technological University 420015, Karl Marx str., 68 Kazan, Russia

gainullin@kstu.ru

Abstract. The paper considers theexperimental-theoretical model devised to determine the regularities of the quasi-stationary electromagnetic field structure of the HFI discharge burning in the inductor of finite dimensions at atmospheric pressure.

1. Introduction

To manage effectively technological processes using the principle of high-frequency inductivegas heating and to optimize them, it is necessary to have reliable information on the structure of the HF-discharge and the distribution of main electromagnetic and thermal parameters in it as well. However, the lack of simple and trustworthy methods to calculate the operation of plasmatrons prevents the development of new plasma technologies.

Though existing methods of operational diagnostics of the HFI-discharge can provide information on the distribution of most of its electromagnetic and thermal parameters, it is actually difficult to maintain measurements within a single cycle when the discharge is burning and its characteristics do not change for a long time. It is due to the fact that usually different types of measurements require a different set of equipment and time for its installation and debugging.

In this regard, the task has become urgent to create an experimental-theoretical model that would require a minimum number of measured quantities, so that it would be possible to calculate all remaining characteristics of the discharge with a sufficient degree of accuracy.

2. Method of solution

Taking into account the harmonic nature of the change in magnetic and electric fields over time, the system of Maxwell equations in the case of a cylindrically symmetric quasi-stationary electromagnetic field of a high-frequency induction discharge of the finite length for the main field components can be represented in the following form [1-3]:

$$\frac{\partial E_\varphi}{\partial z} = -\frac{\omega}{c} H_r \sin\left(\varphi_{H_r} - \varphi_{E_\varphi}\right);$$

$$\frac{\partial \varphi_{E_\varphi}}{\partial z} = \frac{\omega}{c} \frac{H_r}{E_\varphi} \cos\left(\varphi_{H_r} - \varphi_{E_\varphi}\right);$$

$$\frac{\partial H_z}{\partial z}=-\frac{1}{r}\frac{\partial}{\partial r}(rH_r)\cos\left(\varphi_{H_z}-\varphi_{H_r}\right)-H_r\frac{\partial\varphi_{H_r}}{\partial r}\sin\left(\varphi_{H_z}-\varphi_{H_r}\right);\qquad(1)$$

$$\frac{\partial\varphi_{H_z}}{\partial z}=\frac{1}{H_z}\left[\frac{1}{r}\frac{\partial}{\partial r}(rH_r)\sin\left(\varphi_{H_z}-\varphi_{H_r}\right)-H_r\frac{\partial\varphi_{H_r}}{\partial r}\cos\left(\varphi_{H_z}-\varphi_{H_r}\right)\right];$$

$$\frac{\partial H_r}{\partial z}=\frac{\partial H_z}{\partial r}\cos\left(\varphi_{H_z}-\varphi_{H_r}\right)-H_z\frac{\partial\varphi_{H_z}}{\partial r}\sin\left(\varphi_{H_z}-\varphi_{H_r}\right)+\frac{4\pi}{c}\sigma E_\varphi\cos\left(\varphi_{H_r}-\varphi_{E_\varphi}\right);$$

$$\frac{\partial\varphi_{H_r}}{\partial z}=\frac{1}{H_r}\left[\frac{\partial H_z}{\partial r}\sin\left(\varphi_{H_z}-\varphi_{H_r}\right)+H_z\frac{\partial\varphi_{H_z}}{\partial r}\cos\left(\varphi_{H_z}-\varphi_{H_r}\right)-\frac{4\pi}{c}\sigma E_\varphi\sin\left(\varphi_{H_r}-\varphi_{E_\varphi}\right)\right],$$

where H_z - is the amplitude of the longitudinal component of the magnetic field in the discharge, H_r - the amplitude of the radial component of the magnetic field in the discharge, E_φ - the amplitude of the azimuthal component of its electric field, $\varphi_{H_z}, \varphi_{H_r}, \varphi_{E_\varphi}$ phase angles of these components, respectively.

Analysis (1) shows that in this system of equations, the number of quantities characterizing the field is oneunity greater than the number of equations themselves, i.e. the given system of equations is not closed. Therefore, if we considerone of the quantities characterizing the field as the given one, it is possible to obtain at the output of this system a set of various dependences consisting of electromagnetic quantities and conductivity in the discharge σ. In its turn, using the found conductivity, one can also find the temperature field $T(\sigma)$ in the discharge. Thus, the problem to choose the input quantity to solve the obtained system of equations arises.

If we choose experimentally obtained data for the solution (1) as the input parameter for H_z, two important factors should be taken into account. Firstly, in this case, there is no need to use additional dependencies to determine the field temperatures in the discharge since the change in the magnitude of the magnetic field over the discharge cross section is determined by the absorption of electromagnetic energy in the conducting layer of the gas.Therefore, ceteris paribus, electro-conductivity determines the rate of the magnetic field quantity change along the radius of the discharge. Due to this unambiguous dependence, it is not necessary to resort to additional equations and solve the problem within the framework of Maxwell equation system.

Secondly, to solve the system of equations (1), it is significant to specify boundary conditions apparent for all electromagnetic quantities except H_z and do not require additional experimental specification.

Based on the system of equations (1), conductivity can be calculated by the formula:

$$\sigma(r,z)=-\frac{c}{4\pi}\left(\frac{\frac{\partial H_z}{\partial r}-\frac{\partial H_r}{\partial z}\cos(\varphi_{H_z}-\varphi_{H_r})-H_r\frac{\partial\varphi_{H_r}}{\partial z}\sin(\varphi_{H_z}-\varphi_{H_r})}{E_\varphi\cos(\varphi_{H_z}-\varphi_{E_\varphi})}\right)\qquad(2)$$

If we approximate the amplitude of the longitudinal magnetic field in the discharge H_z by a smoothed cubic spline while solving the equation (2), then conductivity manifests a sharp divergence near the axis of the plasmoid and up to the distance of about one third of the plasma bunch radius. It is obvious that this phenomenon is a direct consequence of $H_z(r)$ incorrect interpolation near zero and, therefore, in this case it is impossible to use the finite number of its experimentally measured values by cubic splines to solve the problem to reconstruct the amplitude H_z as a continuous function of the radial coordinate r.

To find out true causes of erroneous conductivity calculations in the HFI-discharge, it is required to analytically study its structure and, especially, the behavior of all its main characteristics including $H_z(r)$ in the paraxial region of the plasmoid.

A two-dimensional mathematical model was constructed in works [4,5] to investigate the structure of the quasi-stationary electromagnetic field of the HFI discharge; it became a further development of the known model of constant conductivity of the HFI – J.J. Thomson's discharge laborated for one-dimensional case [6].

Within the framework of the constructed model, relations were obtained to calculate the amplitude of the longitudinal component of the magnetic field $H_z(r,z)$, radial component of the magnetic field $H_r(r,z)$, and the azimuthal electric field $E_\varphi(r,z)$, and phase angles of these quantities near the plasmoid axis at values r, close to zero as well:

$$H_z^a(r,z) = H_z^a(0,0)I_0(br)\cos(bz)\times\left\{1+\frac{8\pi^2\omega^2\sigma^2{}_{cp}}{b^4c^4}\left(\frac{2I_0(br)-2-\frac{b^2r^2}{4}}{I_0(br)}\right)^2 - \right.$$

$$\left. -\frac{16\pi^2\omega^2\sigma^2{}_{cp}}{b^4c^4}\left(\frac{I_0(br)-1-\frac{b^2r^2}{4}}{I_0(br)}\right)\right\}; \tag{3}$$

$$H_r^a(r,z) = H_z^a(0,0)I_1(br)\sin(bz)\times\left\{1+\frac{8\pi^2\omega^2\sigma_{cp}^2}{b^4c^4}\left(\frac{I_1(br)-\frac{br}{2}}{I_1(br)}\right)^2\right\}; \tag{4}$$

$$E_\varphi^a(r,z) = \frac{\omega}{bc}H_z^a(0,0)I_1(br)\cos(bz\times\left\{1+\frac{8\pi^2\omega^2\sigma^2{}_{cp}}{b^4c^4}\left(\frac{I_1(br)-\frac{br}{2}}{I_1(br)}\right)^2\right\}; \tag{5}$$

$$\varphi_{H_z}(r,z) = \frac{\pi}{2}+\frac{4\pi\omega\sigma_{cp}}{b^2c^2}\left(\frac{I_0(br)-1}{I_0(br)}\right); \tag{6}$$

$$\varphi_{H_r}(r,z) = \frac{\pi}{2}+\frac{4\pi\omega\sigma_{cp}}{b^2c^2}\left(\frac{I_1(br)-\frac{br}{2}}{I_1(br)}\right); \tag{7}$$

$$\varphi_{E_\varphi}(r,z) = \frac{4\pi\omega\sigma_{cp}}{b^2c^2}\left(\frac{I_1(br)-\frac{br}{2}}{I_1(br)}\right). \tag{8}$$

IOP Conf. Series: Journal of Physics: Conf. Series **927** (2017) 012016 doi:10.1088/1742-6596/927/1/012016

In these relations the constant $b = \frac{1}{L}\arccos\frac{H_z(0,L)}{H_z(0,0)}$, where L is the plasmoid length, I_0, I_1 is modified Bessel function of the zero and first order, respectively.

Experiments to measure magnetic field were conducted on a test stand based on a high-frequency plasmatron of 60 kW that operates at a frequency of 1.76 MHz, plasma-forming gas is air. To work in conditions of thermal discharge, a special water-cooled magnetic probe was used; it allows to have a long-term operation at high temperatures peculiar to this type of discharge.

Amplitude measurements of the longitudinal component of the magnetic field H_z were conducted at two charges of plasma-forming gas - *9 m³ /hour* and *13 m³ /hour*. As a result, radial distributions were obtained in different sections along z (Fig. 1) [7-9]; $z = 0$ denotes the plane of the plasmoid central section; the coordinate z is measured from this section downstream; magnetic probe step: *r=0,4 cm, z=1 cm*.

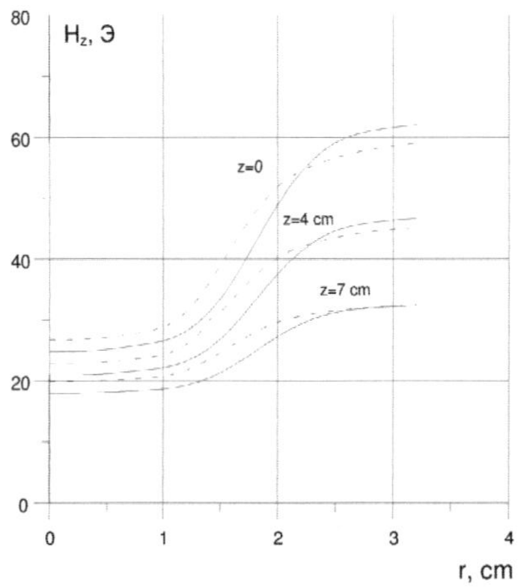

Fig. 1. Radial distribution H_z in different sections of the inductor at the expense of the plasma-forming gas: —— - Q_1=*9 m³/hour*; - - - - - Q_2=*13 m³/hour*

3. Results and conclusion

The system of equations (1) was solved numerically by a finite-difference method with respect to unknown quantities H_r, E_φ, φ_{H_z}, φ_{H_r}, φ_{E_φ} and σ. In this case, the parameters of magnetic and electric fields in the near-beam axis of the plasmoid were approximated by means of the relations (3) - (8). We used amplitude measurements of the longitudinal component of the magnetic field H_z as an input parameter. Figures 2-5 present some results obtained by means of a numerical model in the form of conductivity dependency graphs in the discharge $\sigma(r)$, eddy current density $j = \sigma E_\varphi$, volume density of power included into the discharge $W = \frac{1}{2}\sigma E_\varphi^2$ and temperature distribution for three sections of the HFI discharge, starting from its central section downwards the stream for two different discharge rate of plasma-forming gas – *9 m³ /hour* and *13 m³ /hour*.

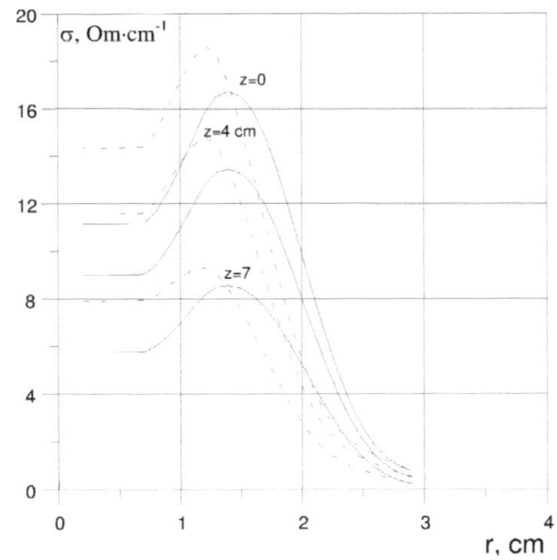

Fig. 2. Radial distribution of conductivity in different sections of the inductor at the discharge rate of the plasma-forming gas:
—— - Q_1=9 m^3/hour; - - - - - Q_2=13 m^3/hour

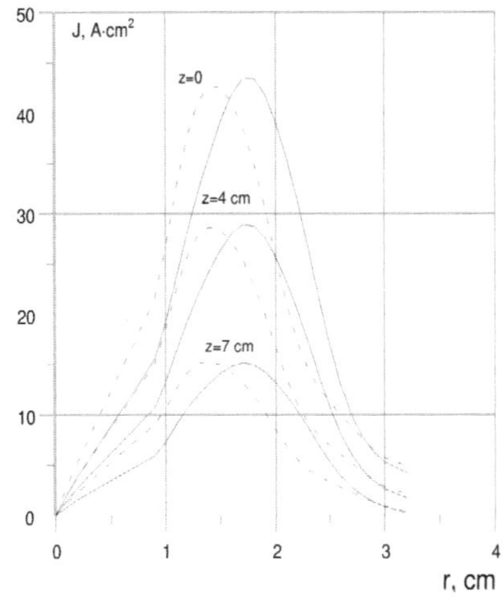

Fig. 3. Radial density distribution of the eddy current in different sections of the inductor at the discharge rate of the plasma-forming gas:
—— - Q_1=9 m^3/hour; - - - - - Q_2=13 m^3/hour

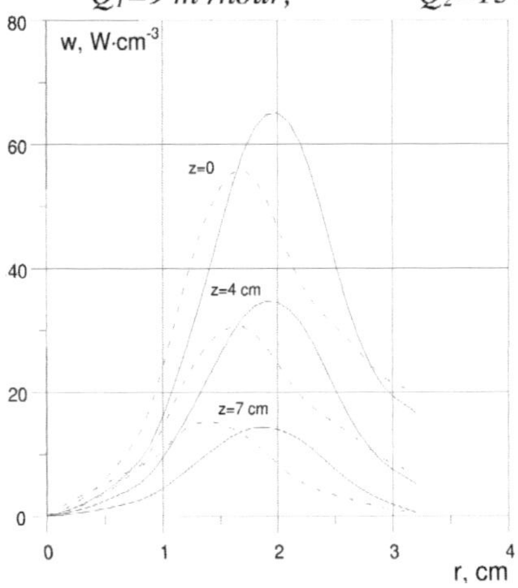

Fig. 4. Radial distribution of the volume density of the power included into the discharge in different sections of the inductor at the discharge rate of the plasma-forming gas:
—— - Q_1=9 m^3/hour; - - - - - Q_2=13 m^3/hour

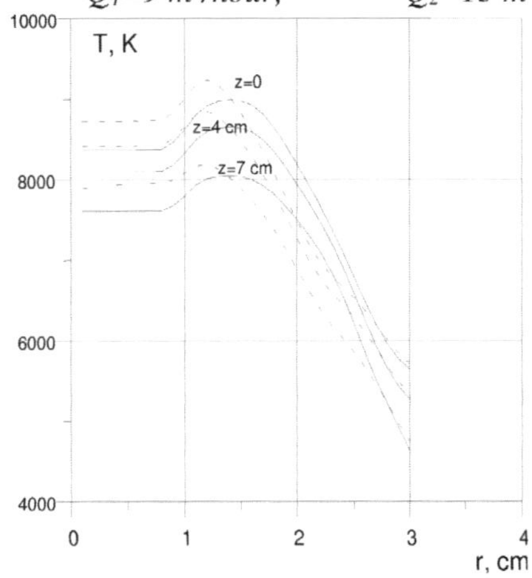

Fig. 5. Radial distribution of temperature in different sections of the inductor at the discharge rate of the plasma-forming gas:
—— - Q_1=9 m^3/hour; - - - - - Q_2=13 m^3/hour

Conclusions

The combination of calculated and experimental research methods makes it possible to study the internal structure and properties of the high-frequency induction discharge of atmospheric pressure more profoundly, as well as to investigate the influence of the discharge rate of the plasma-forming gas on the distribution of its basic electromagnetic and thermal characteristics. Obtained results can be used to develop and optimize various types of power plants using the principle of high-frequency inductivegas heating.

IOP Conf. Series: Journal of Physics: Conf. Series **927** (2017) 012016 doi:10.1088/1742-6596/927/1/012016

References

[1] Gainullin R 1992 *Izvestiya VUZov Fizik.* No 6 p 121
[2] Gainullin R 1995 *Inzhenerno-Fizicheskii* Zhurnal **68** No 2 p 248
[3] Gainullin R 1995 *Engineering Physics and Thermophysics* **68** No 2 p 215
[4] Gainullin R 2007 *Applied Physics* No 3 p 54
[5] Gainullin R 2008 *Engineering Physics* No 2 p 61
[6] Thomson J 1926 *Philos. Mag.* vol 2 No 9 p 674
[7] Gainullin R 2007 *Applied Physics* No 3 p 61
[8] Gainullin R 2008 *Izvestiya VUZov. Problemi energetiki* No 1 p 54
[9] Gainullin R 2008 Applied Physics No 5 p 44

IOP Conf. Series: Journal of Physics: Conf. Series **927** (2017) 012017 doi:10.1088/1742-6596/927/1/012017

Electrophysical and heat properties of the discharge with liquid anode

D N Mirkhanov, Al F Gaisin, A I Kuputdinova, R A Mukhametov

Faculty of Physics and Mathematics, Kazan National Research Technical University named after A.N.Tupolev - KAI, Kazan, 420111, Russia

E-mail: mirhanov@list.ru

Abstract. The low-temperature plasma of the gas-discharge was analyzed that generates between the jet and dropping liquid anode and metal cathode at the ambient pressure. The study results of electrophysical properties and heat processes in the discharge combustion area are presented.

1. Introduction

The profound data have been obtained in the field of discharge generation between solid electrodes [1-3]. Currently, discharges are of higher concern to generate in the ambient air between the dropping jets of conductive liquids and the metal electrode [4-12]. This technology of gas discharge formation is attractive for both fundamental and applied researches. Discharge forms and types, electrophysical and spectral properties, plasma composition, plasma component distribution by energy and concentration are of high interest in the field of fundamental researches. Physical properties of discharge generation along the jet and at the border of electrode interaction are attractive. Hydrogasdynamic and heat processes are of concern that emerges within the electrode spacing in the zone of discharge generation. In addition, discharge of this type is relevant in view of practical application in various industries. Aircraft and mechanical engineering sectors face problems on local formation of item inner and outer surface micro-pattern.

2. Experiment

The gas discharge between the jet and dropping liquid anode and metal cathode was studied at the pilot plant with the direct current source and controlled DC voltage up to 4000 V at the rated current of up to 10A.

The following research methods were applied to address the tasks specified:

1. Gas discharge current and voltage oscillations were analyzed by the digital oscillograph GDS – 806 S.
2. The infrared scanner FLIRA6500SC was used for the electrode surface thermography. The data obtained were processed by the ALTAIR Software v5.91.010.

Combustion of the gas discharge within the electrode spacing was identified at the voltage of U=600-1000B, current I=1.2-4.8A, jet length l_c=30 mm, ambient pressure p=⟦10⟧^5 Pa, a copper plate d_a=1 mm was taken as the cathode.

Gas discharge current and voltage oscillations were analyzed by the digital oscillograph GDS – 806 S. It was identified that the discharge between jet and dropping liquid anode and copper cathode burns as impulses with the current pulsation frequency ν=30-70 Hz (Fig. 1).

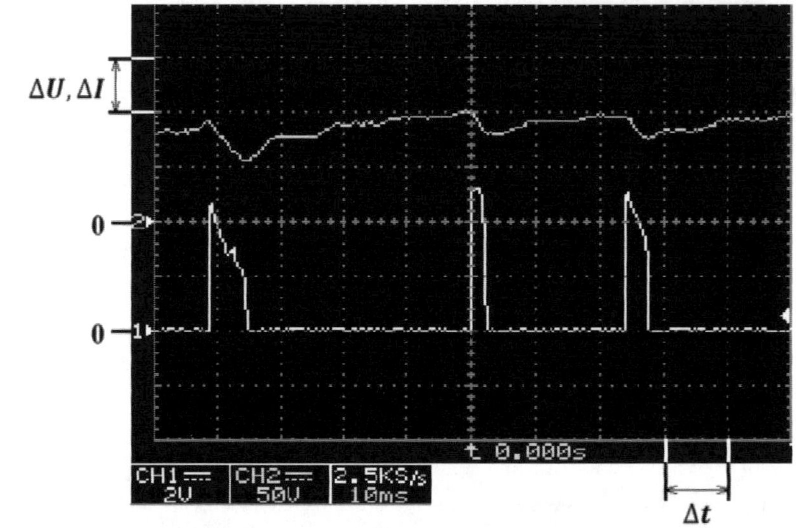

Figure 1. Oscilloscope pattern of the gas discharge current and voltage oscillation: ΔU=500B, ΔI=2A, Δt=10 ms;

This is due to the fact that the electrolyte is intensely vapors generating local steam and gas zones when the potential is supplied to electrodes in the area of the electrolyte and copper electrode contact. Where the environment in the steam and gas shell complies with ionization process conditions, the disruptive discharge generates therein followed by abrupt rise of temperature and pressure values. This causes shock wave generation that result in spreading of the steam and gas shell and jet delivery. In turn, it results in higher resistance in the steam and gas shell followed by the discharge termination, the circuit breaks and the current drops to 0. This process regularly occurs.

Electrode surface thermography in the area of gas discharge combustion was performed by using the infrared scanner FLIRA 6500 SC. The data obtained were processed by the ALTAIR Software v5.91.010.

Based on surface temperature logs (Fig. 2) in the area of discharge combustion, the temperature along the jet and dropping anode varies $T_k \approx 46$-69°C. The temperature of the copper cathode (Fig. 3) is $T_a \approx 54$-69°C.

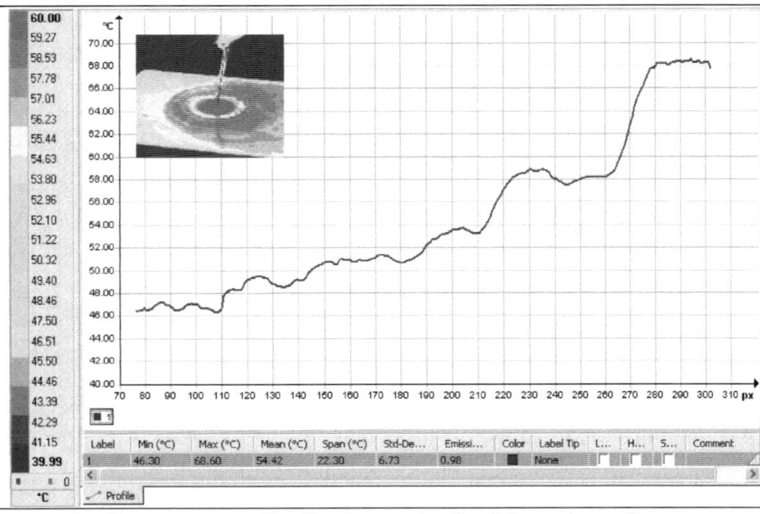

Figure 2. Thermograph of temperature distribution at the liquid anode surface

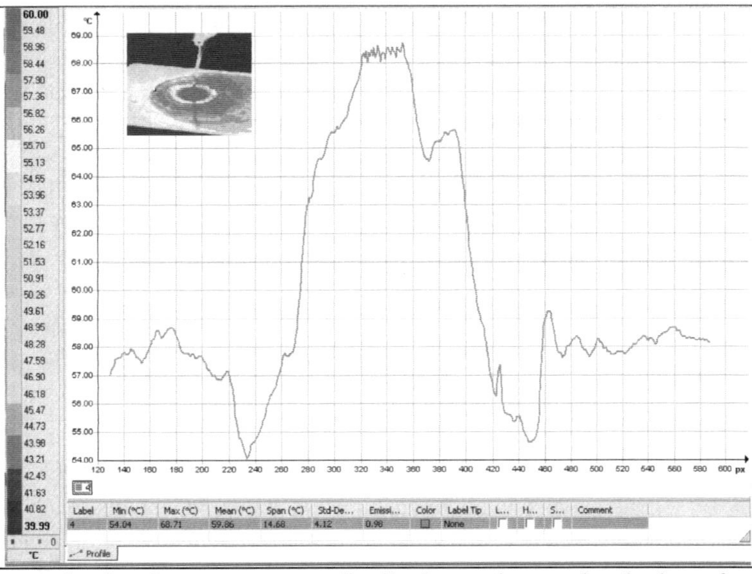

Figure 3. Thermograph of surface temperature distribution at the metal cathode surface.

The temperature reaches T_max=69°C in the area of discharge combustion. It is identified that when the gas discharge is burned, the temperature along the jet and dropping anode drops. This is explained by the fact that the heat input, prior to the disruptive discharge, is due to the electrocaloric effect. Upon the disruptive discharge, the heat input to the jet reduces against the decline in total resistance and drop of jet voltage.

3. Results

1. Peculiarities of discharge are identified to generate between the jet and dropping liquid anode and metal cathode.
2. The discharge is identified to generate as the current impulses with the current frequency ν=30-70 Hz.
3. Thermographs of temperature distribution of liquid anode and metal cathode surface are obtained.

Acknowledgments
This work was supported by state contract regarding resolution 220 of the Ministry of Education and Science of the Russian Federation, project no. 14.Z.50.31.0023.

References

[1] Reiser Yu.P., Physics of Gas Discharge (Edition 3, updated and revised). Dolgoprudny, *2009. - P. 734.*

[2] Reiser Yu.P., Schneider M.N., Yatsenko N.A. High-Frequency Capacitive Discharge: Physics.Experimental Technique. Appendixes. M.: Nauka, *1995. – P. 320.*

[3] Fortov V.E., Son E.E., Gaysin F.M., Bromberg L., Son K.E., O John Khe, I Khe Iong. Plasma Technology (in Korean) Moscow Institute of Physics and Technology, KOFST, *2006. –P. 135.*

[4] Gaisin A.F. 2013 High Temperature **6** 863 – 866

[5] Akishev Yu.S., Grushin M.E., Karalnik V.B., Monich A.E., Pnkin M.V., Trushkin N.I., Kholodenko V.P., Chugunov V.A., Zhirkova N.A., Irkhina I.A., Kobzev E.N. Plasma Physics. *2006. V. 32. No. 12. P. 1142.*

[6] Gaisin A.F., Sarimov L.R. 2011 Plasma Physics Reports **6** 535 – 540

[7] Dautov G.Y., Dautov I.G, Fayrushin I.I, Kashapov N.F. 2013 Journal of Physics: Conference Series 1 012014

[8] Kirko D.L., Savelov A.S., Vizgalov I.V. Russian Physics Journal. 2013. *V. 55. No. 11. P. 1243-1247.*

[9] Machala Z., Jedlovsky I., Chladekova L., Pongrac B., Giertl D., Janda M., Sikurova L., Polcic P. *Eur. Phys. J. D. 2009. V. 54. P. 195.*

[10] Pongrác B., Machala Z. *IEEE Trans. Plasma Sci. 2011. V. 39. P. 2664.*

[11] Kim H.H., Teramoto Y., Negishi N., Ogata A., Kim J.H., Pongrác B., Machala Z., Gañán-Calvo Alfonso M. *J. Aerosol. Sci. 2014. V. 76. P. 98.*

[12] Lu Y., Xu S.F., Zhong X.X., Ostrikov K., Cvelbar U., Mariotti D. *EPL. 2013. V. 102. No. 1. P. 15002.*

IOP Conf. Series: Journal of Physics: Conf. Series **927** (2017) 012018 doi:10.1088/1742-6596/927/1/012018

Discharge between the jet and dropping liquid cathode and metal anode

Al F Gaisin[1], E E Son[2], N F Kashapov[3]

[1]Faculty of Physics and Mathematics, Kazan National Research Technical University named after A.N.Tupolev - KAI, Kazan, 420111, Russia
[2] Joint Institute for High Temperatures of the Russian Academy of Sciences, Moscow, 125412, Russia
[3] Kazan Federal University, Kazan, 420008, Russia

E-mail: almaz87@mail.ru

Abstract. Low-temperature gas-discharge plasma is investigated that is generated within the electrode space between the jet and dropping liquid cathode and metal anode at ambient pressure. The results of plasma composition, electrophysical parameters, electron concentration are presented, the discharge types are specified.

1. Introduction

In parallel with the study of discharges between solid electrodes [1-3], of high concern are discharges generated by direct or alternate current fields within the electrode spacing where a single or both electrodes are the liquids. This type of discharge is of high concern for both the sphere of fundamental studies and the discharge applications in various sectors of production [4-12]. Such industries as aircraft engineering, mechanical engineering, metalwork set urgent tasks to treat metal and item surfaces of different physical nature to improve their surface properties. At the same time, industries are up against such tasks as the surface combined treatment as well as the local section modification of item surfaces. To date, various surface machining technologies are applied for materials and items, namely mechanical, chemical, electrochemical and plasma techniques.

Low-temperature gas-plasma discharge is the perspective technology of surface local treatment that is burnt between the jet and dropping liquid cathode and metal anode (item to be treated) within a wide range of pressure values. "The method is based on the impact of surged electric discharges that generate in the steam-plasma skin formed around the item immersed in the electrolyte. Unlike the electrochemical finishing in acidic solutions, electrolyte-plasma technology allows using environment friendly aquatic solutions of low-concentration salts" [13].

2. Experiment

The gas discharge between the jet and dropping liquid cathode and metal anode was studied at the pilot plant (Fig.1) with the direct current source and controlled DC voltage up to 4000 V at the rated current of up to 10A, where: 1 – electrolytic cell; 2 – electrolyte solution; 3 – metal plate for electropositive potential supply; 4 – gas discharge combustion area; 5 – metal cathode; 6 – jet and dropping liquid anode.

The following research methods were applied to address the tasks specified:

1. The process of gas discharge combustion was recorded by the digital photo- and video devices SONY FDR-AX33.

2. Gas discharge current and voltage oscillations were analyzed by the digital oscillograph GDS – 806 S.

3. The plasma emission spectrum of the HF capacitive discharge was defined by the fibre-optic spectrometer PLASUS EC 150201 MC with the goniosight to fix light rays within the wavelength range of 195 to 1105 nm. The spectral resolution of the device δλ is 1 nm. The composition and concentrations of plasma electrons of HF capacitive discharge was evaluated with no reference to the particular area. The spectrum was interpreted by the line identification correlating the test spectrum against the NIST database.

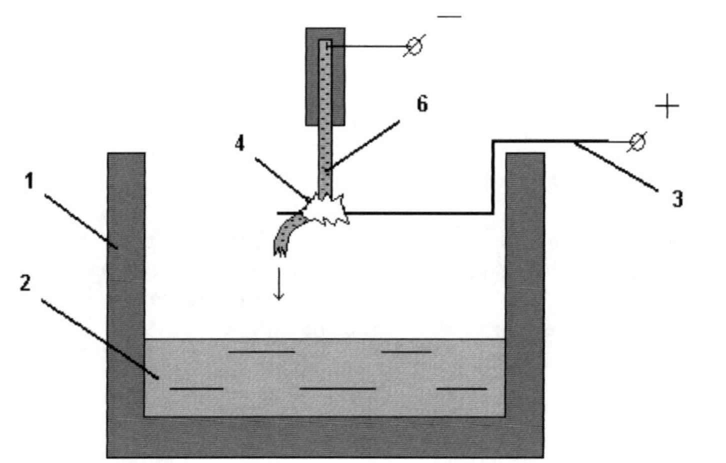

Figure 1. Pilot plant block scheme

Combustion of the gas discharge within the electrode spacing was identified at the voltage of $U = 250 - 1000\text{B}$, current $I = 0.7 - 2.5\text{A}$, jet length $l_c = 30$ mm, ambient pressure $p = 10^5 Pa$, the current pulsation frequency $\nu = 30 - 100Hz$ $(Fig.\ 2)$, a copper plate $d_a = 1\ mm$ was taken as the anode. The discharge is generated in two system areas: 1 – at the border between the jet and dropping liquid cathode and copper anode, 2 – in the area of jet thinning, between the electrolyte drops as formed. The discharge generated in the area of jet thinning is due to that the potential supplied to electrodes results in the potential difference in the area between drops that emerge and disruptive discharge generates between them.

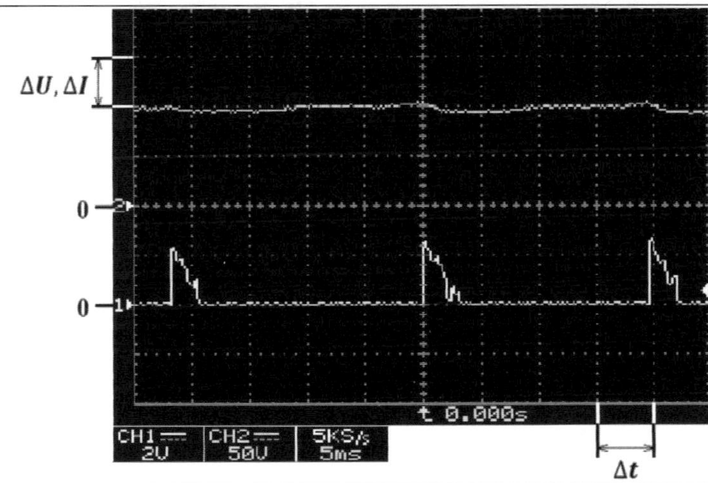

Figure 2. Oscilloscope pattern of the gas discharge current and voltage oscillation

The method of optical spectroscopy was applied to evaluate the plasma composition and electron concentration of the gas discharge. The device-based spectral spread (Fig. 3) was verified by the potassium atom K I - 766.2 nm. In both spectra, the minimal width of single lines was $\Delta\lambda g$ = 1 nm which is taken as the instrumental width. The emission spectrum analysis of the gas discharge indicates that different elements are available in the plasma test area: oxygen O I, hydrogen H I, natrium Na I, potassium K I, zinc Zn I, calcium ions Ca II, zinc ions Zn II, as well as nitrogen molecules N2+ (C-B) and hydroxyl OH (A-X). To define the electron concentration in the plasma of HF capacitive discharge, contours of hydrogen lines were analyzed in the Balmer series (H_α, H_β, H_γ). To calculate the electron concentration n_e, the width of Voigt profile of hydrogen line was defined. The half-width of lines at the half-height was $\Delta\lambda f$ = 1.4 nm for H_α, $\Delta\lambda f$ = 1 nm for H_β, and $\Delta\lambda f$ = 1.2 nm for H_γ. The value of n_e by H_β line was not calculated since its half-width appeared to be equal to the minimal width of single lines $\Delta\lambda_g$ = 1 nm. In view of device-based spreading in assumption of the Voigt profile of the line to be registered, the Lorentz profile constituent of the H_α line is equal to $\Delta\lambda L$ = 0.661 nm, and that for H_γ line is equal to $\Delta\lambda L$ = 0.34 nm. Values $\Delta\lambda L$ conditioned by the linear Stark effect correspond to electron concentrations of $n_e \approx 10^{15} - 10^{16} cm^{-3}$ order.

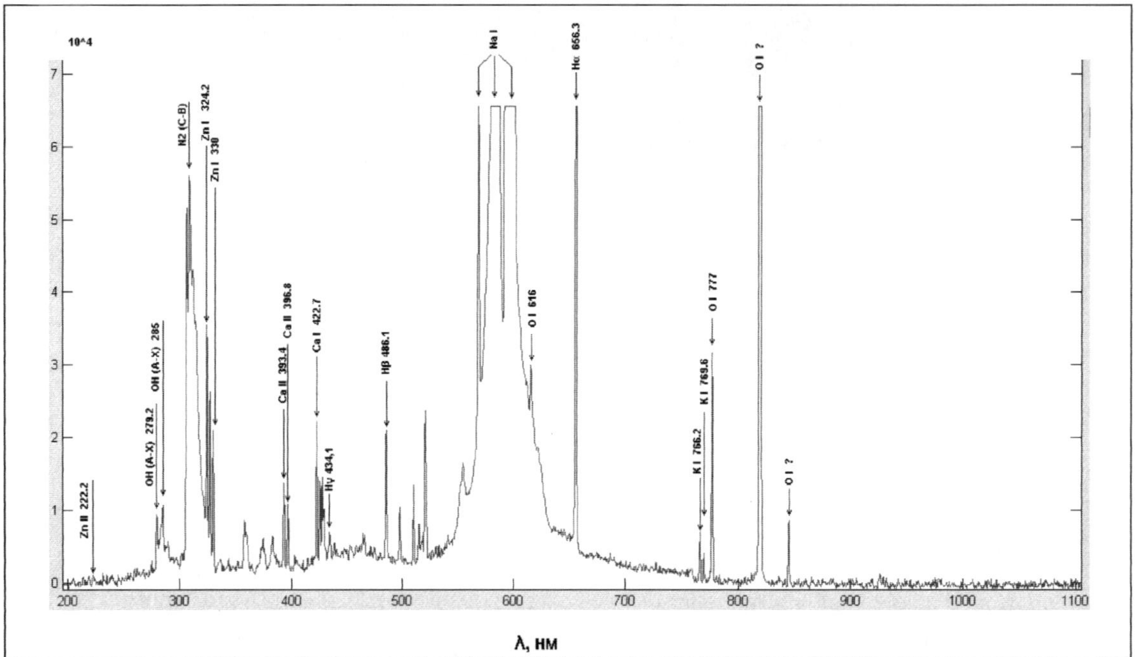

Figure 3. Gas discharge emission spectrum with identified spectral lines.

3. Results

1. The discharge combustion between the jet and dropping liquid cathode and metal anode was identified in two system areas: 1 – at the border between the jet and dropping liquid cathode and copper anode, 2 – in the area of jet thinning, between the electrolyte drops as formed.
2. The discharge current pulsation was defined within the range of I=0.7-2.5A, frequency ν=30-100Hz.
3. The electron concentration n_e≈〖〖10〗^15 - 10〗^16 〖см〗^(-3) was identified.

Acknowledgments

This work was supported by state contract regarding resolution 220 of the Ministry of Education and Science of the Russian Federation, project no. 14.Z.50.31.0023.

References

[1] Reiser Yu.P., Physics of Gas Discharge (Edition 3, updated and revised). Dolgoprudny, 2009. - 734 p.

[2] Reiser Yu.P., Schneider M.N., Yatsenko N.A. High-Frequency Capacitive Discharge: Physics.Experimental Technique. Appendixes. M.: Nauka, 1995. – 320 p.

[3] Fortov V.E., Son E.E., Gaysin F.M., Bromberg L., Son K.E., O John Khe, I Khe Iong. Plasma Technology (in Korean) Moscow Institute of Physics and Technology, KOFST, 2006. -135 p.

[4] Gaisin A.F. 2013 High Temperature **6** 863 – 866

[5] Akishev Yu.S., Grushin M.E., Karalnik V.B., Monich A.E., Pnkin M.V., Trushkin N.I., Kholodenko V.P., Chugunov V.A., Zhirkova N.A., Irkhina I.A., Kobzev E.N. Plasma Physics. 2006. V. 32. No. 12. P. 1142.

[6] Gaisin A.F., Sarimov L.R. 2011 Plasma Physics Reports **6** 535 – 540

[7] Dautov G.Y., Dautov I.G, Fayrushin I.I, Kashapov N.F. 2013 Journal of Physics: Conference Series **1** 012014

[8] Kirko D.L., Savelov A.S., Vizgalov I.V. Russian Physics Journal. 2013. T. 55. № 11. C. 1243-1247.
[9] Machala Z., Jedlovsky I., Chladekova L., Pongrac B., Giertl D., Janda M., Sikurova L., Polcic P. *Eur. Phys. J. D. 2009. V. 54. P. 195.*
[10] Pongrác B., Machala Z. IEEE Trans. Plasma Sci. 2011. V. 39. P. 2664.
[11] Kim H.H., Teramoto Y., Negishi N., Ogata A., Kim J.H., Pongrác B., Machala Z., Gañán-Calvo Alfonso M. *J. Aerosol. Sci. 2014. V. 76. P. 98.*
[12] Lu Y., Xu S.F., Zhong X.X., Ostrikov K., Cvelbar U., Mariotti D. *EPL. 2013. T. 102. № 1. C. 15002.*
[13] I.S, Kulikov, S.V. Vashenko, A.Ya. Kamenev. Electrolyte and plasma treatment of materials. – Minsk: Belarus. *Navuka* [Science], 2010, – P. 232.

Modeling of glow discharge in a gas flow

I G Galeev and T Ya Asadullin

Kazan National Research Technical University named by A.N.Tupolev
420111 Russia Kazan, K.Marx str., 10

galeev.physics@kstu-kai.ru

Abstract. The discharge plasma of positive column in electronegative gas flow has been described by two-dimensional integro-differential system of equations written in the approximation of “narrow channel”. An efficient algorithm of solving this system of equations is suggested. In this work an implicit method is used of solution with selection gradient of the electric field along the channel. The simulation of discharge characteristics was conducted under various boundary conditions at the inlet of the discharge area and with various full discharge currents.

Improving of volume stability of the glow discharge is an important task for various applications. This problem can be solved by various ways such as: longitudinal and transverse organization of a gas flow [1-3], changing a shape of channel [4], influence of magnetic field [5], suppression of the emerging canals of the discharge contraction [6], redistribution of the electric current in the cathode area [7], a supersonic gas flow [8,9] and etc.

In this paper a theoretical model of the positive column plasma is suggested for the longitudinal glow discharge in the flow of electronegative gas. The gas is pumped throw the sectional electrodes arranged in the beginning and end of the plane channel. The main processes that determine the characteristics of such discharge are diffusion of charged particles, birth and death processes of electrons and ions, convective charge transfer. The ambipolar nature of electron diffusion is disturbed in electronegative gas plasma by fairly strong effect from negative ions. A concentration of negative ions in this discharge may be of the same order and higher than the electron concentration.

An accepted two-dimensional system of equations has been written in the form

$$D_e N\frac{\partial^2}{\partial y^2}\alpha_e+\mu_e\left(\frac{\partial(E_x\alpha_e)}{\partial x}+\frac{\partial(E_y\alpha_e)}{\partial y}\right)-W_x N\frac{\partial\alpha_e}{\partial x}-W_y N\frac{\partial\alpha_e}{\partial y}+$$
$$+(\nu_i-\nu_a)n_e+\nu_d n_- -\beta_{ei}n_e n_+ +q=0, \tag{1}$$
$$D_- N\frac{\partial^2}{\partial y^2}\alpha_- +\mu_-\left(\frac{\partial(E_x\alpha_-)}{\partial x}+\frac{\partial(E_y\alpha_-)}{\partial y}\right)-W_x N\frac{\partial\alpha_-}{\partial x}-W_y N\frac{\partial\alpha_-}{\partial y}+$$
$$+\nu_a n_e-\nu_d n_- -\beta_{ii}n_- n_+ +q=0, \tag{2}$$
$$I=2b_z eE_x\int_0^b(\mu_e\alpha_e+\mu_+\alpha_+ +\mu_-\alpha_-)dy, \tag{3}$$

$$div\vec{j} = 0, \tag{4}$$

$$\vec{j} = e\vec{E}(\mu_e\alpha_e + \mu_+\alpha_+ + \mu_-\alpha_-), \tag{5}$$

$$\alpha_+ = \alpha_e + \alpha_-, \tag{6}$$

$$\rho W_x \frac{\partial W_x}{\partial x} + \rho W_y \frac{\partial W_x}{\partial y} = -\frac{\partial P}{\partial x} + \frac{\partial}{\partial y}\mu\frac{\partial W_x}{\partial y}, \tag{7}$$

$$\frac{\partial}{\partial x}\rho W_x + \frac{\partial}{\partial y}\rho W_y = 0, \tag{8}$$

$$G = 2b_z\int_0^b \rho W_x dy, \tag{9}$$

$$c_p\rho W_x\frac{\partial T}{\partial x} + c_p\rho W_y\frac{\partial T}{\partial y} = \frac{\partial}{\partial y}\lambda\frac{\partial T}{\partial y} + \mu\left(\frac{\partial W_x}{\partial y}\right)^2 + W_x\frac{\partial P}{\partial x} + +(j_e k + j_+ + j_-)E, \tag{10}$$

here marked: ν_i, ν_a, ν_d – frequencies of ionization, attachment and detachment; β_{ei}, β_{ii} – the coefficients of electron-ion and ion-ion recombination, q is the power of external sources of ionization, μ_+, μ_- - mobility of positive and negative ions, $\alpha_-= n_-/N$, n_- - the concentration of negative ions, *k* is a fraction of power going into gas heating, defined by the collisions of electrons with neutrals, b_z, $2b$ – the width and height of the channel.

Equations (1) - (10) are supplemented by boundary conditions:
at the entrance (x =0) - $\alpha_e = \varphi_e(y)$, $\alpha_- = \varphi_-(y)$, $W_x=\psi(y)$, $W_y = 0$, $T =\Theta(y)$;
on the wall of discharge camera - $\alpha_e = 0$, $\alpha_- = 0$, $W_y = W_x= 0$, T = const.
In addition the full electric current *I* and mass gas flow *G* through the channel are set.

By using the plasma quasineutrality of positive column (6) and the continuity equation of electric current (4), the equation for electronic component was rewritten in the form

$$c_a^e\frac{\partial^2\alpha_e}{\partial y^2} - \boldsymbol{W}N\,grad\alpha_e + (\nu_i - \nu_a)N\alpha_e + \nu_d N\alpha_- - \beta_{ei}N^2\alpha_e(\alpha_e + \alpha_-) + q =$$
$$= \left(\frac{\mu_e}{\mu_e + \mu_+}\right)\left[(c_- - c_+)\frac{\partial^2\alpha_-}{\partial y^2} + (\mu_+ + \mu_-)div(\alpha_-\boldsymbol{E})\right]. \tag{11}$$

Here the coefficient $C_a^e =(C_e\mu_+ + C_+\mu_e)/(\mu_+ + \mu_e)$ describes ambipolar diffusion of electrons with positive ions.

The equation (11) without considering the right part coincides practically with the continuity equation of electrons in electropositive gas, recorded in the approximation of the ambipolar diffusion. This equation includes the volumetric processes of birth and death of the charged particles, but its right part determines influence of negative ions that violate the clean ambipolar nature of electron diffusion. The first component on the right side is caused by difference of the diffusion coefficients the negative and positive ions, the second component is due to spatial charge separation and the additional electric field associated with this.

The numerical solution of the above equation system causes severe difficulties due to the strong exponential dependency of the birth and death frequencies of charged particles from the reduced electric field E/N. If the explicit algorithm is used with real fairly coarse grids it can lead to the instability of the computation and divergence. Therefore in this work an implicit method is used of solution that selects gradients of the electric field along the channel. On each step of integration along the x-axis the gas dynamic characteristics are calculated and then electric ones.

The calculation of the electrical characteristics α_e , α_- ,E. The equations (2), (11) can be written in next finite-difference form

$$\frac{c_-}{\Delta y}\left[\frac{\alpha_{-i+1,j+1}-\alpha_{-i+1,j}}{\Delta y_j}-\frac{\alpha_{-i+1,j}-\alpha_{-i+1,j-1}}{\Delta y_{j-1}}\right]-\left[(W_x)_{i,j}N_{i,j}-\mu_-(E_x)_i\right]\frac{(\alpha_{-i+1,j}-\alpha_{-i,j})}{\Delta x_i}-\left[(W_y)_{i,j}N_{i,j}-\mu_-(E_y)_i\right]\times$$
$$\times\frac{(\alpha_{-i,j+1}-\alpha_{-i,j-1})}{2\Delta y}-\mu_-\alpha_{-i,j}\frac{E_{xi+1}-E_{xi}}{\Delta x_i}+\mu_-\alpha_{-i,j}\frac{E_{y_{i,j+1}}-E_{x_{i,j-1}}}{2\Delta y}+$$
$$+(\nu_a)_{i+1,j}N_{i,j}\alpha_{ei,j}-(\nu_d)_{i+1,j}N_{i,j}\alpha_{-i,j}-(\beta_{ii})_{i+1,j}N_{i,j}^2\alpha_{ei,j}(\alpha_{ei,j}+\alpha_{-i,j})=0, \tag{12}$$

$$\frac{c_a^e}{\Delta y}\left[\frac{\alpha_{ei+1,j+1}-\alpha_{ei+1,j}}{\Delta y_j}-\frac{\alpha_{ei+1,j}-\alpha_{ei+1,j-1}}{\Delta y_{j-1}}\right]-(W_x)_{i,j}N_{i,j}\frac{(\alpha_{ei+1,j}-\alpha_{ei,j})}{\Delta x_i}-(W_y)_{i,j}N_{i,j}\frac{(\alpha_{ei,j+1}-\alpha_{ei,j-1})}{2\Delta y}+$$
$$+(\nu_i-\nu_a)_{i+1,j}N_{i,j}\alpha_{ei,j}+(\nu_d)_{i+1,j}N_{i,j}\alpha_{-i,j}-(\beta_{ei})_{i+1,j}N_{i,j}^2\alpha_{ei,j}\times$$
$$\times(\alpha_{ei,j}+\alpha_{-i,j})+q=\left(\frac{\mu_e}{\mu_e+\mu_+}\right)\left\{\frac{(c_--c_+)}{\Delta y}\left[\frac{(\alpha_{-i+1,j+1}-\alpha_{-i+1,j})}{\Delta y_j}-\frac{(\alpha_{-i+1,j}-\alpha_{-i+1,j-1})}{\Delta y_{j-1}}\right]+\right.$$
$$\left.+(\mu_-+\mu_+)\left[\alpha_{-i,j}\frac{E_{xi+1}-E_{xi}}{\Delta x_i}+E_{xi}\frac{(\alpha_{-i+1,j}-\alpha_{-i,j})}{\Delta x_i}+\frac{(\alpha_-E_y)_{i,j+1}-(\alpha_-E_y)_{i,j-1}}{2\Delta y}\right]\right\}, \tag{13}$$

Here marked with: $\Delta y=(\Delta y_j+\Delta y_{j-1})/2=\Delta y_{j-1}(\delta+1)/2,\ \delta=1{,}1,$ The frequency of ionization, recombination coefficient are presented by the first two members of rows

$$(\nu_i)_{i+1,j}=(\nu_i)_{i*,j}+\left(\frac{\partial\nu_i}{\partial E}\right)_{i*,j}\Delta E_i+\frac{1}{2}\left(\frac{\partial^2\nu_i}{\partial E^2}\right)_{i*,j}(\Delta E_i)^2+\ldots,\quad \beta_{i+1,j}=\beta_{i*,j}+\left(\frac{\partial\beta}{\partial E}\right)_{i*,j}\Delta E_i+\frac{1}{2}\left(\frac{\partial^2\beta}{\partial E^2}\right)_{i*,j}(\Delta E_i)^2+\ldots,\cdot$$

$(\nu_i)_{i*,j}=f_1(E_i,N_{i+1,j})$, $\left(\frac{\partial\nu_i}{\partial E}\right)_{i*,j}=f_2(E_i,N_{i+1,j})$. For the attachment, detachment coefficients it is similar.

Further a solution is sought in the form

$$\alpha_{ei+1,j}=\gamma_{i+1,j}+\Delta E_i\omega_{i+1,j},$$
$$\alpha_{-i+1,j}=\xi_{i+1,j}+\Delta E_i\eta_{i+1,j}. \tag{14}$$

After substituting this expressions into (12) и (13) and equating the members with ΔE_i and without, equations for determining the values $\gamma_{i+1,j}$, $\omega_{i+1,j}$, $\xi_{i+1,j}$, $\eta_{i+1,j}$ can be obtained as

$$\frac{A_j}{\Delta y_{j-1}}\gamma_{i+1,j-1}-\left[A_j\left(\frac{1}{\Delta y_j}+\frac{1}{\Delta y_{j-1}}\right)+\frac{(W_xN)_{i,j}}{\Delta x_i}\right]\gamma_{i+1,j}+\frac{A_j}{\Delta y_j}\gamma_{i+1,j+1}==-\left\{\frac{(W_xN\alpha_e)_{i,j}}{\Delta x_i}-(W_yN)_{i,j}\frac{(\alpha_e)_{i,j+1}-(\alpha_e)_{i,j-1}}{2\Delta y}+\right.$$
$$+(\nu_i-\nu_a)_{|*}(N\alpha_e)_{i,j}+(\nu_d)_{|*}(N\alpha_-)_{i,j}-(\beta_{ei})_{|*}(N^2\alpha_e(\alpha_e+\alpha_-))_{i,j}+q_{i,j}-\left(\frac{\mu_e}{\mu_e+\mu_+}\right)\times\left\{\frac{B_j}{\Delta y_{j-1}}\xi_{i+1,j-1}-\right.$$
$$\left.\left.-\left[B_j\left(\frac{1}{\Delta y_j}+\frac{1}{\Delta y_{j-1}}\right)-(\mu_-+\mu_+)\frac{E_{xi}}{\Delta x_i}\right]\xi_{i+1,j}+\frac{B_j}{\Delta y_j}\xi_{i+1,j+1}+(\mu_-+\mu_+)\left[\frac{(\alpha_-E_y)_{i,j+1}-(\alpha_-E_y)_{i,j-1}}{2\Delta y}-\frac{(E_x\alpha_-)_{i,j}}{\Delta x_i}\right]\right\}\right\}, \tag{15}$$

$$\frac{A_j}{\Delta y_{j-1}}\omega_{i+1,j-1}-\left[A_j\left(\frac{1}{\Delta y_j}+\frac{1}{\Delta y_{j-1}}\right)+\frac{(W_xN)_{i,j}}{\Delta x_i}\right]\omega_{i+1,j}+\frac{A_j}{\Delta y_j}\omega_{i+1,j+1}==-\left\{\frac{\partial(\nu_i-\nu_a)}{\partial E}_{|*}(N\alpha_e)_{i,j}+\frac{\partial\nu_d}{\partial E}_{|*}(N\alpha_-)_{i,j}-\right.$$
$$-\frac{\partial\beta_{ei}}{\partial E}_{|*}(N^2\alpha_e(\alpha_-+\alpha_e))_{i,j}-\left(\frac{\mu_e}{\mu_e+\mu_+}\right)\left\{\frac{B_j}{\Delta y_{j-1}}\eta_{i+1,j-1}-\left[B_j\left(\frac{1}{\Delta y_j}+\frac{1}{\Delta y_{j-1}}\right)-(\mu_-+\mu_+)\frac{E_{x_i}}{\Delta x_i}\right]\eta_{i+1,j}+\frac{B_j}{\Delta y_j}\eta_{i+1,j+1}+\right.$$
$$\left.\left.+(\mu_-+\mu_+)\frac{(\alpha_-)_{i,j}}{\Delta x_i}\right\}\right\}, \tag{16}$$

$$\frac{G_j}{\Delta y_{j-1}}\xi_{i+1,j-1}-\left[G_j\left(\frac{1}{\Delta y_j}+\frac{1}{\Delta y_{j-1}}\right)-\frac{\left(\mu_- E_{x_i}-(W_x N)_{i,j}\right)}{\Delta x_i}\right]\xi_{i+1,j}+\frac{G_j}{\Delta y_j}\xi_{i+1,j+1}=-\left\{-\frac{\left(\mu_- E_{xi}-(W_x N)_{i,j}\right)}{\Delta x_i}(\alpha_-)_{i,j}+\right.$$
$$+\mu_-\frac{(\alpha_- E_y)_{i,j+1}-(\alpha_- E_y)_{i,j-1}}{2\Delta y}-(W_y N)_{i,j}\frac{(\alpha_-)_{i,j+1}-(\alpha_-)_{i,j-1}}{2\Delta y}+(\nu_a)_{|*}(N\alpha_e)_{i,j}-(\nu_d)_{|*}(N\alpha_-)_{i,j}-$$
$$\left.-(\beta_{ii})_{|*}\left(N^2\alpha_-(\alpha_e+\alpha_-)\right)_{i,j}\right\} \tag{17}$$

$$\frac{G_j}{\Delta y_{j-1}}\eta_{i+1,j-1}-\left[G_j\left(\frac{1}{\Delta y_j}+\frac{1}{\Delta y_{j-1}}\right)-\frac{\mu_- E_{x_i}-(W_x N)_{i,j}}{\Delta x_i}\right]\eta_{i+1,j}+\frac{G_j}{\Delta y_j}\eta_{i+1,j+1}=$$
$$=-\left\{\mu_-\frac{(\alpha_-)_{i,j}}{\Delta x_i}+\frac{\partial\nu_a}{\partial E}_{|*}(N\alpha_e)_{i,j}-\frac{\partial\nu_d}{\partial E}_{|*}(N\alpha_-)_{i,j}--\frac{\partial\beta_{ii}}{\partial E}_{|*}\left(N^2\alpha_-(\alpha_-+\alpha_e)\right)_{i,j}\right\}. \tag{18}$$

Here marked with:

$$A_j=\frac{c_e}{\Delta y},\ B_j=\frac{c_- - c_+}{\Delta y},\ G_j=\frac{c_-}{\Delta y},\ (\nu_i)_{|*}=\nu_i(E_i,N_{i+1}).$$

The equations (15) – (18) are supplemented with boundary conditions: $\gamma_{i,1}=\omega_{i,1}=\xi_{i,j}=\eta_{i,j}=0$ - on the wall, $\gamma_{i,N_y-1}=\gamma_{i,N_y+1}$ etc. - on the plane of symmetry. First the equation (17), (18) are solved and then the obtained results are used for the right parts (15), (16) to solve the equations for the unknown γ и ω.

The conservation equation of total electric current (3) was represented in finite-difference form by using the formula of trapezes and then used to determine the increment of electric field intensity ΔE_i for the i-th step of integration. Substitution of the expressions for α_e и α_- (14) into (3) gives a quadratic equation for E_{i+1} , which easily is solved. By knowing the increment $\Delta E_i = E_{i+1} - E_i$ it is easily to determine values $(\alpha_e)_{i,j}$ и $(\alpha_-)_{i,j}$. The transverse component of the electric field E_y can be obtained by integration of equation (4) on the y-coordinate in the range from b to y

$$E_y(x,y)=\frac{1}{(\mu_e+\mu_+)\alpha_e+(\mu_-+\mu_+)\alpha_-}\left\{(c_+-c_e)\frac{\partial\alpha_e}{\partial y}+(c_+-c_-)\frac{\partial\alpha_-}{\partial y}++\frac{\partial}{\partial x}E_x\int_y^b[(\mu_e+\mu_+)\alpha_e+(\mu_-+\mu_+)\alpha_-]dy\right\}.$$

Modeling of glow discharge in the air flow. As it is known air is the electronegative gas, there electrons effectively attaché to the oxygen molecules and form negative ions.

Modeling of positive column of the glow discharge in a stream of air was conducted with the following constants, parameters and dependencies of the frequencies from the reduced electric field E/N: $\mu_e = 1{,}5\ 10^{24}$ (V м s)$^{-1}$, $\mu_+ = \mu_- = 6{,}4\ 10^{21}$(V· м · s)$^{-1}$, $C_+ = C_+ = 1{,}5\ 10^{20}$ (м · s)$^{-1}$, $C_e = 1{,}0\ 10^{24}$ (м · s)$^{-1}$, $b_y = 10^{-2}$ м, $\nu_i = N\cdot(3{,}17\cdot 10^{-16}\cdot \exp(-6{,}4\cdot 10^{-19}\cdot N/E) + 4{,}01\cdot 10^{-15}(-8{,}4\cdot 10^{-19}\cdot N/E))$, $\nu_d = N\cdot 10^{-23}$ 1/s, $\nu_a = N\cdot 10^{-16}\exp(-2{,}83\cdot 10^{-19}\cdot N/E)$ при $E/N < 8\cdot 10^{-20}$ V · м^2 [10], $\beta_{ei} = 2\cdot 10^{-13}$ м^3/s, $\beta_{ii} = 3\ 10^{-13}$ м^3/s, $\nu_a = N\cdot 3{,}17\cdot 10^{-18}\cdot \exp(-1{,}31\cdot 10^{-19}\cdot N/E)$ при $E/N > 8\cdot 10^{-20}$ V · м^2 , k =0,15, I =60 мA, G = 3 g/s, $b_z = 5\ 10^{-2}$ м.1.

Modeling of discharge characteristics was conducted for case with a main discharge supported between two rod electrode systems situated at the beginning and end of the discharge chamber. At the upper electrode system the pre-ionization was organized by the additional discharges between each pairs of next rod electrodes. Meanwhile the upper electrode system was also as cathode for the maim discharge.

The distributions of electron and ion concentration were calculated in various sections in the positive column of a longitudinal glow discharge in a plane channel of the discharge chamber. The simulation of discharge characteristics was down under various boundary conditions at the inlet of the discharge area, and with various full discharge currents.

The electron concentration distribution n_e was adopted substantially uniform at the initial cross section, that depends on not independent character of the main charge. The accepted design of upper electrode system of pre-ionization leads to nonuniform ionization of the passing gas. There the electron concentration maximums correspond to the cathode locations of the auxiliary discharges. The computations reveal that the n_e distributions are aligned with the distance from the initial discharge cross-section due to the diffusion of charged particles, the absolute value of the electron concentration decreases along the flow due to diffusion, volume recombination of the particles and attachment.

This work has been done with the assistance of a grant from the Ministry of education and science of the Russian Federation 3.6564.2017/8.9.

References

[1] Galeev I G and Asadullin T Ya 2017 *J. of Phys. Conf. Ser.* **789** 012012
[2] Galeev I G and Timerkaev B A 1988 *High Temperature* **25** Is.5 613-19
[3] Galeev I G and Timerkaev B A 1986 *J. of Eng. Phys.* **51** Is. 3 1112-16
[4] Galeev I G, Tikhomirov I A and Asadullin T Ya 1995 *Rus. Phys. J.* **38** Is.10 1036-39
[5] Galeev I G and Asadullin T Ya 2017 *J. of Phys. Conf. Ser.* **789** 012003
[6] Galeyev I G, Asadullin T Ya and Timerkaev B A 2016 *J. of Phys. Conf. Ser.* **669** 012001
[7] Galeev I G and Asadullin T Ya 2016 *J. of Phys. Conf. Ser.* **669** 012015
[8] Galeev I, Goncharov V, Timerkaev B 1992 *High Temperature* **30** 439.
[9] Saifutdinov A I, Timerkaev B A and Zalyaliev B R 2016 *High Temperature* **54** Is. 5 632-38
[10] Aleksandrov I L, Vysikailo F I, Islamov R Sh, Kochetov I V, Napartovich A P and Pavlov V G 1981 *Rus. Thermophysics of high temperatures* **19** Is.3 485-90

Nonlinear Spatial Absorption of Alfven Waves by Dissipative Plasma

M B Gavrikov[1] **and A A Taiurskii**[1]
[1]Keldysh Institute of Applied Mathematics of the Russian Academy of Sciences, Miusskaya sq., 4, Moscow, 125047, Russia

E-mail: nadya_p@cognitive.ru, tayurskiy2001@mail.ru

Abstract. A mathematical model is proposed for explaining anomalous heating of the solar corona by Alfven waves generated in lower, much cooler, layers of the Sun. Numerical analysis shows that the absorption of Alfven waves occurs at wavelengths of the order of skin depth, in which case neither the classical nor Hall's MHD equations are applicable. Therefore, our research was based on the equations of two-fluid magnetohydrodynamics that fully take into account the inertia of the electrons. This investigation reveals some important features of the absorption process, in particular: the depth of the heated layer of the solar corona is finite and the parameters of the Alfven waves stabilize to a quasi-stationary regime whose parameters can be found.

1. Introduction

We consider nonlinear absorption of plane Alfven waves by incompressible homogeneous dissipative plasma. The results of this investigation are used here for studying the mechanism of anomalous heating of the solar corona up to $\sim 6\cdot 10^{6}\ {}^{\circ}\mathrm{K}$, due to the decay of Alfven waves generated in the lower, much cooler, layers of the Sun [1]. On the other hand, these results demonstrate the possibilities of the two-fluid electromagnetic hydrodynamics of plasma [2], since the absorption of Alfven waves caused by dissipation occurs at distances of the order of skin depth $\sim c/\omega_{\mathrm{p}}$ (ω_{p} is plasma frequency), in which case neither the classical nor Hall's MHD is applicable. In this paper, we formulate a mathematical problem describing nonlinear spatial absorption of a plane Alfven wave falling on the fixed boundary of dissipative incompressible completely ionized two-component plasma that fills up a half-space. It is shown that: (i) bremsstrahlung "locks" the Alfven wave, so that it penetrates the dissipative plasma only to a finite depth; (ii) the parameters of the decaying Alfven wave stabilize (in time) to a quasi-stationary regime whose parameters can be found by solving a certain boundary value problem on the half-line for a special system of ordinary differential equations.

2. Equations of Electromagnetic Hydrodynamics (EMHD) of Plasma

The EMHD equations of incompressible homogeneous plasma have the form

$$\mathrm{div}\mathbf{U} = 0, \quad \rho = \mathrm{const}, \quad \frac{\partial \rho \mathbf{U}}{\partial t} + \mathrm{Div}\Pi = \mathrm{Div}\mathrm{P},$$

$$\mathbf{E} + \frac{c^2\lambda_+\lambda_-}{4\pi\rho}\mathrm{rotrot}\mathbf{E} = \frac{\mathbf{j}}{\sigma} - \frac{1}{c}[\mathbf{U},\mathbf{H}] + \frac{1}{\rho}\mathrm{Div}\mathrm{W},$$

$$\frac{1}{c}\frac{\partial \mathbf{H}}{\partial t}+\mathrm{rot}\mathbf{E}=0,\quad \mathrm{div}\mathbf{E}=0,\quad \mathbf{j}=\frac{c}{4\pi}\mathrm{rot}\mathbf{H},\quad \mathrm{div}\mathbf{H}=0,$$

$$\rho_\pm c_\mathrm{p}^\pm\left[\frac{\partial T_\pm}{\partial t}+\mathrm{div}T_\pm\mathbf{v}_\pm\right]=\mathrm{div}\left(\chi_\pm\nabla T_\pm\right)+\mathrm{tr}\left(\Pi_\pm \mathrm{D}_\pm\right)+\frac{m_\mp}{m_\Sigma}\frac{j^2}{\sigma}\pm b\left(T_- - T_+\right)-p_\mathrm{T}^\pm, \tag{1}$$

where Π is the momentum flux tensor, P is the viscous stress tensor, W is the "hall terms" tensor,

$$\Pi=\Pi^\mathrm{h}+\Pi^\mathrm{p}+\Pi^\mathrm{c},\quad \mathrm{P}=\Pi_*^{(\mathrm{c})}+\Pi^{(\mathrm{U})},$$

$$\mathrm{W}=(\lambda_- - \lambda_+)(\Pi^\mathrm{p}+\Pi^\mathrm{c})+(\lambda_- p_+ - \lambda_+ p_-)\mathrm{I}_3+\lambda_+\lambda_-(\mathbf{jU}+\mathbf{Uj})-\Pi_*^{(\mathrm{U})}-\Pi^{(\mathrm{c})}, \tag{2}$$

$$\Pi^\mathrm{h}=\rho\mathbf{UU}+p_\Sigma \mathrm{I}_3,\ \Pi^\mathrm{p}=\frac{H^2}{8\pi}\mathrm{I}_3-\frac{\mathbf{HH}}{4\pi},\ \Pi^\mathrm{c}=\lambda_+\lambda_-\frac{\mathbf{jj}}{\rho}.$$

Here and in what follows, the subscripts "$\pm$" refer to electrons and ions, $\lambda_\pm=m_\pm/e_\pm$, $\lambda_\Sigma=\lambda_+ + \lambda_-$, $p_\Sigma=p_+ + p_-$, $m_\Sigma=m_+ + m_-$, $\rho=\rho_+ + \rho_-$, $\mathbf{U}=(\rho_+\mathbf{v}_+ + \rho_-\mathbf{v}_-)/\rho$, I_3 is the three-dimensional identity tensor, σ is the plasma conductivity, $p_\mathrm{T}^\pm$ are the losses in bremsstrahlung of electrons and ions, respectively, $c_\mathrm{p}^\pm=\mathrm{k}/((\gamma-1)m_\pm)=\mathrm{const}$, k is the Boltzmann constant. The viscous stress tensors (with plasma incompressibility taken into account) have the form

$$\Pi^{(\mathrm{U})}=2\mu_\Sigma \mathrm{D}^\mathrm{U},\ \Pi_*^{(\mathrm{U})}=2\mu_* \mathrm{D}^\mathrm{U},\ \Pi^{(\mathrm{c})}=2\mu^* \mathrm{D}^\mathrm{c},\ \Pi_*^{(\mathrm{c})}=2\mu_* \mathrm{D}^\mathrm{c},\ \Pi_\pm=2\mu_\pm \mathrm{D}_\pm, \tag{3}$$

where $\mathrm{D}^\mathrm{U}=\mathrm{def}\mathbf{U}$, $\mathrm{D}^\mathrm{c}=\mathrm{def}(\mathbf{j}/\rho)$, $\mathrm{D}_\pm=\mathrm{def}\mathbf{v}_\pm$ are strain tensors, $\mu_\Sigma=\mu_+ + \mu_-$, $\mu_*=\lambda_-\mu_+ - \lambda_+\mu_-$, $\mu^*=\lambda_-^2\mu_+ + \lambda_+^2\mu_-$, $\mu_\pm$ are hydrodynamic viscosities of electrons and ions. The hydrodynamic parameters of electrons and ions can be expressed in terms of $\mathbf{U}$, ρ, $\mathbf{j}$ and have the form $\mathbf{v}_\pm=\mathbf{U}\pm(\lambda_\mp/\rho)\mathbf{j}$, $\rho_\pm=(\lambda_\pm/\lambda_\Sigma)\rho$. The transport coefficients σ, $\mu_\pm$, $\chi_\pm$, b and bremsstrahlung have been calculated by the following formulas:

$$\mu_+=3.44\cdot 10^{-18}\left(\frac{m_+}{m_-}\right)^{1/2}T_+^{5/2},\quad \mu_-=1.857\cdot 10^{-18}T_-^{5/2},\quad \sigma=0.906\cdot 10^7 T_-^{3/2},$$

$$b=9.044\cdot 10^{38}\left(\frac{m_-}{m_+}\right)^3\frac{\rho^2}{T_-^{3/2}},\quad \chi_-=3.657\cdot 10^{-5}T_-^{5/2},\quad \chi_+=6.436\cdot 10^{-5}\left(\frac{m_-}{m_+}\right)^{1/2}T_+^{5/2}, \tag{4}$$

$$p_\mathrm{T}^+=0,\quad p_\mathrm{T}^-=8.044\cdot 10^{32}n_\mathrm{i}n_\mathrm{e}T_-^{1/2}\ \mathrm{erg}/(\mathrm{s}\cdot\mathrm{cm}^3), \tag{5}$$

where $T_\pm$ are measured in °K, and n_i, n_e in cm^{-3}.

3. Alfven Waves in EMHD

In the dissipation-free planar case, planar Alfven waves are exact solutions of the EMHD equations (1) – (5) and have the form

$$U_\perp=u(t)\exp(\mathrm{i}\kappa x),\ H_\perp=h(t)\exp(\mathrm{i}\kappa x),\ E_\perp=e(t)\exp(\mathrm{i}\kappa x),\ T_\pm=\mathrm{const},\ \rho=\mathrm{const},\ \mathrm{U}_x=0, \tag{6}$$

where $\kappa>0$ is arbitrary and the complex notation is used: $U_\perp=U_y+\mathrm{i}U_z$, $H_\perp=H_y+\mathrm{i}H_z$, $E_\perp=E_y+\mathrm{i}E_z$. Here, $H_x=\mathrm{const}$ and $e(t)$ is expressed through $u(t)$, $h(t)$ as follows:

$$e(t)=\left(\frac{\mathrm{i}H_x}{c}u(t)+\frac{\mathrm{i}\kappa\Lambda \mathrm{v}_\mathrm{A}}{\omega_\mathrm{p}}h(t)\right)\left(1+r^2\right)^{-1},\quad r=\frac{\kappa c}{\omega_p},\quad \Lambda=\sqrt{\frac{\lambda_+}{\lambda_-}}-\sqrt{\frac{\lambda_-}{\lambda_+}},\quad \mathrm{v}_\mathrm{A}=\frac{H_x}{\sqrt{4\pi\rho}}. \tag{7}$$

The functions $u(t)$, $h(t)$ have the form

$$u(t)=C_1\exp(\mathrm{i}\omega_+t)+C_2\exp(\mathrm{i}\omega_-t),\quad h(t)=\frac{(4\pi\rho)^{1/2}}{\kappa \mathrm{v}_\mathrm{A}}\{C_1\omega_+\exp(\mathrm{i}\omega_+t)+C_2\omega_-\exp(\mathrm{i}\omega_-t)\}. \tag{8}$$

Here $\omega_\mathrm{p}=(4\pi\rho)^{1/2}(\lambda_+\lambda_-)^{-1/2}$ is the plasma frequency, C_1, C_2 are arbitrary complex constants. The transverse component of the current density, $j_\perp=j_y+\mathrm{i}j_z$, also varies according to the law $j_\perp=j(t)\exp(\mathrm{i}\kappa x)$, and $j_x=0$, where $j(t)=-\frac{\kappa c}{4\pi}h(t)$. Finally,

$$\omega_\pm=\omega_\pm(\kappa)=\frac{\kappa \mathrm{v}_\mathrm{A}}{2}\left\{\frac{r\Lambda}{1+r^2}\pm\left[\frac{r^2\Lambda^2}{(1+r^2)^2}+\frac{4}{1+r^2}\right]^{1/2}\right\}. \tag{9}$$

It is convenient to invert formula (9):

$$\kappa=\kappa(\omega)=\frac{\omega_\mathrm{p}|\omega|}{c}\left(\frac{\mathrm{v}_\mathrm{A}^2\omega_\mathrm{p}^2}{c^2}+\frac{\omega_\mathrm{p}\mathrm{v}_\mathrm{A}}{c}\Lambda\omega-\omega^2\right)^{-1/2}, \tag{10}$$

where $-\omega_\mathrm{c}^+<\omega<\omega_\mathrm{c}^-$ for $H_x>0$, $-\omega_\mathrm{c}^-<\omega<\omega_\mathrm{c}^+$ for $H_x<0$, and $\omega_\mathrm{c}^\pm=|H_x|(\lambda_\pm c)^{-1}$ are cyclotron frequencies of electrons and ions.

4. Statement of the Spatial Absorption Problem

Consider a planar Alfven wave running from left to right in the domain $x<0$ and arriving at the dissipative plasma boundary $x=0$, the plasma filling up the half-space $x>0$. Further propagation of the Alfven wave in the domain $x>0$ is accompanied by its absorption, which is the subject of this study. From the results of the preceding section, it follows that for the longitudinal magnetic field $H_x<0$, the frequency of the incident Alfven wave varies on the interval $-\omega_\mathrm{c}^-=H_x/(\lambda_-c)<\omega<0$, and if $H_x>0$, then its frequency varies on the interval $-\omega_\mathrm{c}^+=-H_x/(\lambda_+c)<\omega<0$. Here we consider only the first case. The plasma in the region $x>0$ at the initial instant is assumed to be magnetized, stationary, homogeneous, and isothermal:

$$U_\perp|_{t=0}=0,\quad H_\perp|_{t=0}=0,\quad T_\pm|_{t=0}=T^0,\quad U_x|_{t=0}=0,\quad \rho|_{t=0}=\mathrm{const}, \tag{11}$$

where the longitudinal magnetic field $H_x=\mathrm{const}$ in the domain $x>0$ and the constant ρ are the same as in the domain $x<0$, from which the Alfven wave comes. The values of the dissipative plasma parameters on the boundary $x=0$ coincide with the values of the Alfven wave parameters on the same boundary, and therefore, in view of (6) – (10) with $C_2=0$, $C_1=U_0$, we have

$$U_\perp\big|_{x=0} = U_0 \exp(\mathrm{i}\omega t), \quad H_\perp\big|_{x=0} = U_0 \frac{\sqrt{4\pi\rho}}{\kappa v_A}\omega \exp(\mathrm{i}\omega t), \quad T_\pm\big|_{x=0} = \text{const},$$

$$E_\perp\big|_{x=0} = \frac{\mathrm{i}U_0}{1+r^2}\left(\frac{H_x}{c} + \Lambda\sqrt{\lambda_+\lambda_-}\,\omega\right), \quad j_\perp\big|_{x=0} = -U_0 \frac{\omega c \rho}{H_x}\exp(\mathrm{i}\omega t). \tag{12}$$

where U_0 is an arbitrary complex constant and $\kappa = \kappa(\omega)$ has the form (10). At infinity, the dissipative plasma parameters coincide with the unperturbed plasma parameters:

$$U_\perp\big|_{x=+\infty} = 0, \quad H_\perp\big|_{x=+\infty} = 0, \quad T_\pm\big|_{x=+\infty} = T^0, \quad E_\perp\big|_{x=+\infty} = 0, \quad j_\perp\big|_{x=+\infty} = 0, \tag{13}$$

In the case of plane symmetry $(\partial/\partial y = \partial/\partial z = 0)$, system (1) – (5) admits a quasi-stationary solution of the form

$$U = U_\perp = U(x)\exp(\mathrm{i}\omega t), \quad H = H_\perp = H(x)\exp(\mathrm{i}\omega t), \quad E = E_\perp = E(x)\exp(\mathrm{i}\omega t),$$

$$j = j_\perp = j(x)\exp(\mathrm{i}\omega t), \quad T_\pm = T_\pm(x), \tag{14}$$

where ω is the frequency of the incident Alfven wave. Substituting this solution into (1) – (5), in the case of plane symmetry, we obtain a boundary value problem for the amplitudes $U(x)$, $E(x)$, $T_\pm(x)$. In dimensionless form, this problem is reduced to the following system of ordinary differential equations on the half-line $[0,+\infty]$:

$$\mathrm{i}\rho\omega U - \frac{H_x}{\omega}\frac{\mathrm{d}^2E}{\mathrm{d}x^2} - \frac{\mathrm{d}}{\mathrm{d}x}\left(\mu_\Sigma \frac{\mathrm{d}U}{\mathrm{d}x}\right) + \frac{\mathrm{i}\xi}{\rho\omega}\frac{\mathrm{d}}{\mathrm{d}x}\left(\mu_* \frac{\mathrm{d}^3E}{\mathrm{d}x^3}\right) = 0,$$

$$E - \left(\frac{\xi^2}{\rho} - \frac{\mathrm{i}\zeta\xi}{\omega\sigma} - \frac{\xi\Lambda H_x}{\omega\rho}\right)\frac{\mathrm{d}^2E}{\mathrm{d}x^2} - \mathrm{i}H_x U + \frac{\xi}{\rho}\left(\mu_* \frac{\mathrm{d}U}{\mathrm{d}x}\right) - \frac{\mathrm{i}\xi^2}{\omega\rho^2}\frac{\mathrm{d}}{\mathrm{d}x}\left(\mu_* \frac{\mathrm{d}^3E}{\mathrm{d}x^3}\right) = 0,$$

$$\frac{\mathrm{d}}{\mathrm{d}x}\left(\chi_\pm \frac{\mathrm{d}T_\pm}{\mathrm{d}x}\right) + \frac{m_\mp}{m_\Sigma}\frac{\xi\zeta}{\omega^2\sigma}\left|\frac{\mathrm{d}^2E}{\mathrm{d}x^2}\right|^2 \pm \frac{\zeta}{\xi}C_0\rho^2\frac{T_- - T_+}{T_-^{3/2}} +$$

$$+\frac{T_\pm^{3/2}}{R_\pm}\left[\left|\frac{\mathrm{d}U}{\mathrm{d}x}\right|^2 + \frac{\lambda_\mp}{\lambda_\pm}\frac{\xi^2}{\omega^2\rho^2}\left|\frac{\mathrm{d}^3E}{\mathrm{d}x^3}\right|^2 \pm 2\frac{\xi}{\rho\omega}\sqrt{\frac{\lambda_\mp}{\lambda_\pm}}\,\mathrm{Re}\left(\mathrm{i}\frac{\mathrm{d}U}{\mathrm{d}x}\frac{\overline{\mathrm{d}^3E}}{\mathrm{d}x^3}\right)\right] - \frac{(1\mp 1)}{4}\frac{\xi_\mathrm{T}}{\xi}Z^2\rho T_-^{1/2} = 0 \tag{15}$$

with the boundary conditions

$$T_\pm(0) = T^0, \quad U(0) = U_0, \quad E(0) = \frac{\mathrm{i}U_0}{1+\kappa^2\xi^2}(H_x + \xi\Lambda\omega), \quad E'(0) = -\frac{\rho U_0\omega^2}{\kappa H_x},$$

$$T_\pm(+\infty) = T^0, \quad U(+\infty) = 0, \quad E(+\infty) = 0, \quad E'(+\infty) = 0, \tag{16}$$

where ω is the frequency of the incident Alfven wave and κ is its wave number, $\kappa = \kappa(\omega) = |\omega|\sqrt{\rho(H_x^2 + H_x\Lambda\xi\omega - \xi^2\omega^2)}$, ξ, ζ, ξ_T are similarity numbers,

$$\xi = \frac{\ell_\mathrm{c}}{L_0} = \frac{c\sqrt{\lambda_+\lambda_-}}{\sqrt{4\pi\rho_0}L_0}, \quad \zeta = 15\frac{(4\pi\rho_0)^{5/2}}{H_0^4}\frac{ce^3}{m_+^2}, \quad \xi_\mathrm{T} = 21.825\cdot\frac{\rho_0^{3/2}}{H_0^2}\frac{10^7 e^5}{c^2\hbar m_+ m_-},$$

Here, $\hbar = 1.05\cdot10^{-27}\,\mathrm{erg\cdot s}$ is the Planck constant, $\ell_\mathrm{c} = c/\omega_p$ is the skin depth, L_0, ρ_0, H_0 are the respective characteristic scales of length, density, and magnetic field strength.

In (15), we have $\mu_\Sigma = \mu_+ + \mu_-$, $\mu_* = (\lambda_- / \lambda_+)^{1/2}\mu_+ - (\lambda_+ / \lambda_-)^{1/2}\mu_-$, $\mu^* = (\lambda_- / \lambda_+)\mu_+ + (\lambda_+ / \lambda_-)\mu_-$, $\mu_\pm = T_\pm^{5/2} / R_\pm$, $\sigma = \sigma_0 T_-^{3/2}$, $\chi_\pm = C_\pm(\xi / \zeta)T_\pm^{5/2}$, and the constants σ_0, C_0, $R_\pm$, $C_\pm$ have the form $\sigma_0 = 2.59$, $C_0 = m_- / m_+$, $R_+ = 1.1(\zeta / \xi)(m_- / m_+)^{1/2}$, $R_- = 2.05(\zeta / \xi)$, $C_- = 2.11 m_+ / m_-$, $C_+ = 3.7(m_+ / m_-)^{1/2}$. Numerical analysis shows that the solution of the initial boundary value problem (1) – (5), (11) – (13) for the EMHD system in the case of plane symmetry stabilizes to the stationary solution (14) with the amplitudes being the solution of the boundary value problem (15), (16). The solution of the boundary value problem (15), (16) was found by the stabilization method.

5. Some Calculation Results

For typical parameters of the solar corona, $\rho_0 = 10^{-12}\,\mathrm{g/cm^3}$, $H_0 = 1\,\mathrm{G}$, $Z = 1$, we have $\zeta = 3\cdot 10^3$, $\xi_T = 3$, $\ell_c = c/\omega_p = 1\,\mathrm{cm}$. In order to take into account small-scale processes, we set $L_0 = \ell_c$, $\xi = 1$. For these parameters of the background dissipative plasma, we have $v_0 = v_A = 2.8\cdot 10^5\,\mathrm{cm/s}$, $t_0 = L_0 / v_A = 3.45\cdot 10^{-6}\,\mathrm{s}$, $T_0 = 10^3\,°\mathrm{K}$, the Debye radius is $r_D = (kT_0)^{1/2}\left(2\pi^{1/2} e n_0^{1/2}\right)^{-1} \cong 2\cdot 10^{-4}\,\mathrm{cm}$, $n_0 = \rho_0 / m_+$, $m_+ = 2m_p$. Consider the calculation results for $U_0 = 1$. The temperature stabilization is represented in figure 1. The electromagnetic field components have the form of "decaying sine curves" shown in figure 2. Our estimates show that for $|U_0| \cong 50$, the temperature of the solar corona goes up to $\sim 10^6\,°\mathrm{K}$.

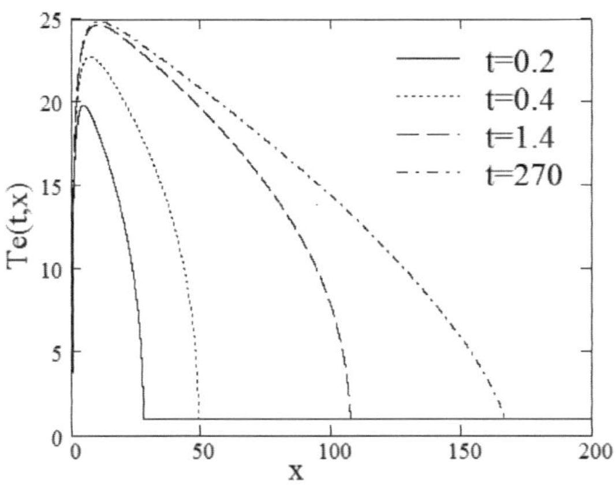

Figure 1. Distribution of the temperature of electrons at different times for $U_0 = 1$.

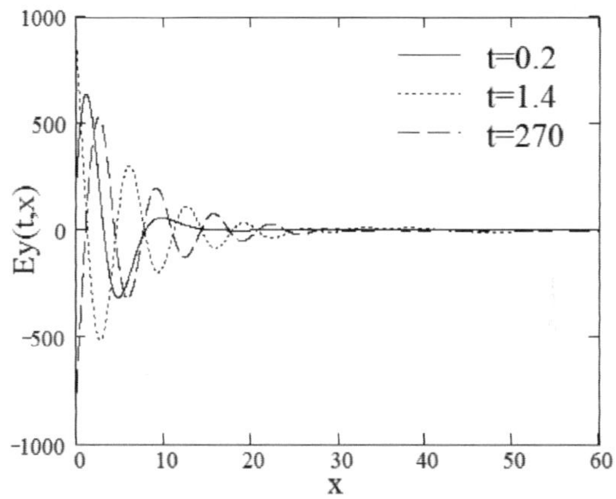

Figure 2. Distribution of the electric field intensity at different times for $U_0 = 1$.

Acknowledgments
This work has been supported by the Russian Science Foundation: Project No. 16-11-10278.

Bibliography

[1] McIntosh S W, Pontien B P, Carlsson M, Hansteen V, Boerner P and Goossens M 2011 *Nature* **475** pp 478-480
[2] Gavrikov M B and Sorokin R V 2009 *Fluid Dynamics* **43** 6 pp 977-989

Site of maxima of conductivity, temperatures, density of the current and specific capacity of the thermal emission in the HFI-discharge

A Gerasimov, A Kirpichnikov, F Sabirova and R Gainullin

Kazan National Research Technological University 420015, Karl Marx str., 68 Kazan, Russia

gerasimov@kstu.ru

Abstract. On the basis of theoretical analysis of distributions of the conductivity, current density and specific power of heat release in the high-frequency induction discharge, a law of crowding of maxima of these values has been established.

1. Introduction

In electro-plasma processes and power plants using high-frequency induction (HFI) plasma, the discharge zone is the main process zone. Information on distribution of main parameters of the discharge (conductivity, temperature, current density, specific power of heat release) and position of maxima of these values gives an opportunity to determine optimal conditions for heating the initial material and to manufacture a high-quality product.

The results of works [1]-[7] allowed us to suggest that for each section inside a plasmoid of the HFI discharge, the following inequality holds true

$$r_1<r_2<r_3, \qquad (1)$$

where $r_1=r(\sigma_{max})$, $r_2=r(j_{\varphi\ max})$ and $r_3=r(W_{max})$ are the radial coordinates corresponding to the maxima of physical quantities within the parentheses.

In these relations σ is the conductivity of the discharge, $j=\sigma E_\varphi$ is the eddy current density, E_φ is the strength of the azimuth electric field, and $W=\frac{1}{2}\sigma E_\varphi^2$ is the power input to the discharge per unit volume averaged over a period of the high-frequency field.

Figure 1 shows dependences of the conductivity of the discharge $\sigma(r)$, eddy current density $j=\sigma E_\varphi$, bulk density of the power input to the discharge $W=\frac{1}{2}\sigma E_\varphi^2$ in various sections of a HFI plasmoid, starting from its central section downstream and illustrating the relation (1). The results are obtained on a stationary semi-commercial unit HFI-11/60 using a procedure described in Refs. [1-7].

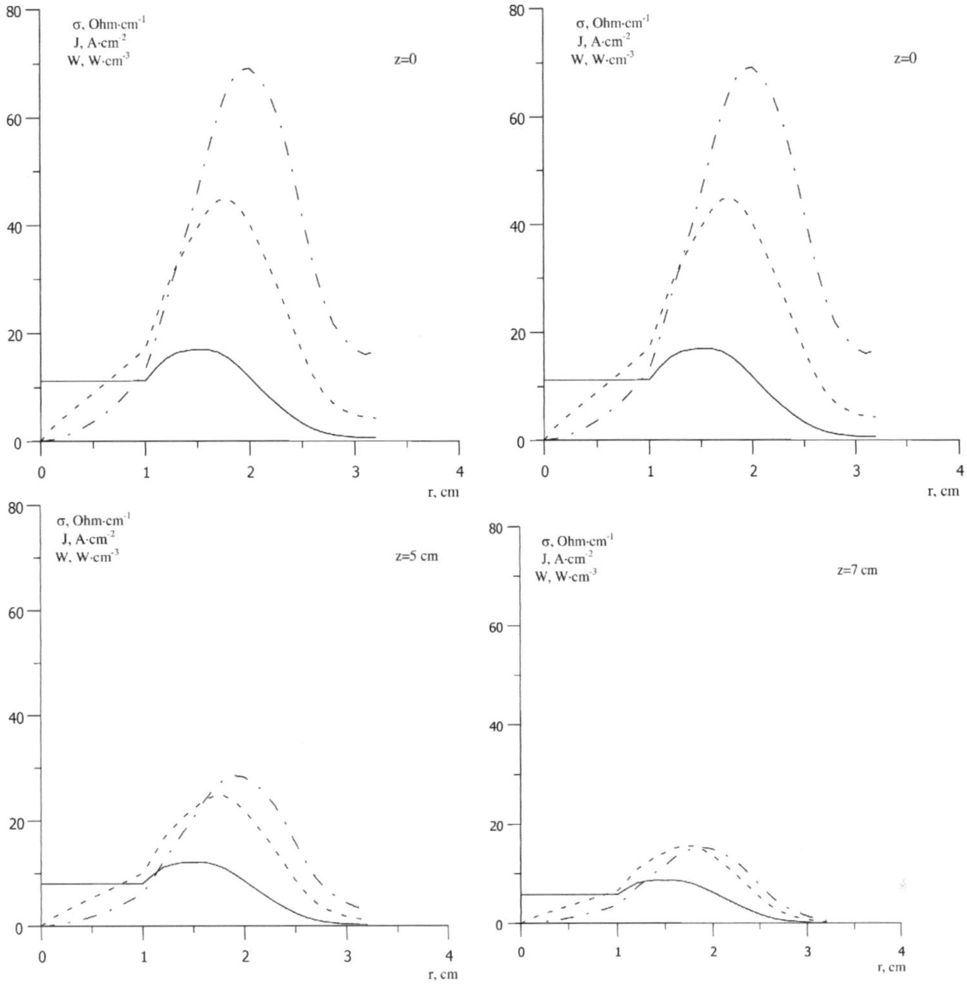

Figure 1 Radial distribution of conductivity σ - ____, eddy current density j - - - - and heat power W -·-·-·-·- for various cross-section of the plasmoid

2. Method of solution

Inequality (1) can be analytically proven as follows. Let us assume that the conductivity of the discharge σ(r,z), current density $j_\varphi(r,z)=\sigma E_\varphi$ and power of heat release $W=\frac{1}{2}\sigma E_\varphi^2$ are the functions, which at each fixed section of the discharge reach their maximum values only once on the interval $0 \le r \le R$. One may prove that in this case the relation $r(\sigma_{max}) < r(j_{\varphi max}) < r(W_{max})$ holds true.

At the point $r_1 = r(\sigma_{max})$ $\left.\frac{\partial\sigma(r)}{\partial r}\right|_{r_1} = 0$, $\frac{\partial\sigma(r)}{\partial r} > 0$ when $r < r_1$, And $\frac{\partial\sigma(r)}{\partial r} < 0$ when $r > r_1$.

Then at the point $r_2 = r(j_{\varphi max})$ we have $\left.\frac{\partial(\sigma E_\varphi)}{\partial r}\right|_{r_2} = \left.\frac{\partial\sigma}{\partial r}\right|_{r_2} E_\varphi(r_2) + \sigma(r_2)\left.\frac{\partial E_\varphi}{\partial r}\right|_{r_2} = 0$.

Since $E_\varphi(r), \sigma(r), \frac{\partial E_\varphi(r)}{\partial r}$ are nonnegative values in the whole volume of a discharge, at the point r_2 $\left.\frac{\partial\sigma}{\partial r}\right|_{r_2} = -\frac{\sigma(r_2)}{E_\varphi(r_2)}\left.\frac{\partial E_\varphi}{\partial r}\right|_{r_2} < 0$, and this means that the point $r_2 = r(j_{\varphi\max})$ is situated at the descending path of the function $\sigma(r)$, i.e., on the interval $r_1 < r < R$. Therefore, $r_2 > r_1$.

At the point $r_3 = r(W_{max})$, in turn, we have:

$$\left.\frac{\partial(\sigma E_\varphi^2)}{\partial r}\right|_{r_3} = \left.\frac{\partial(\sigma E_\varphi)}{\partial r}\right|_{r_3} E_\varphi(r_3) + \sigma(r_3)E_\varphi(r_3)\left.\frac{\partial E_\varphi}{\partial r}\right|_{r_3} = 0.$$

Hence $\left.\frac{\partial(\sigma E_\varphi)}{\partial r}\right|_{r_3} = -\sigma(r_3)\left.\frac{\partial E_\varphi}{\partial r}\right|_{r_3} < 0$, and, consequently, $r_3 > r_2$.

Finally we obtain the relation $r_1 < r_2 < r_3$, which was to be proven.

Thus, one may state that inside the HFI discharge the radial coordinates corresponding to maximum values of the conductivity, eddy current density and power input to the discharge are arranged in increasing order.

Note that using the method applied one can prove a more general result,

$$r_1 < r_2 < \ldots < r_i \ldots < r_n, \tag{2}$$

where $r_i = r\left[\left(\sigma E_\varphi^{i-1}\right)_{max}\right]$; i=1, 2,…, n, and n is integer number that can be made as large as is wished. Inequality (1) is particular case of (2). Therefore, there is a set coaxial cylindrical surfaces inside an HFI discharge that correspond to the maximus of σE_φ^{i-1}.

Under conditions of local thermodynamic equilibrium, the surface of minimum radius $r_1 = r(\sigma_{max})$ of the set of surfaces (2) corresponds to the maximal temperature $T_{max} = T(r_1)$ inside the plasmoid, because the relationship $\sigma = \sigma(T)$ between the temperature and the conductivity of the discharge is unambiguous in this case.

Let us find a law of crowding of r_i points corresponding to maxima of the σE_φ^{i-1} (i=1, 2, 3) values on the r-axis in the direction of the plasmoid periphery, that is first three points r_1, r_2, and r_3, most interesting to technical applications.

Let us examine the function of the form $y_i = \sigma E_\varphi^{i-1}$. The changes of the E_φ value can be approximated by a straight line $E_\varphi = a \cdot r^{i-1}$, while the σ value is chosen in the form

$\sigma = \sigma_{max} \cdot e^{-\left(\frac{r-r_0}{\Delta r}\right)^2}$. Then $y = a \cdot r^{i-1} \cdot \sigma_{max} \cdot e^{-\left(\frac{r-r_0}{\Delta r}\right)^2}$.

In this case, from mathematical point of view, the problem of finding maxima of σE_φ^{i-1} values as a function of the radial coordinate r reduces to the problem of finding maxima of the following function $y = a \cdot r^{i-1} \cdot \sigma_{max} \cdot e^{-\left(\frac{r-r_0}{\Delta r}\right)^2}$, i=1, 2, 3 which can be solved by standard methods of applied analysis.

For the point $r^* = r(f_{max})$, apparently, we have the relation

$$\frac{\partial\left(ar^{i-1}e^{-\left(\frac{r-r_0}{\Delta r}\right)^2}\right)}{\partial r}=a(i-1)r^{i-2}e^{-\left(\frac{r-r_0}{\Delta r}\right)^2}-a\cdot r^{i-1}\cdot\frac{2(r-r_0)}{\Delta r^2}e^{-\left(\frac{r-r_0}{\Delta r}\right)^2}=0$$

or $(i-1)r^{i-2}-r^{i-1}\cdot\frac{2(r-r_0)}{\Delta r^2}=0$, so $r^2-r\cdot r_0-(i-1)-\frac{(\Delta r)^2}{2}=0$.

The solution of this quadratic equation can be written as

$$r_{1,2}=\frac{r_0\pm r_0\sqrt{1+2(i-1)\left(\frac{\Delta r}{r_0}\right)^2}}{2}.$$

On physical grounds the positive sign should be used

$$r^*=\frac{r_0}{2}\left(1+\sqrt{1+2(i-1)\left(\frac{\Delta r}{r_0}\right)^2}\right).$$

At i=i+1 $\tilde{r}^*=\frac{r_0}{2}\left(1+\sqrt{1+2i\left(\frac{\Delta r}{r_0}\right)^2}\right)$. That is

$$\Delta=\tilde{r}^*-r^*=\frac{\sqrt{1+2i\left(\frac{\Delta r}{r_0}\right)^2}-\sqrt{1+2(i-1)\left(\frac{\Delta r}{r_0}\right)^2}}{2}$$

When $i\to\infty$, $\tilde{r}^*(i)-r^*(i-1)\to 0$ or $\tilde{r}^*\approx r^*$, i.e. the points corresponding to maxima of the σE_φ^{i-1} values, should crowd

3. Results and conclusion

Let us estimate the value of this crowding. The estimation procedure is illustrated in Figure 2.

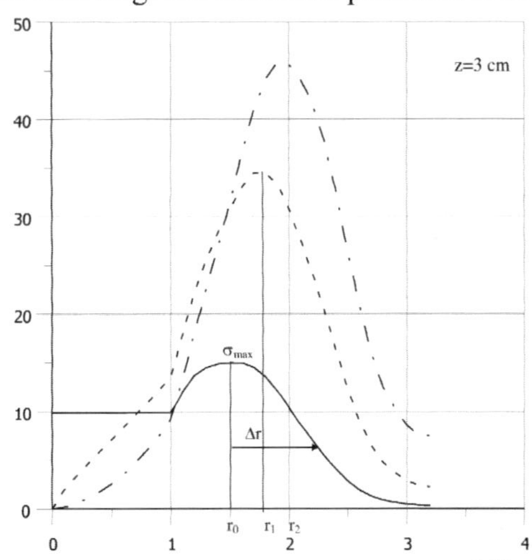

Figure 2 Illustration of a technique of an estimation of value of a crowding

The point $r_0(\sigma_{max})$ corresponding to the maximum of the conductivity value σ is chosen on the plot. In our case, σ_{max}=15.0. The radial coordinate r_0 corresponding to σ_{max} is 1.5 cm. Then a standard half-width of an exponential curve Δr is determined, i.e. at this point the value σ is diminished e-fold from σ_{max}. In our case, Δr=0.8 cm. Then $\frac{\Delta r}{r_0} = \frac{0{,}8}{1{,}5} = 0{,}53$.

Let us estimate the Δ values for the j and W values denoting them as Δ_1 and Δ_2, respectively. As one may see from the figure, $\Delta_1 = r_1 - r_0$ and $\Delta_2 = r_2 - r_1$.

$$\Delta_1 = 1{,}5 \cdot \frac{\sqrt{1+2\cdot 1\cdot (0{,}53)^2} - 1}{2} = 0{,}19,$$

$$\Delta_2 = 1{,}5 \cdot \frac{\sqrt{1+2\cdot 2\cdot (0{,}53)^2} - \sqrt{1+2\cdot 1\cdot (0{,}53)^2}}{2} = 0{,}16.$$

Analogous calculations have been performed for other sections.

The results of Δ_{1calc} and Δ_{2calc} calculations are summarized in Table 1. The Δ_{1emp} and Δ_{2emp} values obtained directly from the plots are also given in the table.

Table 1. The results of calculations values Δ_{1calc} and Δ_{2calc}

Z, cm	0	1	2	3	4	5	6	7
Δ_{1emp}	0,3	0,3	0,3	0,3	0,3	0,3	0,3	0,3
Δ_{1calc}	0,21	0,20	0,20	0,19	0,19	0,19	0,19	0,19
Δ_{2emp}	0,2	0,2	0,2	0,2	0,1	0,1	0,1	0,1
Δ_{2calc}	0,17	0,17	0,16	0,16	0,16	0,16	0,16	0,16

Apparently, the best agreement of empirical data with the calculated ones is observed for the first four plasmoid sections downstream. This fact can be explained by more precise approximation of the amplitude of the azimuth electric field in this area of a plasma bunch.

Procedures described in the present paper can be useful for a broad circle of experts in different fields of low-temperature induction plasma physics and engineering.

References

[1] Kirpichnikov A 1992 *Izvestiya VUZov Fizik.* No 1 p 56
[2] Kirpichnikov A 1992 *Izvestiya VUZov Fizik.* No 11 p 102
[3] Kirpichnikov A 1994 *Izvestiya VUZov Fizik.* No 2 p 77
[4] Kirpichnikov A 1995 *Hight Themperature* vol 33 No 1 p 139
[5] Kirpichnikov A 1996 *Hight Themperature* vol 34 No 6 p 989
[6] Kirpichnikov A 1998 *Izvestiya VUZov Fizik.* No 2 p 125
[7] Gerasimov A 2004 *Izvestiya VUZov Fizik.* No 7 p 65

High Power Argon, Nitrogen Plasma Torches

A Hakki, N Kashapov, K Sadikov
Kazan (Volga-Region) Federal University,
420008, Kremlyovskaya St, bdg.18, Kazan, Russia

E-mail: hakki.adnan@gmail.com

Abstract. The paper describes a high power supply for Argon and Nitrogen plasma torches. A high frequency was used in order to drive the pulse width modulation circuit. The average output current consumption (AOCC) was modified from 20A up to 80A by increasing the pulse width from 2μsec up to 3μsec for Argon gas plasma torches. The (AOCC) was reduced from 70A down to 25A by increasing the pulse width from 6μsec up to 8μsec in the case of Nitrogen gas plasma torches.

Key words. Argon plasma, Nitrogen plasma, Bridge converter, Modulation circuit, MOSFET transistors, LPF

1. Introduction

Bridge converters have many advantages such as a high output power, a small size, a high efficiency factor, a high reliability. Usually the type of the power transistors of the bridge converters is MOSFET because, these kind of transistors are very fast, so they can work as switches in the converter circuit.
In modern power supplies, the MOSFET power transistors can be built on one electronics chip (Module). MOSFET transistors can be derived by very simple circuits. MOSFET driver circuits are connected directly to the pulse width modulation port (PWM) of the used microprocessor. The method of this power supply is described in our paper for feeding plasma torches [1-5].

2. Experimental Setup

The block diagram of plasma torches power supply is shown in Fig.1. The schematic circuit corresponding to Fig.1 is shown in Fig. 2.

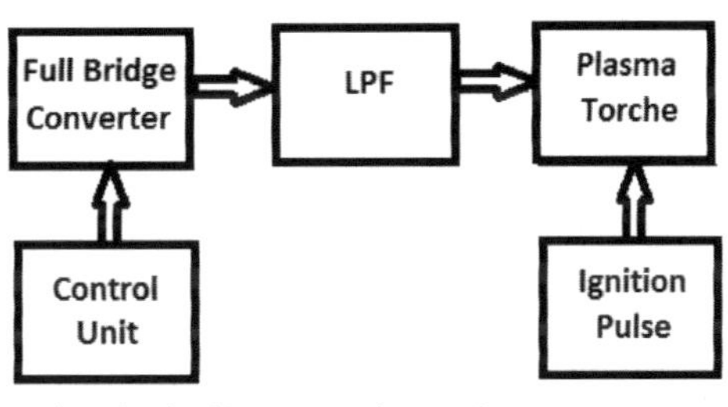

Figure 1. Block diagram of arc plasma power supply

3. Results and discussion

A high frequency high power transformer was used for insulation (T1), a high frequency rectifier and a high frequency filter with two stages were added to prevent the vibration of plasma torches pulses returning to the input circuit, and to limit the load current. An electrical transformer was used in order to sense the output current (T2). A monitor plays an important role to show the output current value while adjusting the duty ratio δ. A feedback voltage circuit was used to generate the voltage Vr, the value of this voltage was in the interval:

$$0V \leq V_r \leq 4V$$

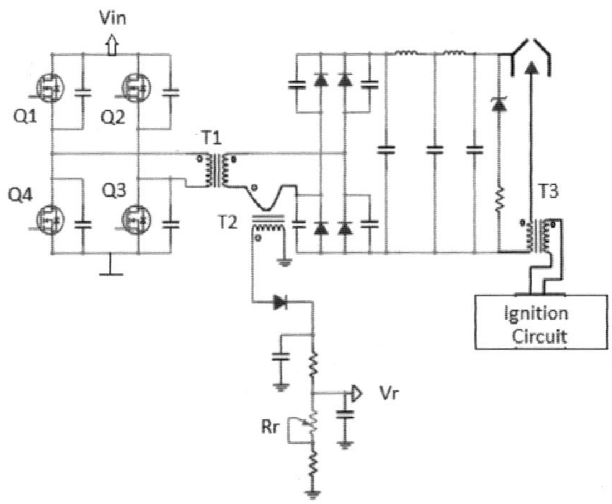

Figure 2. Schematic circuit of power supply

Fg. 3 shows the schematic circuit of MOSFET driver. The output voltage of this circuit are the signals (Q1G, Q1S, Q3G, Q3S). The signal (Q1G, Q1S) drives the transistor Q1 in the bridge converter, the signal (Q3G, Q3S) drives the transistor Q3 in the bridge converter, and these two signals are synchronized. The signal (Q2G, Q2S) drives the transistor Q2 in the bridge converter, the signal (Q4G, Q4S) drives the transistor Q4 in the bridge converter, and these two signals are synchronized too. The ignition circuit generates a high voltage pulse in order to start the plasma process as shown in fig. 2 [8-10].

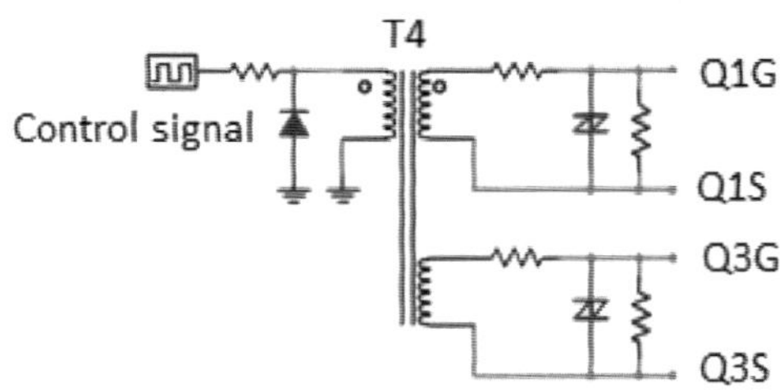

Figure 3. Schematic of MOSFET driver

The output voltage V_o is given by [6-7]:

$$V_o = \frac{1}{T}\int_0^T V_{oi}(t)\,dt$$

$$V_o = \frac{1}{T}\left(2\int_{0}^{T_{on}} \frac{n_2}{n_1} V_d \, dt + \int_{T_{on}}^{T} 0 \, dt \right)$$

$$V_o = 2\frac{n_2}{n_1}\frac{1}{T} V_{IN}\, \delta\, T$$

Where δ is the duty ratio, V_{IN} is the input DC voltage.
In the case of $n = \frac{n_2}{n_1} = 0.5$

$$V_o = V_{IN}.\delta$$

Working frequency

$$f \approx 50 \text{ KHz} \Rightarrow T = \frac{1}{f} = 20\mu sec$$

by changing the period T_{on} (the width of the driving pulses) the output voltage value is regulated.
We can set:

$$\delta_{Max} = 0.40$$

$$T_{onmax} = T.\delta_{Max} = 8\mu sec$$

Delay Time

$$T.D_{min} = \frac{T}{4} - \frac{T_{onmax}}{2} = 1\mu sec$$

Note that during the delay of time T. D , all transistors are in cut off situation. Target efficiency of full bridge converter is 90%, then the input power Assuming at this efficiency is given by:

$$P_{IN} = \frac{P_{out}}{\eta} \approx 9KW$$

In single pulse width modulation control, there is only one pulse per half cycle. In that case, the output voltage is changed by varying the width of the pulse. By varying the duty ratio from 0 to 0.4 the pulse width T_{ON} can be modified from 0μsec up to 8μsec and the average output voltage from 0V to 240V. The switches (Q1, Q2, Q3, Q4) are assumed to be the fast power MOSFET N–Channel transistors. If the current of selected transistors (I_{DS}) is not enough one can collect few transistors in parallel. In this converter Transistor APT12060LVR was used in the converter. Each two transistors were collected in parallel. So, the full bridge converter contains 8 transistors assembled on one aluminum substrate for cooling. The APT12060LVR transistor has the following characteristics:

$$V_{DS} = 1200V, I_{DS} = 20A\,,$$

$$R_{DS(on)} = 0.6\Omega,\ t_r = 13nsec$$

A capacity of (1nF) was added in parallel with each transistor to limit the voltage spikes. Also we added a Schottky diode in parallel with the load to limit the inverse voltage spikes. Table. 1 shows the parameters of the designed converter:

Parameters	The value
V_{IN}	535V
V_{INmax}	600V
V_{INmin}	500V

IOP Conf. Series: Journal of Physics: Conf. Series **927** (2017) 012022 doi:10.1088/1742-6596/927/1/012022

V_{outmax}	240V
f	50KHz
δ_{Max}	40%

Plasma Torches Topology: fig.4 shows the typical plasma torches volt-ampere characteristics.

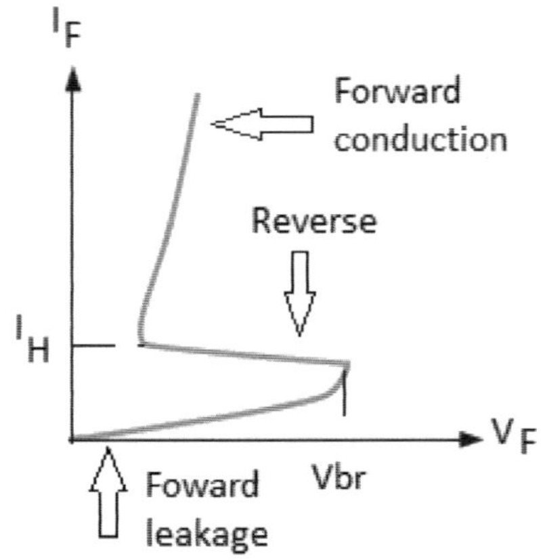

Figure 4. Typical Plasma Torches volt-ampere characteristics

One can see in fig.4 that there are 3 regions: Forward leakage region, Reverse region (negative resistor) and Forward conduction region (positive resistor). The working region of the Plasma Torches could be in the Forward Conduction region (positive resistor), or in the Reverse region (negative resistor). But it cannot be in the Forward leakage region because in that case the current is not enough.

Forward Conduction Mode (positive resistor)

Fig.5 shows the plasma torches volt-ampere characteristics for Argon (Ar) gas. Notice, if $\delta_i < \delta_j$ And,

$$\delta_i \Rightarrow V_{oi} \Rightarrow I_{oi}$$

$$\delta_j \Rightarrow V_{oj} \Rightarrow I_{oj}$$

$$V_{oi} < V_{oj} \Rightarrow I_{oi} < I_{oj}$$

This means that the load is the positive resistor. For example

$$\delta_1 \Rightarrow V_{o1} = 50V \Rightarrow I_{o1} = 20A$$

$$\delta_2 \Rightarrow V_{o2} = 70V \Rightarrow I_{o2} = 80A$$

$$V_{o1} < V_{o2} \Rightarrow I_{o1} < I_{o2}$$

In that case the behavior of the Argon gas is similar to the positive resistor that means that when the duty ratio increases the output current increases too.

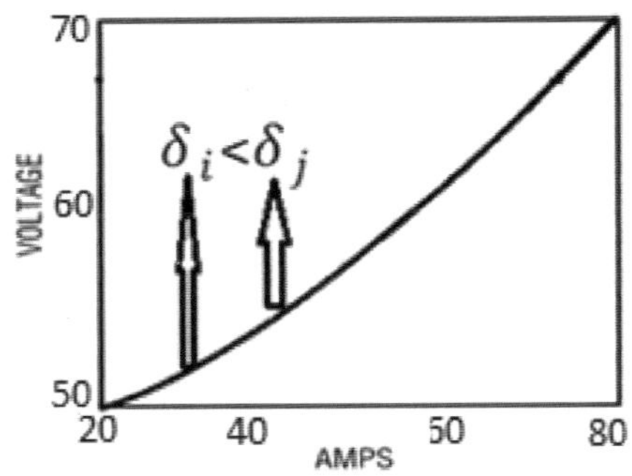

Figure 5. Plasma torches volt-ampere characteristics (Argon)

Reverse Mode (negative resistor)

Fig.6 shows the plasma torches volt-ampere characteristics for Nitrogen (N2) gas. Let us note that if $\delta_i > \delta_j$ and,

$$\delta_i \Rightarrow V_{oi} \Rightarrow I_{oi}$$

$$\delta_j \Rightarrow V_{oj} \Rightarrow I_{oj}$$

$$V_{oi} > V_{oj} \Rightarrow I_{oi} < I_{oj}$$

Then the load is the negative resistor. For example

$$\delta_1 \Rightarrow V_{o1} = 120V \Rightarrow I_{o1} = 25A$$

$$\delta_2 \Rightarrow V_{o2} = 110V \Rightarrow I_{o2} = 70A$$

$$V_{o1} > V_{o2} \Rightarrow I_{o1} < I_{o2}$$

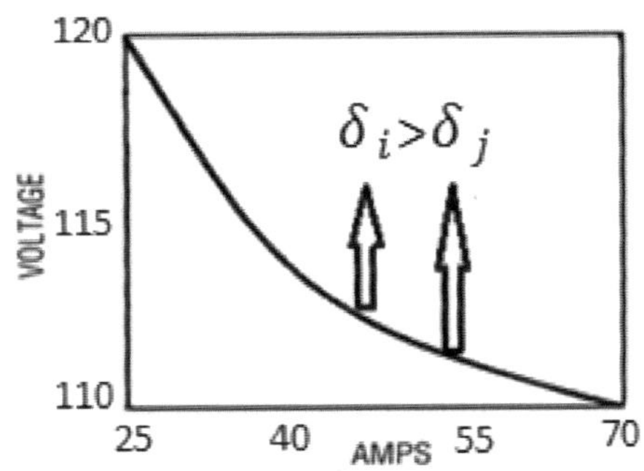

Figure 6. Plasma Torches volt-ampere characteristics (Nitrogen)

So, in this case the behavior of the Nitrogen gas (N2) is similar to the negative resistor that means that when the duty ratio increases the output current decreases.

4. Conclusion

The minimum output current consumption of Argon plasma torches was 20A at pulse width 2μsec, by increasing the pulse width up to 3μsec, the maximum output current consumption was 80A. The minimum output current consumption of Nitrogen plasma torches was 25A at pulse width 8μsec, by decreasing the pulse width down to 6μsec, the maximum output current consumption was 70A, so in short, Argon plasma torches considered as a positive load, and Nitrogen plasma torches considered as a negative load.

References

[1] Poki Chen, Yu-Kang Lo, Kuan-Cheng Chiu, "Intelligent high efficiency controller for 2-kW interleaved series resonant converters" IEEE Transactions on Industry Applications, 13564289, DOI: 10.1109/PEDS.2013.6527056, 2013

[2] Hakki. A., Luchkin. A.G, Rahimov. N.F, Sadikov. K.G, Luchkin. G.S, (2017) Journal of Physics: Conference Series, Volume 789, conference 1

[3] Y. C. Chuang et al, "Implementation and Analysis of an Improved Series- Loaded Resonant DC-DC Converter Operating Above Resonance for Battery Chargers," IEEE Transactions on Industry Applications, Vol. 45, pp.1052-1059, 2009

[4] Hakki.A , Sadikov. K, Kashapov. N, (2016) Journal of Physics: Conference Series

[5] D. Aggeler, J. Biela, and J. W. Kolar, "A compact, high voltage 25 kW, 50 kHz DC-DC converter based on sic jfets," in Proc. Twenty-Third Annual IEEE Applied Power Electronics Conference and Exposition APEC 2008, 24–28 Feb. 2008, pp. 801–807

[6] R. Severns (2005). Design of Snubbers for Power Circuits. Available from: http://www.cde.com/tech/design.pdf

[7] Hakki. A, Fayrushin. I, Kashapov. N, (2016) Journal of Physics: Conference Series, 669 (1), art no.012022..

[8] K. Venkatachalam, C. R. Sullivan, T. Abdallah, and H. Tacca, "Accurate prediction of ferrite core loss ith nonsinusoidal waveforms using only steinmetz parameters," in Proc. IEEE Workshop on Computers in Power Electronics, 3–4 June 2002, pp. 36–41

[9] Design of Planar Power Transformers. Ferroxcube, Available from: http://www.ferroxcube.com/appl/info/plandesi.pdf

[10] Hakki. A , Kashapov. N, (2015) IOP Conference Series: Materials Science and Engineering, 86 (1), art. no. 012002

Evolution of the shape of a high-current and moderate energy electron beam at its propagation through D2 at low pressure

Yu S Akishev[1,2], V B Karalnik[1], M A Medvedev[1], A V Petryakov[1], N I Trushkin[1], A G Shafikov[1]

[1] SRC RF TRINITI, 108840, Moscow, Troitsk, Pushkovykh street, 12, Russia
[2] NRNU MEPhI, 115409, Moscow, Kashirskoe shosse, 31, Russia

Abstract. The paper presents the experimental data on the time-spatial evolution of electron beam that is formed and propagates in a dielectric tube of large volume filled with D2 at low pressure of 1-2 Torr. The e-beam is generated by gas discharge in a short gap powered by the stepwise high voltage with amplitude up to 25 kV. Initially, this discharge appears in a strongly over-voltage regime which is unstable and eventually (in several hundreds of ns) transits into abnormal glow discharge at the voltage about of 1.5 - 2.5 kV and the current amplitude of several hundreds of Amps. The every of the mentioned discharge regimes generates the e-beam with energy up to 25 and 2.5 keV respectively. The ability of the over-voltage gap to generate the high- and low-energy e-beams happening one by one can be of great interest for many applications. The features in spatial structures of high-energy and low-energy e-beams in the course of their propagation over long distance through deuterium at low pressure are revealed and discussed.

1. Introduction

Strong over-voltage pulsed discharges at low pressures ($P < 10$ Torr) in a narrow gap ($d < 1$ cm) between the solid cathode and grid anode (so called "open discharges") operate under conditions which correspond to an appearance of run-away electrons which are able to form the electron beam. It means that it is possible to generate the electron beams not only in a vacuum with the use of the e-emissive cathodes but in gaseous atmosphere as well. The open discharges have a high efficiency in generation of the e-beams. Typically, the e-beam current corresponds to 80-90% of the total discharge current. At present the strong over-voltage pulsed discharges are widely used for generation of e-beams with an energy ε ranging from 1 to 70 keV [1]. As a rule, the increase in magnitude of parameters P, d, ε leads to strong diminishing in duration of both the over-voltage discharge and the e-beam generated by gas discharge.

If to speak about practical applications, it necessary to note that the high-energy e-beams (about 70 keV) generate the hard bremsstrahlung X-radiation which is bio-hazard and has a high penetrability. Due to that the installations using the e-beams of high energy (about of 100 keV) represent bio-danger at their service and demand the application of special measures of bio-protection. In contrast, the e-beam of 25 keV in energy does not generate the hard bremsstrahlung X-radiation. This is a reason why the gas discharges powered with voltage up to 25 kV and generating the high-current e-beams with energy up to 25 keV are of great interest.

In this paper the three-electrode system operating in D2 at lower pressure (1-2 Torr) and powered by the stepwise applied voltage with amplitude up to 25 kV is presented. This pulse discharge system exhibits two regimes (strong over-voltage discharge and anomalous glow discharge) and generates the sequence of two e-beams with energy up to 25 and 2.5 keV respectively which happen one by one. The features in spatial structures of high-energy and low-energy e-beams in the course of their propagation over long distance through deuterium at low pressure are revealed and discussed.

2. Experimental setup

Sketch of the used experimental setup is shown in figure 1. The electrical scheme is depicted schematically in Figure 1a. The pulsed over-voltage discharge has been formed in a narrow gap between the solid cathode (1) and the grid anode (2). The distance between these electrodes is equal to 3 mm. This discharge is ignited with a high voltage transferred by high-current thyratron T from the capacitor C1 through the ballast resistor R1 with a resistance of 44 Ohm. There was a complimentary discharge between the cathode and additional grid anode (3). The distance between cathode and the auxiliary anode is equal to 16 mm. The geometrical transparency of each grid was about 70%. The auxiliary low-current and low-voltage discharge was fed by the capacitor C2 through the ballast resistor R2 with a resistance of 1 MOhm and used to form preionization in the gap of the main discharge. High-voltage diode D prevents the transfer of a negative high voltage from capacitor C1 to the low-voltage capacitor C2.The applied high voltage was measured by HV divider PINTEK HVP-39 (1000:1, 40 kV, 200 MHz). The discharge current was measured by a low-inductive shunt with resistance of 0.024 Ohm. All electrical signals were recorded by the digital oscilloscopes such as Tektronix TDS 520, Tektronix TDS 2012 and Tektronix DPO2024.

Side view of a transparent long quartz tube with three-electrode system generating the sequence of two e-beams of high and low energy is shown in figure 1b. This tube was 11 cm in the inner diameter and 30 cm in length. The right-hand flange of a tube had also a quartz window to observe the image of the discharge and record its UV-Visual spectrum. The discharge and e-beam images were taken by digital camera Canon EOS 550 with the exposure time down to 10^{-4} s and multi-frame fast camera equipped with the intensifier with the exposure time down to 50 ns.

Before each experiment, the tube was pumped out up to the pressure of $P=10^{-3}$ Torr and then filled with deuterium of high purity (99.99%) up to the required pressure. To avoid the accumulation in the tube of the impurities happening due to plasma chemistry and inleakage of ambient air, the deuterium was constantly pumped over through a tube with a low gas flow rate.

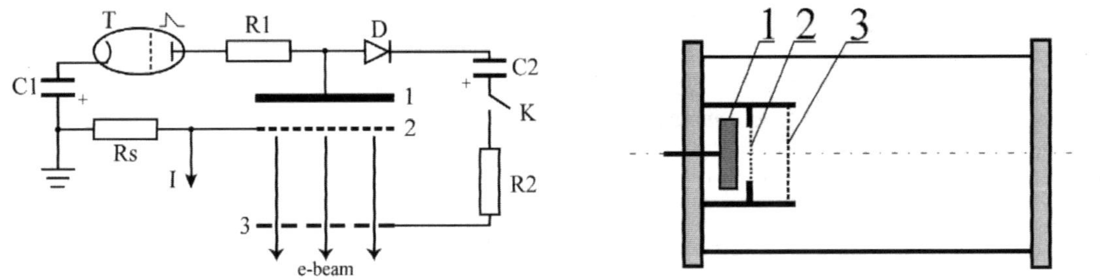

a) b)

Figure 1. a) The electrical scheme of the used experimental setup. 1 - the solid cathode; 2 - the grid anode of the main discharge; 3- the grid anode of the auxiliary discharge; R1=44 Ohm, R2 =1 MOhm, Rs = 0.024 Ohm; C1 = 12.5 nF, C2 = 100 μF, b) The sketch of side view of a transparent long quartz tube with three-electrode system generating the sequence of two e-beams of high and low energy.

3. Experimental results and discussion

The integrated in time images of the e-beam propagating within the tube filled with D2 at pressure of 2 Torr for different applied voltage to the discharge are presented in figure 2. One may see that the e-beam is more or less diffuse at the discharge exit (left-hand side in each photo) but transits to the constricted mode after propagation of some distance from the exit. This effect is more pronounced if the applied voltage and discharge current grow. As the discharge exists in two modes (over-voltage regime and abnormal glow regime) and each of these modes generates an electron beam (high-energy and low-energy e-beam), it is interesting to find out which of these two e-beams has been constricted. In order to do that it is necessary to take the sequence of e-beam images with a short exposure time (50 ns) and compare them with the images of discharge and its current-voltage waveforms at the same moments. The proper information has been collected and is presented in figures 3 and 4. To keep a high spatial resolution, we took the e-beam images separately for the first and second part of its length.

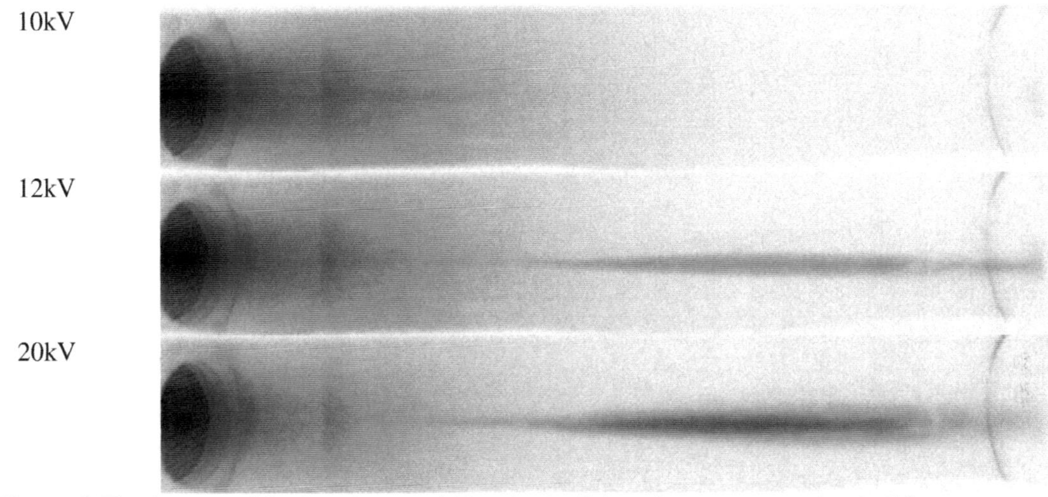

Figure 2. The integrated in time images (side view) of the e-beam propagating in D2 at pressure of 2 Torr for different applied voltage to the main discharge. The voltage of the auxiliary discharge is 1 kV. The e-gun exit is situated at left-hand side in each photo. Exposure time of the every picture is 0.5 s.

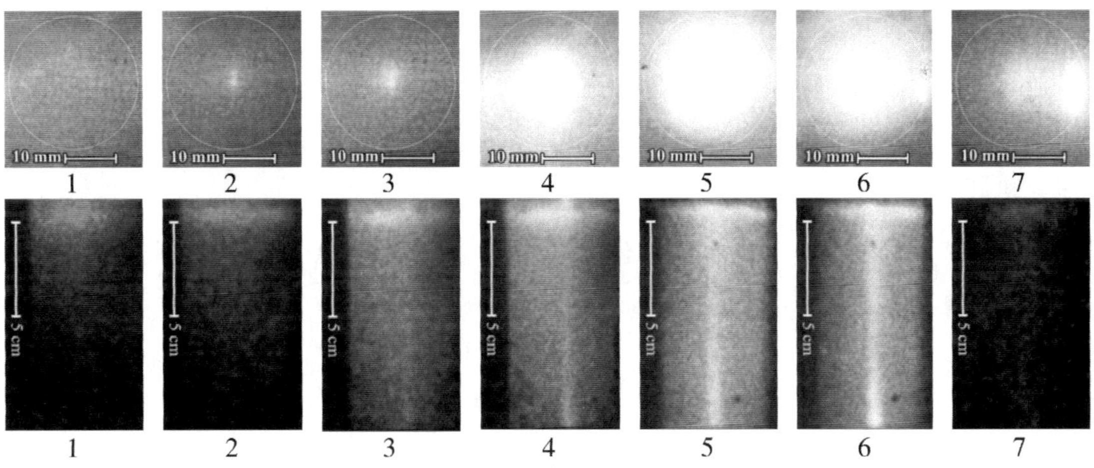

a)

IOP Conf. Series: Journal of Physics: Conf. Series **927** (2017) 012023 doi:10.1088/1742-6596/927/1/012023

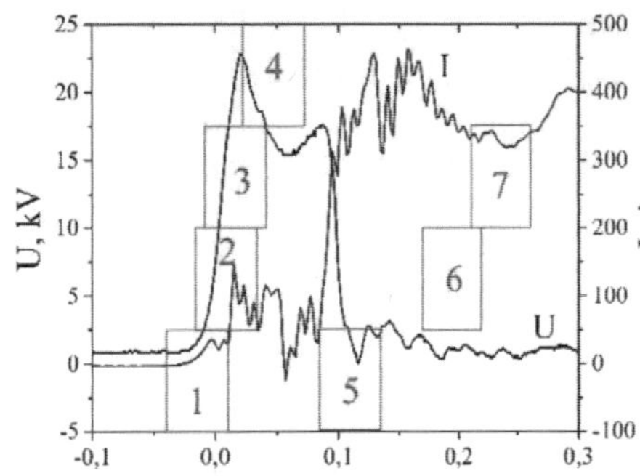

b)

Figure 3. The series of images showing the evolution of both the discharge and e-beam (a) which are correlated with the current-voltage waveform of the discharge (b). The e-beam images correspond only to the first half of the distance travelled by e-beam in the tube. The location on time axis and width of the enumerated squares in figure (b) correlate with the moments and exposure time of the discharge images (upper series) and e-beam images (bottom series) enumerated with the same figures. The voltage amplitude applied to the main discharge is 20 kV. Gas pressure is 2 Torr. Auxiliary voltage is 1 kV.

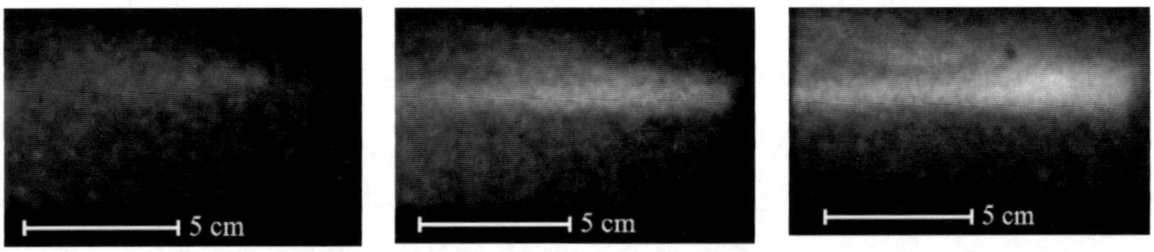

3 4 5

Figure 4. The sequence of the e-beam images corresponding to the second half of the distance travelled by e-beam in the tube. Exposure time of the every shot is 50 ns. The enumeration of the shots correlates with that in figure 3. The discharge parameters are the same as those in figure 3.

Close examination of figures 2-4 reveals that the transverse inhomogeneity of the e-beam which looks like its partial constriction in vicinity of the axis, in fact, is associated at the beginning with a weak transverse inhomogeneity of the over-voltage discharge. However, after the transition of the over-voltage discharge into the transverse uniform abnormal glow regime with a high current but lower voltage, the e-beam partial constriction increases appreciably in spite of the fact that the glow discharge keeps its homogeneity in the cross direction. We suppose that this increase happens due to the pinch-effect produced by the very high amplitude of the e-beam current (about 400 Amps). Eventually, low-energy e-beam disappears after transition of abnormal glow cathode layer into low-voltage pre-arc cathode spot with high current density [2] (see shot #7 in figure 3).

4. Conclusion

In order to generate the pulsed e-beam, the three-electrode system operating in D2 at lower pressure (1-2 Torr) and powered by the stepwise applied voltage with amplitude up to 25 kV was used. It is shown that this pulse discharge system exhibits two regimes (strong over-voltage discharge and anomalous glow discharge with lower voltage) and generates the sequence of two e-beams with energy up to 25 and 2.5 keV respectively. The ability of the over-voltage gap to generate the high- and low-energy e-beams happening one by one opens new opportunities which can be of great interest for many applications. It is revealed that the high-energy e-beam is more homogeneous compared to the low-energy e-beam. Weak expressed partial constriction of the high-energy e-beam in vicinity of the axis is associated with a weak transverse inhomogeneity of the over-voltage discharge. After transition of the over-voltage discharge into abnormal glow regime with a high current, the partial constriction of low-energy e-beam increases appreciably in spite of the fact that the glow discharge becomes more uniform in the cross direction. We suppose that this increased constriction happens due to the pinch-effect produced by the very high current (about 400 Amps) of the low-energy e-beam. Our findings promote more insight into physics of the high-current over-voltage discharge used for generation of the e-beams with energy up to 25 keV.

5. References

[1] Sorokin A and Bokhan P 1994 *Technical Physics Letters* **20** 86
[2] Akishev Yu, Karal'nik V, Kochetov I, Napartovich A and Trushkin N 2014 *Plasma Sources Sci. Technol.* **23** 054013

Acknowledgments

This work has been carried out thanks to full financial support by the Russian Science Foundation (Grant № 16-12-10458).

IOP Conf. Series: Journal of Physics: Conf. Series **927** (2017) 012024 doi:10.1088/1742-6596/927/1/012024

Difference in shape and width of Dα lines in the overvoltage and abnormal glow regimes of open discharge in narrow gap

Yu S Akishev[1,2], V B Karalnik[1], M A Medvedev[1], A V Petryakov[1], N I Trushkin[1], A G Shafikov[1]

[1] SRC RF TRINITI, 108840, Moscow, Troitsk, Pushkovykh street, 12, Russia
[2] NRNU MEPhI, 115409, Moscow, Kashirskoe shosse, 31, Russia

Abstract. Strong overvoltage discharges in a narrow gap between the solid cathode and grid anode (so called "open discharges") are widely used for generation of the high-current electron beams. At low gas pressure (about 1 Torr) we revealed that discharge in the overvoltage gap stressed by stepwise voltage with amplitude up to 25 kV exhibits two regimes which follow one by one and generate the e-beams. The first of them produces the high-energy e-beam but the second generates the low-energy e-beam. The physical properties of these gas discharge regimes have been explored insufficiently. We have done the spectroscopic measurements of these regimes in D2 and found out that the overvoltage regime produces positive ions with high kinetic energy up to several keV but anomalous glow discharge forms the ions with energy of several tens of eV. The existence of high-energy ions in the overvoltage discharge provides a strong increase in the electron secondary emission from the cathode and possibly contributes to the ionization of the gas by fast ions. Our findings promote more insight into physics of the overvoltage discharge generating the high-current e-beams with energy up to 25 keV.

1. Introduction

Strong overvoltage pulsed discharges operating at low pressures (P < 10 Torr) in a narrow gap (d < 1 cm) between the solid cathode and grid anode are often called as "open discharges". The reduced electric field E/N in these discharges reaches of huge magnitude of 10^5 Td and even more. Such electric fields correspond to an appearance of run-away electrons which are able to form the high-current electron beams. The open discharges have a high efficiency in generation of the e-beams. As a rule, the e-beam current corresponds to 80-90% (or higher) of the total discharge current. At present the strong over-voltage pulsed discharges are widely used for generation of e-beams with energy up to 70 keV [1]. However, it necessary to note that the high-energy e-beams (about 100 keV) generate the hard bremsstrahlung X-radiation which has a high penetrability and brings a serious bio-danger. Due to that the installations using the e-beams of high energy (about of 70 keV) represent bio-danger at their service and demand the application of special measures of bio-protection. In contrast, the e-beam of 25 keV in energy does not generate the hard bremsstrahlung X-radiation. This is a reason why the gas discharges powered with the voltage up to 25 kV and generating the high-current e-beams with energy up to 25 keV are of great interest to practical applications. Despite the importance of these discharges for practice, their physical properties have been explored insufficiently.

In this paper, the novel configuration of the pulsed open discharge is presented. We used the three-electrode system operating in D2 at lower pressure (1-2 Torr) and powered by the stepwise applied voltage with amplitude up to 25 kV. In this electrode system, gas discharge exhibits two regimes (strong overvoltage discharge and abnormal glow discharge supported by 25 and 2.5 kV) which happen spontaneously one after another. These regimes were investigated by spectroscopically.

2. Experimental setup

The electrical scheme of the three-electrode system used for generation of the overvoltage discharge is depicted schematically in Figure 1a. The pulsed over-voltage discharge has been formed in a narrow gap between the solid cathode (1) and the grid anode (2). The distance between these electrodes is equal to 3 mm. This discharge is ignited with a high voltage transferred by high-current thyratron T from the capacitor C1 through the ballast resistor R1 with a resistance of 44 Ohm. There was a complimentary discharge between the cathode (1) and additional grid anode (3). The distance between cathode and the auxiliary grid anode is equal to 16 mm. The geometrical transparency of each grid was about 70%. The auxiliary low-current and low-voltage discharge was fed by the capacitor C2 through the ballast resistor R2 with a resistance of 1 MOhm and used to form the preionization of gas in the gap of the main discharge. High-voltage diode D prevents the transfer of a negative high voltage from capacitor C1 to the low-voltage capacitor C2.The applied high voltage was measured by HV divider PINTEK HVP-39 (1000:1, 40 kV, 200 MHz). The discharge current was measured by a low-inductive shunt with resistance of 0.024 Ohm. All electrical signals were recorded by the digital oscilloscopes such as Tektronix TDS 520, Tektronix TDS 2012 and Tektronix DPO 2024.

Figure 1b shows the lay-out of the components of the experimental setup used for spectroscopic measurements. The optical system was focused in such a way in order to collect the light either from the discharge zone including the cathode area or from the exit of discharge zone in the transverse direction. To observe the review UV and visual spectrum of the discharge, the spectrometer Avantes AvaSpec-2048-USB2 was used. Its spectral resolution is as follows: 0.11–0.16 and 0.2–0.3 nm in region of 351–393 and 603–810 nm respectively. To investigate the shape of D_α spectral line in the overvoltage discharge and anomalous glow discharge, the double monochromator equipped with grids was used with a spectral resolution of 0.01 nm. The monochromator recorded the emission of D_α spectral line through each 0.05 nm throughout the width of this line. Based on this information, the spectral line shape was reconstructed. To determine the real shape of the spectral line, the slit function of monochromator was taken into account. The excitation electron temperature in discharge was determined from measurement of the ratio of intensities of the spectral lines of Balmer series.

The three-electrode system was mounted inside the quartz tube. Before each experiment, the tube was pumped out up to the pressure of $P=10^{-3}$ Torr and then filled with deuterium of high purity (99.99%) up to the required pressure. To avoid the accumulation in the tube of the impurities happening due to plasma chemistry and inleakage from ambient air, the deuterium was constantly pumped over through a tube with a low gas flow rate.

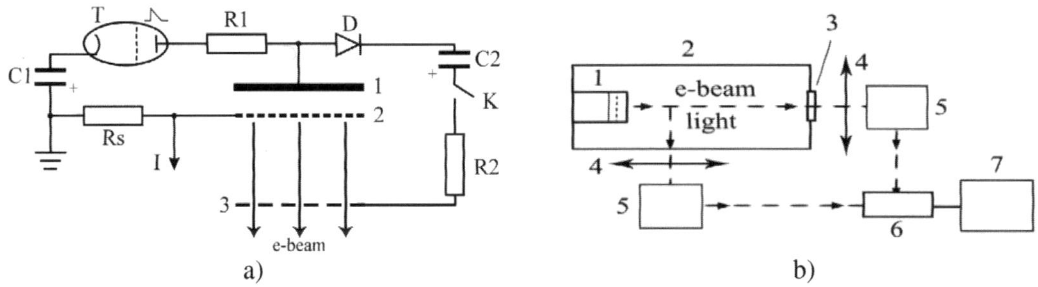

Figure 1. a) The electrical scheme of the three-electrode system generating a high-current e-beam with energy up to 25 keV. 1 - the solid cathode; 2 - the grid anode of the main discharge; 3- the grid anode of the auxiliary discharge; R1 = 44 Ohm, R2 =1 MOhm, Rs = 0.024 Ohm; C1 = 12.5 nF, C2 = 100 μF. b) scheme of spectral measurements: 1-the e-beam gun, 2-the quartz tube, 3-the output quartz window, 4-the quartz lens, 5-the monochromator or spectrometer, 6-PMT, 7-the oscilloscope.

3. Experimental results and discussion

The typical current-voltage waveform of the overvoltage discharge in short gap filled with D2 at pressure P=2 Torr and excited by the stepwise voltage with amplitude of 20 kV is shown in figure 2. Note that the high-voltage regime with the current growing up to 120 A is unstable and exists only up to 180 ns and then abruptly transits into abnormal glow regime with the discharge current about 300 Amps. Figure 3 presents the review UV-Visual spectra of the discharge light collected from the gap including the cathode area (a) and in transverse direction from zone close to the discharge exit (b), i.e. excluding the cathode area.

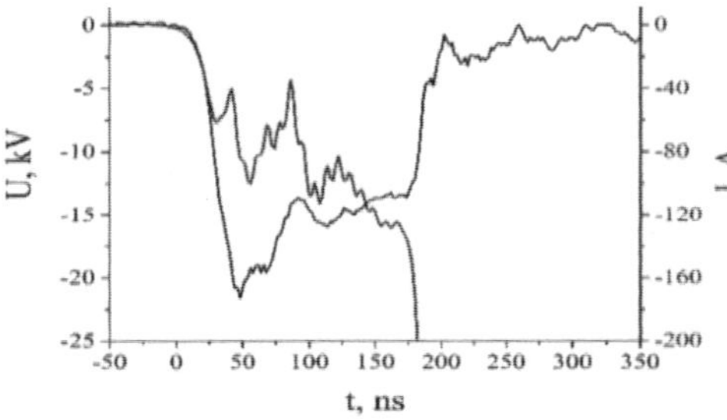

Figure 2. The example of typical current-voltage waveform of the overvoltage discharge in D2.Gas pressure is P=2 Torr, the amplitudes of the applied stepwise and auxiliary DC voltages are 20 and 1 kV. The high-voltage regime exists up to 180 ns and then abruptly transits into abnormal glow regime.

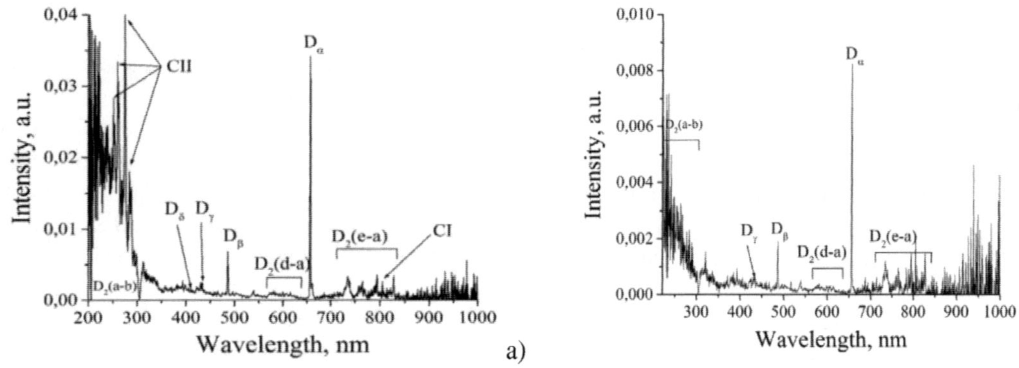

Figure 3. The review UV-Visual spectra of the discharge light collected from the gap including the cathode area (a) and in transverse direction from zone close to the discharge exit (b), i.e. excluding the cathode area. D2, P=2 Torr, U=20 kV.

One may see in figure 3 that gas discharge emits not only the atomic lines of Balmer series D_α, D_β, D_γ, D_δ but molecular bands D_2(a-b), D_2(d-a), D_2(e-a) as well. Note the intensities of all lines emitted from the discharge zone including the cathode area (a) exceed by the factor about 4 the lines emitted from the zone excluding the cathode area (b). Besides, there are two features in spectrum (a): 1) there are no spectral lines of the cathode material (Fe) proving that there is no cathode local melting. It means that there are no high-current and hot arc spots on the cathode; 2) there are spectral lines of carbon in the cathode region - we suppose they appear due to the decarburization of steel that occurs without its melting but if local heating exceeds 700°C.

The emission of spectral lines of Balmer series in the anomalous glow regime is produced predominantly by plasma located in vicinity of the anode. Spectral information about intensities of the D_α, D_β, D_γ, D_δ lines (see figure 4a) similar to that presented in figure 3a was used to determine so called the excitation temperature T_{exc} of electrons in plasma of anomalous glow discharge. This temperature is determined from Boltzmann plot for Balmer series in accordance with this formula:

$$kT_{exc} = -\frac{E_i - E_j}{\ln\left(\frac{I_i\lambda_i}{g_iA_i}\cdot\frac{g_jA_j}{I_j\lambda_j}\right)}$$

where E_i, E_j and I_i, I_j are the energies of i and j levels and the intensities of the lines from these levels; g_i, g_j, A_i, A_j, λ_i and λ_j are the statistical sums, Einstein coefficients and the wave lengths corresponding to transitions from these levels. The obtained data for T_{exc} vs gas pressure P are presented in figure 4b.

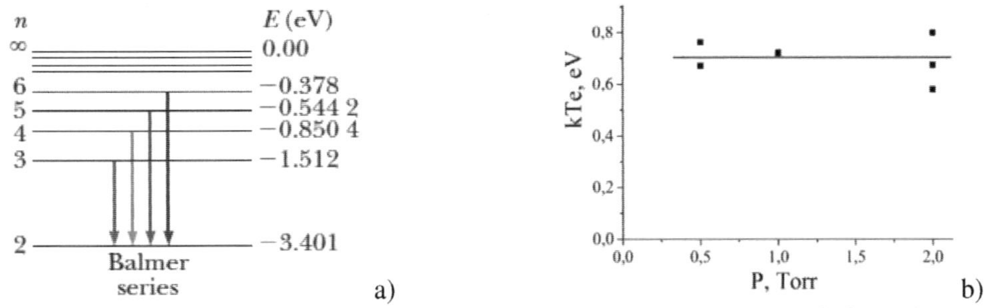

Figure 4. The scheme of Balmer series (a), D_α: (n=3 → n=2); the obtained dependence of the excitation temperature T_e of electrons in plasma near anode vs gas pressure (b). Gas D2, U = 20 kV.

The number density of electrons n_e in D2 plasma of abnormal glow open discharge at P=2 Torr was estimated with use of Saha equation for Te ≈ 0.7 eV. We have got that n_e is equal approximately to $3x10^{13}$ cm^{-3}. This magnitude in a reasonably concordance with the estimation obtained from the known current density and electron drift velocity in plasma of abnormal glow regime.

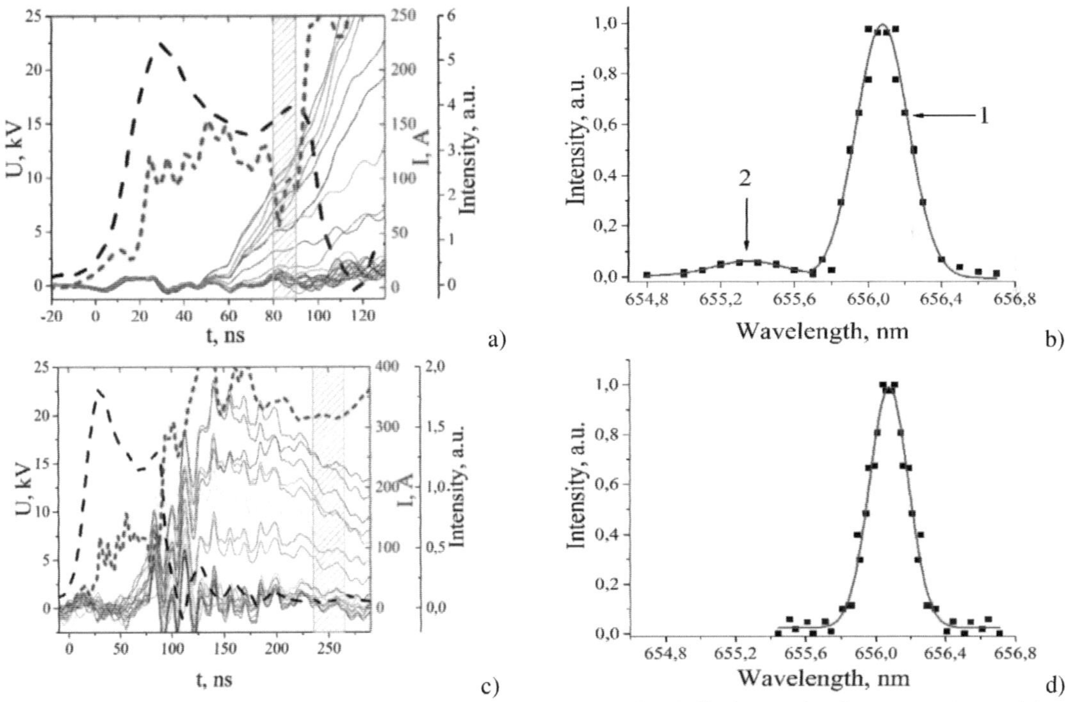

Figure 5. The current-voltage waveforms of the open pulsed discharge in the overvoltage (a) and abnormal glow (c) regimes and the shapes of D_α spectral lines (b) and (d) corresponding to (a) and (c) regimes. D2, P=2 Torr, U=22 kV. Average current in regimes (a) and (c) is 120 and 320 Amps. Spectral lines 1 and 2 in (b) correspond to the emission of fast excited atoms formed due to: (1) the charge exchange of fast ions and (2) impact the cathode by fast ions (see figure 6a).

All experimental profiles were well approximated with Gauss formula. It means that Doppler effect makes the main contribution to the formation of the line broadening. Note that the shape of D_α

emission in the overvoltage regime is very unusual for gas discharge plasma at moderate electric fields. In the case of super high the reduced electric field $E/N \approx 10^5$ Td, D_α emission is characterized by the existence of a wing corresponding to emission of very fast excited D atoms (figure 5b). As a rule, such wing is observed in the emission produced by the impact of high-energy ion beams to solid surface (see figure 6a and paper [2]). Based on this, we can state that the wing in figure 5b is formed due to the impact of fast positive ions with the cathode. It leads to neutralization of some ions and their reflection in a form of fast and excited D atoms. However, the symmetrical profile of D_α line is formed by emission of fast excited atoms happened mainly due to the charge exchange of fast ions and other reactions [3] listed in figure 6b.

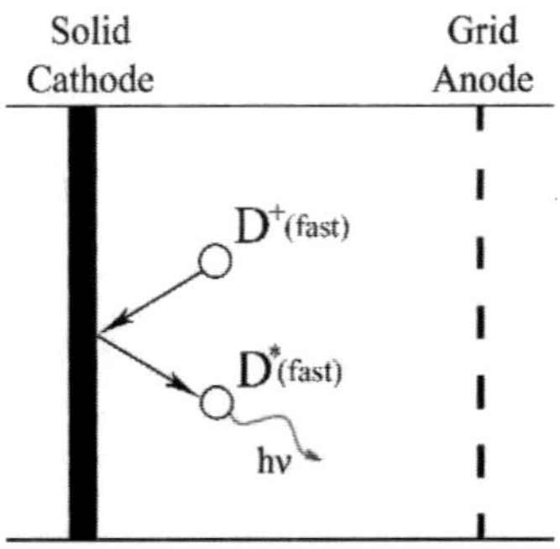

a)

$D^+(fast) + D_2\ (slow) \rightarrow D^*(fast) + D_2^+$
$D_2^+(fast) + D_2 \rightarrow D^*(fast) + D + D_2^+$
$D_3^+(fast) + D_2 \rightarrow D^*(fast) + D_2 + D_2^+$
$D(fast) + D_2 \rightarrow D^*(fast) + product$
$D_2(fast) + D_2 \rightarrow D^*(fast) + D + product$
$D^+(fast) + D \rightarrow D^*(fast) + D^+$
$D_2^+(fast) + D \rightarrow D^*(fast) + D + D^+$
$D_3^+(fast) + D \rightarrow D^*(fast) + D_2 + D^+$
$D(fast) + D \rightarrow D^*(fast) + product$
$D_2(fast) + D \rightarrow D^*(fast) + D + product$

b)

Figure 6. a) the sketch clarifying the appearance of fast excited D atoms due to impact of high-energy ions with the cathode; b) the set of gas-phase reactions producing the fast excited D atoms in plasma.

In the overvoltage regime, fast run-away electrons quickly leave the gap. Therefore the space charge in the gap is formed predominantly by positive ions. In this case, the electric field strength increases towards the cathode. Because of that the ions in vicinity of the cathode will have maximum kinetic energy and produce the fast excited atoms practically with the same energy. Close examination of the spectroscopic data like those presented in figure 5 has led us to the conclusion that the overvoltage open discharge fed with the voltage of 20-25 kV produces the positive ions with kinetic energy up to 3 keV (or higher) at the cathode and fast excited atoms in the bulk of the gap due to the charge exchange with the energy of 0.3±0.1 keV. In the case of abnormal glow regime, the kinetic energy of fast excited atoms is estimated not higher of 100 eV.

4. Conclusion

Spectroscopic investigation of the open discharge powered with a high voltage up to 25 kV in D2 at gas pressure of 2 Torr revealed that the overvoltage regime of this discharge produces the positive ions with kinetic energy up to 3 keV (or higher) at the cathode and fast excited atoms in the bulk of the gap due to the ion charge exchange with the energy of 0.3±0.1 keV. In the case of abnormal glow regime, the kinetic energy of fast excited atoms is estimated not higher of 100 eV.

5. References

[1] Sorokin A and Bokhan P 1994 *Technical Physics Letters* **20** 86
[2] Thomas E 1980 *Progress in Surface Science* **10** 383
[3] Barbeau C and Jolly J 1990 *J. Phys. D: Appl. Phys.* **23** 1166

Acknowledgments

This work has been carried out thanks to full financial support by the Russian Science Foundation (Grant № 16-12-10458).

Improvement of Technological Properties of a Vegetal Tanning Agent in Gas Discharge Plasma

A K Khairullin[1], E F Voznesensky[1], G R Rakhmatullina[1], A M Sabirov[1], N V Tikhonova[1]

[1]Kazan National Research Technological University, 68 Karl Marx street, Kazan 420015, Russian Federation

E-mail: beckhenkel@mail.ru

Abstract. The article considers the possibility of modification of the vegetal tanning agent quebracho in the plasma of a radio-frequency induction discharge at low pressure. It is established that plasma treatment leads to a decrease in the size of colloid fractions and a decrease in the molecular weight, while the functionality of the vegetal tanning agent is preserved. Application of the obtained product in the process of leather retanning allows improving its physical, mechanical and consumer properties.

1. Introduction

Natural tannin-containing vegetal extracts, such as oak, mimosa, quebracho, have a good binding ability, therefore they are used to structure collagen fibers in the production of the leather. However, vegetal tanning agents are significantly inferior to toxic chromium as to their ability to penetrate in the dermis, because of the large molecular weight and the formation of large aggregates in solution.

It is proposed to activate the dry tanning vegetal extract of quebracho in the conditions of a radio-frequency induction (RFI) discharge at low pressure in order to improve its technological properties. The treatment of the plant extract was carried out on the equipment described in the source [1], the following parameters of plasma treatment were used: plasma-forming gas – argon, pressure in the working chamber 30 Pa; consumption of argon 0,04 g/s; energy of plasma ions 10–15 eV; the ion current density 20 A/m^2; the dry material was injected into the discharge with a stream of plasma-forming gas and trapped after processing.

2. Experimental studies

Technological properties of vegetal tanning agent mainly depend upon the ability of molecules to diffuse in the dermis structure, which is determined by the molecular weight and the size of colloid fractions in solution.

A comparison of the infrared spectra of the control and experimental samples does not reveal any significant differences, there are small fluctuations in the intensity of absorption bands. On this basis, one can conclude that treatment does not significantly affect on the content of active groups of the sample and does not cause its complete destruction.

It was experimentally established that the plasma modification leads to a decrease in the moisture content of the sample from 5,85 % to 0,45 %.

IOP Conf. Series: Journal of Physics: Conf. Series **927** (2017) 012025 doi:10.1088/1742-6596/927/1/012025

When the control and experimental samples are dissolved in the water, a certain deterioration of the solubility is observed in the experimental sample, probably due to dehydration in conditions of radio-frequency plasma. The solution of the experimental sample has a visually more intense colour. The initial sample forms a less coloured turbid solution.

The values of the intrinsic viscosity of solutions of the experimental sample are 2,7 times lower than those of the control sample. This may indicate a decrease in the molecular weight due to partial destruction in conditions of radio-frequency plasma.

The size distribution of colloid fractions was determined using the ZetaPALS 90Plus Nanoparticle Size Analyzer (Brookhaven, USA) (Fig. 1).

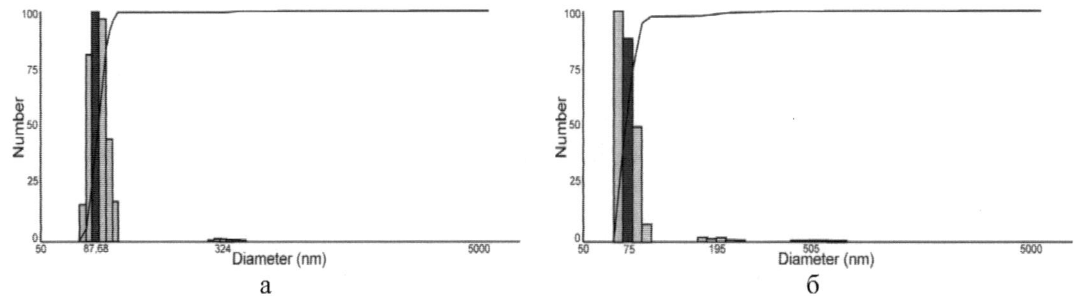

Fig. 1. Histograms of the particle size distribution in the solution of quebracho samples:
a – control, b – experimental

The results show that the size of tanning agent colloid fractions decreases from 68,94–98,13 nm and 367,33–502,88 nm in the control sample to 48,37–68,94 nm and 260,28–405,29 nm in the experimental sample.

The molecular weight distribution was determined using a MALDI UltraFlex III TOF/TOF mass spectrometer (Bruker Daltonik GmbH, Bremen, Germany).

MALDI mass spectra are obtained in a linear mode. The Nd: YAG laser (λ = 266 nm) was used. Data was processed using FlexAnalysis 3.0 (Bruker Daltonik GmbH, Bremen, Germany). Positively charged ions were fixed. A metal target was used. 2,5-Dihydroxybenzoic acid (DHB) or para-nitroaniline (p-NA) was used as a matrix. 0,5 μl of a 0,1 % solution of the matrix in CH_3CN and 0,5 μl of the sample were successively applied to the target and evaporated (Fig. 2, 3).

The molecular weight range using the DHB matrix is 200–1000 for the control and experimental samples. However, in the samples, two predominant fractions can be identified ($I > 400$ arb. units): 311,16; 545,13 in the control and 255,23; 375,20 in the experimental.

The range of molecular weights using the p-NA matrix is 300–1000 in the control, 400–1000 in the experimental sample. In the samples, it is also possible to identify the predominant fractions: 685,78 ($I > 1250$ arb. units); 686,74 ($I > 500$ arb. units) in the control and 360,41 ($I > 400$ arb. units) in the experimental. So, the MALDI mass spectra show a change in the dispersity of molecules, a decrease in the molecular weight of the predominant fraction of the vegetal tanning agent in the conditions of RFI treatment.

To study the diffusion of vegetal extract molecules, the retanning process is performed with control and experimental samples in laboratory conditions. The retanning process of the semi-finished sheepskin using the modified tanning agent should be carried out at pH = 4; in this case the heat resistance of the leather increases by 9 °C; the strength limit – by 15 %; the relative elongation – by 50 %. The obtained leather also has improved properties: softness and uniform colour of the surface.

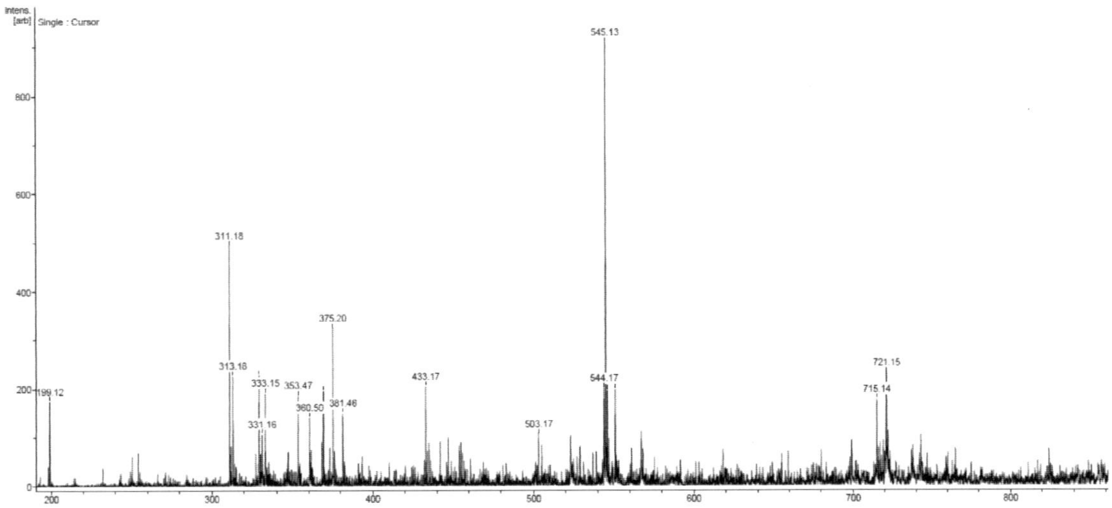

a

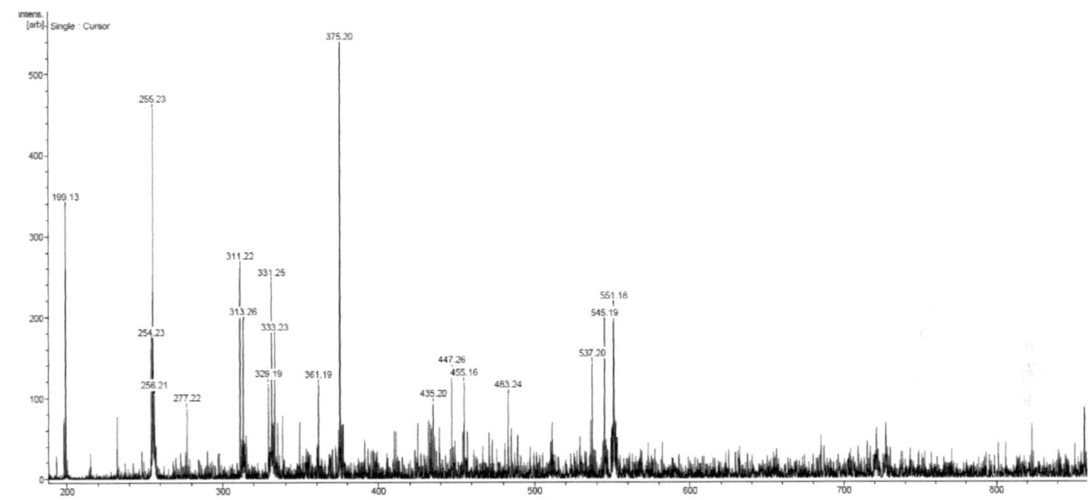

b

Fig.2. MALDI mass spectra of quebracho samples using the matrix DHB:
a – control, b – experimental

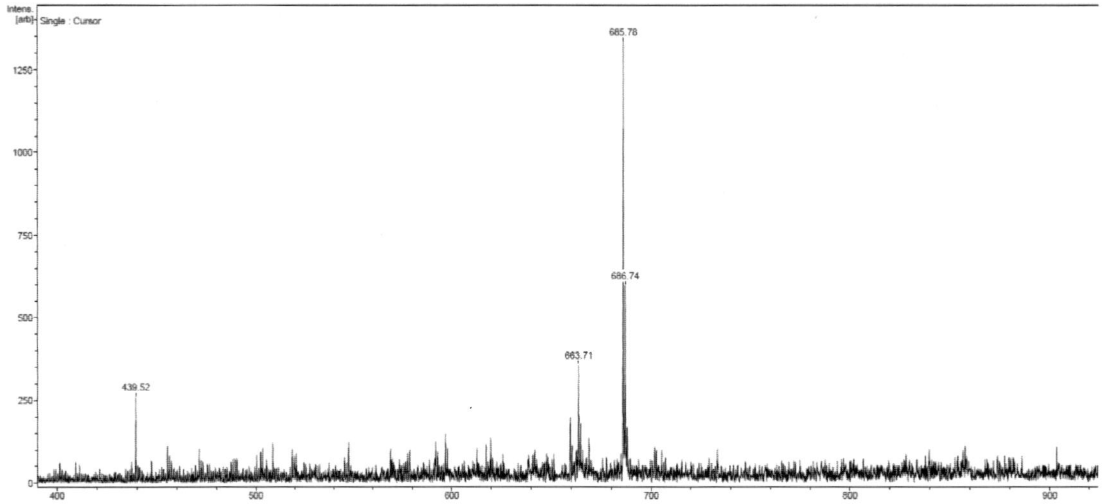

a

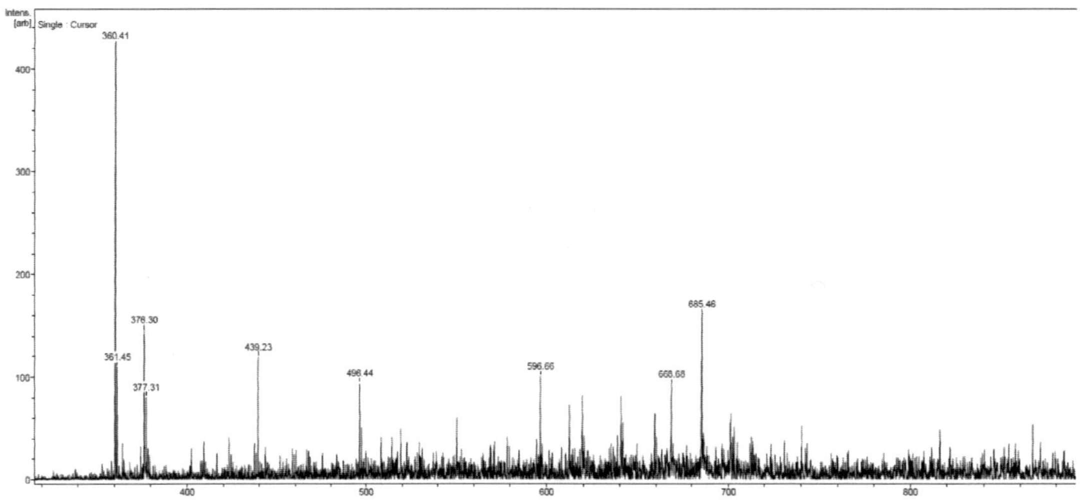

b

Fig.3. MALDI mass spectra of quebracho samples using the matrix p-NA:
a – control, b – experimental

3. Conclusion

Thus, the results indicate the production of light fraction of the vegetal tanning agent with improved technological properties from the plant extract in conditions of low-pressure RFI-plasma.

Acknowledgments

The plasma processing was performed on radio-frequency induction installation of Kazan National Research Technological University.

References

[1] Abdullin ISh, Zheltukhin VS, Kashapov NF. Radio-frequency plasma-jet treatment of materials at low pressures. Theory and practice of application. Publishing house of Kazan State University, 2000.

Materials surface smoothing to sub-nanometer level of roughness by argon cluster ion beam

N G Korobeishchikov, I V Nikolaev and M A Roenko

Novosibirsk State University, Pirogova str. 2, Novosibirsk, 630090, Russia

E-mail: korobei@ci.nsu.ru

Abstract. In this study, the influence of the accelerated argon cluster ion beam on the topography of stainless steel and optical glass surfaces has been investigated. The possibility of smoothing the material`s surface to the sub-nanometer level of roughness is shown. The dose dependencies of the root-mean-square roughness and power spectral density function are obtained. It is demonstrated that by the cluster ion treatment the effective roughness decreases at different ranges of spatial frequencies. Anisotropic structure of the treated material effects on the obtained surface roughness.

1. Introduction

To date, there are many different applications for potential technologies running on a sub-nanometer level, which, in turn, requires new parameters of materials that do not match their initial properties. The surface roughness is one of the most important physical characteristics of the material. It is assumed that thanks to the newest technologies, atomically smooth surfaces will be obtained without disturbing the structure and physicochemical contamination of the material being processed. One of the most promising methods of surface polishing is smoothing with the help of cluster ion beams formed from supersonic gas jets. Pollution of the processed surface is completely excluded due to the use of high vacuum condition and non-reactive gas, for example, argon. So-called lateral sputtering of the material occurs due to the interaction of the large number of the cluster particles with approximately the same amount of the target particles, which makes it possible to smooth out the surfaces of various materials [1, 2].

2. Surface topography evolution by cluster ions influence

2.1. Experimental details

The experiments were performed on equipment, a detailed description of which is given in [3]. In this work, we used the cluster ion beam formed from a supersonic argon jet [4]. The processing was carried out with the mean cluster ion size 1200 atoms/cluster and the cluster ion energy of 22 keV at normal angles of incidence of the cluster ions to the target. The flat samples of the stainless steel having initial mirror surface and the industrial polished optical glass were used. To observe the surface morphology before and after treatment, the atomic force microscope (AFM) Ntegra Prima HD was used. The AFM measurements were carried out at different scan sizes from 10 ×10 μm to 40 ×40 μm with a resolution of 1024 × 1024 pixels.

2.2. Stainless steel smoothing

The results of polishing by argon cluster ions of the stainless steel are illustrated in Figure 1, which shows the 3D images of surface at different treatment stage. The scan area was 20 × 20 μm, the maximum irradiation dose was 2.8×10^{17} ion/cm^2. The average surface roughness R_q was determined from four measurement points. There is a noticeable smoothing of surface irregularities. At each processing stage, the root-mean-square roughness R_q decreases by more than 2 times.

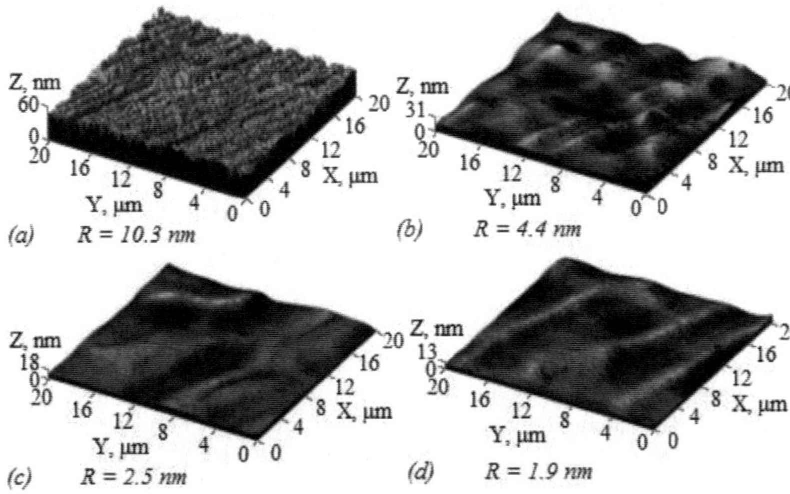

Figure 1. AFM images of the stainless steel surface before and after cluster ion polishing: (a) initial sample, (b) 1st treatment, (c) 2nd treatment, (d) 3hd treatment.

Figure 2 shows the dependences of the average roughness R_q of the stainless steel surface on the irradiation dose for different size of the scan area. It can be seen that of the surface roughness is a significantly decreased at all scales. At scan area of less than 10 ×10 μm the R_q roughness decreases to 0.7 nm and shows a tendency to further decrease with increasing dose. At the same time, at a characteristic size of the scan area above 10 × 10 μm, a restriction of the roughness R_q at a level of 2-5 nm was obtained. As is known, stainless steel is an alloy of several chemical components: iron, chrome, nickel, etc. Obviously, the chemical elements are distributed in stainless steel in the form of micro-granules, have different physical hardness, respectively, different sputtering coefficients. This leads to a limitation of the decrease in roughness in the polishing of the surface of stainless steel by cluster ions at low spatial frequencies.

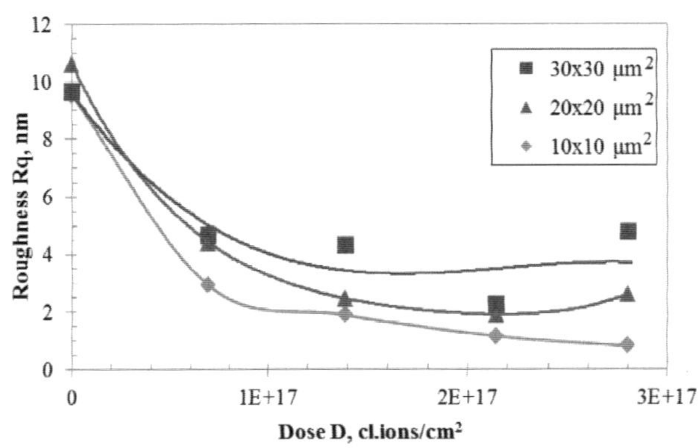

10

Figure 2. Surface roughness of stainless steel as a function of the irradiation dose at different scan area.

More complete information about the surface roughness can be obtained by used the power spectral density (PSD) function, which describes the height distribution depend on the spatial frequency, taking into account lateral roughness values. The spatial frequency v is defined as the reciprocal of the distance between the measurement points on the surface profile [5].

Figure 3 shows the PSD functions for stainless steel before and after treatment for scan area 20×20 μm^2. The sharp peaks of the original curve indicate periodically repeating values of the heights of the surface profile relative to the mean value. These peaks correspond to scratches from the previous mechanical polishing of the sample. After processing the surface roughness is reduced at all spatial frequencies: in the middle frequency region ($\nu \sim 1$ μm^{-1}) by 3 orders of magnitude, in the high frequency ($\nu \sim 10$ μm^{-1}) by 2 orders of magnitude, in the low frequency ($\nu \sim 0.1$ μm^{-1}) is less than 1 order. The PSD function after 3rd treatment differs little from the 2nd treatment.

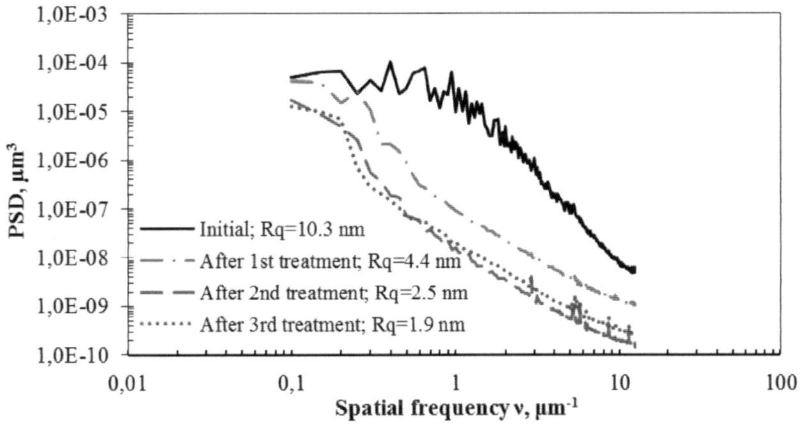

Figure 3. PSD functions of stainless steel surface roughness before and after cluster ion polishing.

2.3. Optical glass smoothing

Also, the experiments were performed with samples of industrial optical glass LK having the diameter of 25 mm and the thickness of 10 mm without preliminary treatment. 3D AFM images of the optical glass surface before and after cluster ion polishing are shown in Figure 4. The scan area was 10×10 μm. The total height difference decreased from 20 nm to 9 nm after processing. Before processing, the surface of the optical glass had a granular structure with a characteristic size of about 1 μm, but after processing such a structure is not observed.

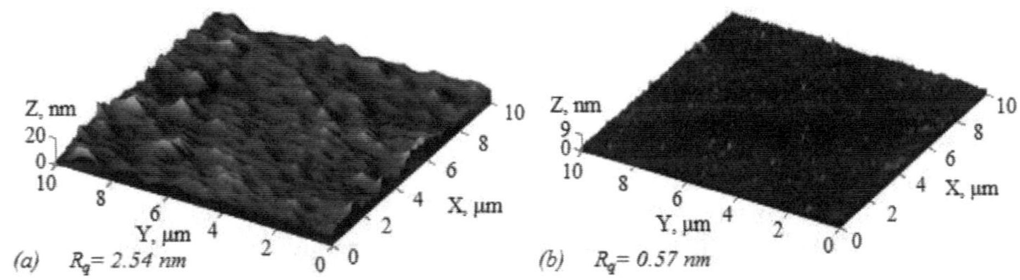

Figure 4. AFM images of the optical glass surface: (a) before processing, (b) after processing.

PSD functions and R_q values of the optical glass surface initial and after processing by argon cluster ions are shown in Figure 5. The surface roughness of optical glass is reduced at all spatial frequencies after processing: in the high-frequency ($\nu \sim 10\ \mu m^{-1}$) range is less than 1 order, in the low-frequency ($\nu \sim 0.2$-$0.3\ \mu m^{-1}$) by 2 orders of magnitude. The PSD function is reduced by 1 order of magnitude in the mid-frequency range ($\nu \sim 1\ \mu m^{-1}$) but there is a pronounced peak on the "after treatment" curve. Presumably, this peak characterizes the frequency of periodic low-dimensional scratches, which remained from preliminary mechanical polishing of the sample. Comparing Figure 3 and Figure 5, it can be noted that the PSD function of the glass decreases by 2 orders of magnitude after one treatment, whereas the PSD function of the steel decreases by 1 order after 2-3 treatments only.

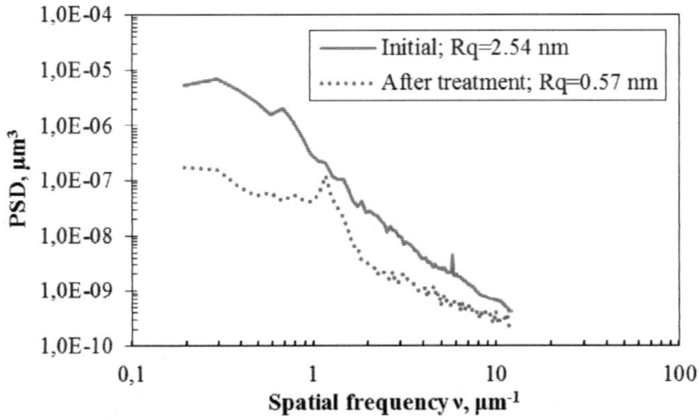

Figure 5. PSD functions of optical glass surface before and after cluster ion polishing.

3. Summary

The influence of accelerated argon cluster ion beam on the surface topography of stainless steel and optical glass was studied. The root-mean-square roughness R_q of stainless steel is reduced by 5.4 times, and the optical glass – 4.5 times to the sub-nanometer level. For the stainless steel surface, the roughness is maximally reduced at medium and high spatial frequencies, while for the glass surface roughness decrease at low spatial frequencies. It has been found that the limiting surface roughness obtained by cluster ions treatment can be restricted by the anisotropic structure of the material being processed.

Acknowledgments

This work was supported by the Ministry of Education and Science of the Russian Federation (Grant No 11.1402.2017/4.6) and the Russian Foundation for Basic Research (Grant No 17-01-00512). The experimental results were obtained by the use of the facilities of the Shared Equipment Center NSU "Applied Physics".

References

[1] Yamada I 2014 *Appl. Surf. Sci.* **310** 77
[2] Popok V N 2011 *Mater. Sci. Eng. R* **72** 137
[3] Korobeishchikov N G, Kalyada V V, Skovorodko P A, Shmakov A A, Khodakov M D, Shulschenko G I, Voskoboynikov R V, and Zarvin A E 2015 *Vacuum* **119** 256
[4] Korobeishchikov N G, Zarvin A E, Madirbaev V Z, and Sharafutdinov R G 2005 *Plasma Chem. Plasma Proc.* **25** 319
[5] Martínez J F G, Nieto-Carvajal I, Abad J, and Colchero J 2012 *Nan. Res. Lett.* **7** 174

IOP Conf. Series: Journal of Physics: Conf. Series **927** (2017) 012027 doi:10.1088/1742-6596/927/1/012027

Electromagnetic launcher for heavy projectiles

A V Kozlov, A V Kotov, V P Polistchook, A V Shurupov and M A Shurupov

Joint Institute for High Temperatures of the Russian Academy of Sciences, Izhorskaya13 Bldg 2, Moscow 125412, Russia

E-mail: polistchook@mail.ru

Abstract. In this paper, we present the electromagnetic launcher with capacitive power source of 4.8 MJ. Our installation allows studying of the projectile acceleration in railgun in two regimes: with a solid armature and with a plasma piston. The experiments with plasma piston were performed in the railgun with the length of barrel of 0.7-1.0 m and its inner diameter of 17-24 mm. The velocities of lexan projectiles with weight of 5-15 g were in a range of 2.5-3.5 km/s. The physical mechanisms that limit speed of throwing in railgun are discussed.

1. Introduction

Until now, the record speeds of projectiles throwing with weight of 1 g in electromagnetic launcher (EML) were about 6.5 km/s. These values are two to three times less than the expected [1]. The substantial erosion of the barrel walls under the action of extreme heat loads, which results in increase of total accelerated mass [1], is considered to be the main obstacle to achieving design projectile speeds.

In the work [2], it was suggested that the speed of projectile in EML is limited by the value of sound velocity in plasma of railgun channel. The sound velocity, estimated for typical working conditions of EML, do not exceed 7 km/s [2]. This fact indirectly confirms the fairness of the hypothesis [2]. The sound velocity in plasma increases with decreasing medium molecular weight of heavy particles μ_m and with increase of the plasma temperature. The sound velocity in the railgun barrel can achieve $10 - 12$ km/s if the plasma temperature is of about 30 kK and its medium molecular weight is of about 7 atomic units [2]. Light plasma ($\mu_m \sim 7$ a.u.) can be created with the use of lithium initiator. Obtaining the high temperature of plasma is a more complex problem, since it requires a very high rate of energy input in the railgun. By estimations, in this case the growth rate of current in EML must be above 20 kA/μs. It is extremely hard to provide dynamic strength of the barrel under such extreme regimes of acceleration.

The study of some processes during high-speed impact requires a relatively low speed of projectiles, at the level of $4-5$ km/s, but with their weight of more than 10 g. At such speeds of projectiles the limit, which is associated with the sound velocity in the channel plasma, can be overcome due to the use of the low-molecular weight initiators. Earlier we explored acceleration in EML the projectiles with weight of ~2 g up to velocities of about 5 km/s. In work [3] we described the results of these experiments and also a numerical simulation of the targets destruction due to the high-speed impact. This paper presents the results of our research aimed to resolve of the task of the throwing of the relatively heavy projectiles with weight above 10 g.

2. Experimental set

We have created an EML with capacitive energy source (CES) of $E_0 = 4.8$ MJ when it is fully charged to 6 kV. CES was sub-divided into seven modules, each with its own independent triggering system and delay timer. Installation allowed studying the projectile acceleration in railgun in two regimes: with a solid armature and with a plasma piston (PP). Energy of CES is sufficient for throwing of projectile with a mass mb of 0.3 kg at a speed of 3 km/s.

This paper describes the results of experiments with PP, where we used three modules of CES with nominal stored energy E_0 of 1.2 MJ. The length of the railgun barrel was 0.7 – 1.0 m, its internal diameter was 17 – 24 mm. Electrodes were made of copper, and insulators were made of lexan. After assembly, the channel was placed into the power housing, which was made of stainless steel.

The weight of lexan projectile was 5 – 15 g; length of the heaviest projectile was of 32 mm. The discharge in railgun was triggered by using of graphite disk, which provided the initial contact between the electrodes. The application of graphite initiator with weight of about 1 g allowed producing of the comparatively "light" plasma on the initial part of the channel. In the experiments [3] we used as the initiator the lithium plate, which weight was of about 10% of projectile weight. The sound velocity in lithium plasma is higher than in carbon plasma. Additionally, lithium requires the significantly less energy to create the PP in comparison with carbon. On the other hand, the lithium initiator explodes faster than graphite initiator that creates more difficulties in the retaining of impulse pressure in the channel breech.

We equipped the railgun with diagnostic system, consisted of: Rogowski coil - to measure breech current, transducers based on current transformers to determine breech and muzzle voltage, magnetic probe (MP) - to control the motion of PP along the channel. MP had form of 7 mm wide loop, padded along the insulation wall of the channel. After departure from the channel, the motion of projectile and its collision with target was controlled by high-speed videography (100,000 frames per second). The target represented itself duralumin disk with thickness of 60 mm and diameter of 150 mm. The time of projectile collision with target was determined by contact sensor.

3. Experimental results and discussion

In our experiments maximum throwing speed (3.5 km/s) was obtained for the projectile with weight of 5 g. Previously, when throwing lighter projectile ($m_b = 1.3$ g) we have achieved the speed of 5.4 km/s [4]. In experiments [3] we used CES with nominal stored energy of 0.8 MJ.

Figure 1 shows the oscillograms of the current in railgun I, breech voltage V_b, muzzle voltage V_m and the signal from MP V_p. Energy of CES with charging voltage of 3.6 kV was $E_0 = 0.68$ MJ. In this test the projectile with weight of 14.3 g was accelerated in the channel with length of 1 m up to the speed of $u_m = 3.0$ km/s, its kinetic energy E_k was 65 kJ.

Neglecting the voltage drop at the electrodes, it follows from the equation of circuit that the rate of change of magnetic flux and consequently projectile acceleration is proportional to the difference of input and output voltage $(V_b - V_m)$. This difference sharply increased to ~ 0.45 kV (Figure 1) through $\sim 20\,\mu$s after the start, then it changed more smoothly with achieving maximum value of 0.9 kV at time $t = 370\,\mu$s. The sharp rise in muzzle voltage V_m and drop of signal from MP V_p at $t_e = 620\,\mu$s is explained by the departure of projectile from the channel.

If we neglect the loss of momentum due to the friction between projectile and channel walls and disregard an increase of throwing mass due to its erosion, the acceleration of projectile (a) and its velocity (u_t) is defined from equations [1]:

$$a = \frac{L'}{2m_b} I^2(t) \tag{1}$$

$$u_t = \frac{L'}{2m_b} \int I^2(t')dt' \qquad (2)$$

In these equations $L' \sim 0,65\,\mu$Hn/m is inductance gradient of the railgun.

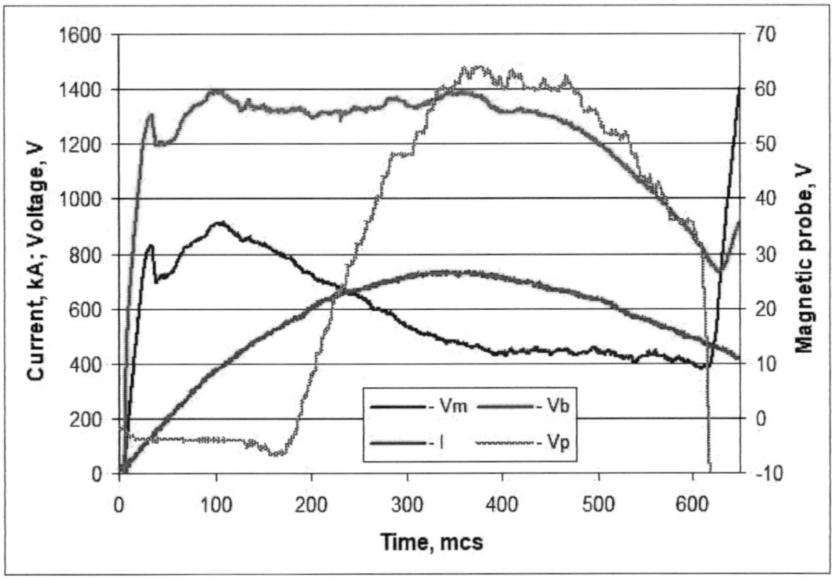

Figure 1. Oscillograms of current and voltages

From equation (1) and Figure 1 it follows that the maximum projectile acceleration at the maximum current (t_m =350 μs, I_m = 730 kA) was $1.6 * 10^7$ m/s^2, and the peak plasma pressure p_m was 0.53 GPa. In this experiment the average values of acceleration and plasma pressure were approximately three times less than the maximum values, and were respectively $5 * 10^6$ m/s^2 and 0.16 GPa. Plasma pressure in the final stage of acceleration (t_e=620 μs) was 0.2 GPa.

In this experiment the net energy efficiency of launcher, i.e. the ratio of E_k to E_0 was 9.5%. Theoretical speed of projectile u_t was calculated using the formula (2) and oscillogram of the railgun current, which is presented on Figure 1. In the moment of the projectile departure from the channel its theoretical velocity reached 4.3 km/s, which corresponds to projectile kinetic energy E_t =132 kJ.

The difference between E_t and E_k is a mainly estimation of the sum of the friction forces work of projectile on the channel walls and energy losses due to the increase of kinetic energy of the PP. In this experiment the sum of energy losses reached to 67 kJ. By the time of the projectile departure from the channel, the charge of CES was about 10% of the initial, so almost all electromagnetic energy, remaining in the circuit, proved to be stored in the inductance. When the projectile left the railgun, the current I was equal to 460 kA (Figure 1), and magnetic energy E_m =0.17 MJ. Thus, about 56% of the energy stored in CES mainly was spent on creation of PP and heating of electrodes.

In this experiment the real projectile velocity was ~70% of the calculated one. Our results may clarify one possible mechanism of loss of momentum projectile. It can be seen from Figure 1 that at time t_b = 290 μs on the voltage oscillogram of MP there is a "plateau" lasting of about 20 μs. This peculiarity is being registered during the stage of increasing current, therefore at moment t_b= 290 μs the braking of current jumper occurs. It follows from the equation (2) that at the moment t_b = 290 μs the calculated speed of projectile was of about 1.5 km/s.

Figure 2 shows the central part of the channel, where the projectile velocity by estimations increased from 1.5 to 2 km/s. At the surface of the insulator there are many "notches", the

Figure 2. The central part of the channel

similar damages with smaller size and not so numerous also present on the electrode surface. We did not observe the similar damages near the entrance of the channel; near the exit of the channel the intensity of such walls damages was noticeably less. The time of projectile passing through the zone of damaged walls in central part of the channel corresponds to the time of braking of current jumper.

Many researchers have observed the damages on inner walls of railgun, similar to those shown in Figure 2. For example, in work [4], when the projectile velocity was of about 1.5 km/s, the damages like gouging appeared on copper electrodes of railgun with duralumin armature. The characteristic size of these damages was of about 10 mm. In work [5], we have explained the occurrence of these injuries due to the development of resonant deformation of barrel (RDB) under the action of moving front of high pressure.

Effect of RDB appears when the speed of the high-pressure front becomes comparable with the sound speed in the wall of the channel. In the real design of railgun the analysis of RDB is a very complex problem. In the channel with thin cylindrical walls RDB develops when the velocity of high-pressure front reaches one of the two critical speeds [6]. The first critical speed corresponds to bending resonance of the channel; the second critical speed causes an axial vibration resonance. Occurrence of RDB results in creation of dynamic asperities on the channel surface. The projectile collides with these asperities, and the explosive ejection of walls material into the channel takes place. It causes the drastic slowdown of PP (current jumper); additionally, the temperature of the plasma and its electrical conductivity considerably decreases. In work [5] we have shown that the effect of RDB explains qualitatively many observed features of railgun operation when the projectile velocity is above $2-3$ km/s. In particular, a sharp braking of PP can cause a power surge at the entrance to the channel, and as a result, the appearance of breakouts and shunt discharges in this part of the channel. Similar current structures significantly reduce the efficiency of EML.

The height of the dynamic asperities that occur on the surface of the channel due to RDB increases with local pressure, therefore the degree of damage to the walls near the exit of the channel is noticeably smaller than in its central part.

Figure 3 shows the duralumin target after the experiment under consideration. After the experiment the diameter of the target was reduced to 100 mm. The diameter of the crater on the target and its depth were 52 mm and 26 mm respectively. Figure 3 also shows the projectile with weight of 15 g, which is similar to projectile used in this experiment. For comparison, in our experiments [3] the lexan projectile with weight of 2.6 g and velocity of 4.7 km/s embossed the cork with diameter of 30 mm in duralumin target with 40 mm thick.

Figure 3. Duralumin target after shot

4. Conclusion

The results of experiments on the acceleration in railgun with plasma piston the projectiles with weight of $5 - 15$ g up to velocity of $2.5 - 3.5$ km/s are presented. It was concluded that resonant deformation of barrel under action of moving front of high pressure is one of the main sources of loss of projectile momentum in railgun.

References

[1] Shvetsov G A *et al* 2004 *Proc. Inter. Seminar Gydrodinamika vysokikh plotnostey energii* (Novosibirsk: IGIL) pp 283-304
[2] Shurupov A V *et al* 2004 *Proc. 12th Inter. Symp Electromagnetic Launch Technology* pp 65-69
[3] Fortov V E *et al* 2006 *Inter. J. of Impact Engineering* **33** pp. 254-263
[4] Fair H D 2004 *Proc. Inter. Seminar Gydrodinamika vysokikh plotnostey energii* **23** (Novosibirsk: IGIL) pp 256-266
[5] Kozlov A V *et al* 2008 *13th Inter. Conf. on Equations of State for Matter* (Chernogolovka: IPCP) p 260
[6] Jones P and Bhuta P G 1964 *J. of Applied Mechanics* **31** pp 105-111

IOP Conf. Series: Journal of Physics: Conf. Series **927** (2017) 012028 doi:10.1088/1742-6596/927/1/012028

Modification of an aluminum electrode in a surface dielectric barrier discharge plasma

A V Lazukin [1,3], I V Selivonin[1,2], I A Moralev[2] and S A Krivov[1].

[1] National Research University "Moscow Power Engineering Institute" Krasnokazarmennaya 14, Moscow 111250, Russia
[2] Joint Institute for High Temperatures of the Russian Academy of Sciences, Izhorskaya 13 Bldg 2, Moscow 125412, Russia
[3] JSC «SPA «CNIITMASH», Sharikopodsibnikovskaja 4, Moscow 115088, Russia

E-mail: morler@mail.ru

Abstract. The paper is devoted to the phenomenological study of the aluminium electrode modification in the surface dielectric barrier discharge in ambient air. The morphological changes found on the electrode edge after discharge treatment include the formation and pit-wise erosion of the relatively thick oxide film and redeposition of the oxide vapor in the form of the dendrite structures. These formations are obtained independently on the thickness of the electrode in the range 10-80 μm, with minor changes of the structures morphology. It is shown that in the case of thin foils, intense oxidation of the electrode edge leads to the mechanical separation of the oxidized region from the electrode.

1. Introduction

In the experimental studies of surface dielectric barrier discharge (SDBD) and in a variety of technological problems two basic types of electrode configurations are used: with a pair of dielectric-covered electrodes (coplanar surface discharge) and with one electrode exposed directly to the discharge. The latter class of systems is characterized by higher electric field strengths in the discharge formation region, lower operating voltages amplitudes and assembling simplicity.

Although thin aluminum foils are one of the most common materials used for electrode manufacturing in SDBD, erosion of the aluminum electrodes in SDBD is not described in the literature. Most of the work devoted to changes in electrodes geometry, cell parameters during SDBD operation are devoted to modification of the dielectric barrier [1,2].

Atmospheric pressure SDBD develops as a set of separate pulsed microdischarges, each having its own electrode spot at the edge of the exposed electrode. The parameters of the microdischarge pulses are close to ones obtained for the pulsed corona discharge in air. In this regard, one can expect some similarity of electrode processes in SDBD and corona discharges.

The erosion of the pin electrodes in the corona discharge was studied in [3-7], in particular, works [3,4,8] describe to the morphological changes that occur on aluminum electrodes. It is shown that in the case of aluminum, the formation of an oxide film on the electrode has a significant effect on both the erosion process and the dynamics of the discharge. The main processes responsible for changing the shape and morphology of the pin electrode was shown to be the formation of breakdowns and cracks in the oxide film, as well as redeposition of the electrode material to the surface.

Besides the fact that the morphological processes on the electrodes have a significant effect on the discharge cell lifetime, they can have an unexpected effect on other applied aspects of the SDBD-based devices. Thus, in the plasma biological applications, in cases when the object of influence is directly in the discharge zone, the formation of a cloud of charged particles of the nanometer range near the aluminum electrode can affect the results and variability of biological studies. Significant thermal loadings on a limited part of an electrode, in presence of an intensely oxidizing atmosphere, may be interesting from the point of view of fast tests to reveal local defects in metals intended for work under aggressive conditions.

2. Experiment details and analysis techniques

In a present study an asymmetric barrier discharge cell with a linear strip electrode was used [8]. The pair of electrodes, one connected to the high voltage and other grounded, were separated by a 1 mm thick ceramic barrier (alumina ceramics 94% Al2O3, ε=10). The thickness of the foil electrodes used in the study was 10, 20, 35, and 80 μm. Foils were glued onto the surface of the cleaned ceramic surface with an alcohol-based adhesive with a thickness of the order of 2-4 μm. The electrode was exposed in the SDBD for 40 hours. The discharge was created by a sinusoidal voltage of 3 kV RMS at a frequency of 25 kHz. The shape and amplitude of the voltage were controlled by an oscilloscope TDS 2014 (Tektronix) with a high-voltage divider P6015A (Tektronix).

At the end of the electrode exposure (5 iterations for 8 hours), the discharge cell was examined by various microscopy methods. Evaluation of changes in the shape of the electrode edge was carried out on a scanning confocal laser microscope Olympus LEXT OLS4000 with a resolution of at least 200 nm, and also with an electron microscope Nova NanoSem 650 with a resolution of up to 2 nm. The length of the electrode edge exposed in the discharge was 40 mm. The studied region with a typical size of 1.8 x 1.8 mm was chosen in the middle of this span. Before exposure, the electrode was marked at some distance from the edge, that allowed for accurate positioning of the inspection zone during the whole test and examination cycle.

3. Results and discussion

It is found that during the discharge operation the structure of the electrode edge changes is subjected to strong morphological changes. The edge of the electrode is covered with a porous oxide layer; in the immediate region where the electrode spots are observed, a black band is formed, consisting of groups of bound grains. The formation of the black band is supposed to be associated with the transition of aluminum oxide from alpha to gamma state. As a result of return of the erosion material, dendrit-like structures are formed on the electrode surface. The greatest density these deposits reach at some distance from the edge, where a "wall" with a height of up to 20 μm is formed. Detailed structure of the oxide layer was studied using an electron microscope. It was found that the surface of the oxide layer in the region directly exposed to the microdischarges electrode spots is covered by small holes (200 nm in diameter and less) (Fig. 1).

Similar structures were observed in a negative corona discharge, where their formation was associated with the breakdown of the dielectric film at the time of the formation of the next Trichel pulse. Discharge operation at long exposures leads to the significant stresses appearing in the oxidized strip. As a result, in the case of thin foils (10 μm) by the end of 40 hours of the discharge operation, the oxidized part of the electrode is detached from its bulk.

The images of the electrode edge for 20 μm, 35 μm and 80 μm thicknesses are shown in Fig. 2. For the comparison, the left fragment shows a 35 μm electrode before discharge operation. The size of the imaged field is 250 to 650 μm. With respect to the difference in the electrodes height, the total length of the modification zone (the total linear length of the black band and oxide side) remains practically unchanged and is equal to nearly 280 μm.

IOP Conf. Series: Journal of Physics: Conf. Series **927** (2017) 012028 doi:10.1088/1742-6596/927/1/012028

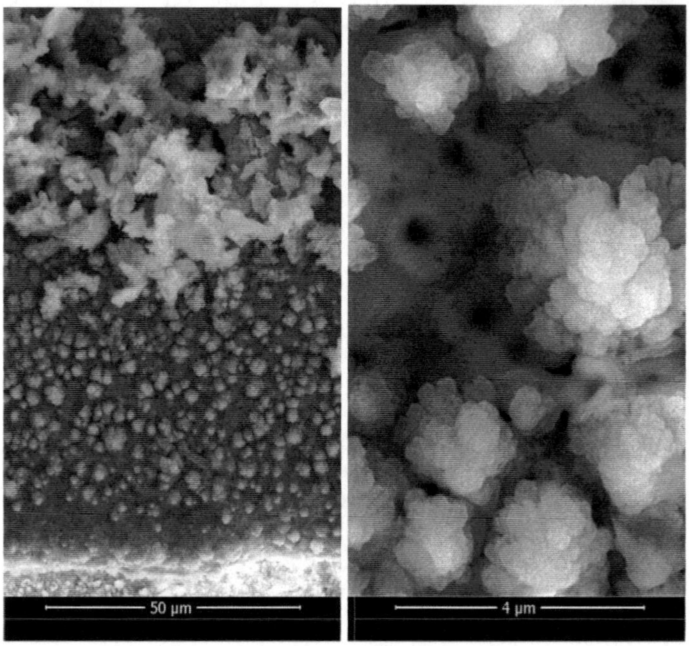

Figure .1 SEM images of 10 µm electrode edge

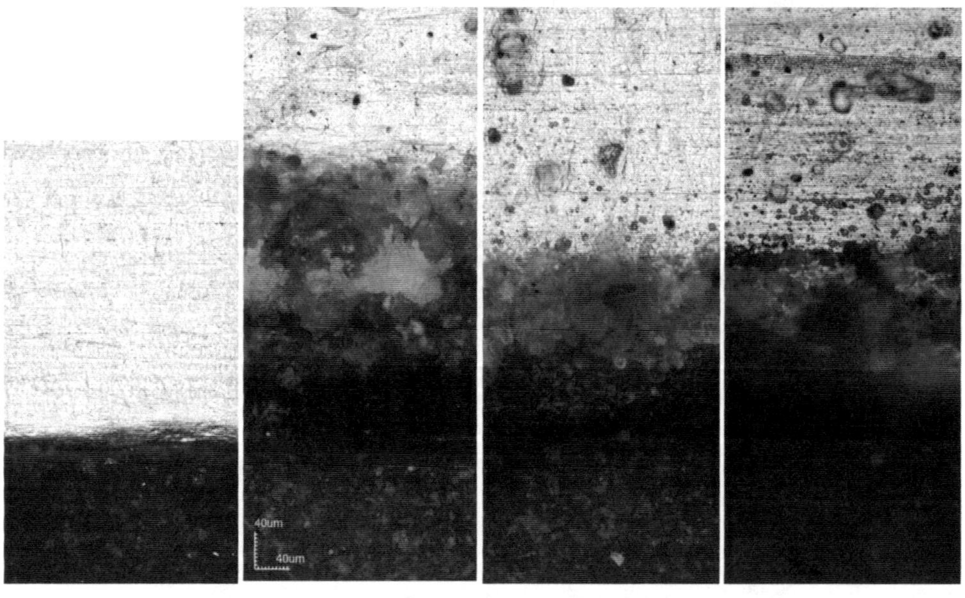

Figure 2. Left to right: 35 µm electrode prior to the exposure; electrodes after 40 hours exposure at 25kHz, 3kV RMS: 20 µm; 35 µm; 80 µm;

Modifications of the electrodes that arise in the process of the surface discharge operation should to cause significant changes in the characteristics of the surface discharge. These changes are assumed to be an object of further studies.

4. Conclusions

Aluminum foil electrodes, regardless of thickness, are subject to considerable modification under the influence of a surface dielectric barrier discharge. The nature of these modifications is similar to the erosion damage that occurs in the negative corona discharge: the formation and point-like breakdown of an oxide film, the formation of craters and local erosion, the redeposition of erosion material onto the electrode. At long exposures, the destruction of the electrode on the form of oxidized layer was detected (separation of the black strip).

Acknowledgments

This work was partially funded by the Russian Federation Ministry of Education and Science as part of agreement to provide subsidies for the implementation of applied scientific research (Unique identificator RFMEFI57915X0114) and partly by RFBR grant #17-58-16004 as a part of Associated International Laboratory "Kinetics and Physics of pulsed Plasmas and their Afterglow" activity.

Authors would like to acknowledge Dr. Alexey Petrov for the fruitful discussion of the results

References

[1] Sokolova M, Kozlov K, Mitin, A, Tatarenko P 2013 *EPJ Applied Physics* **61**, 2, 24312
[2] Ngo A, Pai K, Jacob J 2017 *48th AIAA Plasma dynamics and Lasers Conf* (Denver; United States) K192609
[3] Laan M, Paris P and Repän V, *Le J. Phys. IV* 1997 **7** C4-259-C4-270
[4] Weissler G L and Schindler M 1952 *J. Appl. Phys.* **23** 844
[5] Amirov R H, Petrov A A and Samoylov I S 2009 *International Journal of Plasma Environmental Science and Technology* **3** 35
[6] Petrov A A, Amirov R H, Asinovskii E I and Samoylov I S 2009 *Journal of Plasma Fusion Research Serie.* **8** 780
[7] Amirov R H, Petrov A A and Samoylov I S 2013 *J. Phys. Conf. Ser.* **418** 12064
[8] Gibalov V I and Pietsch G J 2012 *Plasma Sources Sci. Technol.* **21** 2 24010

Influence of small additives of nitrogen in 1D-modeling of non-uniform microwave discharge of low pressure in hydrogen

Yu A Lebedev, A V Tatarinov and I L Epstein
Topchiev Institute of Petrochemical Synthesis RAS (TIPS RAS), Leninsky Prospect, 29, Moscow, 119991, Russia

E-mail: lebedev@ips.ac.ru

Abstract. A one-dimensional simulation for studying the influence of small additives of nitrogen to the parameters and structure of non-uniform microwave discharge in hydrogen at pressure of 1 Torr has been performed. It is shown that the profiles of the electron density and microwave field are affected by changes both in the EEDF and in the ionic constituent of the plasma. The EEDF influences in a spherical region of the discharge (away from the central electrode). The addition of nitrogen increases the mobility of electrons and leads to reduction of size of the plasma formation. The change in the ion components of plasma (N_2H^+ becomes the main ion) affects the structure of the plasma near the surface of the electrode.

1. Introduction

It has been shown in a number of works that even a small amount of gas additives to the plasma-forming gas can greatly change the properties of plasma. In particular, the gas additives may influence the processes of ionization and excitation of particles and therefore the electric field, which in the self-sustained plasma is determined by the balance of charged particles. In works [1, 2] both experimentally (by means of optical emission spectroscopy) and numerical 1D and 2D modeling the effect of small additions of hydrogen on the electrode microwave discharge at low pressure in nitrogen has been investigated. It is shown that the non-uniformity of discharge leads to the fact that the nature of the influence of hydrogen additives (0-30% of the flow rate) on the intensity of the radiation bands of nitrogen is different in different parts of the discharge. There is also a different influence of hydrogen on the vibrational distribution of nitrogen molecules in $C^3\Pi_u$ state in different parts of the discharge. Raising the level of hydrogen addition changes the ion composition of plasma: the main ion N_4^+ in the spherical region is replaced by a lighter ion N_2H^+. This, according to quasi-static one-dimensional modeling, leads to an increase of the microwave field in the spherical part of the discharge and change of the emission intensity of second positive system of nitrogen. Results of 2D modeling [1] basically coincide with the results of the 1D simulation [2]. Thus, it is shown that the main effect of hydrogen additives on parameters of the nitrogen discharge is associated with changes in the ion composition of plasma. In this work we investigated the influence of small additives of nitrogen in the hydrogen discharge. A similar problem has been solved in [3], where the authors on the basis of 0-dimensional simulation studied the influence of changes of the EEDF on the parameters of the hydrogen discharge due to the addition of nitrogen. In this paper, in the framework of 1D model of inhomogeneous electrode microwave discharge in hydrogen, along with the impact of changes in the

EEDF we examine the influence of changes in ion composition due the introduction of small amount of nitrogen (up to 5 vol.%).

2. Model

Here we use a self-consistent quasi-static model of the microwave discharge in the electrode system with spherical symmetry, which was previously described in detail in [4]. The model includes the equation for the electric field strength in the quasi-static approximation, Poisson equation and balance equations for charged and neutral particles. The contribution of processes involving vibrationally excited molecules in the ground state is carried out, using the well-known analytical expression for the vibrational distribution of the molecules, obtained in the diffusion approximation [5]. The EEDF is determined from the solution of the homogeneous stationary Boltzmann equation.

The processes of direct, stepwise and associative ionization, dissociation, recombination of charged and neutral particles in the bulk and on the wall, ion-molecular reactions, excitation and de-excitation of molecules are taken into account. The balance equations for nitrogen plasma describe the kinetics of charged (e, H_2^+, H_3^+, N^+, N_2^+, N_3^+, N_4^+, N_2H^+), and neutral (H, H(2S,2P), H(3S,3P,3D), $N_2(A^3\Sigma^+_u)$, $N_2(B^3\Pi_g)$, $N_2(C^3\Pi_u)$, $N_2(a'^1\Sigma^-_u)$, $N(^4S)$, $N(^2D)$, $N(^2P)$, NH) particles. We use the kinetic schemes for hydrogen and nitrogen, developed in our earlier works [4, 6, 7]. The processes, which include interaction of particles of both gases, are shown in Table 1. The kinetic model includes the processes of quenching of metastables of nitrogen molecules by hydrogen and the reaction of conversion of hydrogen ions into the ion N_2H^+. The plasma is simulated at a gas pressure of 1 Torr, the microwave frequency of 2.45 GHz and gas temperature of 600 K.

Table 1 Processes, which include interaction of particles of both gases.

	Processes	K, см^3 сек$^{-1}$
1	$N_2^+ + H_2 \Rightarrow N_2H^+ + H$	$K = 1.7\cdot10^{-9}$ [8]
2	$N_4^+ + H_2 \Rightarrow N_2H^+ + H + N_2$	$K = 1.1\cdot10^{-8}$ [8]
3	$N_2H^+ + e \Rightarrow H + N_2 + e$	$K = 3.7\cdot10^{-8}$ [9, 10]
4	$H_2^+ + N_2 \Rightarrow N_2H^+ + H$	$K = 1.95\cdot10^{-9}$ [8,11]
5	$H_3^+ + N_2 \Rightarrow N_2H^+ + H_2$	$K = 1.8\cdot10^{-9}$ [9,12]
6	$N_2(A^3\Sigma_g^+) + H \Rightarrow N_2 + H$	$K = 5.0\cdot10^{-11}$ [9,13]
7	$N_2(A^3\Sigma_g^+) + H_2 \Rightarrow N_2 + H + H$	$K = 2.0\cdot10^{-10}\exp(-3500/T_g)$ [14]
8	$N_2(B^3\Pi_g) + H_2 \Rightarrow N_2(A^3\Sigma_g^+) + H_2$	$K = 2.5\cdot10^{-11}$ [8,15]
9	$N_2(a'^1\Sigma_u^-) + H \Rightarrow N_2 + H$	$K = 1.5\cdot10^{-11}$ [9]
10	$N_2(a'^1\Sigma_u^-) + H_2 \Rightarrow N_2 + H + H$	$K = 2.6\cdot10^{-11}$ [9]
11	$N(^2D) + H_2 \Rightarrow H + NH$	$K = 2.3\cdot10^{-12}$ [9,15]
12	$N(^2P) + H_2 \Rightarrow H + NH$	$K = 2.5\cdot10^{-14}$ [9,16]
13	$H + NH \Rightarrow H + N_2$	$K = 5.0\cdot10^{-11}$ [9,17]

3. Results

The results of the calculations are presented in figures 1-3. Figure 1 shows profiles of the electron density and microwave fields for different percentage of additives of nitrogen.

In the near-electrode region with a strong microwave field one finds small (~0.5%) increase in the values of n_e for cases with nitrogen additives compared to n_e for pure hydrogen. In a spherical plasma region (i.e. extended region adjacent to the narrow near-electrode region of strong microwave field, where the value of the microwave field is about 100 V/cm, see Fig.1*b*) the values of n_e for cases with nitrogen additives on the contrary are lower than n_e for pure hydrogen plasma. As for the microwave field, it is growing in the whole region of the plasma in the case of additives of nitrogen (Fig.1 *b*).

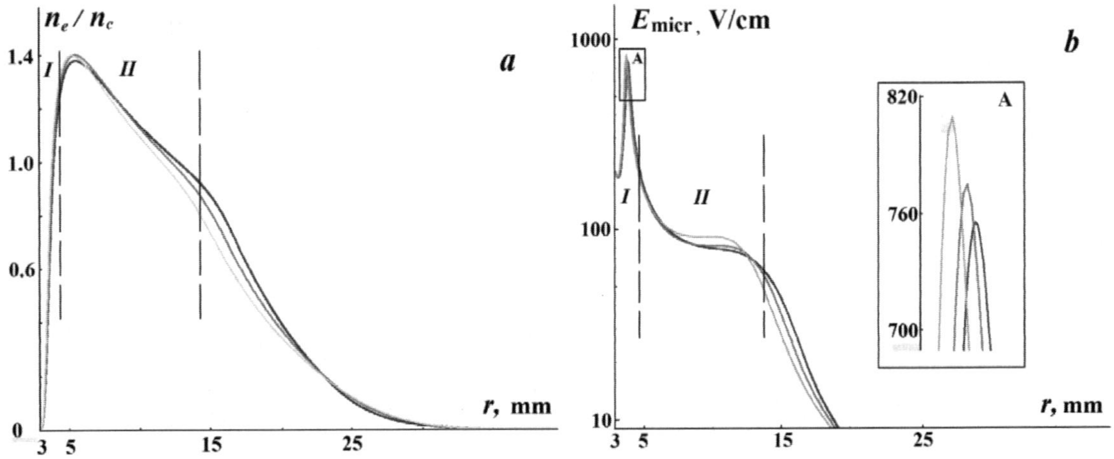

Fig. 1. Radial profiles of the electron density n_e (*a*) and microwave field (*b*). Blue lines – discharge of pure hydrogen; red lines – with 5% nitrogen added; green lines – with 20% nitrogen added. *I* –region of strong microwave field, *II* – spherical region of the discharge, $n_c = 7\times10^{10}$ см$^{-3}$- critical value of plasma density. The inset A in Fig.1*b* shows the region of strong microwave field in an enlarged scale.

Several computational tests have been performed to investigate changes in discharge characteristics, dealt with the introduction of additives of nitrogen.

When making nitrogen additions to the hydrogen discharge, the ion N_2H^+ becomes the main ion (Fig. 2). This ion is much heavier and thus less mobile than the ion H_3^+, which is the main ion in the discharge in pure H_2. This can cause a profile change of the electron density and also change of other plasma properties.

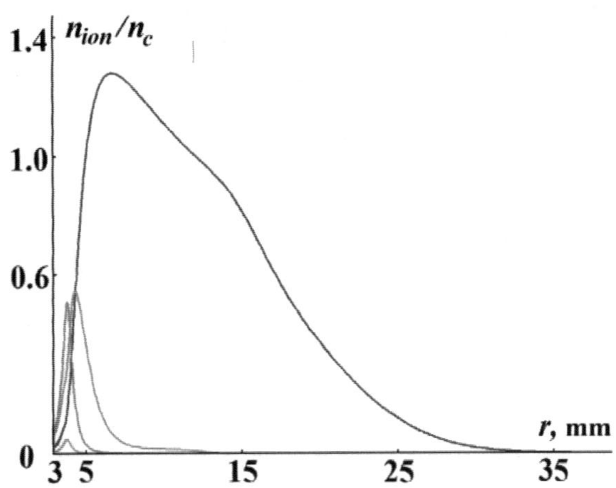

Fig. 2. Radial profiles of ion density for a discharge in hydrogen with 5% nitrogen added: blue line - ion N_2H^+; the dark green line is the ion H_3^+; blue line - ion H_2^+; pink line - ion N_4^+.

To confirm this hypothesis, we made calculations for the mixture with the addition of 5% nitrogen, but with a kinetic scheme which excludes the appearance of the ion N_2H^+. The corresponding profile is shown by the horizontal orange ellipses in figure 3. In the region of strong fields the electron density profile coincides with the profile of a n_e for pure hydrogen discharge, while the rest of the profile coincides with the profile calculated for a 5% additive without changes in the kinetic scheme. This test shows that the presence of the ion N_2H^+ has an effect only in the region of strong microwave fields near the surface of the electrode-antenna.

To determine the effect of changes in the EEDF we made the calculation for the mixture of H_2 with 5% of nitrogen additive, but with the EEDF calculated for the pure hydrogen discharge. As a result, the electron density profile outside the near-electrode region of strong microwave fields almost everywhere coincides with the n_e profile for a discharge in pure H_2 (Fig.3, the vertical purple ellipses). This means that the change in the EEDF due to addition of nitrogen affects the characteristics of plasma in the spherical region, where the magnitude of the microwave field is small.

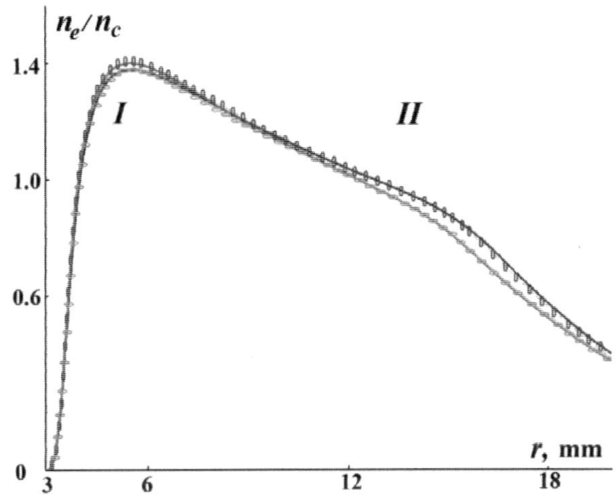

Fig. 3. Radial profiles of the electron density of discharge: blue line - pure hydrogen; red line - with 5% nitrogen added; violet vertical ellipses – with the addition of 5% nitrogen, but with the EEDF, calculated for a discharge of pure hydrogen; horizontal orange ellipses - with the addition of 5% nitrogen, but with a kinetic scheme, which excludes the appearance of ion N_2H^+. *I* – region of strong microwave field, *II* - spherical region of the discharge.

Obviously, a change in the EEDF can affect the reaction rates of mechanisms with the electron impact and transport properties of the electron motion in a DC field due to charge separation, namely the diffusion and mobility coefficients. Our analysis has shown that the change in the EEDF affects mainly the electron mobility. Calculated DC fields of charge separation in plasma are found to be about 1-3 V/cm. For these values of the field the coefficients μ_e for the case with 5% of nitrogen additive are higher (~0.1-0.3%) than the coefficients μ_e for a pure hydrogen discharge. Because the field of charge separation is always directed from the center of the chamber to the electrodes, the electrons drift in this field towards the center of the discharge. Drift flow of electrons to the center of the plasma reduces the diffusion flux of electrons to the electrodes, which leads to a squeeze of the electron profile in comparison with the profile of n_e in pure hydrogen discharge (Fig. 1 *a*).

4. Summary
It is shown that in conditions of strongly inhomogeneous plasma, in case when small additives of nitrogen are added into the hydrogen discharge, the inclusion of processes of ion conversion, along

with a change of the EEDF, leads to a more complex picture of the influence of additives on the plasma parameters than those described in [3]. Changes in the EEDF, in the ion composition of plasma and in transport properties of electrons affect the profile of the electron density and microwave field already at 5% addition of nitrogen. The EEDF influences in a spherical region of the discharge away from the electrode. The addition of nitrogen increases the mobility of electrons and leads to reduction of size of the plasma formation. The change in the ion components of plasma (N_2H^+ becomes the main ion) affects the structure of the plasma near the surface of the electrode.

5. References

[1] Jovovic, J., Epstein I.L., Konjevic N., Lebedev Y.A., Sisovic N. M, Tatarinov A.V. 2012 *Plasma Chemistry and Plasma Processing* **32** 1093
[2] Lebedev Y.A., Mafvludov T.B.., Shakhatov V.A., Epstein I.L. 2010 *High Temp.* **48** 315
[3] Garscadden A., Nagpal R. 1995 *Plasma Sources Sci. Technol.* **4** 268
[4] Lebedev Y.A., Epstein I.L. 2007 *Plasma Physics Reports* **33** 63
[5] Popov M., Ferreira B., Gordiets B. and Osipov A. 2000 *Plasma Kinetics in Atmospheric Gases.* (Berlin: Springer)
[6] Lebedev Y.A., Tatarinov A.V. , Epstein I.L. 2007 *Plasma Sources Sci. Technol.* **16** 726
[7] Lebedev Y.A., Tatarinov A.V. , Titov A.Yu. , Epstein I.L. , Krashevskaya G.V., E.V.Usupova 2014 *J.Phys.D Appl.Phys.* **47** 335203
[8] Tatarova E., Dias F.M., Gordiets B., Ferreira C.M. 2005 *Plasma Sources Sci. Technol.* **14** 19
[9] Gordiets B., Ferreira C. M., Pinheiro M.J., Ricard. A. 1998 *Plasma Sources Sci. Technol.* **7** 363
[10] Amano T. J. 1990 *Chem. Phys.* **92** 6492
[11] Ryan K.R. 1974 J.*Chem.Phys.* **61** 1559.
[12] Hemswworth R.H., Payzant J.D., Schiff H.I., Bohme D.K. 1974 *Chem. Phys. Lett.* **26** 417
[13] Hovis F.E., Whitefield P.D. 1987 *Chem. Phys. Lett.* **138** 162
[14] Herron J.T. 1999 J.Phys.*Chem. Ref. Data* **28** 1453
[15] Piper L.G. 1992 *J.Chem.Phys.* **97** 270
[16] Young R.A., Dunn O.J. 1975 *J.Chem.Phys.* **63** 1150.
[17] Miller J.A., Bowman C.T. 1989 *Prog. Energy and Combust. Sci* **15** 287.

IOP Conf. Series: Journal of Physics: Conf. Series **927** (2017) 012030 doi:10.1088/1742-6596/927/1/012030

Laser ignition of liquid petroleum gas at elevated pressures

E Loktionov, N Pasechnikov, V Telekh

Bauman Moscow State Technical University, Moscow, 105005, Russia

stcpe@bmstu.ru

Abstract. Recent development of laser spark plugs for internal combustion engines have shown lack of data on laser ignition of fuel mixtures at multi-bar pressures needed for laser pulse energy and focusing optimisation. Methane and hydrogen based mixtures are comparatively well investigated, but propane and butane based ones (LPG), which are widely used in vehicles, are still almost unstudied. Optical breakdown thresholds in gases decrease with pressure increase up to ca. 100 bar, but breakdown is not a sufficient condition for combustion ignition. So minimum ignition energy (MIE) becomes more important for combustion core onset, and its dependency on mixture composition and pressure has several important features. For example, unlike breakdown threshold, is poorly dependent on laser pulse length, at least in pico- and to microsecond range. We have defined experimentally the dependencies of minimum picosecond laser pulse energies (MIE related value) needed for ignition of LPG based mixtures of 1.0 to 1.6 equivalence ratios and pressure of 1.0 to 3.5 bar. In addition to expected values decrease, low-energy flammability range broadening has been found at pressure increase. Laser ignition of LPG in Wankel rotary engine is reported for the first time.

1. Introduction

Laser-induced ignition of natural gas applied to internal combustion engines is investigated for a long time. Demonstration of a compact 'laser plug' [1] accelerated research in this field, and particularly, led to a dedicated annual international Laser Ignition Conference appeared. Laser spark benefits as compared to traditional spark plugs are higher compression rate, and possibility of almost any fuel ignition, so lean mixtures burning with lower temperatures could reduce harmful exhausts (NO_x, CH, etc.). No need in electrode and possibility for multi-point, linear or circular ignition can make combustion even more effective. Laser induced combustion wave appears faster and is more stable in time, than electric one, so can be used for ramjets, chemical thrusters, and gas turbines. Need in gas fed engines in general arises not from lower operating costs and harmful emissions only, but from gas direct availability in many remote regions (e.g. Arctic ones), where cost of petrol and diesel delivery is higher than of the original price of product itself.

Laser ignition takes place due to gas optical breakdown followed by plasma and shock wave formation, those lead to deflagration core onset (detonation and autocatalytic reaction are also possible) [2]. Laser breakdown threshold (unlike electric) in gases decreases with pressure increase up to 10s of MPa [3], so smaller ignition energy is needed at higher compression [4]. The latter is helpful for more compact and efficient internal combustion engines design [5]. Lasers could be also rather efficient for heterogeneous [6] and vortical [7] flows ignition.

A lot of data have been published about laser ignition of methane (or compressed natural gas, CNG) [4, 8, 9] and hydrogen [10-12] based mixtures since those are more promising for stationary power plants. However, propane-butane (liquid petroleum gas, LPG) is more widely used in automotive industry because of significant benefits such as higher specific heat release, smaller weight and size of feeding system, longer run between fueling up. Despite this, only one paper is available on propane [13], and we failed to find any on butane or LPG based mixtures laser ignition.
Usually minimum ignition energy (MIE) is experimentally evaluated, and it is found to be significantly higher than for electric ignition, as explained, due to shorter duration and smaller size of the laser impact region [14]. MIE is strongly dependent on air-to-fuel ratio and pressure, and is virtually independent on laser pulse length, at least starting from picosecond [15] to microsecond range. Data available still have a lot of discrepancies, reaching an order of magnitude or even more. Those originate both from multiple experimental conditions features, such as laser focusing [16], gas velocity and temperature [17], etc., and from what the researcher assumes as the MIE: this could be the least energy in a long train of pulses, or the one that leads to a guaranteed ignition [18]. Another significant detail, which stays unclear very often, is weather MIE is defined as an incident or acting energy (incident minus transmitted). The objective of this work was to obtain missing data on MIE of LPG based mixtures at different pressures and to check the obtained results in a model engine.

2. Experimental layout

Single pulses of Nd:YAG laser radiation (1064 nm, 71 ps, LS-2151, Lotis TII) were focused by a 100 mm focal length fused silica lens (25.4 mm dia.). Chamber volume (consisting of standard 40 i.d. vacuum crosses) was vented and pumped down to 1 mbar prior to every measurement. Gas was fed to a partial pressure according to a desired fuel mixture composition, controlled by a membrane vacuum gauge (CTR100, Ceravac). And finally, air was supplied by compressor to the pressure controlled by gauge and a special sensor (ADZ-SIML-20.0, Sensortechnics). To characterize mixture composition we normalized it to 30.5 oxidizer/fuel volumetric (and partial pressure) ratio as stoichiometric r_{stoich} for 6% propane, 30% butane, 64% isobutane (commercial bottle for gas burners). So equivalence ratio was calculated as $\varphi = r_{stoich}/r$. Control photodiode was calibrated with energy meter (30A-P-RP, Ophir) placed instead of the target before each measurement series following a usual procedure. Safety valve opening at 4 bar limited the upper pressure range. Optical breakdown was identified by plasma radiation using a fiber coupled photo multiplying tube (H6780-04, Hamamatsu), which was blind to incident laser radiation.

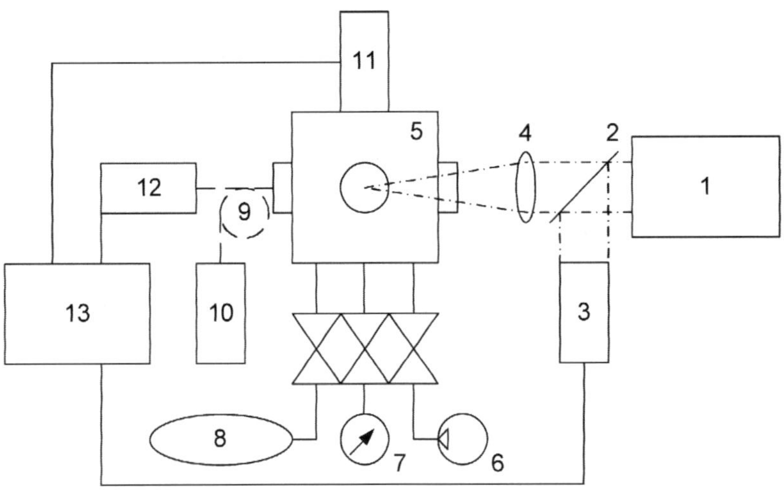

Figure 1. Experimental setup (1 – laser, 2 – beamsplitter, 3 – control photodiode, 4 – lens, 5 – calorimetric bomb, 6 – compressor, 7 – membrane vacuum gauge, 8 – gas bottle, 9 – optical fiber, 10 – spectrometer, 11 – high pressure sensor, 12 – PMT, 13 – oscilloscope)

3. Results and their discussion

We have measured incident MIE, or minimum pulse energy (MPE) as a more practical one, for mixtures of 0.6 to 1.6 equivalence ratio at initial pressure 1 to 3.5 bar, at room temperature. Our laser was unable to supply enough energy for ignition of mixtures with φ less than unity, but it was always far enough for laser breakdown. The results are presented at figure 2. Two trends very important for laser ignition could be seen. MPE, as expected, decreases significantly with pressure increase, same as gas optical breakdown threshold does. Minimum of the dependency drifts towards greater φ (richer mixture), but at the same time curve meander broadens making ignition less sensitive to pulse energy in a wider range of equivalence ratios. Although, data confirming these features can be found in literature for other gases, nothing like this has been shown up to date since very works consider both different pressures and equivalence ratios. Figure 2b also shows that at φ>1.3 MPE dependency on pressure becomes more steep with equivalence ratio increase.

For electric ignition, MIE dependency minimum is known to move towards richer mixtures with fuel molecular weight increase [19]. This is also confirmed for laser ignition at 1 bar: for hydrogen MIE minimum was at φ~0.85 [20], for propane – at φ~1.2 [13], and for dodecane – at φ~3.7 [13]. The optical breakdown threshold was not so sensitive to the mixture composition, obtained value was 1.5 mJ at φ=1, p=1 bar.

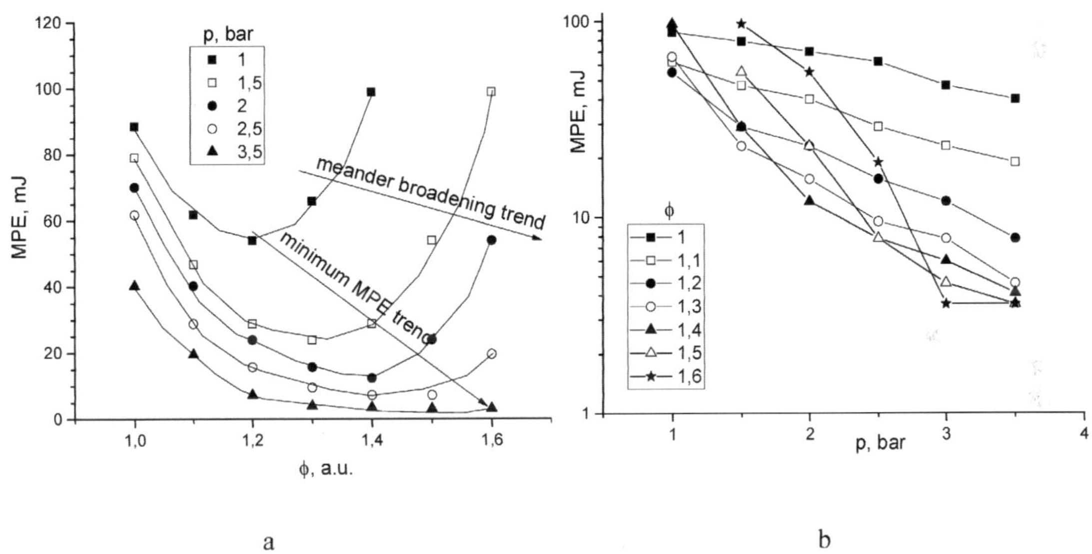

Figure 2. Dependency of MIE on mixture equivalence ratio (a) and pressure (b) at ps-laser impact.

Data for comparison are available for methane, hydrogen and propane at φ=1 and *p*~1 bar mostly. For propane MIE was 0.8 mJ [13], and that is the case when the great difference comes from methods of MIE and MPE measurement. We found in our experiments that 82% of laser radiation was transmitted through the focal region at φ=1 and *p*=1 bar at MPE, at lower energies still sufficient for breakdown up to 98.5% was transmitted through the focal region.

The results we obtained using nanosecond laser pulses did not differ significantly from the discussed above. However, greater pulse energies available from nanosecond lasers allowed igniting mixtures in a broader range of φ and *p*. At mixture pressure of 8 bar, that it usual for ignition in a 2-stroke piston engine, MPE was 2.5 mJ [21]. For methane at 10 bar it is known to be 3 mJ [22]. These values are easily reachable in compact DPSS lasers making those very promising for laser spark plugs design.

We have checked our findings on MPE in a model Wankel rotary engine (49-PI, O.S. Engines) that is normally ignited by glow plug of only 6 mm dia. (type F). For that, we used specially designed laser spark plug based on passively Q-switched Cr/Nd:YAG rod with resonator mirrors coating on edges, end-pumped with a fiber-coupled 50 W laser (FL-FCSB04-50-808, Focuslight). The radiation pulses of 2.5 mJ were focused by 6 mm dia. half-ball lens. Ignition timing was synchronized with shaft rotation. The engine was fed by a rich mixture φ~1.2 and was able to spin up to 15,000 r.p.m. without load. Since engine started deceleration immediately after radiation switch off, glow ignition did not take place.

4. Conclusions

Data on minimum laser pulse energy for butane based LPG) fuel mixtures ignition at elevated pressures have been obtained for the fist time. Also for the first time, we have shown that at pressure increase minimum of ignition energy dependency on air/fuel ratio drifts towards richer mixture. But at the same time, meander of this dependency becomes broader allowing low energy laser ignition in a wider range of equivalence ratios.

Regarding to LPG abundance in transportation, these data are of extreme importance for the development of laser ignited engines. For example, we have shown for the first time possibility of Wankel engine operation with LPG being laser ignited. This can significantly reduce operating costs and harmful emissions for this very compact type of engines.

Acknowledgements
This research has been performed at "Beam-M" facility, following the government task (No.13.6978.2017/BCh) and supported by the grant of the President of Russia (No. 14.Z56.16.5407-MK).

References

[1] Pavel N, Tsunekane M and Taira T 2011 *Opt. Express* **19** 9378-9384
[2] Phuoc T X 2010 Handbook of Combustion, (Weinheim, Wiley-VCH Verlag GmbH) p 95-123
[3] Loktionov E Y, Pasechnikov N A, Protasov Y S, Protasov Y Y and Telekh V D 2015 *J. Appl. Spectrosc.* **82** 607-613
[4] Dumitrache C, Boissiere A, Baumgardner M, Marchese A, Yalin A, Maria A and Roucis J. 2015 *Proc. Laser Ignition Conf.* Argonne, Illinois, Optical Society of America, W3A.4
[5] Morsy M H 2012 *Renew. Sustain. Energy Rev.* **16** 4849-4875
[6] Loktionov E Y and Protasov Y Y 2010 *Doklady Phys.* **55** 313-316
[7] Varaksin A Y, Protasov M V, Romash M E and Kopeitsev V N 2015 *High Temp.* **53** 595-598
[8] Beduneau J-L, Kim B, Zimmer L and Ikeda Y 2003 *Combust. Flame* **132** 653-665
[9] Li X, Yu Y, Yu X, Liu C, Fan R and Chen D 2014 *Opt. Express* **22** 3447-3457
[10] Böker D and Brüggemann D 2011 *Inter. J. Hydrogen Energy* **36** 14759-14767
[11] Weinrotter M, Kopecek H, Wintner E, Lackner M and Winter F 2005 *Inter. J. Hydrogen Energy* **30** 319-326
[12] Morsy M H and Chung S H 2003 *Experim. Therm. Fluid Sci.* **27** 491-497
[13] Lee T-W, Jain V and Kozola S 2001 *Combust. Flame* **125** 1320-1328
[14] Phuoc T X and White F P 1999 *Combust. Flame* **119** 203-216
[15] Ronney P D 1994 *Opt. Eng.* **33** 510-521
[16] Srivastava D K, Wintner E and Agarwal A K 2014 *Opt. Lasers Eng.* **58** 67-79
[17] Griffiths J, Riley M J W, Borman A, Dowding C, Kirk A and Bickerton R 2015 *Opt. Lasers Eng.* **66** 132-137
[18] Xu C, Fang D, Luo Q, Ma J and Xie Y 2014 *Opt. Laser Technol.* **64** 343-351
[19] Lewis B and von Elbe G 1987 *Combustion, Flames and Explosions of Gases*. (Orlando: Academic Press)
[20] Syage J A, Fournier E W, Rianda R and Cohen R B 1988 *J. Appl. Phys.* **64** 1499-1507
[21] Loktionov E Y, Pasechnikov N A and Telekh V D 2017 *J. Phys.: Conf. Ser.* in press
[22] Weinrotter M, Kopecek H and Wintner E 2005 *Laser Phys.* **15** 947-953

Influence of target structure on laser plasma generation efficiency

E Loktionov, V Telekh

Bauman Moscow State Technical University, Moscow, 105005, Russia

stcpe@bmstu.ru

Abstract. Spatial restrictions on heat transfer imposed in foil and sintered powder titanium targets have resulted in sufficient increase of efficiency of laser plasma generation as compared to bulk Ti targets. Especially at low laser fluences, where target material input in plasma is higher than ambient air's. Also, the pattern of momentum coupling coefficient C_m dependency on laser fluence F has changed from convex to concave. Minimum difference in C_m values for bulk and foil targets was 1.68 times, maximum – 1.5 orders of magnitude (always higher for foil). At the impact on sintered porous targets momentum coupling coefficient was lower than for foil, but normalised by mass density, the results were about equal. To our mind, obtained results show that suppression of heat dissipation in porous targets can be same efficient as in foils, but with more benefits for feeding systems and energy efficiency of laser plasma generators.

1. Introduction

Heat transfer in laser irradiated materials defines the amount of energy left for plasma production. The amount of residual energy, i.e. staying in target, normalized to the incident one reaches 0.6 or even more [1]. The difference in laser ablation processes of well (metals) and poorly (polymers) heat conducting materials is well known. The usual parameters for laser ablation efficiency evaluation are specific ablation rate $\Delta m/E$, momentum coupling coefficient C_m, specific impulse I_{sp} and energy efficiency η [2]. The values depend a lot on laser radiation parameters and ambient conditions [3].

To reduce heat losses in metal targets, foils [4, 5] and thin films [6, 7] are often used. However, for plasma generation facilities, foils and films are not always available due to possibility of wrapping or burning through at high laser intensities. Despite even higher ability to reduce heat losses due to lateral transfer also confined, data on laser ablation of porous metals are nearly missing. Although, porous silicon is investigated for electrical micro-thrusters [8, 9]. Laser plasma generators could benefit from using porous targets sintered from powders due to easier powder storage and transfer systems [10], ability for active layer regeneration. This technology is now well developed in 3D-printing [11].

Moreover, porous media provide good abilities for making tailored targets (in a larger scale than multilayered thin films), consisting of different phase materials (impregnated with liquids, gas filled). Better impact localization because of mass transfer restrictions could also be beneficial. Great surface area could be used for laser induced chemical reactions. It is surprising that laser ablation of this class of materials is so under investigated (despite the former is used for the manufacturing of the latter for a long time [12]). By the moment only the applied research in this area could be found in laser cleaning [13].

Polymer powder ablation was studied in [14], but the results were not so good since piling the powder up led to high losses because of splashing (similar to liquids [15]). Compacted (compressed up to 40 MPa) powder ablation led to high mass consumption, perhaps, due to still weak links between grains easily broken by blast wave. The attempt to introduce metallic powder in a polymer matrix also did not give the expected result since grains left the matrix before full evaporation [16]. We suggested that sintered powders should have strong enough links between grains to behave as a bulk material, but with low heat conduction. So the aim of our study was to check if this is going to be true and to compare the results in terms of momentum coupling coefficient with bulk and foil Ti targets.

2. Experimental layout

The sketch of experimental setup is presented at Figure 1. We irradiated targets at ambient conditions in a single pulse mode using frequency doubled Nd:YAG laser (532 nm, 12 ns, up to 80 mJ, Lotis TII LS-2147), spot diameter at target was 224 μm. Incident laser energy was measured by control photodiode calibrated following standard procedure, using energy meter (Ophir 30A-P-RP) placed just after lens instead of a target. Recoil momentum has been measured using a calibrated force sensor (531 mV/N, PCB Piezotronics 209C11). Electric signals were recorded using PC-coupled digital oscilloscope (Aktakom ADS-3114). Force signal was integrated by time to get recoil momentum.

The targets were: 1.5 mm thick Ti sheet (considered as bulk); 0.1 mm thick Ti foil; 0.9 mm thick porous target made of sintered titanium powder (fine gas filter manufactured by VMZ-Techno, 0.26 porosity = non-solid volume / total volume, average pore dia. 8 μm). To avoid targets jumping, those were attached to the sensor by sticky tape. Ablation mass consumption has been measured for 100 pulses (in a scanning mode 0.2 mm/s and 10 Hz pulse repetition) using analytic scales with 10 μg precision (CAS CAUW-120D).

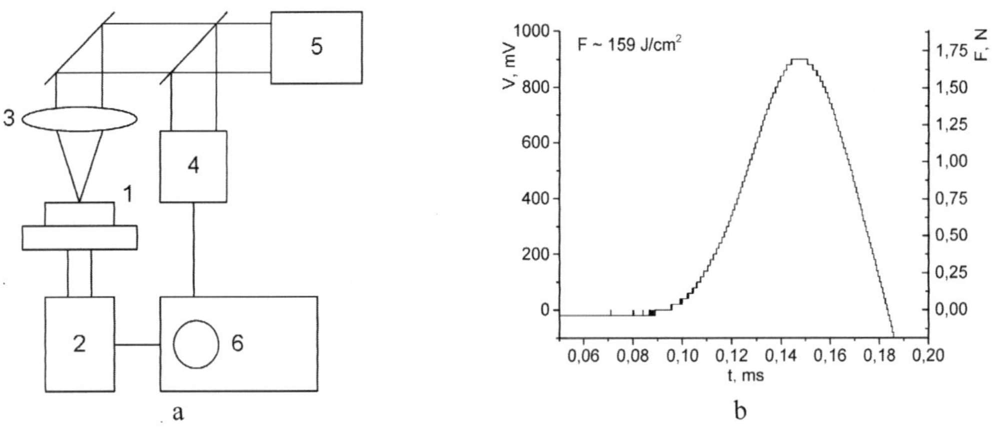

Figure 1. Experimental setup (a) and typical force signal (b):
1 – target, 2 – force sensor, 3 – lens, 4 – control photodiode, 5 – laser, 6 – digital oscilloscope

3. Results and discussion

Data on momentum coupling at laser ablation of titanium are very poor and are unavailable for direct comparison. Most of those are referenced in [17]. The results of our measurements are presented at Figure 2. For the targets with heat transfer restrictions (*2*, *3*) the dependency has quite different pattern as compared to the bulk titanium. Minimum difference in C_m values for bulk and foil targets was 1.68 times at high fluences, maximum – 1.5 orders of magnitude (at lower fluences, always higher for foil). At the impact on sintered porous targets, momentum coupling coefficient was lower than for foil ($C_m(F)$ dependency pattern was identical), but normalized by mass density, the results become about equal. For foil and porous targets, momentum coupling (both values and dependency patter) at low fluences was similar to polymers (which are heat insulators) [18] and metals at femtosecond laser irradiation (when heat affected zone is very small) [19]. Similarity of values obtained for all targets at high intensities can be explained by reduction of residual energy coefficient decreases at laser fluence increase [1] and by increase of ambient air input in recoil momentum, making target properties not so important.

The results presented at Figure 2a are well fitted with a curve $\ln(F/F_a)/F$ [20] corresponding to ablation threshold of F_a = 1 J/cm^2. Those correspond well to the results of [21]. Estimating I_{sp} and η from this results makes it obvious, that at high fluences input of ambient air in recoil momentum becomes dominating since most of laser energy is absorbed before target surface. So ablation exhaust velocity calculation as $\langle v\rangle = C_m/(\Delta m/E)$ is not correct since weight of air involved in thrust generation is sufficient, same as in gas turbine jet engine. Absolute mass removal was nearly asymptotic at fluences greater than 120 J/cm^2, so it looks like some certain amount of target material needs to be ablated for near-surface area saturation. Ablation rates at maximum fluences of 250 J/cm^2 were 4.0 μg/J for bulk target, 2.0 μg/J for foil and 3.1 μg/J for porous. As expected, ablation threshold decreases at heat transfer restriction, and air influence becomes pronounced at lower fluences. This leads to great energy efficiency increase with calculated values far over unity same to shown for laser-induced shockwaves in [22], so target structure becomes not so important fluences higher than ca. 100 J/cm^2. This finding is important for laser shock processing [23] since allows to reduce laser power. Porous layer could be easily coated on almost any surface, and then removed if needed.

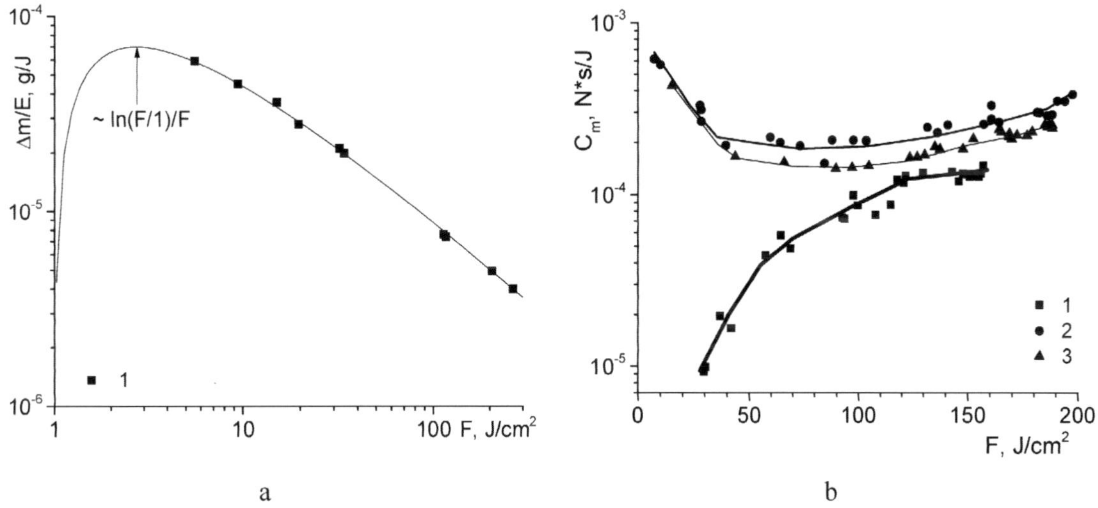

Figure 2. Specific ablation rate (a) and momentum coupling coefficient (b) at impact on Ti targets: 1 – bulk (1.5 mm), 2 – foil (0.1 mm), 3 – porous

Neither target microscope imaging, nor weighing revealed increased mass consumption from a porous media, so our initial suggestion about sufficient strength of inter-grain links was correct. Another influencing parameter is focal spot to grain size ratio. In our case it was ~11, so even at comparatively weak links there should be no large scale destruction. Focal spot is unknown for [14], but grain sizes were 40 μm and 300 μm, so the discussed ratio should have been significantly less. It is still unclear how ablated mass measurement could be affected by oxidizing – ca. 1.7 mm width sky blue stripes (color is characteristic for titanium oxide) left along scans at all targets. Further experiments in vacuum could answer questions regarding to ambient air influence.

4. Conclusions

Spatial restrictions on heat transfer imposed in foil and sintered powder titanium targets have resulted not only in sufficient efficiency of gas-plasma flows generation increase as compared to bulk Ti targets (especially at low laser fluences), but also in change of the pattern of momentum coupling coefficient C_m dependency on laser fluence F from convex to concave.

To our mind, this confirms the role of heat dissipation processes in decrease of efficiency at the impact on bulk targets. In case of impossible use of foil targets, thick porous targets could be used nearly same effectively. Porous media and powder based working media seem to be rather promising for laser plasma generators.

Acknowledgements

This research has been performed at "Beam-M" facility, supported by the Russian Ministry of Education and Science – grant of the President of Russia No. 14.Z56.16.5407-MK.

References

[1] Vorobyev A Y and Guo C 2007 *J. Phys.: Conf. Ser.* **59** 418-423
[2] Loktionov E Y, Ovchinnikov A V, Protasov Y S, Protasov Y Y and Sitnikov D S 2012 *Opt. Spectrosc.* **112** 631-637
[3] Loktionov E Y, Protasov Y S and Protasov Y Y 2014 *Quant. Electron.* **44** 225-232
[4] Horisawa H, Yamaguchi S and Funaki I 2011 *Appl. Phys. A* **104** 935-940
[5] Chen S N, Gregori G, Patel P K, Chung H K, Evans R G, Freeman R R, Saiz E G, Glenzer S H, Hansen S B, Khattak F Y, King J A, Mackinnon A J, Notley M M, Pasley J R, Riley D, Stephens R B, Weber R L, Wilks S C and Beg F N 2007 *Phys. Plasmas* **14** 102701-6
[6] Bedeaux B C, Trott W M and Castaneda J N 2010 *J. Appl. Phys.* **107** 034912
[7] Rodriguez G, Valenzuela A R, Clarke S A and Thomas K A 2006 *Proc. SPIE*, **6261** 62610O-10
[8] Wayne A C, Christopher J M, Raghav R and Sarah B 2015 *J. Micromech. Microeng.* **25** 115022
[9] Wang S, Shen R, Ye Y, Hu Y, Wu L and Yang C 2013 *Proc. SPIE*, **8796** 87960C-6
[10] Loktionov E, Protasov Y and Telekh V 2017 *J. Phys.: Conf. Ser.* **830** 012061
[11] Gu D 2015 *Laser Additive Manufacturing of High-Performance Materials.* (Berlin Heidelberg: Springer-Verlag)
[12] Laiho R and Pavlov A 1995 *Thin Solid Films* **255** 9-11
[13] Sanjeevan P and Klemm A J. 2005 *Proc. 2nd Scottish Conf. Postgrad. Res. Built and Natural Environment (PRoBE)*. Glasgow, Caledonian University, 463-472
[14] Sasoh A, Ogita N and Sinko J E 2010 *Proc. AIP*, **1230** 163-167
[15] Fardel R, Urech L, Lippert T, Phipps C, Fitz-Gerald J and Wokaun A 2009 *Appl. Phys. A* **94** 657-665
[16] Schall W O, Tegel J and Eckel H-A 2005 *Proc. AIP*, **766** 423-432
[17] Loktionov E Y, Protasov Y S and Protasov Y Y 2013 *Opt. Spectrosc.* **115** 758–763
[18] Sinko J E and Phipps C R 2009 *Appl. Phys. Lett.* **95** 131105-3
[19] Loktionov E Y, Ovchinnikov A V, Protasov Y S, Protasov Y Y and Sitnikov D S 2013 *High Temperature* **51** 867-869
[20] Loktionov E Y, Ovchinnikov A V, Protasov Y Y and Sitnikov D S 2011 *Plasma Phys.s Rep.* **37** 1208-1214
[21] Vladoiu I, Stafe M, Negutu C and Popescu I M 2008 *J. Optoelectron. Adv. Mat.* **10** 3177-3181
[22] Loktionov E Y, Ovchinnikov A V, Protasov Y Y, Protasov Y S and Sitnikov D S 2012 *J. Appl. Spectrosc.* **79** 104-112
[23] Ocana J L, Porro J A, Morales M, Correa C, Colon C, Alonso A and Autric M 2012 *Proc. AIP* **1464** 209-218

Improvement of technical purpose materials performance characteristics with the radio frequency low pressure plasma

L Yu Makhotkina[1], V I Khristoliubova[1]

1Kazan National Research Technological University, 420015, Russian Federation, Kazan, 68 Karl Marks st.

valllerrriya@mail.ru

Abstract. The main aim of the work is to solve the actual problem of increasing the competitiveness of tanning products by reducing the prime cost and improving the quality of finished products due to the increased durability of the working elements of tanneries. The impact of the low pressure radio frequency (RF) plasma in the processes of treating for modification of the materials for special purposes is considered in the article. The results of working elements of tanneries and the materials for special purposes sample processing by a RF low pressure plasma are described. As a result of leather materials nano structuring and nano modifying physical, mechanical and hygienic characteristics were increased. Processing of the technical purpose materials allows to increase operational performance of products and extend their lifespan.

1. Introduction

The production of competitive products and the assurance of the release of guaranteed high quality materials of light industry through the introduction of modernized technologies is an integral part of increasing the production efficiency of the textile and light industries. Conditions that ensure the increase of an efficiency in the production can be achieved through technological re-equipment, improving of the quality of raw materials and reagents and carrying out of organizational measures [1-5]. One of the main reasons for the technical and technological backward trend of the leather industry from foreign countries is the low working capacity of the equipment used. A worn and technically obsolete machinery and equipment park is not able to produce a modern assortment of high-quality competitive products. In the leather industry, it is especially important to modernize the equipment for technological operations of planing, breaking and smearing of genuine leathers, which are carried out on planing, milling and breaking machines. At the same time, the quality of planing, skinning and breaking up of leather fabric is determined not only by the properties of the material of the processed semi-finished product, but also by the performance characteristics of the tools (spiral, disk, center, and knife-edge knives), knife-surface grinding regimes along the generatrix, and parameters of the grinding device. In this case, in connection with this, it is necessary to have a scientific justification for the sharpening of the surface of the knives along the generatrix and the optimal design of the grinding device. Recently, complex studies of the system "grinding machine - knife shaft, disk (spiral) knives - a processed semi-finished product" are aimed at providing the required quality of the produced leather while optimizing the physical and mechanical characteristics of tools, increasing the resistance of spiral and disk knives, which, in turn, determines the performance characteristics of the quality of the milling and planing equipment and the productivity of the machinery of production.

Problems of improving the quality of tanning materials during mechanical operations have been considered for a long time [6-8]. A review and analysis of the studies carried out in this direction show that it is necessary to create applied methods for increasing the quality and durability of the working elements of tanneries.

The work is aimed at solving the actual problem of increasing the competitiveness of tanning products by reducing the prime cost and improving the quality of finished products due to the increased durability of the working elements of tanneries.

2. Results and discussion

To investigate the effect of the interaction of a RF low pressure plasma with the material surface of tanneries equipment the experiment was performed in an inert gas plasma, argon, 0.06 g/s and in a mixture of a plasma chemical gas, methane, 0.004 g/s and argon, 0.06 g/s. The working pressure in the chamber is 24-26 Pa. For the deposition of a diffusion protective coating it is necessary to conduct preliminary cleaning and surface activation in a RF plasma of a capacitive discharge. Surface treatment with argon ions with an energy of 60-80 eV made it possible to reduce the surface roughness parameter from 0.4 to 0.19 μm, which in turn increases the quality of the workpiece itself.

Characteristics of a plasma RF generator: power consumption 3.5 kW, output frequency 13.56 MHz. In order to concentrate the electric field near the surface an additional negative potential of -20 V was applied to the product. The processing time was ~ 40 minutes in total: 20 minutes in argon, 20 minutes in a mixture of gases, inert and plasma chemical.

To determine the physical and mechanical properties the microhardness, roughness, elastic modulus and elastic recovery coefficient a scanning nano-hardness meter "NanoScan-3D" was used [9]. Hardness measuring is implemented on the analysis of the dependence of a load when indenter is pressed into the surface of the material on the depth of introduction of the indenter. This method is based on the ISO 14577. The indentor of Berkovich type is used which is a triangular diamond pyramid with an angle at the apex of about 142°. During the dynamic indentation the indentor is pressed into the surface of the sample at a constant speed, when the specified load is reached, the indenter is retracted in the opposite direction. The values of the load and the corresponding displacement of the indenter are recorded. Feature: piezoelectric resonance cantilever with tuning fork with high flexural stiffness of the console (~ 2·104 N / m).

Change of the properties such as hardness, Young modulus and coefficient of restitution before and after RF low pressure plasma treatment can be seen below along the depth of the formed diffusive coating (Fig. 1.a-1.c).

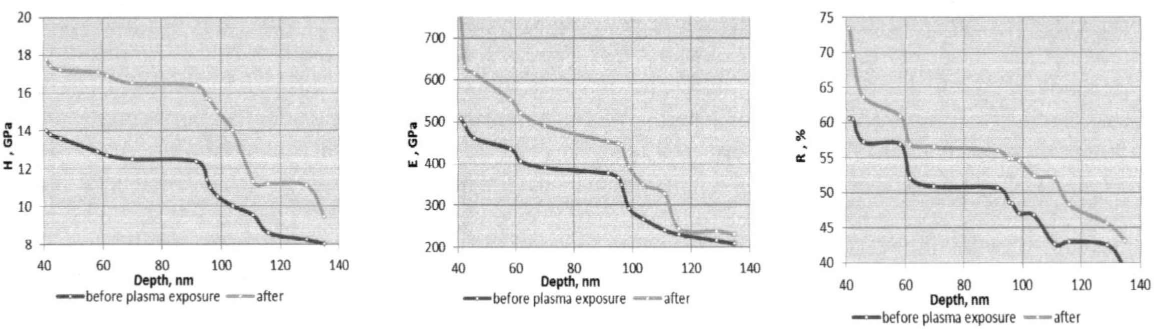

a b c

Fig. 1. Changes of physical and mechanical properties of tool steel before and after RF low pressure plasma treatment along the depth of the metal sample: a – hardness, b - Young modulus, c - coefficient of restitution.

Characteristics of the center and disc milling drills made of tool steel before and after the treatment were changed as follows: surface microhardness of a control knife 4,25±0,5 GPa, knife with a diffusive coating 15±4 GPa, roughness parameter of the knife edge is Ra, 2,50±0,5 mkm, knife with a diffusive layer 0,5±0,1 mkm, groove radius of a cutting edge of a control knife 36±0.5 mkm, knife with a diffusive layer on a surface 38±0.5 mkm.

Studies have shown that strengthening of the working elements of machining tanneries such as center and milling drills with a protective coating of carbides and nitrides with a strengthening phase makes it possible to increase the resistance of working elements to wear and the time of stable processing of semi-finished products, and also improves the quality of finished leather materials in the batch by reducing mechanical defects in leather. According to the test results, it is established that the relative increase in the area of hides during the cutting with experimental knives is more smoothed and ranges from 30 to 40%, and for serial knives from 10 to 45%. Evaluation of the cutting quality was carried out by an organoleptic method and it was established that the sheepskin leather fabric achieves sufficient plasticity in all directions with an increase in area from 25 to 40%, with a relative reduction in area below 25% a repetition of the hides is necessary.

The research of the properties changes of leather samples treated with low pressure RF plasma was carried out on an experimental samples using two technologies: nanostructured samples (genuine leather of a cattle, which at the final stage of finishing production was treated with RF plasma of reduced pressure in the mode Wp = 1.5 kW, τ = 7 minutes), nanomodified samples (genuine leather of a cattle at the final stage of fatliquoring on the top size and dyeing, at step of the fixer deposition coating film two-fold biocidal agent of silver nano solution of 0.2% was injected and at the finishing production the samples were treated by RF plasma in a reduced pressure mode Wp = 1.5 kW, τ = 7 minutes).

For the investigation of the surface and cutting edge structure of leather control samples and treated with the low temperature RF plasma scanning electron microscope Carl Zeiss evo-40 INCAx-sight was used. The range of the accelerating voltage of the electron microscope during the examining of the surfaces and cuts of control leather samples processed in low temperature RF plasma and modified samples in low temperature RF plasma with silver - 0.2-30 kV. The working current range is 0.5 pA - 5 mkA.

The physical and mechanical properties of leather treated by low temperature RF plasma and modified by silver are determined by the properties of the fibers and their mutual arrangement. The microstructure of the cut-off of the control leather samples is shown in fig. 2.a – 2.e, the cut-off of the leather samples of cattle treated with low temperature RF plasma - in fig. 3.a – 3.e, the cut-off of the leather samples of cattle treated with low temperature RF plasma with silver - in Fig. 4.a – 4.e. All the images were made at magnification of 100, 200, 500, 1000, 2000 times. The edge of the untreated sample has a homogeneous hummocky surface, where the character of the arrangement of collagen beams and fibers is observed. The average angle of inclination of fiber bundles to the surface of the dermis varies in different parts of the dermis. Interfascicular spaces evenly distributed over the area of the cut are clearly visible. After the interaction of the plasma the regularity of the interweaving of the structural elements of the leather is not disturbed and the angle of the interweaving of the fiber bundles that affects the strength, wear resistance and plasticity of the leather is not changed. However, after low temperature RF plasma treatment the fibrous structure separates. This makes the leather to become softer and more viscous, the penetration depth of the dyes increases. Samples treated with silver and low temperature RF plasma acquire a soft, uniform surface due to deeper penetration of silver. This kind of treatment improves the bactericidal properties the leather.

Thus, as a result of plasma treatment, the maximum changes of a leather properties of a cattle occur. This is explained by the fact that the plasma treatment is volumetric and allows the plasma to be effected throughout the volume [10,11].

IOP Conf. Series: Journal of Physics: Conf. Series **927** (2017) 012032 doi:10.1088/1742-6596/927/1/012032

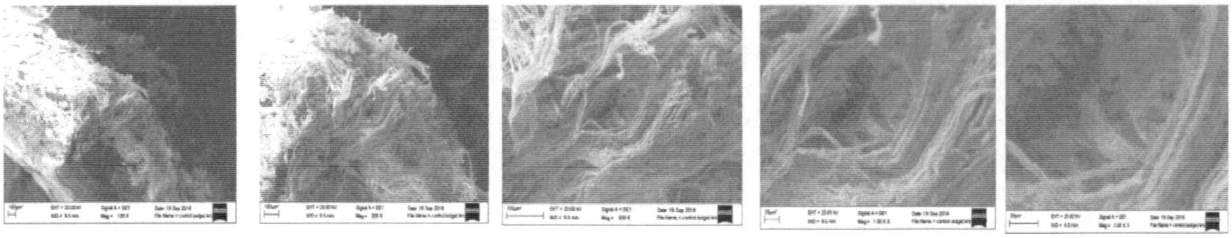

a b c d e

Fig. 2. Images of the control sample of the cattle leather: a - x100, b - x200, c - x500, d - x1000, d - x2000.

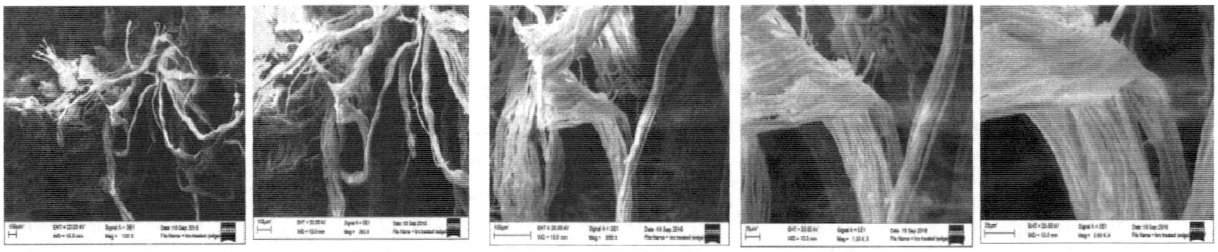

a b c d e

Fig. 3 Images of the cattle leather sample treated by low temperature RF plasma: a - x100, b - x200, c - x500, d - x1000, d - x2000.

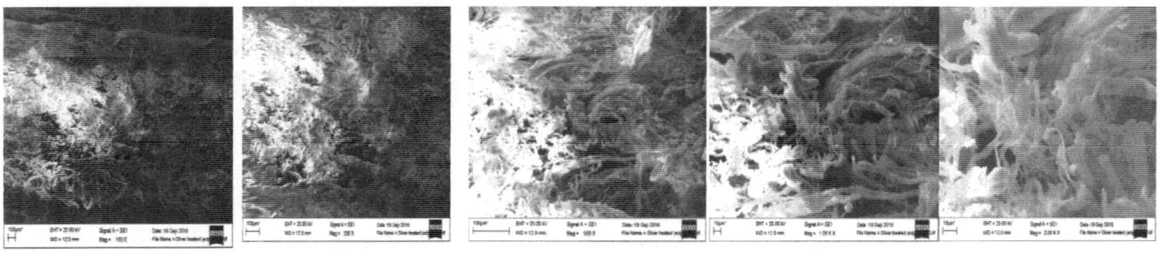

a b c d e

Fig. 4. Images of the cattle leather sample modified by low temperature RF plasma with silver: a - x100, b - x200, c - x500, d - x1000, d - x2000.

As a result of nanostructuring and nanomodifying of materials the physical, mechanical and hygienic characteristics of special technical purposes materials samples are increased. Processing of leather materials in the specified ranges of processing conditions in the low pressure RF plasma can increase the performance of products for special use and extend their service life.

Samples treated with plasma were cut with knives also treated with plasma. The edge part of a sample cut with the treated tool was clear without villi and scoring. These features are very important in the furrier's work when the parts of the products are cut from the skin. No micro cracks are observed on the edge of the sample cut with a knife treated by low temperature RF plasma. The elimination of those cracks is very significant as they lead subsequently to a weakening of the bonding strength of the fibers among themselves and this in turns leads to a decrease in the strength of the material.

3. Conclusions

It is established that processing in a jet RF discharge at a reduced pressure leads to a change in the physical and mechanical properties of metals and their alloys. The effect of the low-pressure RF discharge on the physical and mechanical properties of metals and their alloys depends both on the plasma parameters and on the structure of the material. It has been experimentally established that the change in the physical and mechanical properties of tool steel after exposure to low-pressure jet RF plasma occurs with a change in the phase composition and the parameters of the crystal lattice on the surface. The physical regularities of the change in the properties of the metal are revealed.

Gasification (carbiding) of surface layers of metals and alloys occurs at a depth of up to 500 nm during the processing time up to 40 minutes, which results in an increase in the strength properties, longevity and service life of the products by a factor of 1.5-2. The developed technological processes of plasma gas saturation of steel made it possible to increase the service life of the products by 2-2.5 times in comparison with the untreated ones, to increase the microhardness of the surface by 25%, and the wear resistance by 20%.

As a result of nano structuring and nano modifying of genuine leather materials itself for the manufacturing of footwear uppers physical, mechanical and hygienic characteristics of leather were also increased. Processing of the footwear upper materials allows to increase operational performance of technical products and extend their lifespan. The following application of the new technology of cutting the modified materials with the processed treated tools that has protective diffusion coating will allow to obtain high quality of the finished leather products at the final stage and decrease the ineffective outage time of the equipment in leather industry.

References

[1] Osin Yu N, Makhotkina L Yu, Abutalipova L N and Abdullin I Sh 1998 SEM and X-ray analysis of surface microstructure of a natural leather processed in a low temperature plasma *Vacuum* v 51(2) pp 221-225.

[2] Makhotkina L Yu and Sharifullin S N 2016 Design of special purpose products made of nanomodified collagen-containing materials with radio-frequency discharge *IOP Conference Series: Materials Science and Engineering* v 134, 012020.

[3] Khristoliubova V I, Kashapov N F and Shaekhov M F 2016 Gas and plasma dynamics of RF discharge jet of low pressure in a vacuum chamber with flat electrodes and inside tube, influence of RF discharge on the steel surface parameters *IOP Conference Series: Materials Science and Engineering* v 134, 012017.

[4] Makhotkina L Y and Khristoliubova V I 2015 Variation of footwear wear resistance depending on methods of treatment. *International Journal Of Applied And Fundamental Research* № 1, URL: www.science-sd.com/460-24749 (11.03.2015).

[5] Sirazieva L F, Stepin S N and Makhotkina L Yu 2004 Dispersing agents for emulsion type coating compositions (review) *Lakokrasochnye materialy i ikh primenenie* № 10, pp 25-28.

[6] Makhotkina L Yu 2006 Processing of integrated footwear materials by RF plasma of reduced pressure *Leather and footwear industry* № 5 pp 36-37.

[7] Ladianov V I, Gilmutdinov F Z, Nikonova R M, Kashapov N F, Shaekhov M F and Khristoliubova V I 2017 Effects of low pressure radio frequency discharge on the physical and mechanical characteristics and chemical composition of diffusion coating on a surface of complex configuration details *J. Phys.: Conf. Ser* v 789, 012031.

[8] Makhotkina, L Yu and Khristoliubova V I 2017 Studies of the influence of nonequilibrium plasma thermal exposure on the characteristics of the capillary-porous polymer material *J. Phys.: Conf. Ser* v 789, 012031.

[9] Makhotkina, L Yu, Khristoliubova V I and Khannanova-Fakhrutdinova L.R. 2016 Design of special purpose products made of nanomodified leather *Mathematics Education* 11(6) iejme.2013.134, pp. 1495-1503.

Propagation of a negative streamer over a bubble floating on a water surface

Yu S Akishev[1,2], V B Karalnik[1], M A Medvedev[1], A V Petryakov[1], N I Trushkin[1] and A G Shafikov[1]

[1] SRC RF TRINITI, 108840, Moscow, Troitsk, Pushkovykh street, 12, Russia
[2] NRNU MEPhI, 115409, Moscow, Kashirskoe shosse, 31, Russia

Abstract. The main goal of this work is to clarify in detail the interaction of negative corona and negative streamers with a bubble floating on a water surface. The corona and streamers were initiated with the metallic pin stressed by a step-wise high voltage of a negative polarity. It was shown that at low applied voltage and short distance between pin and bubble a negative corona foregoes an appearance of the streamers. This corona produces the ionic wind which directed towards the top of a bubble and influences both the geometry of a bubble and its life-time. In the case of high amplitude of the applied voltage and long distance between tip and bubble the negative streamer sliding on the bubble surface happens. A difference in the interaction of positive and negative streamers with a water bubble is discussed.

1. Introduction

The non-thermal plasma (NTP) at atmospheric pressure generates plentifully various reactive plasma species like atoms, radicals, excited atoms and molecules, ions, etc [1,2]. This is a reason why NTP has been widely used for initiation or acceleration of the needed bio-chemical processes in a liquid. At present different plasma-liquid systems are developed for activation of many types of liquids including the waste or tap water. Some approaches are based on the use of streamer discharges in ambient air which generate the streamers (thin current filaments) propagating above/over a liquid. Exactly the streamers are enriched with the reactive species. In this case, one may point two key parameters determining the efficiency of plasma activation. One of them is the rate of transfer of reactive species from streamers into the liquid (the higher this rate, the higher efficiency). Another one is the size of a liquid area which tightly contacts with streamers (the larger size of contacting area, the higher efficiency of plasma activation). The best way to satisfy these requirements simultaneously is to form many bubbles (foam) on a water surface and to generate the streamers sliding over the bubble surface. The efficiency of such approach has been demonstrated in paper [3] by the example of positive streamers generated by the pin electrode located above the bubbles. Besides the higher activation efficiency, such approach has an essential advantage from the electrical engineering point of view - because the bubble wall is extremely thin, the formation of streamers on a bubble surface does not require high-voltage pulses with very short (of ns range) rise-time that is necessary in the case of formation of long streamers sliding on a flat surface of a deep water.

In this work, the interaction of negative streamers with a bubble floating on a water surface is studied. The streamers were initiated by the metallic pin stressed by a step-wise high voltage of a negative polarity. There is a fundamental difference in behavior of a discharge in the case of a positive and negative polarity of the pin. In the latter case, a negative corona foregoes an appearance of streamers. This corona exhibits the current pulses (Trichel pulses) and the ion wind which both

IOP Conf. Series: Journal of Physics: Conf. Series **927** (2017) 012033 doi:10.1088/1742-6596/927/1/012033

influence the self-consisted behavior of the discharge-bubble system. The main goal of this work is to clarify in detail the mentioned above features of the interaction between negative streamers and the bubbles floating on a water.

2. Experimental setup

The scheme of the setup used in our experiments is shown in figure1. The sharpened metallic pin of a stainless steel wire of 1 mm in diameter served as a high-voltage electrode initiating the discharge on the floating bubble artificially formed on the liquid surface. The applied voltage amplitude was varied up to -15 kV. The grounded aluminum electrode was covered with a thin layer (about 2–3 mm) of a tap water with the electrical conductivity of 7.5 × 10^{-4} S/cm. Plasma forming gas was ambient air at atmospheric pressure. An isolated bubble floating on flat water was formed by a slow injection of a small amount of air through a narrow nozzle in the center of the aluminum plate. To do that a syringe was used. The diameter of a bubble basis ranged from 10 to 17 mm. A distance between the point electrode and the top of a bubble was varied in the range of 2–10 mm.

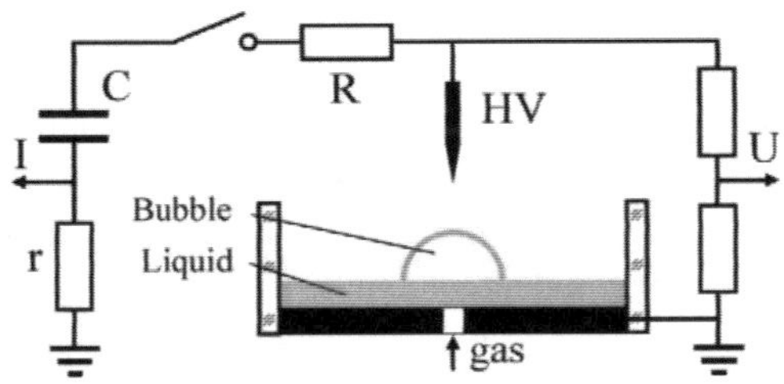

Figure 1. Sketch of the setup for study of interaction of a negative corona and streamer with a bubble.

3. Experimental results and discussion

At small distance h between the tip of HV electrode and bubble surface (h = 1 mm) and under low applied voltage U that is not enough to initiate a negative corona (U = -2 kV), we observed the attraction of the top of a bubble to the high-voltage electrode (see figure 2a and 2b). The rising of the bubble top on 0.23 mm in height happens slowly - approximately in 0.5 s. This effect is attributed with a positive charge on the top of a bubble induced by a negative potential on the high-voltage needle. Note that after a while, the attraction of water leads to accumulation of the water droplet on the HV electrode (see figure 2c).

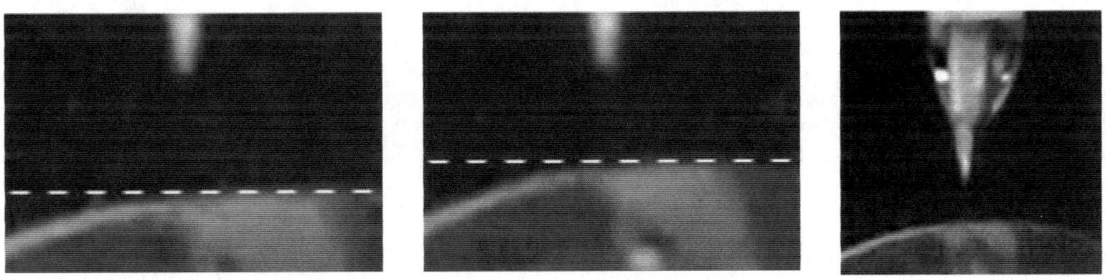

a) b) c)

Figure 2. The images showing the attraction of a bubble wall to the needle at the negative potential U lower than the corona inception voltage (a, b). The dashed lines in shots (a) and (b) mark the position of the bubble top at $U = 0$ and $U = -2$ kV. The image of the needle with small water droplet (c).

If increase the distance h up to 2.3 mm and applied voltage U up to or above of -4 kV, several new things happen. First, attraction of the bubble to the HV needle is practically absent, possibly due to diminishing the induced charge on the bubble surface. Second, a negative corona appears which exhibits pulsing regime (Trichel pulses). Typical example of the current waveform for Trichel pulses is presented in figure 3. In contrast to normal pin-to-plane corona with a metallic flat electrode, the amplitude of Trichel pulses in corona between the pin and a bubble is strongly irregular. We attribute this feature of Trichel pulses to non-stationary and inhomogeneous deformation of the bubble surface by the ionic wind generated by negative corona. The jet of ionic wind is directed from the needle towards the top of a bubble, crumples the bubble wall and makes a relatively deep dint.

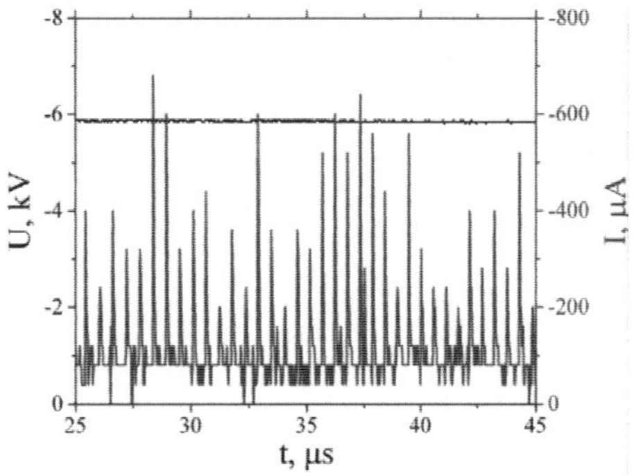

Figure 3. The fragment of the current waveform showing the behavior of Trichel pulses in the negative corona between HV needle and a bubble.

<u>Third</u>, after a while (approximately in 15 ± 1ms) the negative corona is succeeded by the plasma regime when many low-current streamers shunt the gap between HV needle and bubble. This regime eliminates the ionic wind and the bubble restores its initial shape. Plasma regime is unstable and quickly disappears. After that (approximately in 0.4 ± 0.2 s) the negative corona appears again and the ionic wind afresh crumples the bubble wall and makes relatively deep dint. Such alternating can happen several times. However, every time the jet of ionic wind drives water from the bubble top down to his basis, thins the bubble wall at this place and eventually breaks the bubble. The sequence of events described above is presented in figure 4.

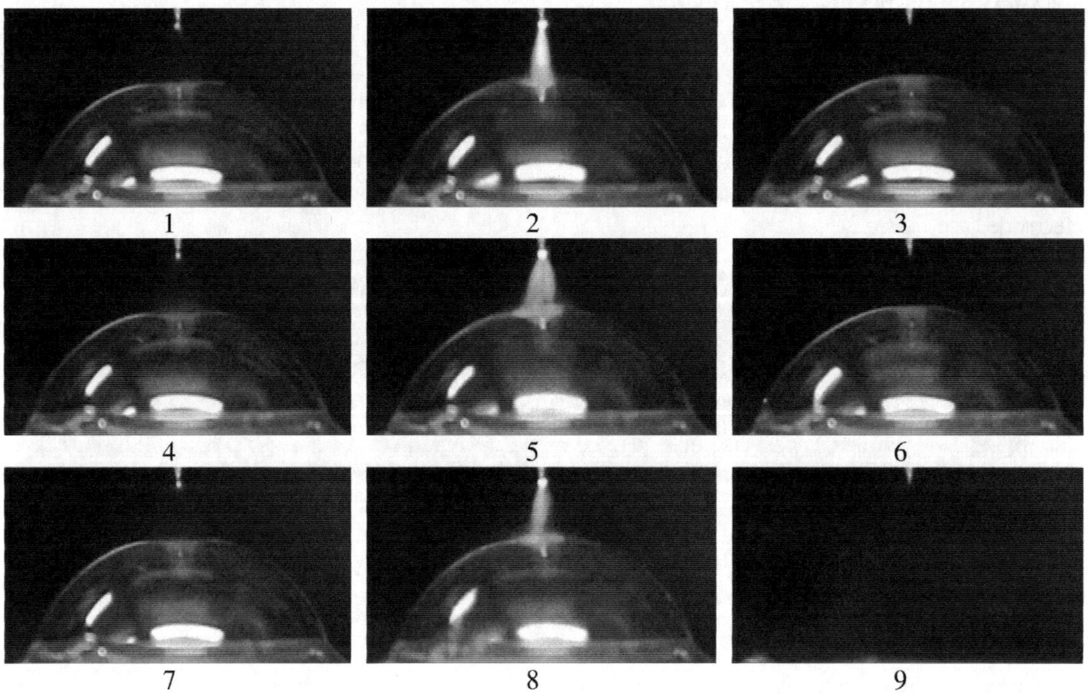

Figure 4. Set of the succeeded images showing the alternating of corona and streamer regimes. The exposure time of each shot is 16.7ms; the applied voltage amplitude is -4kV; the diameter of a bubble basis is 12 mm. Image #9 corresponds to the broken bubble.

As a rule, at the applied voltage lower than 10 kV the negative streamers cover the only small area around the top of the bubble and do not reach its base. However, at higher voltage, the streamers are able to reach the base (see figures 5a and 5b). There is a difference in the spatial structure of the positive and negative streamers propagating from the needle to the bubble. The positive needle induces a single streamer which does not branch and slides on the bubble in a form of alone current filament and reaches the bubble base (see figure 5c).The negative needle induces also a single streamer but this streamer splits itself into several secondary streamers at some distance in front of the bubble. Note the space between primary and secondary streamers and bubble is filled with glow discharge (see figures 5a and 5b). One may conclude therefore that negative polarity of the needle provides the larger area activated by non-thermal plasma at the same operating conditions. The impossibility of a negative streamer to strike perpendicularly to the bubble surface can be explained by the mechanism of his propagation. This mechanism assumes that electrons quickly drifting from the streamer head provide the ionization in front of the head and promote the forward movement of the streamer. It means that in the case of close approach of a streamer to a bubble some amount of electrons can be quickly accumulated on the bubble surface and negatively charge it. Due to that, the streamer will be repelled from the surface and deviate his trajectory from the normal direction.

A probability for the streamer to form the splitting at the turning point depends on the time spent at this point. Because the time diminishes with the applied voltage amplitude (the higher amplitude, the shorter the time), the probability of the splitting strongly diminishes as well (see figures 5a and 5b). Note the breaking of a bubble at low applied voltage happens after several strikes of low-current streamers but high-current streamer breaks the bubble by a single strike. A reason is that the high-current streamer releases higher energy on the bubble, the amount of which is enough to evaporate locally the bubble wall.

a) b) c)

Figure 5. The images of negative (a, b) and positive (c) streamers interacting with a bubble at higher applied voltage. a) U = –14 kV; b) U = –17 kV; U = +14 kV. The diameter of a bubble basis is 13 mm.

4. Conclusion

It was shown that at low applied voltage and short distance between HV needle and bubble, a negative corona foregoes an appearance of the streamers. This corona produces the ionic wind which directed towards the top of a bubble and influences both the geometry of a bubble and its life-time. In the case of high amplitude of the applied voltage and long distance between tip and bubble the negative streamer sliding on the bubble surface happens. A difference in the interaction of positive and negative streamers with a water bubble is shown and discussed.

5. References

[1] Akishev Yu, Balakirev A, Grushin M, Karalnik V, Kochetov I, Napartovich A, Petryakov A and Trushkin N 2015 *IEEE Trans. Plasma Sci.* **43** 745

[2] Korolev Yu, Frants O, Landl N, Geyman V and Matveev I 2007 *IEEE Trans. Plasma Sci.* **35** 1651

[3] Akishev Yu, Arefi-Khonsari F, Demir A, Karal'nik V, Petryakov A and Trushkin N 2015 *Plasma Sources Sci. Technol.* **24** 065021

Acknowledgments

This work was supported by Russian Foundation for Basic Research (grant № 16-02-00613).

Radio-frequency Induction plasma modification effects of disperse systems based on mineral glass

I S Miftakhov, E F Voznesensky, A I Nagmutdinova, A V Trofimov, V A Sisoev, A R Garifullina and I V Krasina

Department of Plasma Technology and Nanotechnology of High Molecular Weight Materials, Kazan National Research Technological University, Kazan, 420015, Russia

fortmayn@mail.ru

Abstract. In the article the possibility of use of high-frequency induction plasma discharge of low pressure as a tool for pre-processing the glass optical elements. Established effects of polishing, improve the adhesion of the glass elements to the polymer binder and the increase of the reflection coefficient.

1. The introduction

Low-temperature nonequilibrium plasma can serve as way for achieving the surface effects of materials modification. During materials treatment surface is bombarded by active particles, such as ions, electrons, plasma-forming gas atoms. As a result, near-surface materials area becomes modified. Given method is ecologically clean due to the absence of active chemical and biological substances which resulting in the absence of effluents and emissions of harmful gases.

Other advantages include short time of treatment the possibility to set accurate parameters for obtaining desirable effects, and wide variety of treated materials [1]. Plasma technologies can be used for obtaining various properties to non-organic materials, including ones consist of mineral glass.

Glass is an amorphous solid that does not have ordered crystal structure, and is obtained by overcooling of a melt [2]. Technical and household glasses are needed to be adequate to increased requirements for its hardness, wear resistance, optical properties and soundproofing. Needed glass properties can be obtained by adding different components during the producing of materials at the silicate formation process [3].

Traditional methods of the surface treatment of end glass product are used for improving decorative and operational properties, and consist in chemical etching with sulfuric and hydrofluoric acids [4-5]. Disadvantages of traditional methods are: low ecological purity of the processes, lengthy duration and necessity of preliminary treatment.

Plasma modification does not change internal structure of the material and provides a wide variety of its effects. In the investigation [6] functional microrelief on the surface of the diffractional optical elements has been created by using radio-frequency induction discharge. Formed elements are a diffractional lattices of phase type. In the investigation [7], cold argon plasma was used for changes in the glass surface. The changes in the chemical structure (silicate system regrouping), surface energy and morphology (microrelief surface formation) under plasma treatment was also discovered. It should be noted that there is a possibility of the large micro-meters roughness removal from production fiber glass surface with negligible weakening [8].The glass surface adhesion increasing without microrelief formation can occur due to the chemically active centers forming on the surface, and because of silicate system regrouping during ion dispersion of the near-surface layer.

2. Materials, methods and equipment

Objects of this investigation are the glass microspheres (GMS) with 50-100 μm dispersion, which are used for road markings as a retroreflective material approached to BS 6088:1981.

We have considered the possibility of the microspheres surface smoothing under the influence of radiofrequency induction low-pressure discharge with inert gas – argon (Ar).

Technical parameters of the treatment were the following: plasma forming gas – argon; gas rate $G_{Ar} = 0,04$ g/s; current intensity on the a generating lamp anode $I_A = 2,5$ A; pressure in the working chamber $P = 40–90$ Pa. GMS were injected into the discharge and with plasma forming gas flow were passed through discharge "core". Behind a plasma generator cut, treated particles were caught by filter-trap. Plasma treatment was made with experimental RF Induction plasma set which has been described in the source [9].

3. Results

Investigation of the surface before and after RF Induction plasma treatment was done by Scanning Electron Microscope (SEM) method and Atomic-Force Microscope (AFM) method. Established plasma effect of surface polishing and microcracks healing is shown with SEM-images (Figure 1)

AFM-images before and after RF Induction plasma treatment are shown at the Figure 2-3. Calculated values of roughness are shown in the Table 1.

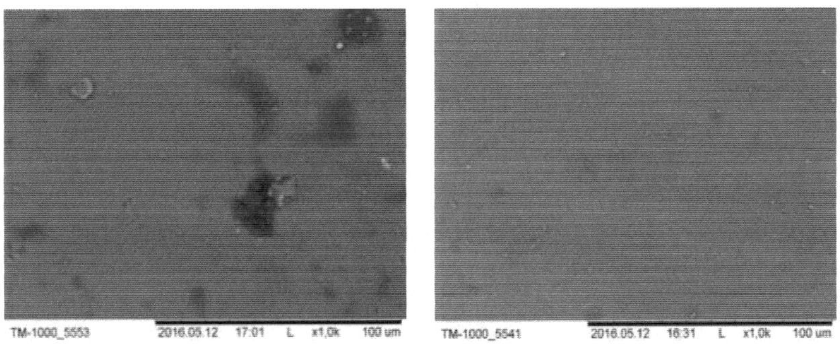

a b

Figure 1. SEM-image of the GMS surface with ×1000 magnification: a – starting sample; b – after RF Induction plasma treatment.

Table 1. Roughness parameters of the sample before and after RF Induction plasma polishing, nm.

Roughness parameter	Starting sample	Sample after RF Induction plasma polishing
Z_{max}, nm	338,732	187,857
R_z, nm	170,145	92,8459
R_a, nm	17,2589	7,48692

* Z_{max} – maximal relief height; R_z – sum of the mean absolute values of the five largest profile ledges and five largest profile valleys within basic length, R_a –arithmetic average of the absolute values of the profile deviation.

At the SEM-images it is shown that surface of the starting microspheres is covered with apprets, pollutes and defects (figure 1 a). Surface of the experimental sample shows less amount of pollutes, mechanical defects are smoothed (figure 1 b), which also is confirmed with AFM data of roughness and with higher magnification.

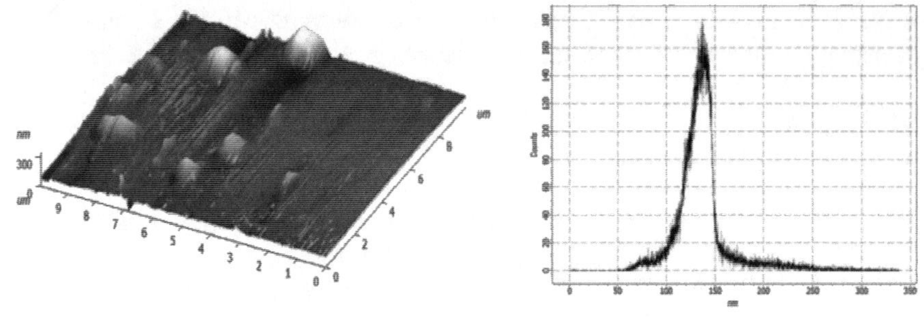

a b

Figure 2. AFM –image of the glass surface without plasma treatment.

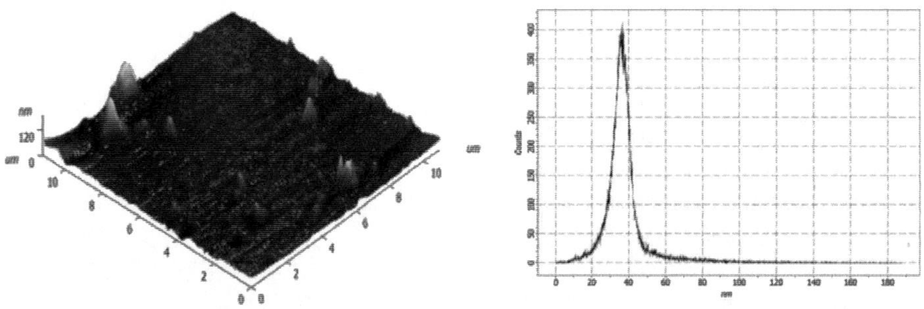

a b

Figure 3. AFM-image of the GMS surface after RF Induction plasma treatment.

Thus, microscope results demonstrate cleaning effect and polish effect of the retroreflective micro spheres surface, and, also, microcracks and microroughnesses healing. It is concluded that such treatment can elevate optical properties of the glass microspheres and its adhesion to polymer binding agent.

Resulting from the specific application of the GMS, the increase of its adhesion to polymer binder under plasma treatment, is a necessity which allows to extend service life of the covering, based on GMS.

In the study, the possibility of RF plasma discharge application has been investigated as an instrument for surface modifying and increasing the adhesion force.

For the wear resistance examination of the retroreflective materials based on GSM we made test samples. For the substrate we used plates of steel with measurements 150×60 mm. Beforehand, plates were degreased and coated with layer of the polymer bonding agent (acryl copolymer) with 0,2-0,3 mm thickness and with layer of the spheres with 0,1-0,15 g/sm^2 consumption. After 48 hours the obtained composite was tested with the use of laboratory test bench for determination surfaces service life under mechanical load. Results are shown in the Table 2.

Table 2. Investigation results of the surfaces service life.

Sample	Mass lost after mechanical impact, g	Mass lost relatively to starting sample %
Treated	1,69	18,76
Untreated	2,5	27,79

Results show that the mechanical resistance of the coating based on plasma modified microspheres has increased 9%.

The destruction of the coating was evaluated by confocal laser scanning microscopy (CLSM) method (Figure 4).

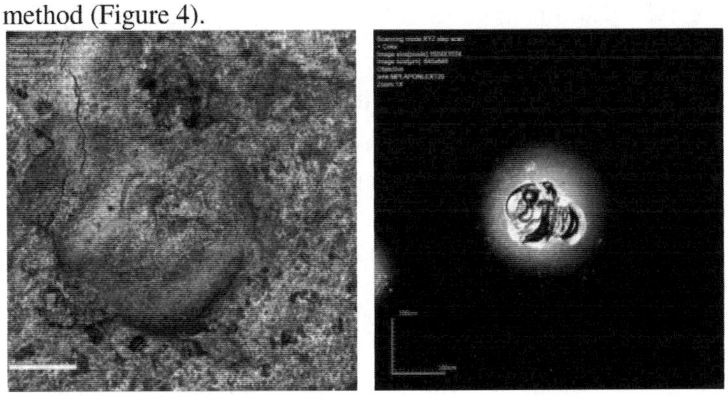

a b

Figure 4 CLSM-images of the retroreflective coatings on the steel plates with acryl copolymer binder a) GSM without plasma treatment; b) GMS modified under plasma treatment.

Modified microspheres after physical and mechanical tests maintain their even distribution.

Microscopy results demonstrate cohesive destruction of the material and of the spheres themselves (Figure 4 b), which provides evidence for adhesion force increasing.

In addition to service life of the coating, important factor for obtained material is reflection coefficient.. Measurements were made with spectrophotometer X-RiteColor DTP22 Swatchbook. Control samples and samples with modified under plasma treatment GMS were investigated and compared. Results are shown at the Figure 5.

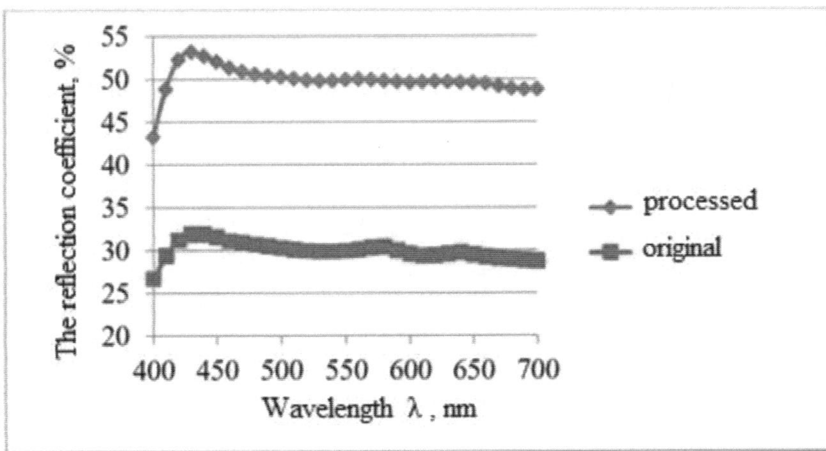

Figure 5 The graph of wavelength against reflectance.

Diagram at the Figure 5 shows that after modification in RF gas discharge, reflective coefficient for silicate spherical materials with 10-60 μm dimensions shifts in the more high areas, which provides the evidence for improvement in optical properties of the GMS after plasma treatment. In terms of obtained data, positive influence of plasma treatment at optical properties of the glass microspheres surface has been established.

4. Conclusions

Thus, it is concluded that RF Induction low-pressure plasma treatment for GMS modification potentially is a highly effective method. During the investigation, positive effect for the roughness smoothness of the surface has been confirmed. Parameters of the roughness were reduced more than 80%, and the exploitation properties, such as adhesion interaction, was increased 66% in comparison with the starting sample, and reflective coefficient was also increased. Therefore, RF Induction plasma treatment can be recommended as a method of GMS preprocessing for retroreflective coatings.

References

[1] Miftahov I S, Voznesensky E F, Abdullin I Sh, Gataullin L 2014 *Ionnaya polirovka opticheskih materialov amorphnogo stroeniya v usloviyach VCH-razryada ponizhennogo davleniya* [Ion polishing of the amorphous optical materials by RF low pressure discharge] Kazan (*Vestnik tekhnologicheskogo universiteta* vol 23) pp. 73-75 (In Russian)

[2] Gutnikov S I , B I Lozoriak, A N Seleznev, 2010. *Steklyannie volokna: posobie dlya studentov* [Glass fibers: tutorial for students] Moskow MSU. – 53 p. (In Russian)

[3] Kitaygorodskiy I I, Kachalov N N, Vargin V V, Evstrop'ev K S, 1961 *Technologiya stekla* [Technology of glass] Moskow *(gosudarstvennoe izdanie literaturi po stroitel'stvu I stroitel'nim materialam)* – 612 p. (In Russian)

[4] Temnikova N V, 2009 *Razrabotka sposobov uluchsheniya ekspluatacionnih charakteristik float-stekla* [Development of ways for improvement exploitation properties of the float-glass during non-stop manufacturing process: diss. thesis for PhD degree (05.17.11)] / Moscow, Russian Chemical Technological University named after D.I. Mendeleev,– 18 p. (In Russian)

[5] Eitel W, 1954 *Phizicheskaya chimiya silikatov* [Physical chemistry of the silicates] / University of Chicago Press, – 1592 p. (In Russian)

[6] Odinokov S B, Sagaletian G R, Goncharov A S, 2012. *Experimental'noye issledovanie prcessa plazmochimicheskogo travleniya stekla pri izgotovlenii difraktsionnih i gologramnih opticheskih elementov* [Experimental investigations in the plasma-chemical process of the glass etching during producing diffraction and hologram optical elements] / MSTU named after N.A. Baumen (*nauka i obrazovanie vol 5)* pp. 391-410. (In Russian)

[7] Miftahov I.S., Voznesensky E F, Abdullin I Sh, Fadeev A O, Gataullin L 2015 *Primrnrnie plazmi VCH-razryada ponizhennogo davleniya v processah polirovki opticheskih materialov polycrystallicheskogo stroyeniya* [Applications of the RF low-pressure plasma in the polishing processes of the optical polycrystal materials] Kazan (*Vestnik tekhnologicheskogo universiteta* vol 11) pp. 144-146. (In Russian)

[8] Krishnamurthy V and Ihab L Kamel 1989 *Journal of Materials Science* **24** 3345–3352

[9] Abdullin I Sh, Jeltuhin V S, Kashapov N F, 2000. *Visokochastotnaya plazmenno-struynaya obrabotka materialov pri ponizhennih davleinyah. Teoriya I praktika primeneniya.* [Radio-frequency plasma-stream treatment of the materials under low pressure. Theory and practice of the application] Kazan: Kazan St. Un. publ. house,. – 348 p. (In Russian)

The change in physical and mechanical properties of technical fabric by means of plasma treatment

S Ilyushina[1], I Krasina[2], A Minyazova[3], M Antonova[4], A Parsanov[5] and B Eworo[6]

[1] Associate professor, the Department of Technology of Chemical, Natural Fibers and Products, Kazan National Research Technological University, Kazan, Russia
[2] Head of the Department of Technology of Chemical, Natural Fibers and Products, Kazan National Research Technological University, Kazan, Russia
[3] Graduate student the Department of Technology of Chemical, Natural Fibers and Products, Kazan National Research Technological University, Kazan, Russia
[4] Associate professor, the Department of Technology of Chemical, Natural Fibers and Products, Kazan National Research Technological University, Kazan, Russia
[5] Associate professor, the Department of Technology of Chemical, Natural Fibers and Products, Kazan National Research Technological University, Kazan, Russia
[6] undergraduate student, the Department of Technology of Processing Plastics And Certification Materials, Kazan National Research Technological University, Kazan, Russia

strelfy@mail.ru

Abstract. The properties of a technical cushion fabric modified by a low-temperature plasma are studied. It established that the treatment with plasma of low pressure allows us to give anti-adhesive properties of technical fabrics with improvement of physical and mechanical properties.

1. Introduction

The manufacturing of technical products is the fastest growing in the textile industries in the world's leading countries. The basis for intense increase in volumes of production of technical textiles and expansion of spheres of its application is active in researching activities, which are directed considerable for financial resources and in which appears numerous innovative developments [1].

Technical fabrics are widely used in various sectors of the economy. In contrast to fabrics of a household purpose, they are characterized by higher breaking load, surface density and thickness. The performance of technical textiles is due to their purposes. These properties include water - and air permeability, conductivity and electrified, thermo-, light - and chemical resistance, bond strength with the coating, the linear shrinkage from the effects of temperature stability of the relaxation properties when working under stress not exceeding above 10% of the breaking loads, and others [2].

However, it should be noted that the appearance of industrial fabrics does not impose such strict requirements compared to that of household fabrics. Technical fabrics are widely used for production of conveyor belts, rubber hoses, flat and V-belts; soft furnishings of the seats and the interior lining of cars; for awnings and covers, materials, shells, balloons and inflatable hangars, folding garages, tourist tents, automobile and aviation tires; as a filter partitions. Technical fabrics coated with rubber or polyvinyl chloride, used in manufacturing of containers for storage and transportation of liquids and bulk goods, inflatable boats, life jackets and rafts on ships, flexible fences (skirts) air-cushion vessels, straps for lowering and lifting of pipes of large diameter for construction of oil and gas pipelines, pipes for air supply in mines [2].

A significant proportion of fabrics for technical purposes used as cushioning materials for storage of uncured rubbers preform, rubber profiles and tire cord prior to assemble and vulcanize.

During operation of the cushioning materials, experience various static and dynamic effects: the dynamic stretching and bending at the seam and rolling, abrasion surface material, and static electricity. The main reason for the decline is the life of the strip is the loss of its dimensional stability. Consequently, in the process of multiple seaming raw rubber blanks on their surface scarring. In addition, increased abrasion of the surface of the cushioning material leads to the penetration of oxygen to the uncured rubbers and reducing their consumer properties.

2. The objective of study

As the cushioning materials used cotton and combined fabrics: calico, muslin, percale, chafer and chafer containing lavsan and cotton fibers, To reduce the adhesion of cushioning materials to the raw rubber blanks require additional processing costly release formulations or develop completely new reliable materials.

In this case, to regulate adhesion, physical and mechanical characteristics, and reduce the cost of technical fabrics for mechanical rubber goods are urgently use to develop new formulations of technical fabrics or the modification of their surface.

Non-equilibrium low-temperature plasma changes the adhesive properties of the material surface depending on the parameters of the modification. In connection with this interest, the impact on non-equilibrium low-temperature plasma on the fabrics for technical purposes, containing natural and synthetic fibers.

The surface and strength properties of technical fabrics can be adjust by changing the parameters of low-temperature plasma treatment [3].

The objective of the study selected technical cushioning fabric ChLH containing cotton and polyester fibers (51% cotton, 49% polyester), manufactured by company "Krez", Elabuga. Insulating fabric cushioning ChLH used in the tire industry between steel cord and rubber, therefore, is to make the necessary anti adhesive properties of the technical fabrics to the rubber.

3. Methods and means of research

Processing of technical fabrics was performed in an industrial plasma installation, the process parameters that were changed within the following limits: the voltage of the anode (Ua) from 3 to 7.5 kV; anode current (Ia) from 0.3 to 0.7 A; the processing time (t) from 1 to 10 minutes. The Plasma gas: gas mixture argon / propane-butane in the ratio of 70%÷30%. The value of plasma gas flow and pressure in the vacuum chamber was G=0.04 g/s and P=26.6 Pa, respectively, and installed in the early works [4, 5] as optimal for the processing of chemical fiber materials.

To evaluate the surface properties of the fabric selected standard methods for determination of capillarity, water absorption and contact angle of wetting.

4. Results of experiments and discussion

The experiment dependence on the values of the capillary modes plasma treatment of tissue ChLH presented in Table 1.

According to table 1, the lowest elevation of tissue fluid after the plasma-forming gas treatment in argon / propane-butane (70% ÷ 30%) was observed for the parameters: Ia = 0,5 A, Ua = 4,5 kV, t = 3 min, P = 26,6 Pa.

Then, we carried out experimental evaluation of the contact angle of the wettability of the surface of the fabric, The result of the measurement of the equilibrium-wetting angle was considered to be the arithmetical average values of wetting angle six drops of water. Therefore, found that after processing the cushioning tissue in the environment of argon/propane-butane wetting angle is increased by 46% compared to a control sample.

Table 1. The influence of non-equilibrium low-temperature plasma processing on tissue capillarity (t = 3 min, P = 26.6 Pa; G = 0,04 g/s, argon / propane-butane (70%÷30%)).

Processing modes	Ia, A	0,2	0,3	0,3	0,5	0,5	0,5	0,7	0,7	0,8	Not treated
	Ua, kV	4,5	3	6	2,5	4,5	6,5	3	6	4,5	
Capillarity, mm		86	77	63	102	5	89	70	83	8	105

In the next phase of the research was determined water absorption capacity of technical fabrics, after plasma modification, the cushioning tissue water absorption decreased by 20%.

The decrease in the water absorption of technical fabric after the action of the plasma flow of a high-frequency capacitive discharge of reduced pressure can occur both due to splitting and changes in the internal structure of the processed materials, both by changing the free energy of the surface of the pores and capillaries, and by restructuring the nature of the functional bonds [6, 7].

Due to the fact that the textile material is exposed to temperature during the manufacturing of rubber products, the effect of the plasma modification on the thermal shrinkage of the technical fabric was determined. The results of the experimental data showed that after the modification of the technical fabric in a high-frequency capacitive discharge of reduced pressure, the values of thermal shrinkage in hot air are reduced by 28%.

Reducing the shrinkage of the tested materials when exposed to the plasma stream of the high-frequency capacitive discharge of low pressure occurs, firstly, due to the shrinkage effect of the plasma: recombination of ions and bombardment of the surface with them, and secondly, the thermal effect of the plasma leads to increased of kinetic energy of molecules and atoms and weakening of intermolecular bonds, which contributes to the resumption of the relaxation process. Defects caused by the formation of a "buried" layer of a plasma-forming gas atoms,give rise to stretching stresses, which prevent shrinkage.

Study of changes in the adhesive properties of technical fabrics after plasma modification was performed by the method of determining the bond strength rubber-cord (H-method).

Tests to determine the strength of the bond between the rubber and the cloth consisted in pulling the cord from a rubber sample, which, when placed on a flat surface, was shaped like the letter H. Samples were cured at 150 ° C for 20 minutes in a mold. The mold has mutually perpendicular channels, into which strips of rubber compound and threads of textile cord laid. Then samples were determined under normal conditions for 24 hours, and then tested for bond strength on a breaker RM-50.

According to the experimental data, after the plasma treatment, the strength of bonding with the rubber for the lining fabrics, the bond strength decreased by 1.5 times.

Investigation of the strength properties of the cushion fabric showed some improvement in strength characteristics. In Figure 1 is a graph of the fracture curves of samples of technical fabric.

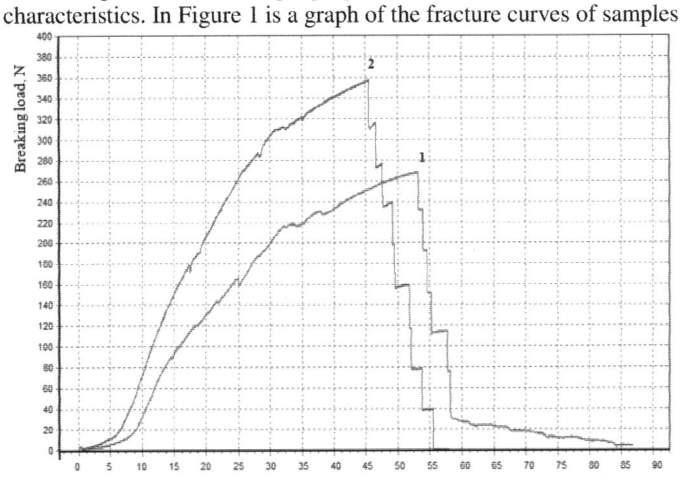

Figure 1. The nature of the destruction curve of the cushion fabric:
1 - without plasma exposure;
2 - the sample modified in an optimum mode.

From figure 1 it is seen that after plasma modification in optimum conditions, the breaking load for cushioning fabric increased by 33%.

This can be explained by the fact that in the case of processing in the environment of argon-propane-butane, there is a probability of grafting of monomeric units and fragments of the molecules of propane-butane to the resulting free radicals, the formation of additional bridges and cross-links, which greatly reduces the number of free radicals after processing, results in a significant increase in strength and impart anti-adhesion properties.

5. Conclusion

Thus, a study of surface properties showed that for the cushion fabric in the optimal mode (I_a = 0.5 A, U_a = 4.5 kV, t = 3 min, P = 26.6 Pa argon / propane-butane in the ratio 70% ÷ 30%) drop in capillarity by 10 times, water absorption by 20%, wetting angle increased by 46%, thermal shrinkage decreased by 28%, breaking load increased by 33%, and bond strength with rubber decreased by 1.5 times.

Analyzing the obtained results, it can be concluded that plasma treatment of technical cushioning fabric in the plasma-forming medium argon / propane-butane allows increasing the strength and anti-adhesion properties of the fabric.

References

[1] Eneke M 2011 *Technical textiles are a market with immense potentialities* vol 3, Ed A Sergeenkov (Moscow: Industry reports) pp. 19-21

[2] Kerimov S 1994 *Production of technical fabrics: (equipment and technology)* (Moscow: Legprombytizdat) p. 237

[3] Ilyushina S and Sergeeva E 2011 *Modification of technical tissues by a nonequilibrium low-temperature plasma* Int. conf. of Physics of high-frequency discharges (Kazan: KSTU) p. 286

[4] Shaehov M 2006 *Physics of high-frequency discharge of low pressure in the processes of capillary-porous and fibrous materials* (Kazan) p. 350

[5] Sergeeva E 2010 *Regulation of the properties of synthetic fibers, yarns, fabrics and composite materials on their basis with the help of nonequilibrium low-temperature plasma* (Kazan) p. 363

[6] Khammatova V 2006 *Regulation of the molding ability of textile materials using plasma technologies* (Kazan) p. 316

[7] Ilyushina S and Sergeeva E 2011 *Change in water absorption of polyester and polyamide technical fabrics under the influence of nonequilibrium low-temperature plasma* conf. of Innovations in Chemistry: Achievements and Prospects (Kazan: KNITU) p.127

Electrical properties of titanium nitride films synthesized by reactive magnetron sputtering

W M Mohammed, A I Gumarov, I R Vakhitov, I V Yanilkin, A G Kiiamov, S S Kharintsev, S I Nikitin, L R Tagirov and R V Yusupov

Kazan Federal University, Kremlevskaya str. 16a, 420008 Kazan, Russia

E-mail: Roman.Yusupov@kpfu.ru

Abstract. Reactive dc magnetron sputtering was employed to produce thin films of titanium nitride using titanium metallic target, argon as the plasma gas and nitrogen as the reactive gas. A set of the films was studied deposited on the Si, fused silica and crystalline (001) MgO substrates with various deposition conditions. The films deposited on the Si and SiO_2 substrates are polycrystalline while deposited at slow rate to the heated to 600°C MgO substrate are highly epitaxial according both to XRD and LEED data. Electrical resistivity of the films was measured by means of the four-probe van der Pauw method.

1. Introduction

Titanium nitride (TiN) is a material that for a long time is used in many technological, medical and scientific fields due to its outstanding properties. Recently, it has been found that TiN thin films may serve also as the plasmonic materials that can hold the temperatures much higher than the classic plasmonic elementary compounds like gold, silver or copper [1]. Owing to its outstanding physical properties such as high hardness, modulus of elasticity, thermal expansion coefficient and a transition to superconducting state [2, 3], TiN is an attractive material for application in various photoelectric devices [4, 5]. Also, some important features of titanium nitride such as good wear resistance, high corrosion resistance, low friction coefficient, high conductivity and a bright, golden color enabled it to be used in protective and functional coatings on cutting tools, bearings, etc [6]. Due to high electrical conductivity and thermal stability of TiN film, TiN coating has recently been led to electrodes in microelectronic devices [7]. Titanium nitride is well known for lustrous golden color [8] which is useful for decorative applications. Golden color is a result of the high reflectance of TiN at the red end of the visible part of spectrum with low reflectance near the ultraviolet region. This makes it a potential coating for IR-reflective back contacts material and a solar energy absorber suitable for solar cells [9]. In addition, titanium nitride could replace the polycrystalline silicon and be used as a gate electrode in MOS devices [10]. It may be also used in biomedical materials because of its noncytotoxical nature [11].

In this paper, the deposition of titanium nitride films on magnesium oxide, silicon and fused silica substrates by reactive dc magnetron sputtering under the different conditions of deposition will be described. Also, we report the electrical properties of TiN thin films produced by reactive magnetron sputtering.

2. Experimental part

2.1. Synthesis

Titanium nitride (TiN) films were deposited by the reactive dc magnetron sputtering with the power of 30-150 W on the cleaned silicon, fused silica and crystalline magnesium oxide substrates. Titanium metal target with a purity of 99.99% was used for deposition. Before the deposition, the pressure was set to $5 \cdot 10^{-3}$ mbar inside the chamber with desired argon and nitrogen gas partial pressures. The deposition conditions are shown in Table 1.

Table 1. Conditions employed for the preparation of TiN films

Parameter	**Value**
Power	30, 50, 100, 150 W
Deposition time	30, 60, 120, 180 min
N_2 concentration	20, 10, 5, 1, 0.99 %
Sub. temperature	25, 500, 600 °C
Base pressure	$5*10^{-3}$ mbar

2.2. Characterization

In situ XPS measurements were performed in the ultrahigh vacuum chamber (base pressure ~ $3 \cdot 10^{-10}$ mbar) equipped with the Mg-K_α X-ray source operated at 12.5 kV and 250 W, and a Phoibos-150 hemispherical energy analyzer (all from SPECS). The crystal structure and orientation were tested with Bruker D8 Advance X-ray diffractometer with the use of Cu-K_α ($\lambda = 1.5418$ Å) radiation in the Bragg–Brentano geometry with the scanning rate of 0.002 degree per second in the range of 2θ angles from 17° to 82° and a step of 0.0153°. Surface roughness of the deposited films was measured using atomic force microscopy (NTEGRA-Spectra system by NT-MDT). Low energy electron diffraction (LEED) was utilized to determine the near-surface crystalline structure of the films and of the MgO substrate surface. The resistivity measurements were performed using the van der Pauw four-point probe approach.

After deposition of titanium nitride films, they were analysed by LEED in situ and by X-ray diffraction (XRD) ex situ. Figure 1 shows the XRD pattern of the TiN films on MgO (001) substrates. In figure 2 the LEED patterns are presented of the bare MgO (001) substrate and of the deposited onto it TiN film. Both figure 1 and figure 2 indicate a highly textured epitaxial growth of TiN on MgO due to an almost ideal lattice matching and identical crystal structures of MgO and TiN; lattice constants are respectively 4.212 Å and 4.240 Å [12] ($\delta a/a = 0.66\%$).

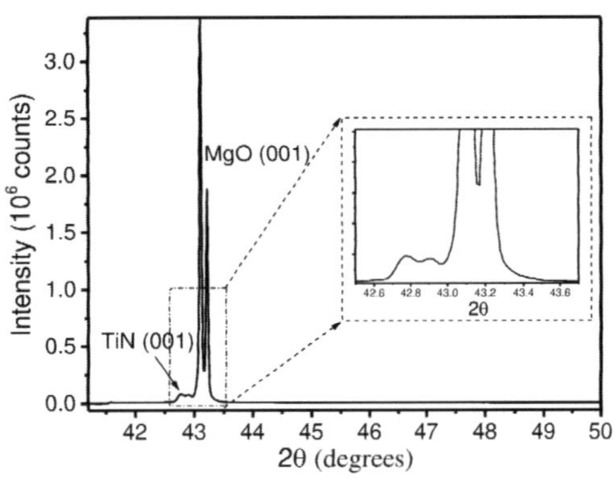

Figure 1. XRD pattern of the 86 nm thick epitaxial TiN film deposited on (001) MgO.

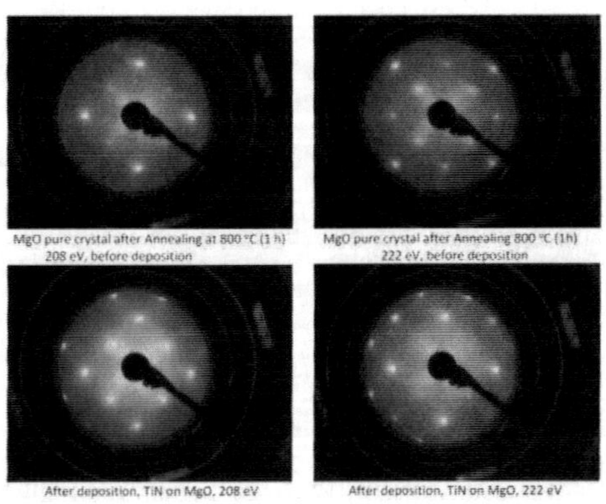

Figure 2. Surface structure of MgO substrate annealed at 800°C before deposition (top) and TiN/MgO after deposition (bottom).

XPS spectra of Ti 2*p* and N 1*s* for titanium nitride films are shown in figure 3. No trace of oxygen was found in the in situ collected XPS data of the as deposited films. However, the presence of oxygen is revealed immediately as soon as the films are exposed to the air. From the spectrum of Ti 2*p*, we can observe the two peaks that reflect the spin degeneracy of the 2*p* level ($2p_{3/2}$, $2p_{1/2}$) with the binding energy of 455 eV and 460.8 eV, respectively. These values indicate the TiN formation. Similarly, the spectrum of N 1*s* with the binding energy of 397.2 eV also indicates the TiN formation [13]. Analysis of the AFM topography of the films deposited onto the epi-polished MgO gives the mean surface roughness of the in situ annealed MgO substrate of about 0.4 nm, and of the deposited TiN film onto it 0.8 nm; it is known that the substrate annealing is more practical and effective method for rehabilitating the surface of the substrate [14].

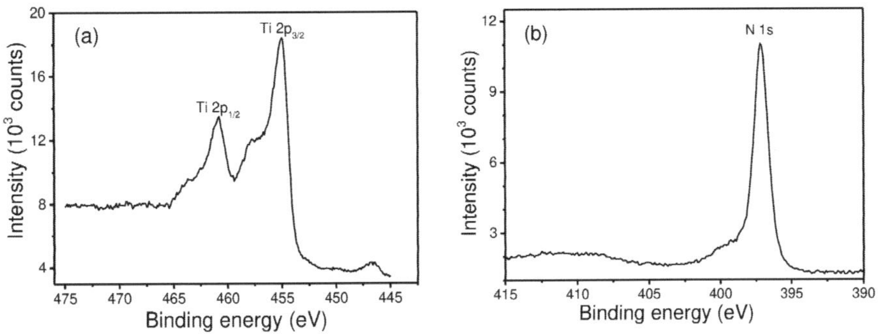

Figure 3. XPS spectra of as-deposited TiN film: (a) Ti 2*p*-lines and (b) N 1*s*-line.

2.3. *Resistivity measurements*

The resistivity measurements were performed using the van der Pauw four-point probe system. As shown in figure 4, the shape of the sample is square with ohmic electrodes connected to the four corners according the van der Pauw method. The resistivity was calculated from the measured R_A and R_B values as follows [15, 16]:

$$e^{-\pi\frac{R_A}{R_S}} + e^{-\pi\frac{R_B}{R_S}} = 1, \qquad (1)$$

where R_S is the sheet resistance; this can be written as follows

$$\rho = \frac{\pi d}{\ln(2)} \frac{R_A+R_B}{2} \cdot f. \qquad (2)$$

Here ρ is the resistivity, d is the thickness and f is a factor depending on the values of R_A and R_B . The last can be expressed as:

$$f = \left(\frac{R_A-R_B}{R_A+R_B}\right)^{2k_1} - \left(\frac{R_A-R_B}{R_A+R_B}\right)^{4k_2} \qquad (3)$$

where k_1 and k_2 are the constants:

$$k_1 = \frac{\ln(2)}{2},\ k_2 = \left(\frac{(\ln(2))^2}{4} - \frac{(\ln(2))^3}{12}\right) \qquad (4)$$

One of the problems that had to be solved was a development of the way to create the resistive connection which was not the case for, e.g., aluminum or gold wire ultrasound bonding. We have found that the best choice among the available to us was the indium/gallium alloy which prompts ohmic contact as shown in figure 5. The changes in resistivity as a function of deposition conditions such as nitrogen flow rate, substrate temperature and thickness were studied as it is shown in Table 2.

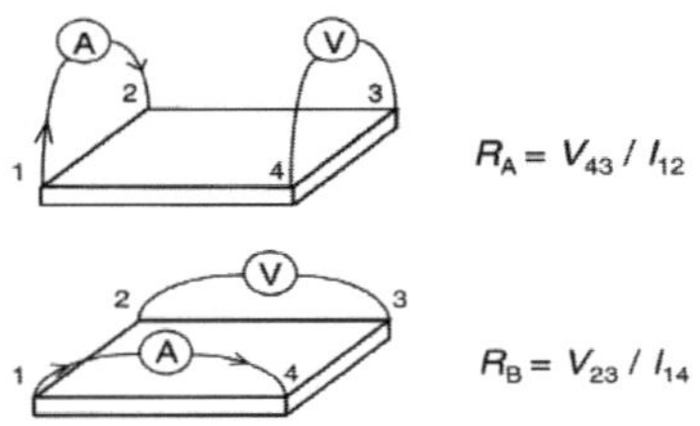

Figure 4. Two ways of an electrical connection to the films for measurements of the resistivity by van der Pauw technique.

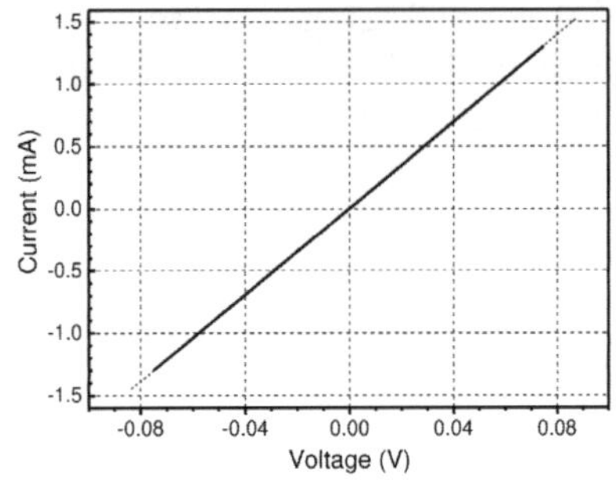

Figure 5. I-V curve of the films with the indium/gallium alloy bonded electric contacts; dashed line is a linear fit to the data.

Table 2. Values of TiN films resistivity for various deposition conditions.

Films / conditions	TiN/Si				TiN/SiO_2			TiN/MgO			
N_2 concentration (%)	100	20	20	10	0	20	20	20	5	1	0.99
Substrate temp. (°C)	25	500	500	500	25	500	500	500	500	500	600
Thickness (nm)	110	150	150	108	110	150	60	90	36	86	13
Resistivity (μΩ·cm)	130	26.1	15.4	5.1	16.7	12.4	15.3	12.4	8.3	2.12	11.2

Clearly, the resistivity of the deposited films lies in a wide range and may differ by two orders of magnitude. Typical value of the film resistivity lies in the range of 10-15 μΩ·cm, close to the bulk one. The highest value was obtained for the film deposited on the substrate held at room temperature in the pure nitrogen atmosphere. This is most probably due to the formation, at least, partial, of the dielectric Ti_3N_4 film. The lowest resistivity value among the series was observed for the highly epitaxial TiN film on the in situ annealed (001)-MgO substrate. The observed reduction may be related to the vanishing carrier scattering on the crystalline domain walls or inter-grain boundaries.

3. Conclusions

Titanium nitride films have been prepared in different conditions by reactive dc magnetron sputtering. Formation of the stoichiometric oxygen-free TiN phase is shown by XPS. A highly epitaxial growth of TiN film on the MgO (001) substrate is observed due to the perfect matching between TiN and MgO lattices. The resistivity of the films was measured by van der Pauw technique; low values of resistivity were obtained for deposited films on MgO (001) due to the epitaxial growth.

4. Acknowledgements

The Program of Competitive Growth of Kazan Federal University supported by the Russian Government is gratefully acknowledged. Synthesis and analysis of the films were carried out at the PCR Federal Center of Shared Facilities of KFU. The work of R.V.Yusupov was funded by the subsidy No. 3.7704.2017/6.7 allocated to Kazan Federal University for the state assignment in the sphere of scientific activities.

References

[1] Naik G V, Schroeder J L, Ni X, Kildishev A V, Sands T D and Boltasseva A 2012 *Opt. Mater. Express* **2** 478
[2] Pierson H O 1996 *Handbook of refractory carbides and nitrides: properties, characteristics, processing, and applications* (New York: Noyes Publications) p 193
[3] Stone D S, Yoder K B and Sproul W D 1991 *J. Vac. Sci. Technol.* **9** 2543

[4] Dimitriadis C A, Lee J I, Patsalas P, Logothetidis S, Tassis D H, Brini J and Kamarinos G 1999 *J. Appl. Phys.* **85** 4238
[5] Kadelec S, Musil J and Vyskocil J 1992 *Surf. Coat. Technol.* **54–55** 287
[6] Jin Y, Kim Y G, Kim J H and Kim D K 2005 *Journal of the Korean Ceramic Society* **42** 45
[7] Talyansky V, Vispute R D, Ramesh R, Sharma R P, Venkatesan T, Salamanca-Riba Y X, Li L G, Wood M C, Lareau R T, Jones K A and Iliadis A A 1998 *Thin Solid Films* **323** 37
[8] Niyomsoan S, Grant W, Olson D L and Mishra B 2002 *Thin Solid Films* **415** 187
[9] Guo S Y, Shafarman W N and Delahoy A E 2006 *J. Vac. Sci. Technol.* **24** 1524
[10] Lima L P B, Moreira M A, Diniz J A and Doi I 2012 *Phys. Status Solidi C* **4** 1427
[11] Serro A P, Completo C, Colaço R, Santos F, Lobato da Silva C, Cabral J M S, Araújo H, Pires E and Saramago B 2009 *Surface and Coatings Technology* **203** 3701
[12] Zgrabik C M and Hu E L *Opt. Mater. Express* 2015 **5** 2786
[13] Mishra S K, Kumar R, Sreemany M and Pathak L C 2015 *Journal of Materials Engineering and Performance* **24** 5013
[14] Zhang Q Y, Shaibo J, Ju J, Liu M, Cheng S, Ma Y J, Xiao J Y, Jiang X N and Ma C Y 2016 *Crystal Growth & Design* **16** 4272
[15] Matsumura T and Sato Y 2010 *Journal of Modern Physics* **1** 340
[16] Rietveld G, Koijmans C V, Henderson L C, Hall M J, Harmon S., Warnecke P and Schumacher B 2003 *IEEE transactions on instrumentation and measurement* **52** 449

IOP Conf. Series: Journal of Physics: Conf. Series **927** (2017) 012037 doi:10.1088/1742-6596/927/1/012037

Experiments on phase transitions in three-dimensional dusty plasma under microgravity conditions

V I Molotkov[1], V. N. Naumkin[1], A. M. Lipaev[1], D. I. Zhukhovitskii[1], A. D. Usachev[1], V. E. Fortov[1], H. M. Thomas[2]

[1]Joint Institute for High Temperatures of the Russian Academy of Sciences, Izhorskaya 13 Bldg 2, Moscow 125412, Russia

[2]Institute of Materials Physics in Space (DLR-MP) German Aerospace Center, 51147 Cologne, Germany

E-mail: molotkov@ihed.ras.ru

Abstract. Complex (dusty) plasmas are composed of weakly ionized gas and charged microparticles and represent the plasma state of soft matter. Due to the "heavy" component, microparticles, and the low density of the surrounding medium, the rarefied gas and plasma, it is necessary to perform experiments under microgravity conditions to cover a broad range of experimental parameters which are not available on ground. The investigations have been performed onboard the International Space Station (ISS) with the help of the PK-3 Plus laboratory. This laboratory was mainly built to investigate the crystalline state of complex plasma, the so-called plasma crystal, its phase transitions and processes in multi-particle mixtures. The crystal–liquid phase transition was obtained in large three-dimensional isotropic dusty plasma system. Observations of a transition of the dusty plasma system state due to the particle charge reduction and due to variations of the plasma component density are presented.

1. Introduction

Complex plasmas are plasmas containing small solid particles, typically in the micrometer range, the so-called microparticles. These are dusty plasmas which are specially prepared to study fundamental processes in the strong coupling regime on the most fundamental (kinetic) level, through the observation of individual microparticles and their interactions. Many interesting phenomena can be studied starting from small two-dimensional (2D) and three-dimensional (3D) clusters [1], to larger 2D and 3D systems where collective effects play a dominant role [2].

In laboratory conditions, the microparticles are heavily affected by the force of gravity. Under microgravity conditions, e.g. on the International Space Station (ISS), gravity is negligible. Therefore, the particles are pushed out of the strong electric field region close to the electrodes due to their negative charge and can form large, more or less, homogenous particle clouds in the bulk of the discharge. Under these conditions, weaker forces like the ion drag force and the interparticle interactions become important and often dominate the motion and structure formation in complex plasma.

The laboratory, PK-3 Plus, has been improved considerably compared to the PKE-Nefedov and has been equipped with new diagnostic tools [3]. It has provided the next important step

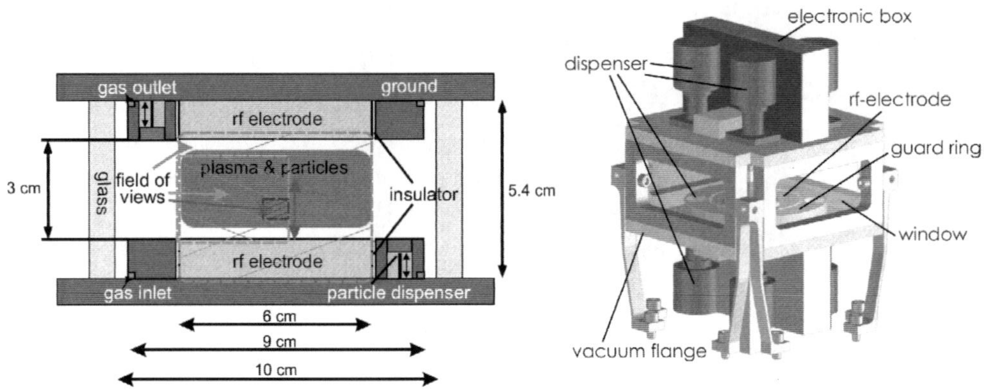

Figure 1. The sketches show the 2D (left) and 3D (right) view of the PK-3 Plus plasma chamber.

in research of complex plasmas under microgravity conditions. The obtained results are unique and opened up a new interdisciplinary research directions, in particular, related to the field of soft condensed matter. The purpose of this paper is to present some new results obtained using the PK-3 Plus laboratory onboard ISS.

The PK-3 Plus setup has a well-balanced symmetrically driven rf-electrode system which provides a homogeneous distribution of the plasma with identical sheaths near both electrodes. This is necessary for a homogeneous distribution of the microparticles under microgravity conditions. A cross-sectional and perspective schematic of the microgravity setup, the PK-3 Plus chamber, is shown in figure 1. Technical details of the setup can be found in [3].

2. Results on phase transitions in 3D complex plasmas

2.1. Phase transition by manipulating the neutral gas pressure

We performed experimental investigations of the fluid-solid phase transitions in large 3D complex plasmas under microgravity conditions. These phase changes were driven by manipulating the neutral gas pressure. Detailed analysis of complex plasma structural properties allowed us to quantify the extent of ordering and accurately determine the phase state of the system. Evaluation of various freezing and melting indicators gave further confidence regarding the phase states. It was observed that the system of charged particles can exhibit melting upon increasing the gas pressure, in contrast to the situation in ground-based experiments where plasma crystals normally melt upon reducing the pressure [4].

2.2. Phase transition due to the particle charge reduction

We performed experiments to demonstrate a change of the structural properties of the dusty plasma system due to a change of the particle charge. There have been used two-species complex plasma with one species composed of small (1.55 μm) and the other composed of big (14.9 μm) particles. The experiments have been carried out in argon gas at a pressure of 10 Pa. The particles are SiO_2 spheres with a diameter 1.55 μm, the big particles are melamine formaldehyde spheres. The particles of different size do not mix. The smaller particles form an inner cloud close to the discharge centre (with the central areas free of particles, see figure 2). The big particles form an outer cloud surrounding and confining the smaller one, as it is seen in figure 2. The bigger particles can be added to the system. This effectively increases the strength of the confinement and compresses the small particle system.

IOP Conf. Series: Journal of Physics: Conf. Series **927** (2017) 012037 doi:10.1088/1742-6596/927/1/012037

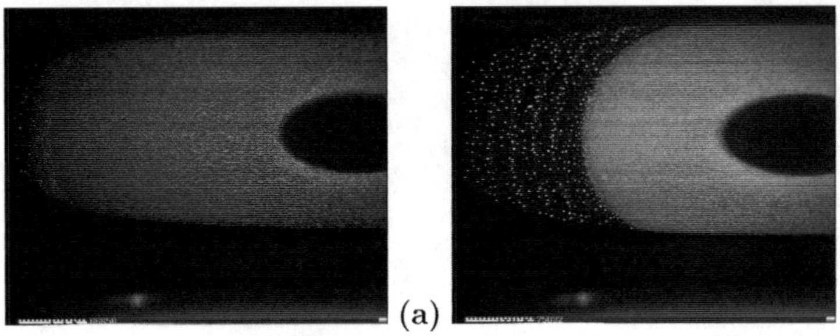

Figure 2. Video images of dusty plasma system: **(a)** initial system; **(b)** compressed system. Snapshots are obtained by quadrant camera.

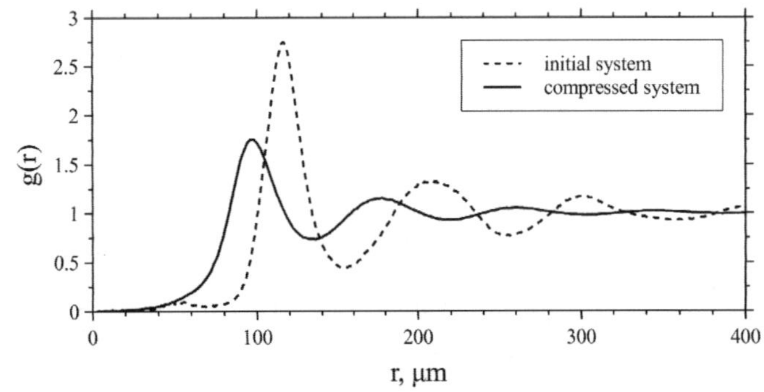

Figure 3. Change of pair correlation function due to compression by bigger particles.

Figure 3 demonstrates a change of the pair correlation function for the initial dusty plasma system (only small particles are present) and for the compressed system (with an addition of bigger particles that increases a confinement of the small particles system). Table 1 presents the results of the calculation of some parameters of the dusty subsystem: interparticle distance Δ obtained from the pair correlation function, dust particle density n_p obtained from the processing of the video images received by the high resolution camera, particle charge Q calculated using the orbital motion limited approximation for the experimental plasma parameters.

For the analysis of the dusty plasma system state we use the Raveche-Mountain-Streett criterion R that is the ratio of the first nonzero minimum of $g(r)$ to the first maximum [5]. Note that in the case when R exceeds 0.2 the dusty plasma system is in the fluid state and in the case when R is less than 0.2 the system is in the well ordered (crystal) state. It is seen from Table 1 that R changes from the small value (0.04) pointing to the high ordering in the subsystem (plasma crystal) to the high value (0.21) pointing to melting of the plasma crystal. Thus, it is a demonstration of the phase transition due to a drop of the absolute magnitude of the particle charge when their density increases.

Table 1. Parameters of the dusty subsystem.

Δ (μm)	n_p (10^5 cm^{-3})	Q (e)	R
125	5.10	2000	0.04
90	12.0	807	0.21

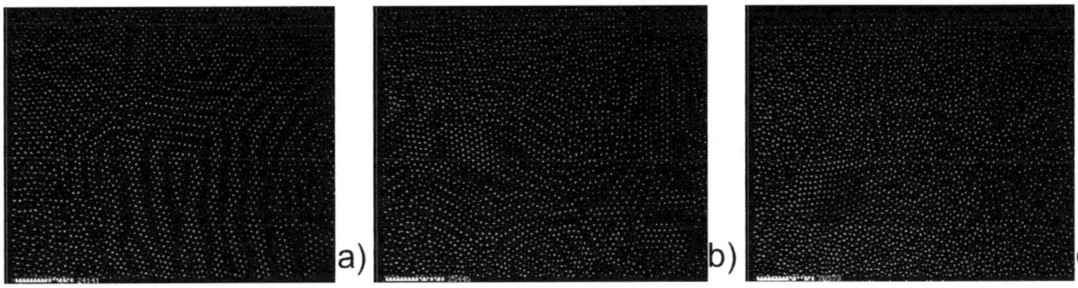

Figure 4. Video images of the restored dusty plasma system for the argon pressure 20 Pa. a) $U_{\text{eff}} = 19.0$ V; b) $U_{\text{eff}} = 16.3$ V; c) $U_{\text{eff}} = 13.1$ V.

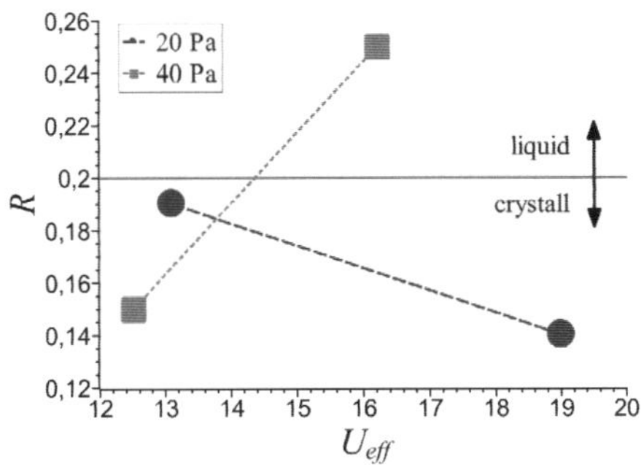

Figure 5. Change of the criterion R on the interelectrode voltage at different pressures.

2.3. Phase transition by changing the plasma component density

Several experiments have been performed to study peculiarities of the crystal-liquid phase transition in the 3D dusty plasma system when varying the voltage between electrodes of the plasma chamber. The change of the interelectrode voltage leads to a change of the plasma component density and one of the important parameters determining a behavior of the dusty plasma system, namely, a screening length λ. In the experiments monodisperse SiO_2 particles of 1.55 μm diameter were used.

The procedure of the experiments was the following one. The radio-frequency discharge was generated at the given argon pressure and at the given interelectrode voltage. Then a formation of an ordered structure took place. Figure 4 presents video images of the dusty plasma system obtained by the high resolution camera at the argon pressure 20 Pa at different values of the interelectrode voltage ($U_{\text{eff}} = 19.0$, 16.3 and 13.1 V respectively). It is clearly seen a decrease of the ordering level when rf voltage at the electrodes is decreasing, that is, when the ion density is decreasing.

The video images for the dusty plasma system for the argon pressure 40 Pa show that in this case the ordering level of the system is changing in the direction opposite in comparison with the previous case.

To interpret the results obtained we made evaluations of the parameters of the dusty plasma systems observed using the pair correlation functions determined on the results of a scanning of the each system. The experimental correlation functions give a possibility to evaluate a level of the dusty plasma system ordering and to determine an interparticle distance.

Figure 5 shows a change of the criterion R when the interelectrode voltage changes, that is, the plasma component density changes. It is seen that a character of the change of the ordering state at two different values of the argon pressure has quite an opposite behavior. At the argon pressure 20 Pa an ordering of the system decreases when there is a reduction of the voltage U_{eff}:

Table 2. Parameters of the dusty plasma systems.

P (Pa)	$U_{\rm eff}$ (V)	n_i (cm^{-3})	λ (μm)	Δ (μm)	k	e^{-k}	Q (e)	U (eV)
20	19.0	8×10^8	43	131	3.05	0.047	1.6×10^3	1.34
20	13.1	5×10^8	54	122	2.26	0.1	8.8×10^2	0.96
40	16.2	9.0×10^8	40	136	3.4	0.033	1.9×10^3	1.28
40	12.5	6.5×10^8	47.5	136	2.86	0.057	1.4×10^3	1.19

the dusty plasma system melts at a reduction of the plasma component density. In the case of the argon pressure 40 Pa when the interelectrode voltage increases the dusty plasma system ordering increases as well.

To obtain values of the plasma component density (in the first turn, the ion density) we use the results of calculations on the base of the SIGLO-2D model [3]. The evaluation of plasma parameters was made without taking into account dust particles. The obtained ion densities for pressures 20 Pa and 40 Pa are presented in table 2. In the table there are also the values of the screening length λ. The table 2 contains also values of the interparticle distances and the screening parameter k, where $k = \Delta/\lambda$, Δ is interparticle distance. The values of the exponential multiplier e^{-k} in the interparticle interaction energy $U = (Q^2/\Delta)e^{-k}$, where Q is microparticle charge are presented in the table as well.

The presented evaluations of the interparticle interaction energy do not explain the character of the ordering state change with the pressure. We need the more detailed theoretical analysis that is a task of the future work.

CONCLUSION

The presented results show several examples of phase transitions in the 3D complex plasma system under microgravity conditions obtained with the help of the PK-3 Plus laboratory onboard the International Space Station.

The manipulation of the neutral gas pressure allows researchers to observe the fluid-solid phase transitions in the large 3D complex plasma The system of charged particles can exhibit melting upon increasing the gas pressure.

Compressing the dusty subsystem of small particles by an addition of the bigger microparticles to the system we have demonstrated the phase transition due to a drop of the particle charge.

The other quite a new observation concerns a study of transitions of the dusty plasma system state due to variations of the ion density. It was demonstrated a rather complicated character of the dependence of the system order upon the ion density.

Acknowledgments

The work is supported by the Russian Science Foundation, grant No. 14-12-01235.

References

[1] Fortov V E and Morfill G E (eds) 2010 *Complex and Dusty Plasmas: From Laboratory to Space* Series in Plasma Physics (Boca Raton, FL: CRC Press)
[2] Fortov V *et al.* 2005 *Phys. Rep.* **421** 1–103
[3] Thomas H M *et al.* 2008 *New J. Phys.* **10** 033036
[4] Thomas H M and Morfill G E 1996 *Nature* **379** 806–809
[5] Khrapak S A *et al.* 2012 *Phys. Rev.* E **85** 066407

Modification of polycarbonate surface in oxidizing plasma

A A Ovtsyn, S A Smirnov, T G Shikova, I V Kholodkov

Ivanovo State University of Chemistry and Technology, Sheremetyevskiy ave., 7, Ivanovo, 153000, Russia

E-mail: moose2004@inbox.ru

Abstract. The properties of the surface of the film polycarbonate Lexan 8010 were experimentally studied after treatment in a DC discharge plasma in oxygen and air at pressures of 50-300 Pa and a discharge current of 80 mA. The contact angles of wetting and surface energies are measured. The topography of the surface was investigated by atomic force microscopy. The chemical composition of the surface was determined from the FT-IR spectroscopy data in the variant of total internal reflection, as well as X-ray photoelectron spectroscopy. Treatment in the oxidizing plasma leads to a change in morphology (average roughness increases), an increase in the surface energy, and the concentration of oxygen-containing groups (hydroxyl groups, carbonyl groups in ketones or aldehydes and in oxyketones) on the surface of the polymer. Possible reasons for the difference in surface properties of polymer under the action of oxygen and air plasma on it are discussed.

1. Introduction

The surface of polymers generally has low surface tension and high chemical inertness, and usually not the best adhesion properties and wettability. Consequently, modification of the surface of polymers plays an important role in improving such processes as: wettability and adhesion to coatings during staining and printing, when working with biomaterials and certain types of composite materials. Therefore, the main methods for modifying the surface of polymers are: plasma treatment, ultraviolet irradiation, laser and chemical oxidation, introduction into the flame, and a number of others [1]. Processing in plasma provides ample opportunities for modifying the polymer surface. A lot of works are devoted to the improvement of polymer-polymer and metal-polymer adhesion, as well as to dyeing and printing on polymers [2, 3]. Polycarbonates (PCs) have become widespread in the industry due to their excellent transparency, high strength and resistance to high temperatures.

2. Experimental setup

For the experiment we used the setup described in [4]. Samples of film polycarbonate (PC) of the brand Lexan 8010 were placed along a generatrix on a thermostated wall of a cylindrical glass reactor 3 cm in diameter in the zone of the positive column. The pressure range of the plasma-forming gas is 50 - 300 Pa. The linear velocity of the gas flow was 30 cm / s. The discharge current is 80 mA. The processing time for the polymer is 180 s.

To determine the surface energy, the "neutral drop" method was used, consisting in placing two drops of different liquids (water and glycerin) with a known surface tension on the surface of the material under study. The contact angle of wetting was determined by digital photography.

The topography of the PC surface was investigated by atomic force microscopy using a "Solver 47 Pro" microscope. The scanning was carried out in a semi-contact mode.

The surface of the polymer was investigated by ATR-FTIR spectroscopy. A Nicolet type spectrophotometer of the Avatar-360 type was used. The ATR element was a zinc selenide crystal, the angle of incidence of the ray was 42 °, with a single reflection, a signal accumulation mode was used based on the results of 32 scans, and the resolution was 2 cm^{-1}. The chemical composition of the polycarbonate surface was determined by X-ray photoelectron spectroscopy (XPS). The source was AlKα (with energy 1486.6 eV) with a multichannel chemospheric electrostatic analyzer Phoibos 100. The obtained spectra were corrected for aliphatic carbon at 285 eV.

3. Results and discussion

Processing of polycarbonate films in plasma leads to an increase in the wettability of the surface: the contact angles of polymer wetting with water and glycerin decrease, the surface energy (σ) increases. The increase in the surface energy of the sample is due to an increase in its polar component (σ^p). The dispersion component (σ^d) changes insignificantly as a result of the action of the plasma (Table 1). Processing in oxygen plasma allows to achieve a better effect.

Table 1. Contact angles of wetting and surface energy of the polycarbonate films

Treatment conditions	Contact angles, degree		Surface energy, mJ/m^2		
	θ (H_2O)	θ (glycerol)	σ^d	σ^p	σ
Initial	81	70	20.3	10.2	30.5
Air, 100 Pa	36	47	10.5	48.2	58.7
Oxygen, 100 Pa	30	45	10.1	52.0	62.1

The study of samples by ATR-FTIR spectroscopy showed that as a result of processing in plasma, the greatest changes occur in the range of stretching vibrations of the O-H bond (3200-3500 cm^{-1}) and the stretching vibrations of the C = O bond in various functional groups (1625-1860 cm^{-1}) (figure 1).

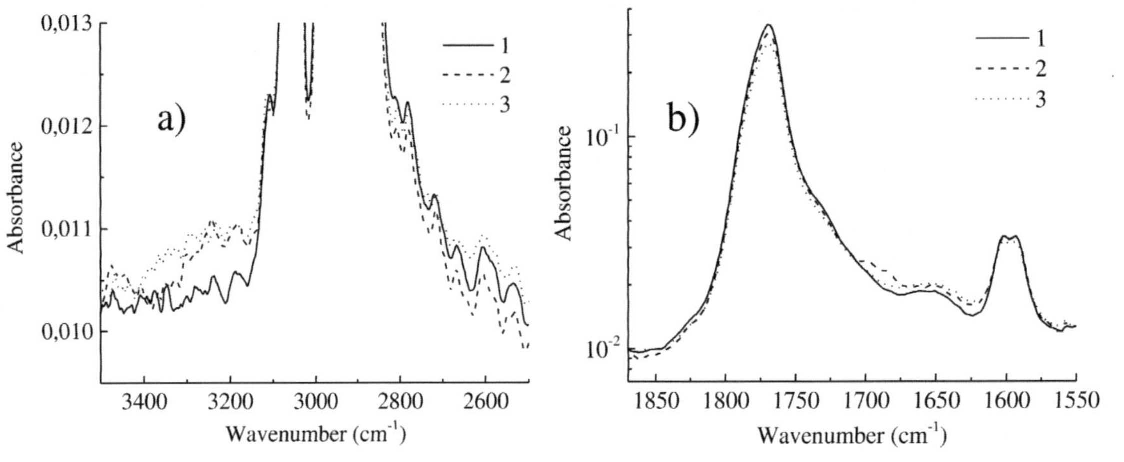

Figure 1. IR spectrum of polycarbonate films. 1 - initial sample; 2-air processed in plasma; 3 - oxygen treated in plasma. The gas pressure is 100 Pa. a) -O-H (3200-3500 cm^{-1}), b) -C = O (1625-1860 cm^{-1}).

The effect of oxygen and air plasma leads to a decrease in the optical density at the maximum of the C = O bond absorption band in the carbonate bridge (1770 cm^{-1}) and to an increase in absorption due to O-H bond vibrations (3220 cm^{-1}), C = O bond vibrations of ketones or aldehydes (1695 cm^{-1}) and in oxyketones (1650 cm^{-1}) [5]. A similar change in the IR spectrum of polycarbonate during thermal-oxidative degradation is reported in [6]. Changes in the spectrum in the region of 1660 cm^{-1} are attributed to the possible formation of quinoid carbonyl groups. The paper presents possible mechanisms for the formation of ketone and aldehyde groups, which are represented as a chain process initiated by the detachment of the hydrogen atom in the methyl group.

Figure 2 (a, b) shows the results of measurements of optical densities at the indicated wave numbers assigned to the optical density at a wave number of 1500 cm^{-1}, corresponding to the maximum of the absorption band of the C-C bond vibrations in the aromatic ring, and reduced to the untreated sample.

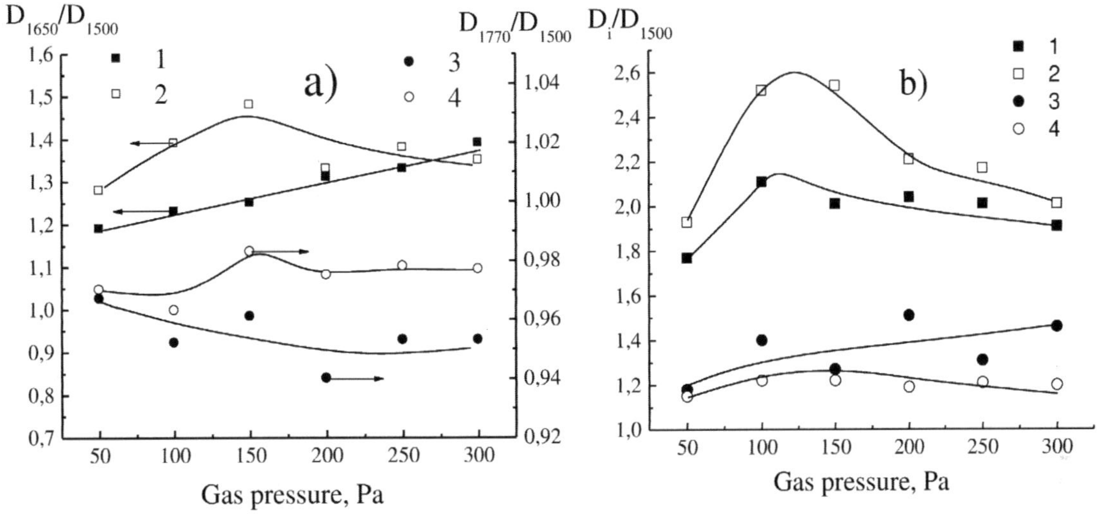

Figure 2. Dependence of the optical density on the gas pressure in the air and oxygen plasma. (a) 1- 1650 cm^{-1}, air plasma; 2 - 1650 cm^{-1}, oxygen plasma; 3 - 1770 cm^{-1}, air plasma; 4 - 1770 cm^{-1}, oxygen plasma. (b) - 3220 cm^{-1}, air plasma; 2 - 3220 cm^{-1}, oxygen plasma; 3 - 1695 cm^{-1}, air plasma; 4 - 1695 cm^{-1}, oxygen plasma.

The decrease in the optical density at the wave number of 1770 cm^{-1} is due to the destruction of the carbonyl groups that make up elementary unit of the polymer. When exposed to air plasma, the decrease is more significant than when processed in oxygen plasma. In the latter case, there is also a higher optical density, and, consequently, the concentration of hydroxyl groups and carbonyl groups in oxyketones (1650 cm^{-1}), but the concentration of carbonyl groups in ketones or aldehydes (1695 cm^{-1}) is lower than in air plasma .

The effect of gas pressure on the change in the concentration of the functional groups under consideration depends on the type of plasma-forming gas. For a plasma of oxygen, non-monotonic dependences with an extremum at 100-150 Pa are characteristic. In the air plasma, there are increasing dependences, except for the concentration of carbonyl groups in the carbonate bridge, which decreases with increasing pressure.

The results of IR spectroscopy as a whole are consistent with the results of the surface investigation by the XPS method, which also show the destruction of some functional groups of the polymer and the formation of others (figure 3 and Table 2). The concentration of functional groups on the polymer surface is determined by the ratio of the rates of their formation and destruction. It was shown earlier [4] that in the interaction of an oxygen-containing plasma with polymers, the main active plasma factor reacting with the sample is $O(^3P)$ oxygen atoms. These particles initiate the polymer plasmolysis process, which, like the thermal oxidative degradation process, is radical-chain. The fluxes of $O(^3P)$ atoms to the polymer surface in the oxygen plasma are almost three times higher than in the air plasma. This leads, on the one hand, to a higher rate of rupture of the initial bonds in the polymer and the formation of free radicals. On the other hand, these radicals in the oxygen plasma with higher rates react with atoms and oxygen molecules, forming both oxygen-containing groups existing in the sample, and new ones. As a result, under the influence of oxygen plasma, the concentration of the carbonyl groups of the carbonate bridge with respect to the initial sample decreases to a lesser degree than in the air plasma. The concentration of newly formed groups in the oxygen plasma as a whole is also higher. As a result, the degree of oxidation of the polymer surface under the action of oxygen plasma is greater than in the air plasma.

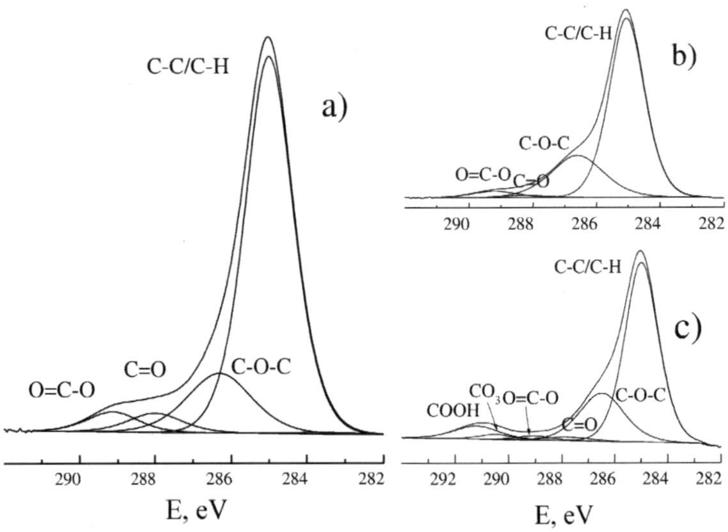

Figure 3. X-ray photoelectron spectroscopy spectra of C_{1s} for untreated (a), air-treated (b) and oxygen-treated (c) polycarbonate. The gas pressure is 100 Pa.

Table 2. Change in the composition of the surface of polycarbonate after treatment in air and oxygen plasma

Sample	Concentration of main groups, %							Concentration of atoms,%	
	$\pi - \pi^*$	C-C/ C-H	C-O-C	C=O	O-C=O	CO_3	COOH	C	O
Untreated	0,5	75,3	15,7	4,3	4,2	0	0	81,50	18,5
Air plasma	0,2	70,9	24,7	1	3,1	0,1	0	82,60	17,4
Oxygen plasma	0	65	23,4	1,8	1,6	1,8	6,4	76,2	23,8

Different character of the absorbance dependences (concentration of functional groups) on the gas pressure at the modification of polycarbonate in an oxygen plasma and air, and also different ratio of concentrations of carbonyl groups in various groupings suggests that, despite the similarity of mechanisms plasma-oxidative degradation of the polymer in an oxygen plasma, the common stage of initiation of the radical chain process, the further development of the chain in the oxygen and air plasma can proceed with various ways.

An increase in the content of polar functional groups on the surface of the polymer as a result of the action of plasma leads to an increase in the polar component of the surface energy and an improvement in the wetting of the sample.

Treatment in plasma leads not only to a change in the chemical composition, but also to the micro relief of the surface layer of the polymer (figure 4). Surface study by the AFM method showed that the average roughness increases from 3.36 nm for the original sample to 19.85 nm and 6.06 nm when exposed to oxygen and air plasma, respectively. A similar situation is observed in [7]. More significant changes in the topology of the surface under the action of oxygen plasma are due to higher polymer etching rates compared to air plasma. An increase in the surface roughness, as well as its oxidation, leads to a decrease in the contact angle of wetting.

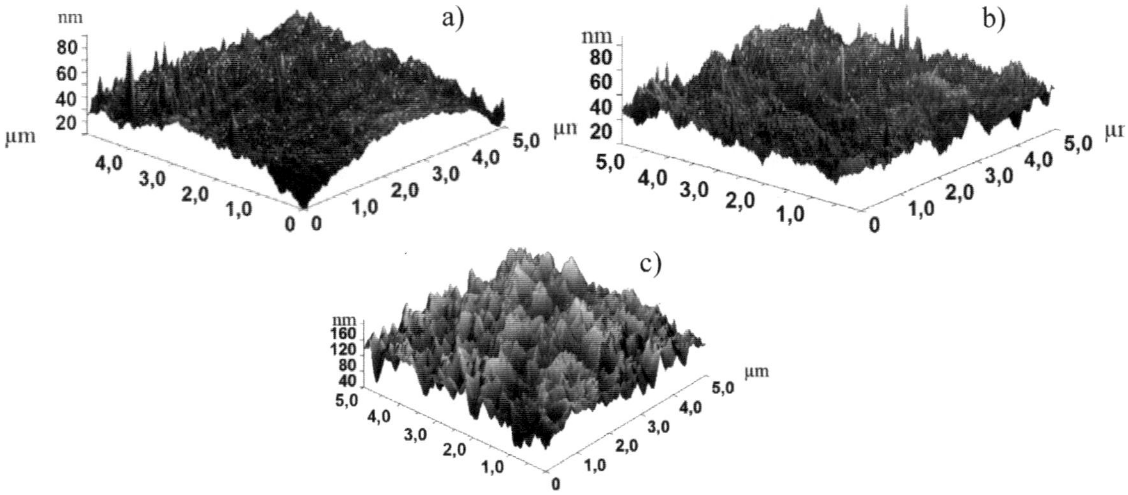

Figure 4. Images of the surface of polycarbonate obtained by atomic force microscopy: (a) - an untreated sample, (b) - processed in air plasma, (c) - processed in oxygen plasma. The gas pressure is 100 Pa.

4. Conclusion

Treatment of polycarbonate in the oxidizing plasma leads to a change in morphology (average roughness increases), an increase in the surface energy, and the concentration of oxygen-containing groups (hydroxyl groups, carbonyl groups in ketones or aldehydes and in oxyketones) on the surface of the polymer. Similarity of mechanisms plasma-oxidative degradation of the polycarbonate in an oxygen and air plasma, the common stage of initiation of the radical chain process, the further development of the chain in the oxygen and air plasma can proceed with various ways.

Acknowledgements

The study was carried out using the resources of the Center for the collective use of scientific equipment of Ivanovo State University of Chemistry and Technology. The work was supported by the Russian Foundation for Basic Research (project No. 16-32-00404).

References

[1] Ozdemir M, Yurteri C U and Sadikoglu H 1999 *Crit. Rev. Food Sci. Nutr.* **39** (5) 457–477
[2] Fu R K Y et al. 2005 *Nucl. Instrum. Methods Phys. Res., Sect. B.* **237** 417–421
[3] Flynn C N, Byrne C P and Meenan B J 2013 *Surf. Coat. Technol.* **233** 108–118
[4] Titov V A, Rybkin V V and Smirnov S A 2009 *High Energ. Chem.* **43** (3) 172–180
[5] Zbinden R 1964 *Infrared Spectroscopy of High polymers* (New York–Academic Press)
[6] Levantovskaya I I, Draluk G V et al 1968 *J. Polym. Sci., Part A: Polym. Chem.* **10** 2488–2494
[7] Cho J S, Choi W K, Jung H J and Koh S K. 1997 *J. Mater. Res.* **12** 277–280

IOP Conf. Series: Journal of Physics: Conf. Series **927** (2017) 012039 doi:10.1088/1742-6596/927/1/012039

The shape of DBD plasma jet striking into the static and quickly moving dielectric and metallic substrate

Yu S Akishev[1,2], V B Karalnik[1], M A Medvedev[1], A V Petryakov[1], N I Trushkin[1] and A G Shafikov[1]

[1] SRC RF TRINITI, 108840, Moscow, Troitsk, Pushkovykh street, 12, Russia
[2] NRNU MEPhI, 115409, Moscow, Kashirskoe shosse, 31, Russia

Abstract. In the case of motionless ambient air, the quickly moving surface involves in the movement the gas adjoining to a surface. This gas-dynamic wall effect can influence both the shape of plasma jet interacting with the moving substrate and the efficiency of plasma modification of the surface to be treated. The issue mentioned above has been studied by an example of the DBD plasma jet striking into the static and fast moving dielectric and metallic substrate. The discs of acrylic plastic and aluminum were used as the static and quickly rotating substrates. The experiments were performed with He and Ar plasma jets generated by sinusoidal DBD operating at atmospheric pressure. It is proved that the change of the shape of plasma jet at the place of its striking into a moving dielectric and metal substrate happens due to the dragging of air by the fast moving surface.

1. Introduction

Non-thermal plasma jets at atmospheric pressure generated by dielectric barrier discharge (DBD) in He and Ar flow are the popular objects to investigation in many publications [1,2]. Such plasma jets are widely used for different applications, in particular, for plasma modification of thermally non-resistant materials. The plasma-chemical modification happens due to interaction of reactive plasma species with the surface. These species are generated in the discharge zone of DBD and transferred to the surface by plasma jet of a small diameter. The reactive plasma species are neutral and charged particles. At atmospheric pressure, the neutral particles are "frozen" into the gas stream forming the plasma jet. The charged particles are ruled predominantly by the electric field in the plasma. The total efficiency of plasma modification is determined by many factors among the most important of them one can note the efficiency in generation of primary reactive species in the discharge zone, their decay and partial transformation into secondary reactive species in the course of transport of reactive species by thin plasma jet to the surface and finally the spreading of primary and secondary reactive species over large area on the surface to be treated. The latter process depends mainly on gas-dynamic and electrodynamic phenomena related to interaction of gas stream with a solid surface and propagation of plasma on the substrate. The pointed above phenomena depend not only on the velocity of jet striking the substrate but on the stray capacitance of the substrate and its state (mobile or motionless) as well.The issues mentioned above have been studied by an example of the DBD plasma jet striking into the static and fast moving dielectric and metallic substrate. The discs of acrylic plastic and aluminum were used as the static and quickly rotating substrates. Additionally, polyethylene film, quartz and ceramic plates were used as the static substrates. The experiments were performed with He and Ar plasma jets generated by sinusoidal DBD operating at atmospheric pressure.

2. Experimental setup

In order to generate a long non-thermal plasma jet at atmospheric pressure we used a sinusoidal DBD in thin quartz tube blown with He and Ar flow. Sketch of the experimental setup is shown in figures 1a and 1c.Inner and outer diameters of the discharge tube were 2.5 and 4.5 mm. Frequency of the powering voltage was fixed and equal to 100 kHz; the amplitude was varied up to 4 kV. In some experiments the inner electrode (thin wire) was grounded and the powering voltage was applied to the outer electrode. Gas flow velocity at the DBD exit is equal to 25-30 m/s. Plasma jet enters the ambient air and strikes the substrate. Dielectric substrates were quartz, acrylic plastic, ceramic and polyethylene. The substrates to be treated were placed on the static or rotating aluminum of 160 mm in diameter, which can be grounded, if it was necessary. A distance from the DBD exit to the disc was 20 mm and 7-10 mm for the case of He and Ar as the plasma forming gases. In the case of rotating disk, plasma jet was directed to the point located of 70 mm away from its center. The disc rotation frequency was varied up to 170 turns per second. Maximum frequency corresponds to the linear velocity of 75 m/s of the disc at the point of the plasma jet striking. Fast rotation of the disc leads to dragging of a thin layer of ambient air adjoining the surface of the disc. Maximum averaged velocity of air measured by Pitot tube located at 0.5 mm from the surface of rotating disc is equal to 25 m/s.

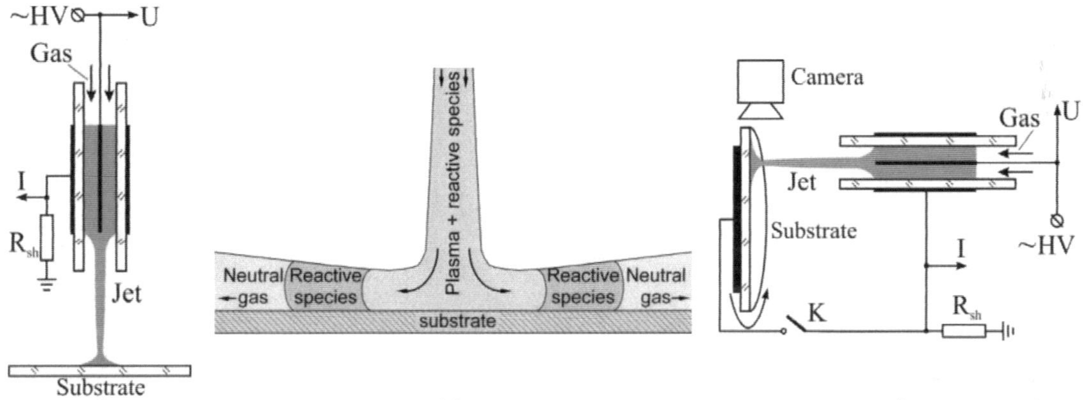

a) b) c)

Figure 1. Sketch of the experimental setup used for generation of a plasma jet by DBD in He and Ar flow and investigation of its colliding with the static (a) and quickly moving (c) dielectric and metallic substrate. The sketch illustrating the difference in radial distribution of plasma, reactive species and neutral gas over surface for the case of plasma jet normal striking the static dielectric substrate (b).

Figure 1b illustrates schematically the difference in the radial distribution of plasma, reactive species and neutral gas on the surface for the case of plasma jet normal striking the static dielectric substrate. This figure shows that the reactive species can be transported by a gas stream to a longer distance compared to the plasma. A reason of that is caused by relatively long time of their life.

3. Experimental results and discussions

3.1. Interaction of plasma jet with the motionless dielectric substrates

Figure 2 demonstrates influence of the electric parameters of DBD on brightness of plasma jet and its spreading on the motionless dielectric substrates. It was found out that the grounding of the inner electrode diminishes appreciably both the length and brightness of plasma jet independently on sort of plasma forming gas. Note that helium as plasma forming gas increases the length of free plasma jet compared to argon in both regimes when the inner or outer electrode is grounded. The dependence of visual diameter of plasma spot on the treated substrate vs the applied voltage is shown in figure 2c.

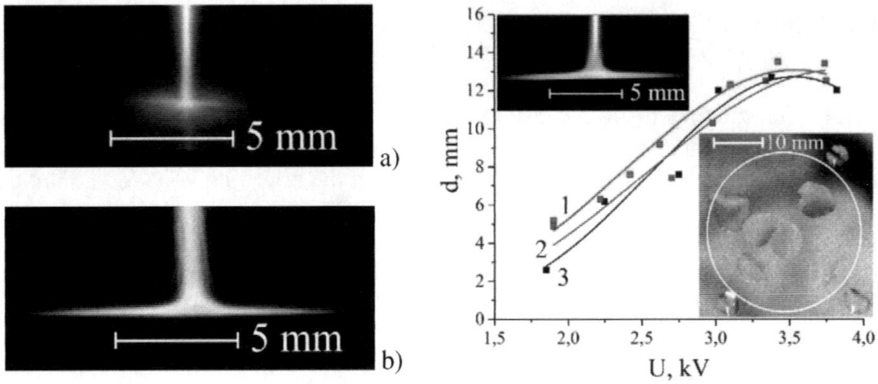

Figure 2. The images of plasma jet striking the polyethylene film (a, b). Ar, V=25 m/s, U=4 kV, h=7.5 mm. a) P=6.3 W, the inner electrode is grounded, the jet brightness is low, the exposure time τ=0.1 s. b) P=5 W, the outer electrode is grounded, the jet brightness is high, the exposure time τ=1 ms. c) The dependence of visual diameter of plasma jet on quartz vs applied voltage. He, V=28 m/s, h=20 mm, the outer electrode is grounded, the jet brightness is high, the exposure time τ=5 ms. Two photos in plot (c) show a large difference between visual diameter of plasma spot on the quartz (top image) and the diameter of the area treated by plasma jet (the circle with the strongly spread water droplets).

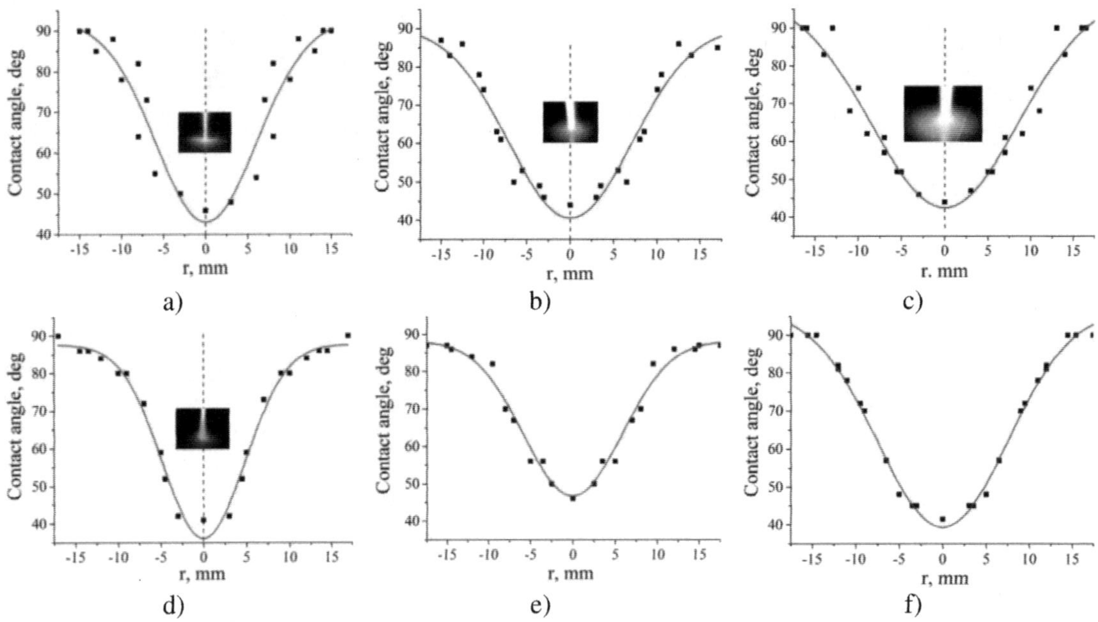

Figure 3. The radial distribution of the contact angle of small water droplets deposited on polyethylene film treated by Ar plasma jet. The images of plasma jet striking the substrate are inserted with the same scale into some plots. The substrate is placed at a distance of 7.5 mm (a)-(c) and 15 mm (d)-(f) from the DBD exit. The inner electrode is grounded (a), (d); the outer electrode is grounded (b, c, e, f). The electric power of DBD: P=6.3 W (a,d); P=4 W (b,e); P=8.5 W (c,f). Treatment time is 2 s.

Figure 3 shows in detail the radial distribution of the contact angle of small water droplets deposited on polyethylene film treated by Ar plasma jet at different experimental conditions. This figure demonstrates the dependence of the treatment effect on a distance from the DBD exit, electric power of DBD and which electrode is grounded. One can see also in this figure that visual diameter of plasma spot is always smaller compared to the diameter of the area treated by plasma jet.

3.2. Interaction of plasma jet with quickly moving the dielectric and metallic substrates

The results on deformation of helium plasma jet on and in vicinity of the moving substrate are presented in figure 4. Note that the shape of plasma spot on a motionless dielectric and metallic surface is not the same - plasma widely spreads on a dielectric surface but concentrates into a small spot on the metallic substrate. Increase in a rotating velocity of the substrate (disc) leads to the dragging of ambient air in the vicinity of the disc surface that in its turn influences the interaction of plasma jet with the surface - the wind induced by the rotation separates the jet from the surface to be treated. This effect is the same for dielectric and metallic substrate and can lead to lowering the efficiency of plasma jet treatment. A positive effect of the wind is the destroying of current spot figure 4 (4c).

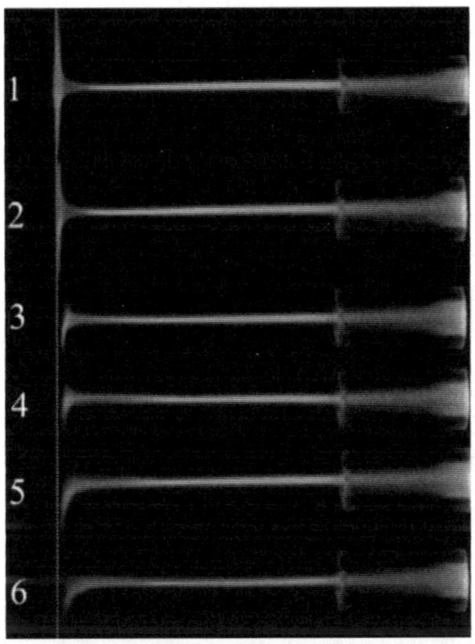

a)

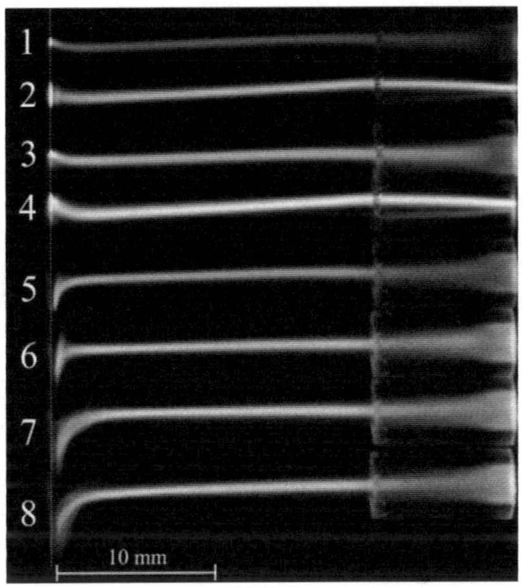

c)

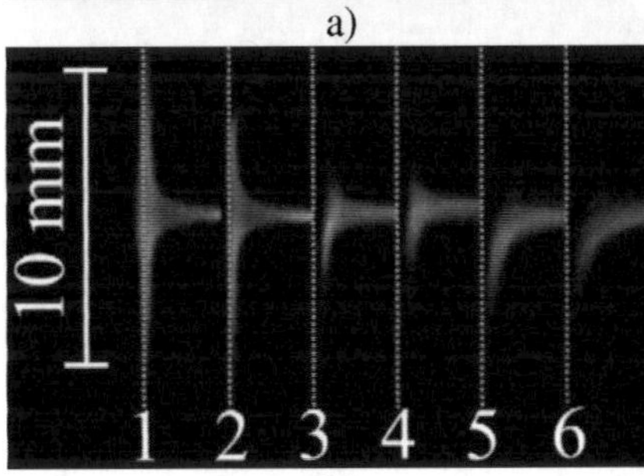

b)

Figure 4. Set of images of He plasma jet striking the substrate of acrylic plastic (a, b) and aluminum (c) rotating with different frequencies F (revolutions per second). Set (b) shows the enlarged image of plasma jet deformation on the substrate and its separation from the surface with increase in F. Dashed line in photos marks the substrate surface. Jet velocity V=28 m/s; P=3.6 ±0.2 W; F=0 (1a, 1c, 2c); F=9 rps (2a, 3c, 4c); F=58 rps (3a, 5c); F=102 rps (4a, 6c); F=139 rps (5a, 7c); F=168 rps (6a, 8c).

Figure 5 presents the results on interaction of argon plasma jet with the rotating substrate. There is a great difference in the behavior of Ar plasma jet interacting with the dielectric and metallic substrate. According to figures 5a, 5b, deformation of argon plasma jet in vicinity of the dielectric substrate grows with increase in rotation frequency but there is no its separation. Ar plasma jet striking the metallic substrate exhibits unexpected behavior. At low electric power $P < 8$ W, plasma jet is diffusive (figure 5d) and interacts with metallic disc like He plasma jet. At higher electric power $P > 8$ W, DBD transits to the regime with alternation of low-current and high-current modes of 8.5 and 27 W respectively (figure 5c), which correspond to diffusive and filamentary plasma jet (figures 5d, 5e, 5f). This phenomenon is attributed to ability of the grounded disc to pass high current to the ground through plasma jet that leads to the development of the ionization instability in plasma jet and its transition into filamentary mode. The wind induced by the disc rotation influences the high-conductive current filaments, which change the angle of their striking the surface (figure 5e, 5f). Note the filaments are strongly twisted. A reason of that is the necessity to increase their total length in order to be able to keep for short time a high voltage drop between the HV electrode and the ground.

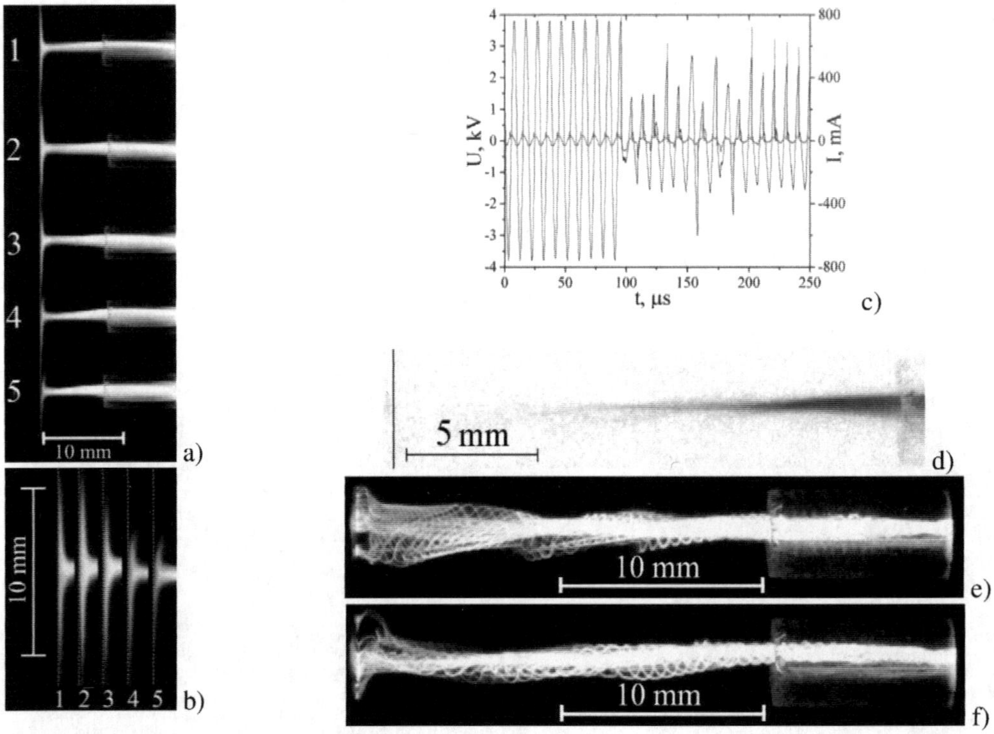

Figure 5. Set of images of Ar plasma jet striking the substrate of acrylic plastic (a, b) and the grounded aluminum disc (d, e, f) rotating with different frequencies F. Set (b) shows the enlarged image of the jet deformation with increase in F. Plot (c) presents the current-voltage waveform of DBD at the alternating regime. Dashed line in photos marks the substrate surface. Jet velocity V=28 m/s; F=0 (1a, 1b, d, e); F=58 rps (2a, 2b); F=102 rps (3a, 3b); F=139 rps (4a, 4b); F=168 rps (5a, 5b, f).

4. Conclusion

It is proved that fast movement of the dielectric and metal substrate leads to change of the shape of plasma jet at the point of its striking. This phenomenon is attributed to gas-dynamic effect – to the dragging of air by the fast moving surface. It was revealed that the wind induced by fast moving substrate can lead to the separation of jet from the surface and lowering the efficiency of plasma treatment. It was found out that Ar plasma can form thin current filaments twisted by a spiral.

References

[1] Akishev Yu, Balakirev A, Grushin M, Karalnik V, Kochetov I, Napartovich A, Petryakov A and Trushkin N 2015 *IEEE Trans. Plasma Sci.* **43** 745

[2] Korolev Yu 2015 *Russian Journal of General Chemistry* **85** 1311

Acknowledgments

This work was supported by Russian Foundation of Basic Research (grant № 17-02-00234).

IOP Conf. Series: Journal of Physics: Conf. Series **927** (2017) 012040 doi:10.1088/1742-6596/927/1/012040

How ionization waves (plasma bullets) in helium plasma jet interact with a dielectric and metallic substrate

Yu S Akishev[1,2], V B Karalnik[1], M A Medvedev[1], A V Petryakov[1], N I Trushkin[1] and A G Shafikov[1]

[1] SRC RF TRINITI, 108840, Moscow, Troitsk, Pushkovykh street, 12, Russia
[2] NRNU MEPhI, 115409, Moscow, Kashirskoe shosse, 31, Russia

Abstract. Plasma jets generated by DBD are not uniform along their length because of existence of so called "plasma bullets" (ionization waves) travelling each half-cycle in the jets. The bullet velocity and the visual shape of plasma jet will change, if plasma jet meets the substrate. This work is devoted to the detail experimental study on influence of the substrate material (dielectric or metal) on the bullet propagation velocity in plasma jet formed by the cylindrical DBD in He flow. It was found out that the conductivity of the substrate has a great impact on the plasma bullets parameters.

1. Introduction

Plasma jet generated by dielectric barrier discharge (DBD) in He flow blown through a narrow dielectric tube is the object of many studies [1,2]. The detailed review of experimental and theoretical investigations devoted to different properties of such plasma jet was done in paper [3]. One of the interesting phenomena observed in DBD plasma jet is so called "plasma bullets" travelling along plasma jet. These "bullets" have been for the first time revealed by M. Teschke et al [4]. The bullets are characterized by higher electric field and higher light emission. A visual length of plasma jet in a free space is determined therefore by a maximum distance traveled by a bullet from the DBD exit. In fact, plasma bullet is the ionization wave, the propagation mechanism of which is similar to the propagation mechanism of the head of normal streamer. One of the important parameters influences the streamer propagation is the displacement current collected by streamer head from the space in front of the streamer. For the case of a sinusoidal DBD in He, it means that the bullet propagation velocity depends not only on parameters of the applied voltage (amplitude and frequency of sinusoidal voltage) but on the configuration of the variable electric field formed around plasma jet in the course of the bullet propagation. If the plasma jet treats the dielectric or metallic substrate placed on different distance from the DBD outlet (in other words, the space in front of the jet is not free), the configuration and amplitude of the variable electric field in front of the moving bullet will change appreciably compared to that for plasma jet in free space. One can possible expect therefore that both the bullet velocity and the visual shape of plasma jet interacting with the substrate will change as well. This work is devoted to the detailed experimental study on influence of the substrate material (dielectric or metal) on both the bullet propagation velocity in plasma jet and shape of the plasma jet contacting the substrate.

IOP Conf. Series: Journal of Physics: Conf. Series **927** (2017) 012040 doi:10.1088/1742-6596/927/1/012040

2. Experimental setup

In order to generate a long non-thermal plasma jet we used a sinusoidal DBD in He flow. Frequency of the applied voltage is equal to 100 kHz; amplitude of the applied voltage was varied up to 4 kV; gas flow velocity at the DBD exit is equal to 25-30 m/s. Plasma jet enters the ambient air at atmospheric pressure and strikes the substrate. Dielectric substrate were quartz ($\varepsilon \approx 4$) and ceramic ($\varepsilon \approx 8.5$) plates with thickness d = 12 and 1.65 mm respectively. Sketch of the experimental setup is shown in figure 1. The treated substrates were placed on the metallic plate which can be grounded, if it was necessary.

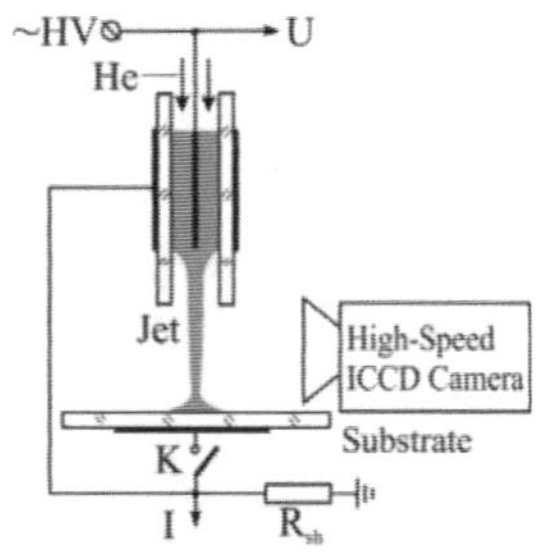

Figure 1. Sketch of the experimental setup used for generation of a plasma jet by DBD in He flow and investigation of the plasma bullets colliding with a dielectric and metallic substrate.

3. Experimental results and discussions

The images of plasma jet either entering free space or striking the dielectric and metallic substrate are shown in figure 2. One can see that plasma jet spreads over a dielectric substrate (the higher applied voltage and ratio ε/d, the stronger spreading) but does not do that on the metallic substrate. A reason is that the plasma spreading over a dielectric substrate is a normal surface dielectric barrier discharge (SDBD) [5]. Indeed, plasma jet plays a role of the conductor transferring the high-voltage potential to the substrate. Due to existence of the stray capacitance between a dielectric substrate and ground, the SDBD over substrate happens. One of the needed conditions for appearance of SDBD is a formation of the tangential electric field on the surface. This condition is not fulfilled for metallic substrate, i.e. SDBD plasma cannot happen on metallic surface. Equivalent electric schemes of DBD corresponding to the images in figure 2 are presented in figure 3. These schemes take into account an existence of stray capacitance of the "plasma jet - substrate" system for different ambient conditions.

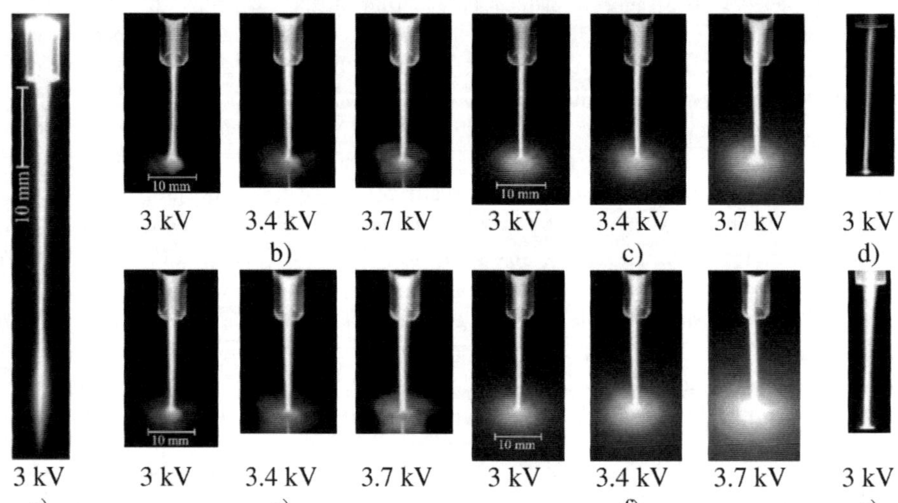

Figure 2. The images of plasma jet under different conditions. Exposure time of each shot is 0.5 ms. a) free plasma jet, some expansion of the jet at its end happens due to transition of gas jet at this place into a turbulent regime; b), e) quartz substrate placed on non-grounded and grounded metallic plate; c), f) ceramic substrate placed on non-grounded and grounded metallic plate; d), g) non-grounded and grounded metallic plate.

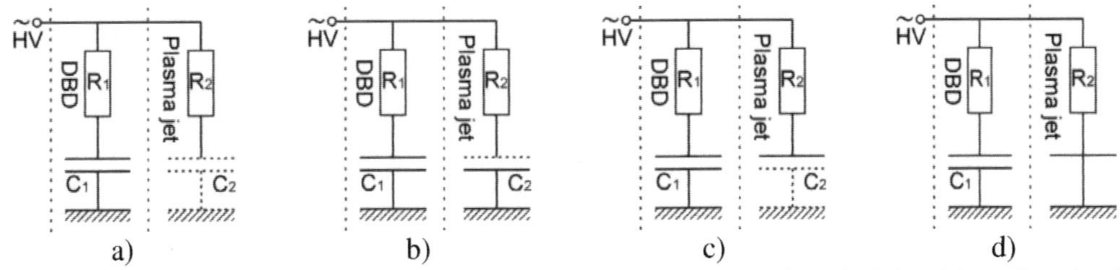

Figure 3. Equivalent electric schemes of DBD with plasma jet. Left and right sides of each picture depict the scheme of DBD and plasma jet.a) jet enters a free space; b) jet impacts a dielectric substrate placed on the grounded metallic plate; c) jet impacts a metallic plate at the floating potential; d) jet impacts a grounded metallic plate. R_1 and R_2 notate the effective resistances of DBD plasma and plasma jet; C_1 and C_2 notate the capacitance of DBD and plasma jet for every configuration a)-d).

The current-voltage waveforms of DBD correlated with the images of plasma jet taken by a fast frame camera at different time moments are presented in figure 4 and figure 5. The enumerated quadrates in the legends of figure 4 show the time position of the corresponding shots presented in figure 5.

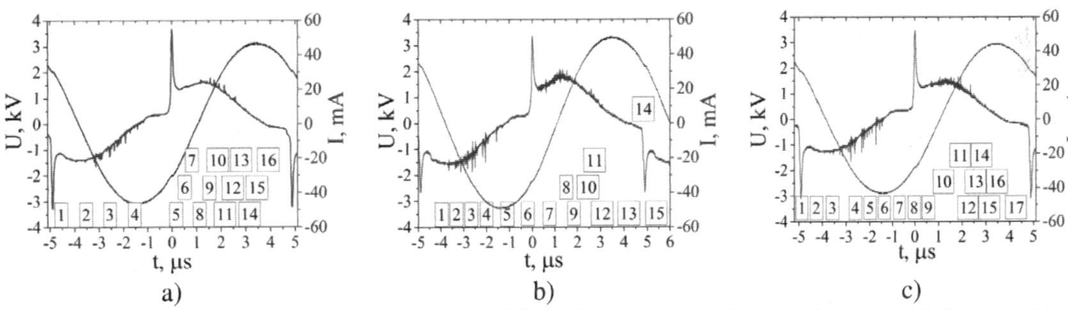

Figure 4. The current-voltage waveforms of DBD in He corresponding to the cases: a) free jet; b) jet impacts the grounded ceramic substrate; c) jet impacts the metallic plate at floating potential.

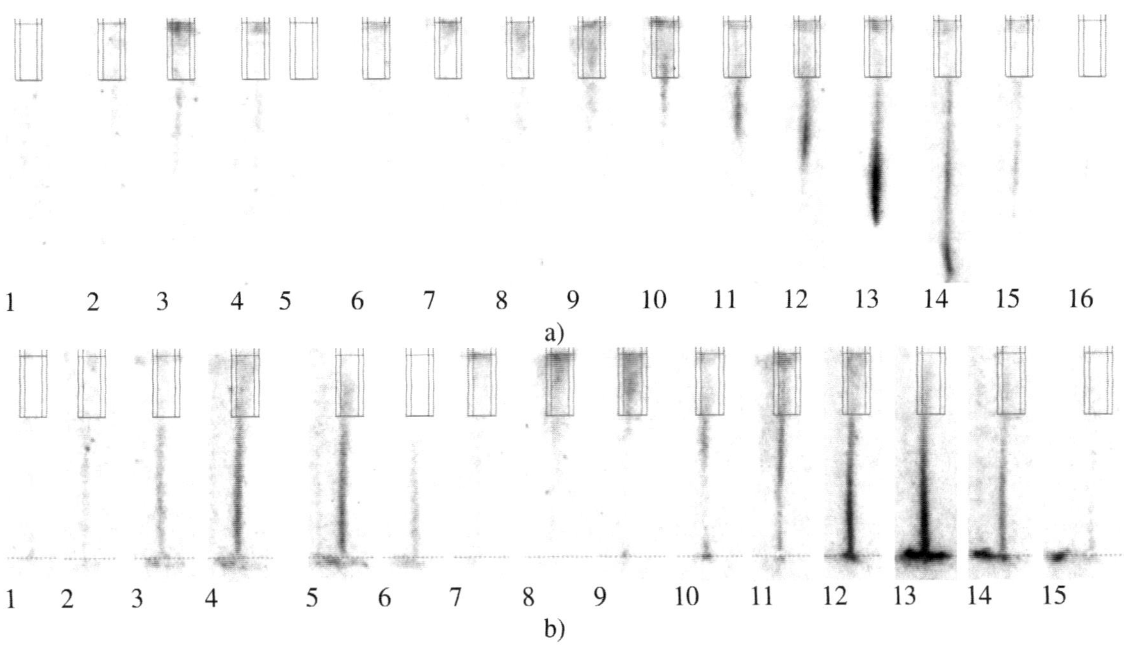

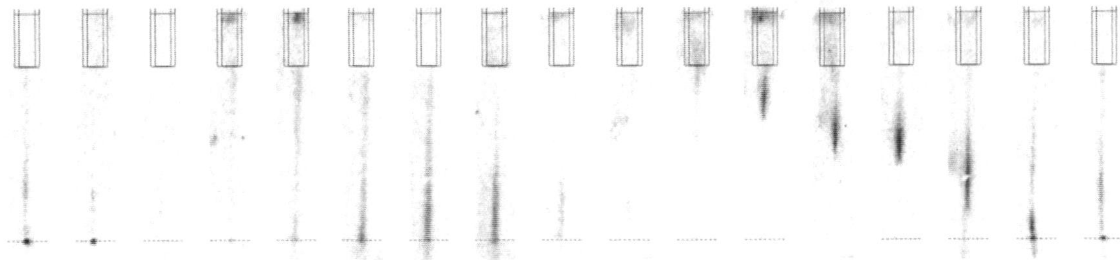

c)

Figure 5. The set of plasma jet images taken by a fast frame camera at different time moments. The time moments of the enumerated shots correspond to time position of the appropriate enumerated quadrates in the legends in Fig.4. a) free jet; b) jet impacts the grounded ceramic substrate; c) jet impacts the metallic plate at floating potential. The exposure time of shots: a) ## 1-4 is 500ns, ## 5–16 is 200 ns; b) ## 1–15 is 200 ns; c) ## 1–17 is 200 ns.

In contrast to figure 2, one can see in figure 4 that the existence of the substrate practically does not influence the current-voltage waveforms of DBD. However, a spatial-temporal behavior of plasma jet and plasma "bullets" strongly depends not only on the existence of substrate but on the material of substrate and its proximity to the ground as well.

In the case of a free jet (figure 5a), the stray capacitance of plasma jet regarding the ground is not so large. In a free jet, the negative "bullets" are practically invisible. A positive plasma "bullet" appears at the exit of the discharge tube not immediately at the moment of the DBD breakdown (shot #5, figure 5a) but about 1 μs later. In general, slow bullet propagation inside the tube correlates with the growth of current but fast bullet propagation in plasma jet correlates with diminishing the current. Fast decay of brightness of the plasma jet (##15, 16, figure 5a) is followed by shortening of his visible length.

In the case of a plasma jet striking a thin grounded ceramic substrate (figure 5b), the stray capacitance of the "plasma jet - substrate" system is large. It leads to that the electric current flowing through plasma jet increases appreciably compared to the case of a free jet. In this case the amount of the electric charge deposited on the substrate in the course of every half-cycle is large and the behavior of "plasma jet - substrate" system is similar to normal DBD. In other words, plasma jet spreading over dielectric substrate can be considered as a single large microdischarge of DBD. The mentioned above results in the following features. The plasma "bullets" are absent in negative half-cycles: the brightness of plasma jet increases simultaneously over the whole length between the substrate and the tube exit (##2-5, figure 5b). The growth of brightness correlates with the current diminishing. Fast decay of the plasma jet brightness (##6, 7, figure 5b) happens also practically simultaneously over the whole length of plasma jet. In a positive half-cycle, the luminosity of plasma jet is forming due to simultaneous movement of two ionization waves quickly propagating from a discharge zone and substrate (##9-11, figure 5b). Maximum luminosity of plasma jet (#13, figure 5b) happens approximately at the moment when the applied voltage is maximum but the discharge current is close to zero. Note that the current flowing to the ground through plasma jet may not be zero at this moment. Decay of the brightness (##14, 15, figure 5b) happens again practically simultaneously over the whole length of plasma jet.

As it is shown in figure 2.g, b, plasma jet does not spread over the metallic substrate and this property does not depend on that there is or not a connection of the substrate with the ground - the connection with the ground leads only to strong increase in both the current flowing through plasma jet and brightness of its luminosity but does not lead to spreading the plasma over the substrate. Figure 5c presents the set of the instant shots of plasma jet interacting with a metallic plate at the floating potential. Interesting that in the negative half-cycle the ionization wave propagates from the substrate towards the discharge tube (##5-8, figure 5c). Possible reason of that is associated with the large charge deposited on the metallic surface at the preceding half-cycle. However, the positive "bullets" propagate in "right" direction from the discharge exit towards the substrate (##11-16, figure 5c).

Close examination of the data similar to those presented in figure 5 allowed us to determine the velocity of the positive plasma bullets vs their distance from the discharge zone. The type of the substrate (dielectric or metal) and its connection with the ground (grounding or floating potential) were the variable parameters of the experiment. The appropriate results are shown in figure 6.

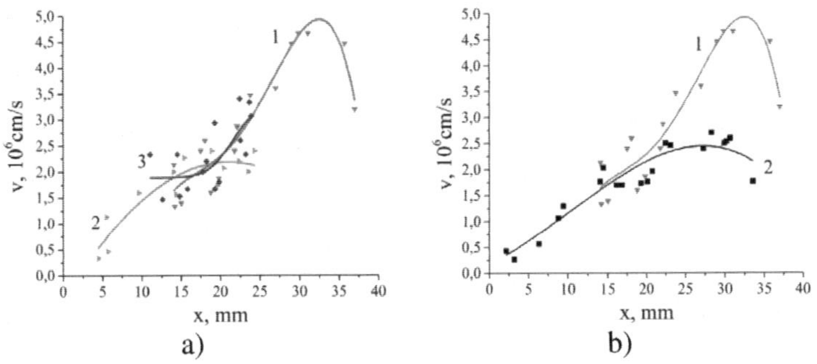

Figure 6. The positive plasma bullet velocity in He plasma jet vs the bullet distance from the discharge zone. a) 1- free jet, 2 and 3 - jet striking the quartz substrate placed on the non-grounded and grounded metallic plate; b) 1 – free jet, 2 - jet striking the non-grounded metallic plate.

One can see in figure 6 that there is a small influence of any substrate on the velocity of positive bullets if the substrate is placed closer than 20 mm from the discharge zone. However, after this distance the existence of the substrate leads to diminishing the bullet velocity propagation in the plasma jet striking the substrate.

4. Conclusion

It was shown that interaction of He plasma jet with the substrate depends on its conductivity – plasma jet striking the dielectric spreads over a substrate (the higher applied voltage and ratio ε/d, the stronger spreading) but plasma jet does not do that on the metallic substrate. It was found out also that there is a small influence of any substrate on the velocity of positive bullets in plasma jet, if the substrate is placed closer than 20 mm from the discharge zone. The existence of any substrate placed at longer distance leads to diminishing the bullet velocity propagation compared to that for free plasma jet.

5. References

[1] Akishev Yu, Balakirev A, Grushin M, Karalnik V, Kochetov I, Napartovich A, Petryakov A and Trushkin N 2015 *Trans. Plasma Sci.* **43** 745
[2] Korolev Yu 2015 *Russian Journal of General Chemistry* **85** 1311
[3] Lu X, Naidis G, Laroussi M and Ostrikov K 2014 *Phys. Rep.* **540** 123
[4] Teschke M, Kedzierski J, Finantu-Dinu E, Korzec D and Engemann J 2005 *IEEE Trans. Plasma Sci.* **33** 310
[5] Akishev Yu, Aponin G, Balakirev A, Grushin M, Karalnik V, Petryakov A and Trushkin N 2013 *J. Phys. D: Appl. Phys.* **46** 464014

Acknowledgments

This work was supported by Russian Foundation of Basic Research (grant № 17-02-00234).

Modification of Surface Properties of Ultrahigh-Molecular – Weight Polyethylene Films by DC Discharge

M Piskarev[1], A Gilman, A Kuznetsov and A Ozerin

Enikolopov Institute of Synthetic Polymer Materials RAS,
ul. Profsoyuznaya 70, Moscow, 117393, Russia

E-mail: mikhailpiskarev@gmail.com, plasma@ispm.ru

Abstract. The methods of surface modification of ultrahigh-molecular-weight polyethylene (UHMPE) by the low-temperature plasma are considered. Installations for treatment of polymers with various types electric discharges as well as the methods for studying the changes occurring on the polymer surface are described. The experimental results are given concerning properties of 100 μm UHMWPE film before and after treatment by the low pressure air dc discharge.The film samples were placed at the anode. The measurements of the peel resistance (T-test according to ASTM 1876-01) of adhesion joint of modified UHMWPE with polyimide and polytetrafluoroethylene films were obtained using polyurethane adhesive.

1.Introduction

Ultrahigh-molecular-weight polyethylene (UHMWPE) (molecular weight of 1.5×10^6 to 6×10^6) is characterized by high mechanical properties, resistance to corrosive chemicals, a low water uptake, a low coefficient of friction, and high abrasion resistance. These properties determine the use of UHMWPE in many areas of technology and, since recently, in medicine and biology, including the manufacture of artificial joints and implants for surgery and orthodontics. Currently, UHMWPE is manufactured under various brand names: Polymin™ SK (BASF, Germany), Polystone M™ (Roechling, Germany), Tivar™ (Quadrant, Belgium), Tecafine PE10™ (Ensinger, Germany), Okulen 2000™ (SP-Plast, Finland), GUR™ (Ticona, Germany), Chirulen™ (Poly-Hi Solidur, Germany). UHMWPE is produced also by Goodfellow (United Kingdom) and Braskem (Brazilian Chemicals).

Note that the UHMWPE surface is hydrophobic and for many applications, for example, in lining, its contact and adhesive properties require a substantial improvement. It is known that low-temperature plasma treatment is currently one of the most technically promising and environmentally friendly methods for surface modification of polymer materials. This method makes it possible to alter the surface properties of a nanosized polymer layer without changing its bulk characteristics [1–6].

In this paper we consider some methods for modification the surface of UHMWPE films by low-temperature plasma and data of adhesion properties of the obtained films [5, 7, 8].

The goal of this study is also the examination of adhesion properties of UHMWPE films using the T-peel test for the contact of this film with PTFE and PI films and Polyurethane adhesive.

2. Methods of plasma modification of UHMWPE films

To modify UHMWPE films, low- or atmospheric-pressure discharge excited over a wide current frequency range, including the utility (50 or 60 Hz), medium (20–90 kHz), radio (RF, 13.56 MHz) and the microwave (MW, 2.45 GHz) frequencies and DC discharge are used [8]. The treatment is carried out using both laboratory setups, and industrial units. The schematics of the most commonly used installations are shown in Figures 1 and 2.

As an example, Figure 1 shows a schematic of an experimental dielectric barrier discharge (DBD) setup with an industrial ac power supply [9]. Glass-coated copper electrodes (*4*, *6*) are arranged in the working chamber (*5*) at a distance of 7 mm from each other. The upper electrode (*4*) is connected to a 50-Hz high-voltage power source (*3*), and the lower electrode (*6*) is grounded. A sample of UHMWPE film was placed on the lower electrode; Ar, He, N_2, or air was used as a plasma gas; and methyl methacrylate was the precursor for the thin film of another chemical nature deposited on the sample surface by plasma polymerization.

Figure 2 shows a schematic of an APC 2000 industrial unit (Sigma Technologies Int.) in which a process for modifying UHMWPE films by treatment in a 90-kHz discharge in air was studied [10]. The working gas was a mixture of He (*3*) and water vapor (*1*, *2*) fed directly through hollow electrodes (*4*) in a certain ratio. A 1-mm thick sheet was mounted on a rotating electrode (*6*) with an Al_2O_3 dielectric coating and treated by DBD with an input power from 0.861 to 2.58 W/cm^2 for 0.4–40 s.

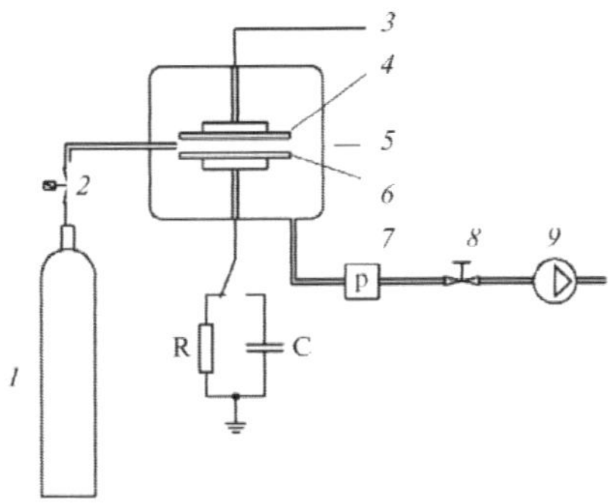

Figure 1. Schematic of an ac DBD reactor: (*1*) process-gas cylinder, (*2*) metering valve, (*3*) high-voltage power supply (50 Hz), (*4*) upper electrode, (*5*) working chamber, (*6*) grounded lower electrode, (*7*) pressure gauge, (*8*) vacuum valve, and (*9*) backing pump [9].

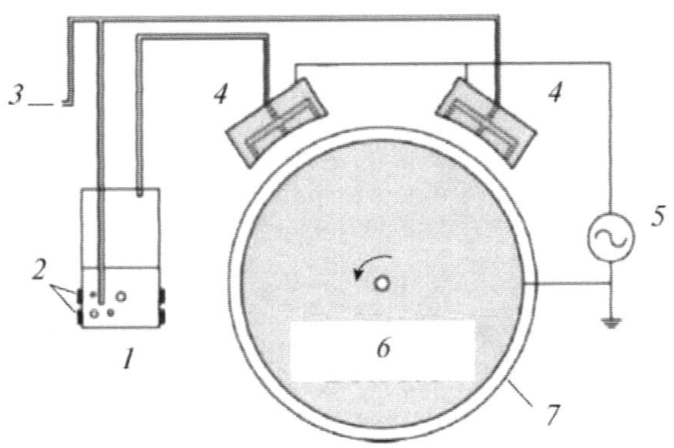

Figure 2. Schematic of the commercial APC 2000 atmospheric pressure discharge plasma machine: (*1*) evaporator, (*2*) heaters, (*3*) He supply line, (*4*) grounded electrodes with He and water vapor supply, (*5*) discharge power source (90 kHz), (*6*) rotating grounded electrode, and (*7*) dielectric coating [10].

Oosterom et al. [11] used single lap-joint shear testing to determine the adhesive strength for UHMWPE sheets (thickness 1.4 mm). Samples of UHMWPE (Chirulen 1020, Germany) were modified by corona discharge (TIGRES CKG unit, 50 Hz, Germany) or by RF glow discharge (TNO Laboratories, 13.56 MHz, the Netherlands). To make measurements, two specimens of a 110 × 25 × 1.4 mm size were lap-jointed with the adhesive MA300 (ITW Plexus, United Kingdom) to have a contact length of 20 mm. After holding for 24 h, the joint was examined using a Zwick 20 kN testing machine (Germany) at room temperature and crosshead speeds of 3 and 6 mm/min. The shear strength was 0.3 MPa for the initial sheet and increased to 1.06 and 0.93 MPa for the samples modified by the corona and the RF glow discharge, respectively.

Rodriguez-Santiago et al. [10] investigated surface changes using the ASTM 1876-01 T-peel test. Goodfellow UHMWPE films (75 μm) were modified by helium–water vapor DBD discharge using a SigmaTechnologies model APC 2000 (90 kHz) atmospheric pressure plasma system. Two specimens of a 15.2 cm×30.5 cm×75 μm size were bonded with the DevThane 5 polyurethane adhesive (Devcon, the United States). The sample prepared in this manner was rolled down with a roller for leveling and held for 24 h at room temperature. For measurements, 30.5 cm×2.5 cm strips were cut, and tests were performed using an MTS Synergie electromechanical testing machine at a crosshead speed of 0.4 cm/s. For the film modified over 15 s, the peel resistance increased tenfold from 0.25 to 2.5 N/cm.

3. Adhesion properties of UHMWPE films modified by DC discharge

Samples of the UHMWPE film of 100 μm thickness (Russia) were used in the study. The procedure for film modification by DC discharge and the installation were detailed in [12].

The samples of 35×40 mm were placed at the anode and treated at a working plasma gas air, pressure of 5–45 Pa and a discharge current of 50 mA for 10–60 s. Filtered atmospheric air with humidity of 50% was used as a working gas. The diameter of the electrodes is 180 mm, so the area of the dielectric sample is approximately 5% of the electrode area. In this case there was not any difference between the glow discharge with sample or without one. The discharge in this case is homogeneous.

The adhesive strength of the polymers studied was evaluated using a T-peel testing configuration according to ASTM 1876-01 [5, 10, 13]. The T-peel test is a test designed to determine the peel resistance (**A**) of adhesive bonds between flexible adherends. Standard Test Method for Peel Resistance of Adhesives (T-Peel Test) according ASTM 1876-01 is shown in Figure 3.

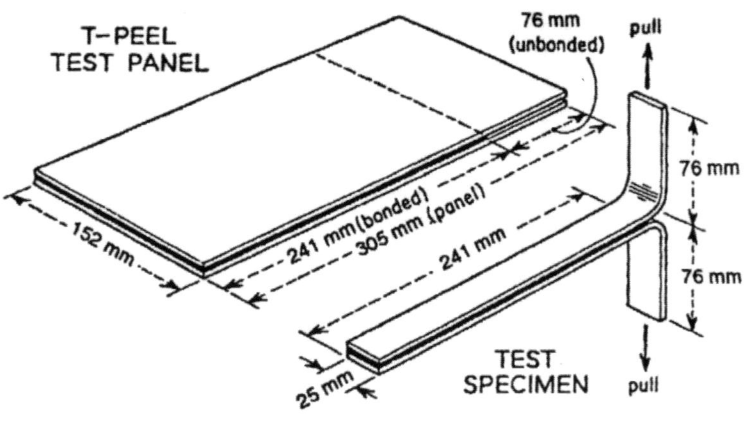

Figure 3. Standard Test Method for Peel Resistance of Adhesives (T-Peel Test) according ASTM 1876-01. Test Panel and Test Specimen [13].

The test specimens consisted of two 15.2 cm×30.5cm films bonded together with a urethane-based adhesive (PU). The PU adhesive is a solution of polyurethane rubber in acetone and ethyl acetate. The adhesive was applied to the adhesive surfaces of 2 films, dried at 90° C for 5 min, then these surfaces were combined and held under a press at a pressure of 3 kg/cm^2 and T = 100° C for 1 min. The cured specimens were then cut into strips (see Figure 3) and tested. The peel tests were conducted on a Hounsfield H1K tensile machine at a speed of 100 mm/min. The result of the measurements is averaging over 10 tests.

Adhesion between pairs of UHMWPE/PTFE and UHMWPE/PI films was also studied. Samples of the PTFE film (60 μm, "Plastpolymer", Russia) and PI (50 μm, PM-1, GOST-6-10-121-85, Russia) were used in the study.

The experimental values of the T-peel resistance (**A**) measured immediately after the plasma treatment according ASTM 1876-01 are given in the table 1.

Table 1. Experimental data on the peel strength (**A**)

Sample	Adhesive	**A**, N/m	
		Initial sample	Plasma treated
UHMWPE/ UHMWPE	PU	12 ± 1	470 ± 30
UHMWPE /PI		12 ± 1	457 ± 31
UHMWPE /PTFE		10 ± 1	280 ± 24

From the data given, it can be seen that the initial films have very low adhesion. Processing by DC discharge at the anode leads to a multiple increase in the adhesion of the film pairs indicated in the table. The **A** value increases approximately 40 times for pairs of UHMWPE/UHMWPE and UHMWPE/PI, and 30 times for the UHMWPE/PTFE pair.

According to the data of X-ray photoelectron spectroscopy data [9,10] the processes of modifying UHMWPE films by oxygen and air plasmas results in formation in the surface layer of oxygen-containing functional groups. Apparently, increase in adhesion strengh A (table 1) is due to the interaction of these polar groups (hydroxylic, carboxylic, fluoroanhydride and others) with the active groups of adhesive.

According to the data of X-ray photoelectron spectroscopy [9,10] the process of UHMWPE modifying by oxygen and air plasmas results in formation in the film surface layer of oxygen-containing functional groups. Apparently, increase in adhesion strengh A (table 1) is due to the interaction of these polar groups (hydroxylic, carboxylic, fluoroanhydride and others) with the active groups of adhesive.

4. Conclusion

Thus, in present paper on the basis of experiments conducted, we have demonstreated that the treatment of UHMWPE by the low themperature plasma, and more specifically, by the low pressure dc discharge is technologically effective and environmentally friendly method to improve its adhesion properties.

References

[1] *Encyclopedia of Low-Temperature Plasma: Introductory Volume IV.* Fortov V E, Ed., 2000 (Moscow: Nauka)
[2] *Polymer Surface Modification: Relevance to Adhesion*, Mittal K L, Ed., 2009, vol. 5 (Boca Raton: CRC)
[3] *UHMWPE Biomaterials Handbook*, Kurtz S, Ed., 2009, 2nd ed. (Philadelphia: Academic)
[4] Friedrich J, *The Plasma Chemistry of Polymer Surfaces:Advanced Techniques for Surface Design*, 2012 (Weinheim: Wiley–VCH)

[5] Pocius A V *Adhesion and Adhesive Technology: AnIntroduction*, 2nd ed., 2002 (Munich: Carl Hanser)

[6] Yablokov M Yu, Sokolov I V, Malinovskaya O S, Gilman A B and Kuznetsov A A, *High Energy Chem.* 2013 **47** 32

[7] Ebnesaijjad Ed S, Ebnesaijjad C 2014 *Surface Treatment of Materials for Adhesive Bonding* (Norwick: William Andrew Publ)

[8] Gilman A, Piskarev M, Kuznetsov A and Ozerin A *High Energy Chem.* 2017 **51**136

[9] Aziz G, Cools P, de Geyter N, Declercq H, Cornelissen F and Morent R *Biointerphases* 2015 **10** 029502

[10] Rodriguez-Santiago V, Bujanda A A, Stein B E and Pappas DD *Plasma Process. Polym.* 2011 **8** 631

[11] Oosterom R, Ahmed T J, Poulis J A and Bersee H E N , *Med. Eng. Phys.* 2006 **28** (4) 323

[12] Demina T S, Drozdova M G, Yablokov M Yu, Gaidar A I, Gilman A B, Zaytseva-Zotova D S, Markvicheva E A, Akopova T A and Zelenetskii A N *Plasma Proc. Polym.* 2015 **12** 710

[13] ASTM 1876-2001, *Standard Test Method for Peel Resistance of Adhesives (T-Peel Test)*

Arc ignition at heating of graphite by fixed current

V P Polistchook, I S Samoylov, R Kh Amirov and V I Kiselev

Joint Institute for High Temperatures of the Russian Academy of Sciences, Izhorskaya13 Bldg 2, Moscow 125412, Russia

E-mail: polistchook@mail.ru

Abstract. Arc ignition after the destruction of graphite samples under prolonged heating by electric current was described. Evidences of liquid film formation on the graphite surface at a temperature of 3.3 kK were presented.

1. Introduction

An occurrence of arc after destruction of conductor, which is heated by electric current, is usually the result of reaching the melting temperature. In work [1] using this method the melting graphite in argon atmosphere was studied and parameters of the triple point of carbon were estimated: T_{tp}=4 kK, p_{tp}=10.5 MPa. Other researchers, who used pulsed heating of graphite by electric current or laser radiation, got a higher temperature of the triple point: T_{tp}=4.8 kK, p_{tp}=11 MPa [2]. At present, these parameters of the triple point of carbon are generally accepted; however the results of the studies [3-5] directly contradict this point of view.

According to the phase diagram of carbon, which was suggested in work [3], the graphite transforms into carbyne at a temperature above 2.6 kK. The time of phase transition "graphite-carbyne" depends on the temperature: at the temperature of 2.8 kK graphite entirely transforms into carbyne during 10^5 s, and at the temperature of 3.8 kK this process lasts for a few seconds [3]. Thus, the melting temperature of graphite should depend on the duration of heating. According to work [3], the triple point in the "vapor-carbine-liquid" system exists at the temperature of 3.8 kK and a pressure of 20 kPa. The evidences of formation carbyne and other carbon phases after prolonged graphite exposure at temperature of 3.0-3.8 kK were presented in works [4-7]. The experiments on melting graphite when heating process lasted from 1 to 10^3 s were described in work [5]. The parameters of the triple point of carbon (T_{tp}=4 kK, p_{tp}=0.1 MPa), which were obtained in work [5], are close to those that were given in work [3].

In work [6], we described the arc that arose after graphite sample exposure at a temperature of about 3 kK for duration of about 10^3 s. According to work [6] at a temperature of about 3.3 kK, the liquid film was observed on the graphite surface. This paper presents the study of ignition stage of this arc.

2. Experimental set

We used installation that has been described in works [6, 7]. The samples were made from rods of spectral graphite with diameter of 6 mm. In the central part of the rod its diameter was decreased up to 3 mm on the length of 20 mm. Resistivity of used graphite at room temperature was ρ_0=11 Ohm×μm; its density was 1.7 g/cm^3; the total impurities in graphite was less than 0.01%. Samples that fixed in molybdenum holders were heated by DC in argon atmosphere.

We measured the current through the sample and the voltage drop on it. The voltage on the sample in transient regimes was recorded using oscillograph adapter AKTAKOM. The process of sample destruction was controlled by videography at a speed of 25 frames/s and exposure time of 0.5 ms. The sample temperature was measured by luminance pyrometer IS-140 at the wavelength of 0.9 μm with 0.9 mm spatial resolution and temporal resolution 10 ms. The true graphite temperature was determined using the data of emissivity factors [6].

3. Experimental results and discussion

Heating current (120-190 A) was chosen so that the initial temperature in downtown the grooving T was 2.7 -2.9 kK. During exposure with the fixed current (10-100 minutes) sample temperature and voltage drop on it monotonically increased that is explained by the local growth of graphite resistivity [7]. When the temperature exceeded ~3.2 kK, an "explosion" of sample occurred, and the arc discharge appeared.

In this work the video frames and voltage oscillograms at sample # 175 in the initial time of arc were compared. Argon pressure was 51 kPa. Sample current before the "explosion" was 138 A; after the destruction of the sample and occurrence the arc with voltage of about 20 V the current decreased to 132 A. Before "explosion" the sample temperature increased at the speed of ~1 K/s. When the temperature reached ~ 3.2 kK, the arc discharge arose, and during ~ 10 ms temperature abruptly increased from 3.2 to 3.4 kK. In this experiment the pyrometer determined the temperature of anode part of the sample. During the first 10 s after "explosion" the anode temperature changed mainly in the range of 3.4 to 3.6 kK.

Figure 1 shows video frames of the sample destruction. Figure 1a shows the moment of "explosion"; the cathode part of the sample is to the right. One can see the intense plasma glow near the surface, on the previous frame (40 ms earlier) this glow was absent. Figure 1b, which was filmed 0.76 s after Figure 1a, shows the gap between the cathode and anode of about 0.2 mm wide. In the previous frame the gap was not visible, and through of 0.12 s it disappeared. Figure 1c was obtained 2.2 s after Figure 1a. As one can see, the gap between the cathode and anode is absent. Figure 1d presents the sample 10.8 s after Figure 1a. On this frame the bridge between cathode and anode is still visible; in the next shot it is absent. The anode temperature at this moment was of 3.4 kK.

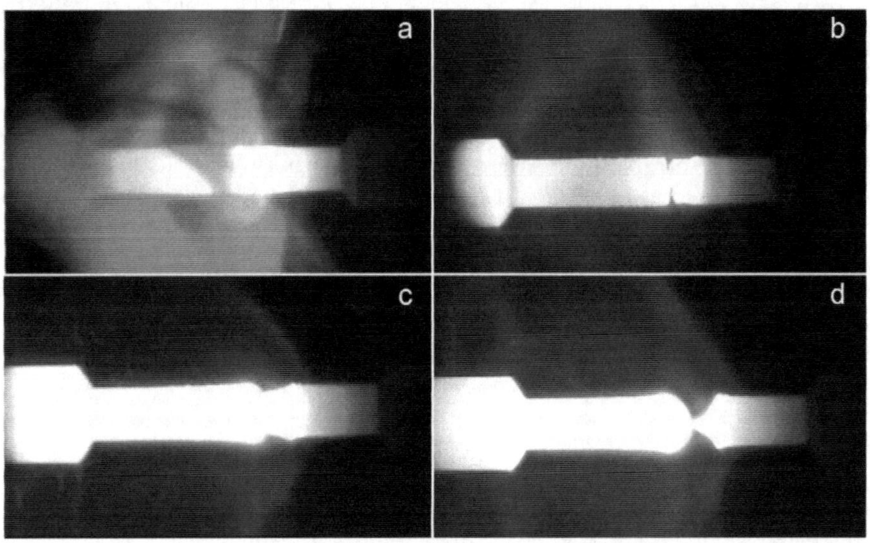

Figure 1. Frames of the sample destruction.

Figure 2 shows the snippets of oscillograms of sample voltage in moments of time corresponding to the frames on Figure 1. Before the "explosion" the voltage drop on the sample was of 12 V. Figure 2a presents the oscillogram in the moment of "explosion", when during 0.5 ms voltage has increased to 35 V. The voltage of about 30 V was maintained for ~ 10 ms, and then it decreased to ~ 26 V. At this moment there was a series of voltage pulses having duration of 10 ms and an amplitude of 35 V; the interval between pulses was of about 60 ms. The pulses begun 0.12 s after "explosion" and lasted about of 0.5 s until the visual gap between cathode and anode part of the sample not appeared.

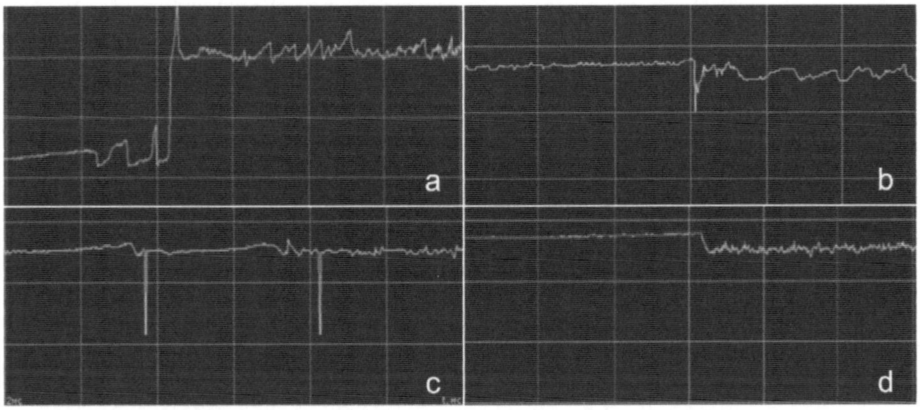

Figure 2. Oscillograms of sample voltage: 2 ms/div, 10 V/div.

Figure 2c shows the oscillogram at the time corresponding to the Figure 1c. The breakdowns on Figure 2c demonstrated that for a short period of time the voltage on the sample tried to restore to the value of 12 V, which was on the sample before "explosion". Decreasing of voltage drop to 12 V could cause short-term extinction of the arc. The anode temperature was decreased to 3.1-3.2 kK in this time.

Figure 2d shows the waveform at the time of the disappearance of the jumper. The destruction of the "bridge" between the electrodes had a little effect on the voltage drop on the sample. This voltage also slightly changed when the initial gap between the electrodes appeared (Figure 1b and Figure 2b).

In this experiment, the final gap between the cathode and the anode was formed approximately 11 s after the surge, i.e. ~10 s after the appearance of the initial gap (Figure 1b). This scenario of the gap formation between the electrodes was often observed in our experiments. Along with this there was another type of sample destruction when the sample was locally heated in area with size of ~3 mm in which there was a gradual decrease in the diameter of the grooving due to the evaporation and ejection of micro particles [6]. In this case, the gap between the cathode and the anode was also installed via ~10 s after the voltage surge. The third kind of the sample destruction when the gap between the electrodes appearing via ~0.5 s after the jump of the voltage was monotonically increased, were observed much less the other two.

The presented results demonstrate that the destruction of the graphite sample after prolonged exposure at high temperature has the appearance of stretched electric "explosion". This "explosion" starts with qualitative change in the character of voltage growth on the sample, where a smooth increase in voltage with a rate of 10 mV/s transforms to a sharp jump in the voltage value of ΔU~20 V (Figure 2a). In the initial time sample integrity is retained, so the power surge is associated with a sharp increase in resistivity. This indicates the rapid growth of content in the sample of carbon phases with high resistivity. Carbyne and diamond-like structures, for example, can play the role of such carbon phases. If one assumes that the longitudinal size of the region with high resistivity is ~5 mm (Figure 1a) then the estimated value of specific resistivity is ρ~250 Ω *μm. That is approximately 20 times higher than the initial resistivity of graphite. This is an estimation of the average over the cross section of the resistivity of the sample material; the local value of the resistivity could be significantly more. The breakdowns shown on Figure 2c can be considered as indirect confirmation of the existence of local regions with high resistivity.

The increase in the resistivity of the sample may indicate an appearance of liquid carbon; according to the hypothesis [3], it is formed by melting the carbon phases with low electrical conductivity.

Apparently, at the initial time of arc burning on the electrode surface there is a competition of two processes – the formation of the melted film and its destruction under the action of the arc. Under the influence of these mechanisms ultimately, the film is formed with a low electrical conductivity, the thickness of which is such that it does not prevent the current transfer.

In work [6] the existence of the melt film on the graphite surface was justified by the peculiarities of the electrodes geometry. First, the anode geometry in the form of mace preserved with a substantial reduction in the anode length. The anode has gained such geometry in time of gap formation (Figure 1d). Second, we often observed the temporal disappearance of the initial gap between the cathode and anode that usually appeared via ~0.5 s after the arc ignition. Both of these features can be explained by the flow of the melt film along the electrode surface under action of the plasma pressure. When this film comes on the colder part of the anode, it solidified, forming a step with height of ~0.1 mm [6].

The appearance of the electrode surface after exposure to the arc also indicates the existence of melt film. When the arc duration was of about 1 s the droplet and hemispherical formations with size of 0.1-0.3 mm were observed on the original surface of electrodes. If the arc existed above 5-10 s, then appearance of the electrode surface in the zone of influence of the arc became a sleek and "melted". Apparently, in the initial moment of time the size of the melt zone was of 0.3-0.5 mm, and then the area of the melting covered the entire working surface of the electrodes.

Earlier we have noted [6] that in our experiments the occurrence of the arc was determined not only by the sample temperature, but also by exposure time at high temperature. This effect is confirmed by the thermogram of sample # 181 presented in Figure 3. In this experiment, the increase of heating current from 160 A to 174 A caused the temperature rise from 3.15 to 3.35 kK. In this regime the movement of "shadows" was observed near the sample surface that resulted in brief decrease of the pyrometer readings [6]. The current of 174 A was maintained for 10 s, and then it was decreased to 130 A. The arc occurred later via 440 s when the current was 142 A. Figure 3 demonstrates that sample temperature at the moment of arc appearance (~3.2 kK) was significantly less than at the regime of short-term overheating (~3.35 kK).

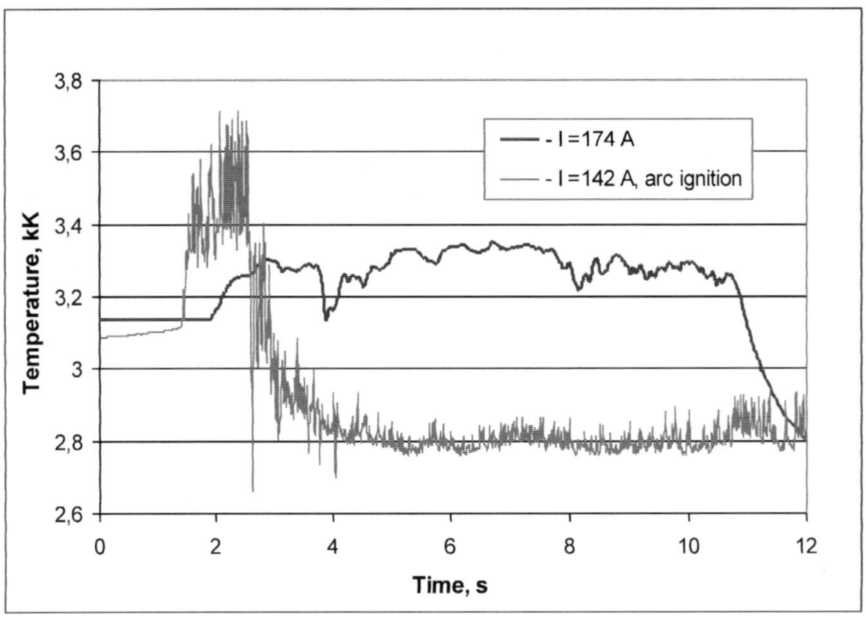

Figure 3. Thermogram of sample # 181.

In our experiments the minimum temperature of graphite, at which its melting was observed, was about of 3.2 kK that corresponds to a brightness temperature of 3.05 kK. Similar "low" melting point of graphite was registered in work [5], wherein the color temperature of the melt film flowing on the sample was equal of 3.04-3.09 kK. However, the authors of work [5] concluded that this temperature was too low for melting, and they measured, in fact, the temperature of solid graphite, along which transparent liquid carbon flowed.

4. Conclusion

The presented results confirm the conclusions of the work [6] that the melt film with thickness of the order of 0.1 mm arises on the surface of graphite electrodes at a temperature of 3.2-3.3 kK. The appearance of the liquid phase of carbon at such "low" temperature is possible only after long, more than 10^2 s, the exposure of graphite at a temperature of about 3 kK. Under these heating conditions the formation of carbon phases with lower, compare to graphite electrical conductivity, takes place. Apparently, the scenario of formation of liquid carbon was close to what was proposed in work [3].

Acknowledgments

The work was supported by the Russian Foundation of Fundamental Research (grant 06-08-00056a). Authors gratefully acknowledge Arkadii Itenberg for assistance in conducting experiments and processing of their results.

References

[1] Basset M J 1939 *J. Phys. Radium* **10** pp 217-228
[2] Bundy F P *et al* 1996 *Carbon* **34** pp141-153
[3] Whittaker A G 1978 *Nature* **276** pp 695-696
[4] Whittaker A G and Kintner P L 1985 *Carbon* **23** pp 255 -262
[5] Asinovskii E I, Kirillin A V and Kostanovskii A V 2002 *Phys. Usp.* **45** pp 869-882
[6] Amirov R Kh *et al* 2016 *High Temperature* **54** pp 644-654
[7] Asinovskii E I *et al* 2006 *High Temperature* **44** pp 396- 405

IOP Conf. Series: Journal of Physics: Conf. Series **927** (2017) 012043 doi:10.1088/1742-6596/927/1/012043

The study of low temperature plasma of pulse discharge in relation to air cleaning units.

A Ponizovskiy[1], S Gosteev[2] and O Kuzhel[3]

[1] Deputy chief engineer, FMKB "Horizon" CS "SPC Gas Turbine" Salute" Moscow RF
[2] Head of Department, FMKB "Horizon" CS "SPC Gas Turbine" Salute" Moscow RF
[3] Director, FMKB "Horizon" CS "SPC Gas Turbine" Salute" Moscow RF

E-mail: horizontL56@rambler.ru

Abstract. In paper it studied parameters of low-temperature plasma (LTP) used in systems for cleaning waste gas. LTP created by positive nanosecond corona discharges, generated by high voltage pulses with a rise time of 50 ns, duration up to 400 ns, an amplitude up to 90 kV and pulses repetition 50-1000 Hz. in coaxial electrode system with gap space 3-10 cm through which moving air with linear velocity v = 0.01 to 10 m/s.

1. Introduction

Currently being introduced systems for cleaning emissions of harmful gaseous impurities and odors based on the technology of low-temperature non equilibrium plasma (LTP) generated by nanosecond streamer discharge [1,2]. LTP characterized by significant densities and energies of electrons capable of creating in the discharge gap high concentrations of active intermediate particles (atomic oxygen, ions and radicals), which initiate radiation-chemical reactions with harmful molecules and transform them into ecologically harmless gases or aerosols. The basis of LTP treatment systems are high-frequency pulse generators and reactor chamber (RC) with wire or multi-point (MP) corona discharge electrodes, in which the plasma-chemical reactions take place. Due to high volume density charges occurring while generating pulsed corona produces charged finely dispersed aerosols, vapors and solid particles as present in the gases, and produced as a result of conversion. The removal of charged particle from air is carried out in the pause between pulses due to the drift of molecules and particles to the grounded electrode under high DC voltage. The flushing of deposit aerosols from the wall of RC carries out by circulating water. The effectiveness of such systems depends on the ability of the introduction of energy into the flow of gas, so that is of particular relevance to the study of streamers multipoint electrode system, in moving air and high pulse repetition rate. Therefore, unlike many studies streamers carried out in stationary air at a single pulse, we studied the streamer in the frequency mode in the air with ozone, nitrogen oxides and other contaminated admixtures. The results obtained allowed to optimize the design of running a series of plasma treatment installations of the "Corona M", made by FMKB "Horizon", intended for cleaning air of gaseous environmentally harmful ways and fine impurities.

2. Experimental Procedures

The pulses are generated by device collected by the scheme Fitch (PGF) with an odd number of steps (Figure 1) [2]. PGF consist of high-voltage charger (HV charge), a three stage of capacitors C_{st}, two charging inductors L_d, switch (thyratron), transformers TR1 and TR2 and block ignition (C, R_1 R_2). The capacity charge to voltage U_A from HV charger.

Main advantage PGF is that, in the first place, the capacitors are connected directly to the load, generated pulse with an amplitude equal to, $U_{pl} \approx 3 \cdot U_A$ by main one switch, secondly, in the pause between pulses the load have constant potential of U_A. As a result, the unit operates as a converter of impurities and as a standard electrostatic precipitator.

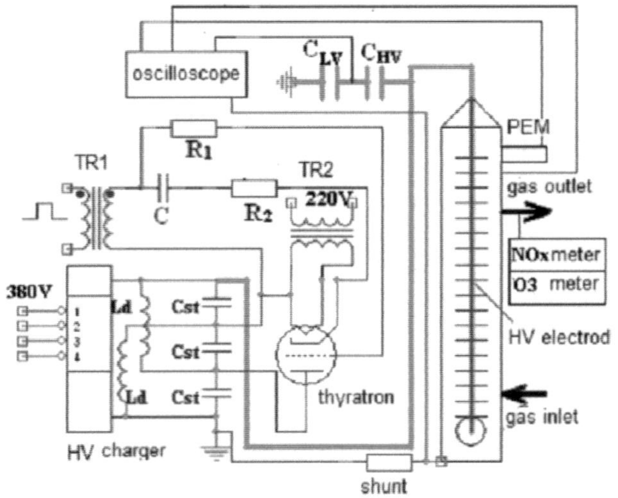

Figure 1. Schematic diagram of the experimental and industrial LTP gas cleaning unit.

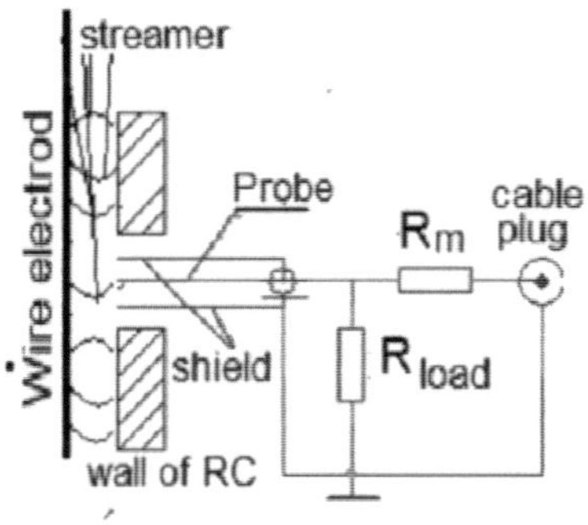

Figure 2. Schematic diagram of the probe measurements of parameters of the streamer.

U_{pl} was measured with a capacitive divider(C_{HV}, C_{LV}), the full discharge current (I_{sum}) was measured with shunt. Concentrations of gaseous impurities, oxygen and ozone gas were measured by analyzers with various types with photo ionization, photometric and electrochemical sensors. LTP was generated in the reaction chamber (RC) with the coaxial electrode system - grounded pipe and central MP or wire electrode. Diameter(D) of RC ranged from 10 to 273 cm, length(L) from 0.3 m to 4 m, electrode spacing (d) d=3-10 cm. Experiments were conducted at air flow rate 0.01-10m/s.

Probe technique for measuring the parameters of the streamer [3] is to measure the current (I_{ST}) through wire probe diameter of 0.5 mm the set flush with the RC wall(Figure 2) loading on resistor R_{load}. This allowed measurement of single streamer current after reaching the probe head streamer.

3. Results

3.1 Investigation of the parameters of streamers probe method.

Using frequency PGF and multichannel oscilloscopes provided evidence of the development of the tape after it crossed the inter-electrode space. Figure 3 shows waveforms of a single streamer current (I_{st}), Isum, U_{pl} and a signal from the photo electron amplifier(PEM) in RC with D=6 cm, L=30 cm and wire electrode.

We believe that the start of streamers (t_{in}) corresponds fracture on the waveform PEM signal. In time streamers is touching the probe and is beginning register of I_{st}. Waveform of I_{st} has a characteristic peak the duration of the t_0 to t_m and then decline with some smooth portion stabilization.

Such character of the current appears to be associated with the neutralization of the head of the streamer (peak) and the deionization of plasma streamer channel (part of the current recession). The time from t_{in} to t_0 is approximately 30 ns that means the velocity of the streamer ~ 10^8 cm/s.

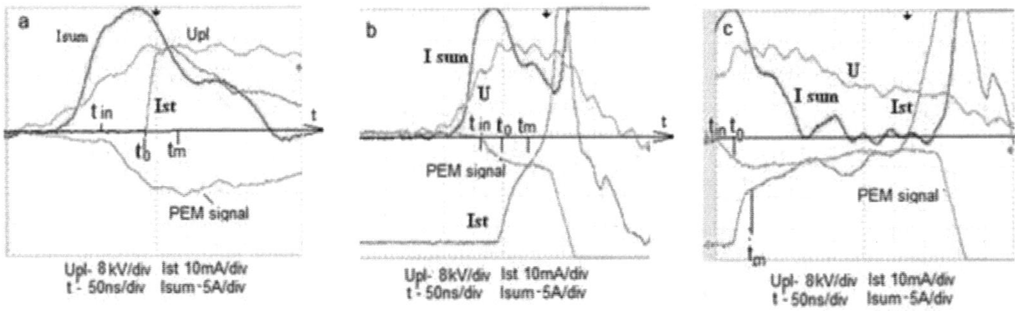

Figure 3. Three scenarios of nanosecond streamer discharges (a) no breakdown; (b) (c) – breakdown.

The development of the streamer after the touch probe has several scenarios. The first one (Figure 3a), associated with the collapse of the plasma channel streamer during the procedure leading to 0.5 μs. Experiments have shown in the intervals d=3-10 cm such a scenario takes place at the middle pulse field strength ($E_{pl}=U_{pl}$ / d) E_{pl} <7 kV/cm.

With increasing E_{pl}>7 kV/cm there are two other scenarios. In the first case, immediately after touching the streamer probe does not decrease, but increases rapidly, leading to a breakdown in the gap $E_{pl}\approx$8 kV / cm (fig.3b). In the second case (fig 3c). I_{st} after touching the probe is maintained stable for hundreds of nanoseconds at 30-50 mA and then increases rapidly. The streamer breakdown occur in the fall of the pulse at $E_{pl}\approx$5 kV/cm. Theoretically streamer breakdown was predicted in [4], but this work suggests that this effect may occur when E≈20 kV / cm. We have shown that the streamer breakdown can occur at three times lower values of E.

3.2 The main factors affecting the efficiency of waste treatment plants

The integral electrical parameters of nanosecond corona and single streamers measurements allowed to explain the dependence of the efficiency of energy transfer into gas from E_{pl}, which directly determines the efficiency of cleaning. Fig. 4 shows the frequency of occurrence of streamers on the probe (N), Q, I_{sum} and efficiency of energy transfer from a PGF into gas (η) as a function of E in RC with D=23.7 cm, d=10 cm, l=2 m. The graph shows that the efficiency means of η occurs in the range of 5 <E <6.5 kV / cm, i.e. when streamers are all inter-electrode distance (frequency of their appearances on the probe increases by an order of magnitude) and charge their heads increases significantly.

The upper limit values of E_{pl} is the field strength at which the streamer breakdown take place. Figure 5 shows W_{in} and I_{sum} , the concentration of ozone kO_3, hour of ozone production (hO_3) and specific energy consumption for ozone production (W/hO3) from the air flow rate (θ). The graph shows that the maximum of efficiency of ozone production take place the transition of flow from laminar to turbulent. The above results showed that the highest efficiency of the plant is provided with the values of the order of E 6.5 kV / cm and duration of pulse ~ 0.5 μs. In this case on the one hand streamers cover the all inter electrode gap and thereby maximize the conversion zone but the other hand do not allow the streamer breakdown occur, whereby instead impurities conversion of nitrogen oxides is generated. Another factor that significantly affects the conversion efficiency is the design of the RC.

Our experiments showed [5] that the energy efficiency of introduction of W increases with increasing distance between HV and ground electrodes. So increasing the diameter of the RC from 30 to 260 mm increases the current from 30 to 200 A at the same pulse voltage gradient E=6.5 kV / cm. This is due not only to an increase in the discharge point that the wire is not more than 200 1/m, and. But D of our RC be limited to 260 mm due to the complexity of designing high-frequency generators for voltage more than 100 kV.

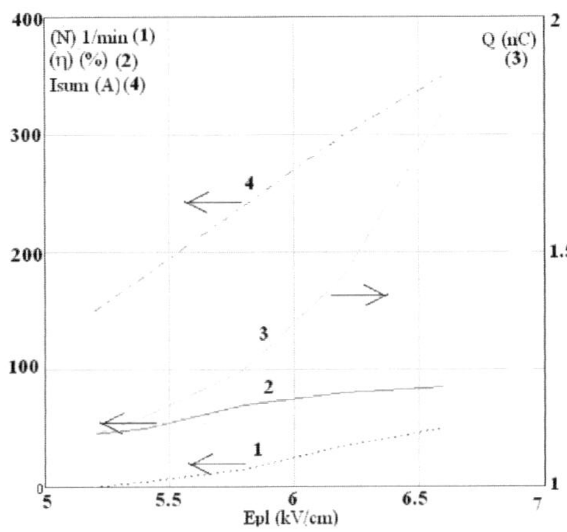

Figure 4. The dependence of the frequency of occurrence of streamers on the probe N(1), efficiency of energy transfer from a PGF into gas (η)(2), Q(3), I_{sum} (4) as a function of E_{pl} (RC - D=25.7 cm, d=9 cm, l=2 m.),

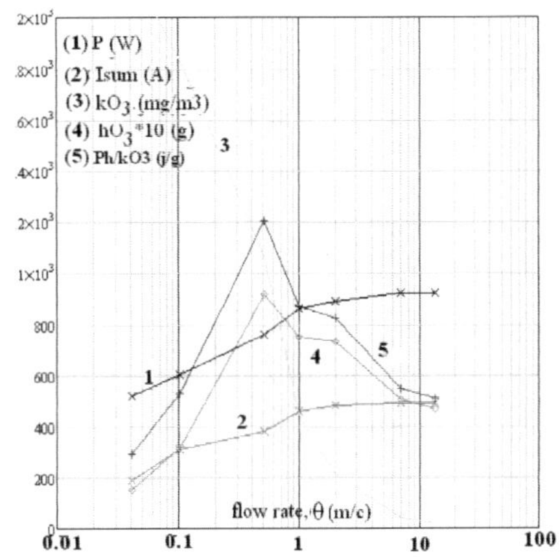

Figure 5 The dependence of the power input in gas W(1), I_{sum}(2), ozone concentration of kO_3(3), hour of ozone production (GO_3) (4) and specific energy consumption for ozone production (W/GO3)(5) from the air flow rate (θ)

Because of the short duration of the pulse, the value introduced into the discharge gap volume energy less than 20 J / m3. Given that the conversion to a concentration of impurities of 100 ppm requires the energy density of at least 10^4 J / m^3, it is obvious that the pulse repetition frequency (f), industrial installations should be much higher than 100 Hz.

4. Installations for waste gas by means of LTP

All of the above have been taken into account in the establishment in FMKB "Horizon" series of installations for cleaning waste gas of contaminants via LTP. Under the scheme of Figure 1 developed and commercially available modular treatment plants ranging from 1 to 8 kW performance (0.3-15)·10^3 m^3/h., Whose photos are shown in figure 6-9. Main specification of units are presented in table 1; results of cleaning efficiency – in table 2.

Table 1. Specification

Name of the parameter	Dimension	Value/characteristic			
		SPC Gas Turbine" Salute" Moscow	Project "Mars 500" Moscow	Pump station Moscow	Pump station Moscow
Gas flow rate max	m^3/hour	5000	200	5000	15000
Power max	kW	6	2	4	6
Pulse voltage max	kV	115	90	90	90

Pulse duration	ns	450	120	250	250
Pulse frequency max	Hz	100-400	100-1000	1000	1000
Reactor chamber		4-tube wet-type	1 tube	4-tube	4-6 tube
Denison	m	1x2.5x5	0.6x1.5x2	1 x 2 x 4	2x1.5x6
Weight	ton	1.5	0.2	.8	1.2
№ of figure		7	9	8	

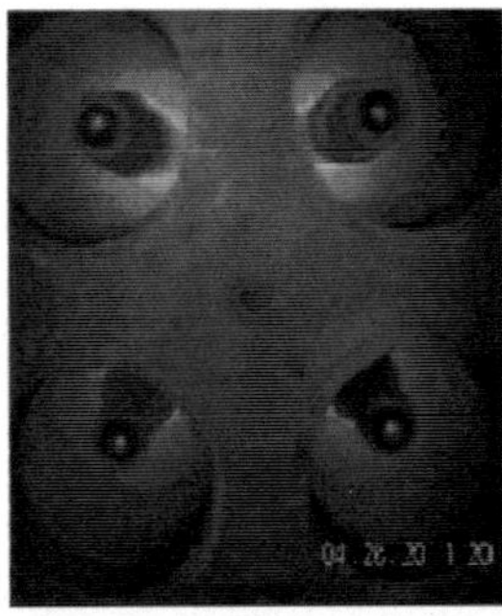
Figure 6. LTP in RC

Figure 7. Unit "KORONA 3" in factory shop

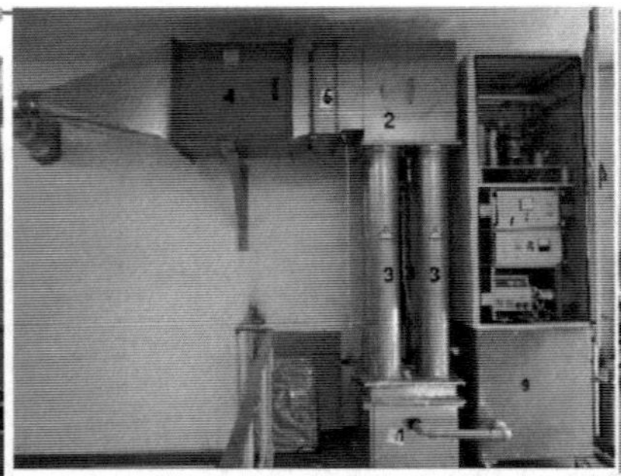
Figure 8. Unit "KORONA M" in pump station

Figure 9. Mobil unit in Project "Mars 500"

Table 2. Cleaning efficiency

Type of pollution	Concentratio mg/m^3	Gas flow rate, m^3/h	Purification degree,%	Specific power, W-h/m^3 mg	Installation site
SO_2	5000	800	70	0.002	Non-Ferrous Metals Plant
NO_X*	3000	36	95	0.006	MSE Corp., USA * with addition of ammonia
Toluene + Benzene	300	50	90	0.13	
Aerosol CrO_3	10	200	99	0.2	" FSUE SIC" Salut
NOx, soot	350 ,15000	50	90	0.016, 10	KIA "Sportage"136kW
HF, aerosol	30	5000	95	0.06	"FSUE SIC" Salut
H2S	150	1000	0.2	0.02	Sewage treatment tank
The smell of sewage		5000	No smell	0.2, **)	**Specific power W-h/m^3
H_2S	16,2	5000	80,4	0.2, **)	
Ammonia	0,55	5000	89,8	0.2, **)	

5. Conclusions

1. The results of experimental studies of plasma parameters generated by nanosecond corona discharge applied to plants for plasma cleaning emissions.
2. For the first time experimentally shown that it is possible streamer breakdown of air gaps of a few centimeters.
3 The factors affecting the efficiency of the treatment plants were determined.
4. The descriptions photos and specification pilot and industrial plants.

IOP Conf. Series: Journal of Physics: Conf. Series **927** (2017) 012043 doi:10.1088/1742-6596/927/1/012043

5. Presented main results of cleaning emissions from industrial and communal enterprises, diesels engine.

References

[1] Masuda S. and Nakao H. IEEE-IAS Annual Conference. 1986 Denver. pp.1173-1182
[2] Ponizovskii A.Z. Ecological Systems and Devices 2007 №11,., S.9-14.9 (In Russian)
[3] Mel'nikov V., Ponizovsky A., Gosteev S., et al *Proceedings of the 19th International Symposium on Plasma Chemistry ISPC19 - 2009,* Bochum, Germany 26-31 July 2009 P1.5.04
[4] Larsson A J 1998 *Phys. D: Appl. Phys.* **31** 1100–1108
[5] Ponizovskiy A. Gosteev S., et al. *Russian Electrical Engineering*, vol.64, No3, 1993, pp.77-83

The kinetics of energetic O^- ions in discharge H_2O plasma

A A Ponomarev[1,2] and N L Aleksandrov[1]

[1]Moscow Institute of Physics and Technology, Dolgoprudny, 141700, Russia

[2]SSC Keldysh Research Center, Moscow, 125438, Russia

E-mail: nick_aleksandrov@mail.ru

Abstract. Using Monte Carlo simulation, the translational relaxation of energetic O^- ions produced by dissociative electron attachment to molecules was studied in water vapor plasmas in a strong electric field. Initial O^- ions are energetic and are more reactive than the ions being in equilibrium with the electric field. Because of this, the energetic O^- ions have a chance to be involved in endothermic reactions prior to energy relaxation of these ions. The probabilities of charge exchange and electron detachment in ion-molecule interactions were calculated versus the reduced electric field. It was shown that several percent of energetic O^- ions produced in water vapor plasma by dissociative electron attachment to H_2O molecules are rapidly transformed to OH^- ions due to charge exchange collisions or are decomposed to release free electrons. The total probability of charge exchange and electron detachment from energetic O^- ions increases up to around 90% when these ions are produced by dissociative electron attachment to O_2 molecules if oxygen is added to water vapour. This means that, in this case, most of energetic ions formed from O_2 molecules are involved in charge exchange and electron detachment reactions prior to their thermalization.

1. Introduction

Non-equilibrium discharge plasmas generated in water vapor, humid air and other H_2O-containing mixtures are important for atmospheric electricity, discharges in the presence of liquid water (discharges with a liquid water electrode, in bubbles and foam) and widely used in technology, with applications ranging from air purification to plasma medicine [1-3]. Therefore, a large body of research is devoted to numerical simulation of the properties of discharge H_2O-containing plasmas taking into account ionization and ion conversion.

Water vapor is an electronegative gas and the formation of negative ions greatly affects the properties of H_2O-containing plasmas. These properties depend not only on the ratio between the density of negative ions and electron density, but on the negative ion composition as well. The density and composition of negative ions in non-equilibrium discharge plasmas are governed by the rates of ion-molecule reactions including ion formation, conversion and destruction. The rates of these reactions in weakly-ionized gases and plasmas in a strong electric field are usually calculated using the non-Maxwellian ion energy distribution and the corresponding reaction cross sections. This distribution is assumed to be in equilibrium with the applied electric field and is controlled by the reduced electric field E/N (N is the gas number density) and by the elastic and inelastic cross sections for ion-molecule interaction.

Translational energy distributions of particles produced in chemical reactions also are non-Maxwellian ones. In many collisional processes, translationally energetic (superthermal) neutral and charged particles are formed. Energetic particles are usually thermalized in collisions with neutral particles. However, the high-energy particles can participate in inelastic collisions that do not occur when the particles are thermalized. As a result, the effective reaction rates change and additional products are formed. This effect has been well studied for energetic atoms formed in chemical reactions (see [4] and references therein). The effect of superthermal particles was also studied for O^- ions produced by dissociative electron attachment to O_2 molecules in oxygen discharge plasmas [5]. It was shown that up to 6% of energetic O^- ions are rapidly converted to O_2^- ions in charge exchange collisions.

The objective of this work was to extend the calculations made in [5] to reactions in water vapor plasmas and in H_2O-containing plasmas. We used the Monte Carlo technique to simulate the translational energy relaxation of energetic O^- ions formed by dissociative electron attachment to molecules. The effect of high-energy O^- ions on charge transfer and electron detachment was numerically studied.

2. The method of simulation

The Monte Carlo technique was used to simulate the energy relaxation of high-energy O^- ions in water vapor in a stationary uniform electric field. We assumed that O^- ions with excessive translational energy were produced by dissociative electron attachment to H_2O and to O_2 when some amount of oxygen was added to the gas.

We used the Monte Carlo technique that is the same as in our previous work [5, 6]. An ion was assumed to be originated with a given initial translational energy. To simulate the motion of O^- ions in water vapor, we used the set of cross sections taken from [7]. This set includes elastic scattering, electron detachment

$$O^- + H_2O \rightarrow e + H_2O_2 \quad (1)$$

and charge exchange

$$O^- + H_2O \rightarrow OH^- + OH. \quad (2)$$

This set of the cross sections was validated in [7] by comparing the calculated ion transport coefficients with measured data.

The initial distribution for O^- ions produced by dissociative electron attachment to H_2O and O_2 molecules was calculated by analogy with the approach used in [5] and averaged over the electron energy distribution. The electron distribution was calculated with the Bolsig+ code and a self-consistent set of electron cross sections for H_2O [8].

3. Calculated results

The kinetic energy of O^- ions produced by electron attachment is in the range 0 – 0.8 eV for the attachment to H_2O molecules and in the range 0.5 – 3 eV for the attachment to O_2 molecules. The initial ion energy distribution changes with time and tends to the ion energy distribution when the ions are in equilibrium with the electric field. In a similar way, the mean ion energy decreases with time and tends to the equilibrium mean value corresponding to a given value of E/N.

The initial ions are energetic and are more reactive in comparison with thermalized ions. The probabilities of endothermic ion-molecule reactions (charge exchange and electron detachment) are elevated when the ions are energetic and tend to equilibrium values with increasing number of collisions. The equilibrium probabilities for endothermic ion-molecule reactions increase with E/N because these processes have energy thresholds and their equilibrium rate coefficients also increase with E/N.

Firstly, let us consider the kinetics of O^- ions formed by electron attachment to H_2O molecules. Figure 1 shows the total charge exchange probability, P_{ce}, for O^- ions in all collisions before their thermalization in water vapor versus E/N. For comparison purposes figure 1 presents the charge exchange probability, $P_{ce\ th}$, for the thermalized ions (in equilibrium with the electric field) after the same number of collisions. The excess charge exchange probability $P_{ce\ ex} = P_{ce} - P_{ce\ th}$ that characterizes the effect of energetic O^- ions on charge exchange collisions is also shown in figure 1.

The probability P_{ce} can reach 11% (for 300 Td), whereas the excess probability for energetic O^- ions reaches a peak value of ≈ 4% for E/N = 200 Td. It may be concluded that, for such values of E/N, the end negative ion products formed by dissociative electron attachment to H_2O molecules are 96% of O^- ions and 4% of OH^- ions. The OH^- ions are generated by charge exchange in collisions of energetic O^- ions with H_2O molecules.

The effect of the energetic O^- ions on charge exchange is much more profound when these ions are formed by electron attachment to O_2 molecules (see figure 2) that can be present in water vapor plasma due to small admixtures. In this case, the excess probability for energetic O^- ions reaches a peak value of ≈ 37% for E/N = 100 Td. These values of E/N are practically important because they are close to the breakdown reduced electric field in water vapor and air.

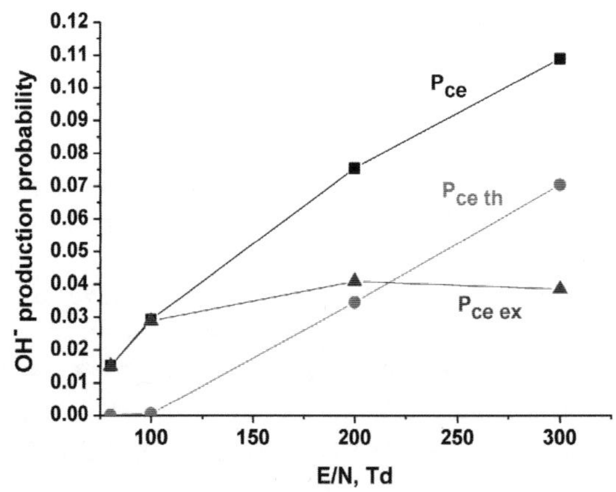

Figure 1. Charge exchange probability during O^- ion energy relaxation in water vapor as a function of E/N: (P_{ce}) the total probability, ($P_{ce\ th}$) the probability for thermalized ions and ($P_{ce\ ex}$) the excess probability of energetic ions. The ions are formed by electron attachment to H_2O molecules.

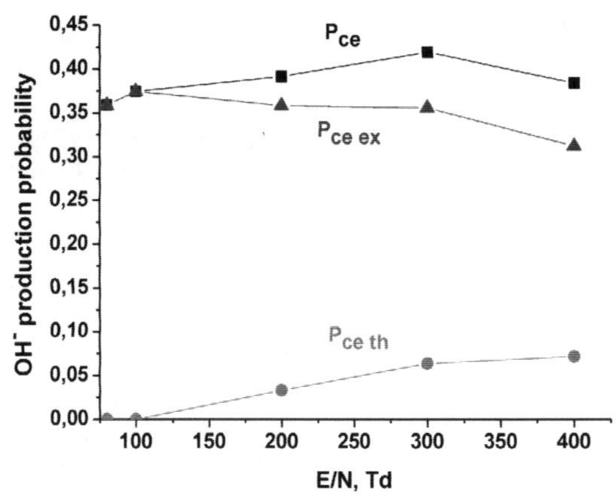

Figure 2. Charge exchange probability during O^- ion energy relaxation in water vapor as a function of E/N: (P_{ce}) the total probability, ($P_{ce\ th}$) the probability for thermalized ions and ($P_{ce\ ex}$) the excess probability of energetic ions. The ions are formed by electron attachment to O_2 molecules.

Figures 3 and 4 show, respectively, the calculated probabilities for electron detachment in collisions between O^- ions and H_2O molecules in water vapor when these ions are produced in electron attachment to H_2O and O_2, respectively. For this reaction, by analogy with the charge exchange reaction, we introduced the total probability (P_d), the probability for thermalized ions ($P_{d\ th}$) and the excess probability for energetic ions ($P_{d\ ex} = P_d - P_{d\ th}$). In the case of electron attachment to H_2O molecules, the probability P_{ce} reaches 8.5% (for 300 Td), whereas the excess probability for energetic O^- ions reaches a peak value of ≈ 2% for E/N = 200 Td. In the case of attachment to O_2 molecules, the effect of the energetic O^- ions on electron detachment is also much more profound. Here, the excess probability for energetic O^- ions reaches a peak value of ≈ 30% for E/N in the range 100 - 400 Td.

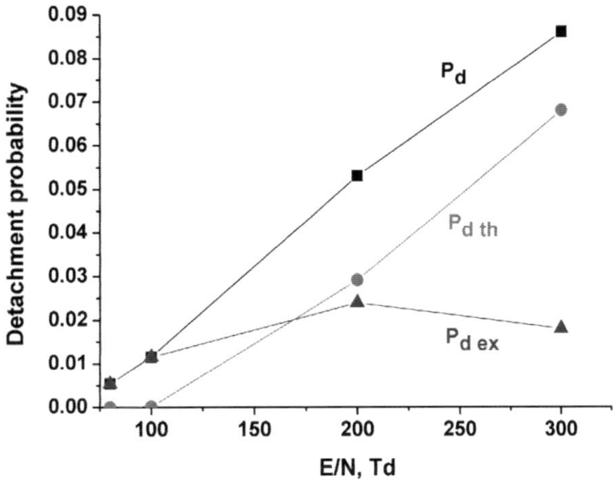

Figure 3. Electron detachment probability during O^- ion energy relaxation in water vapor as a function of E/N: (P_d) the total probability, ($P_{d\ th}$) the probability for thermalized ions and ($P_{d\ ex}$) the excess probability of energetic ions. The ions are formed by electron attachment to H_2O molecules.

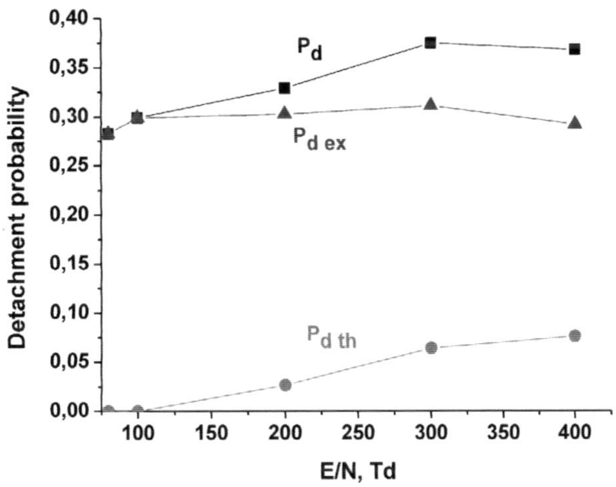

Figure 4. Electron detachment probability during O^- ion energy relaxation in water vapor as a function of E/N: (P_d) the total probability, ($P_{d\ th}$) the probability for thermalized ions and ($P_{d\ ex}$) the excess probability of energetic ions. The ions are formed by electron attachment to O_2 molecules.

4. Conclusions

Using Monte Carlo technique we studied the energy relaxation of energetic O^- ions generated by dissociative electron attachment to H_2O and O_2 molecules in water vapor in a strong electric field. The initial energy of the O^- ions is higher than the mean ion energy. Therefore, these ions can efficiently participate in endothermic reactions. Our calculations showed that the effect of the energetic ions formed by electron attachment to H_2O leads to a 4% increase in the probability of charge exchange and to a 2% increase in the probability of electron detachment prior to ion thermalization. The effect is much more profound for O^- ions formed by electron attachment to O_2 molecules. In this case, up to 37% of the energetic O^- ions are converted to OH^- ions in the charge exchange reaction and 30% of these ions are decomposed in the electron detachment reaction prior to ion thermalization. It may be concluded that only one third of the energetic O^- ions formed by electron attachment to O_2 are thermalized in water vapor plasmas. Other two thirds of these ions are converted to OH^- ions or are decomposed to release free electrons. The effect of energetic O^- ions should be taken into account when simulating the properties of water vapor plasmas.

Acknowledgements

This work was partially supported by the Russian Foundation of Basic Research under the project No. 16-32-00196.

References

[1] Fridman A 2008 *Plasma chemistry* (Cambridge: Cambridge University Press)
[2] Kong M G, Kroesen G, Morfill G, Nosenko T, Shimizu T, van Dijk J and Zimmermann J L 2009 *New J. Phys.* **11** 115012
[3] Bruggeman P J et al. 2016 *Plasma Sources Sci. Technol.* **25** 053002
[4] Aleksandrov N L, Ponomarev A A and Starikovskiy A Yu 2017 *Comb. Flame* **176** 181
[5] Ponomarev A A and Aleksandrov N L 2017 *Plasma Sources Sci. Technol.* **26** 044003
[6] Ponomarev A A and Aleksandrov N L 2015 *Plasma Sources Sci. Technol.* **24** 035001
[7] Stojanovic V, Raspopovic Z, Maric D and Petrovich Z L 2015 *Eur. Phys. J. D* **69** 63
[8] Hagelaar G J H and Pitchford L C 2005 *Plasma Sources: Sci.Technol.* **14** 722

Accurate and approximate methods to calculate capacitance and potential coefficients of two-particle system

M Rodin[1,2], **A Filippov**[1]

[1]State Research Center of Russian Federation "Troitsk Institute for Innovation and Fusion Research", Pushkovykh Street 12, Troitsk, Moscow 108840, Russia
[2]Moscow Institute of Physics and Technology, Institutskiy Pereulok 9, Dolgoprudny, Moscow Region 141700, Russia

E-mail: mikhail.rodin@phystech.edu

Abstract. Accurate and approximate methods to calculate the capacitance and potential coefficients of a system consisting of two macropaticles are considered. A technique of their calculation with sufficient accuracy for physical and technological applications is proposed. Based on them interaction potential and force as functions interparticle distance are obtained.

1. Introduction

Interaction potential plays a valuable role in microparticle coagulation and agglomeration processes. When the capacitance coefficients of a particle system are determined one can calculate the potential in vacuum or uniform dielectric. From this perspective the purpose of the article is a search for a simple method to obtain these quantities for further calculation of the interaction potential and force.

2. Methods of calculating the capacitance coefficients

Let us consider two conducting spheres of radii that in general must not be neglected in the calculation. These macroparticles are situated in vacuum or uniform dielectric in the absence of an external electric field. In case if electric potentials are independent of the angular coordinates, which is also satisfied by spheres situated in plasma or electrolyte, we can define potential and capacitance coefficients [1]

$$\phi_{1s} = S_{11}q_1 + S_{12}q_2, \quad \phi_{2s} = S_{12}q_1 + S_{22}q_2, \tag{1}$$

$$q_1 = C_{11}\phi_{1s} + C_{12}\phi_{2s}, \quad q_2 = C_{12}\phi_{1s} + C_{22}\phi_{2s}. \tag{2}$$

According to [2] (see also [3]) the total electrostatic potential generated by the system of two conducting spheres of radii a_1 and a_2 is given by

$$\phi(r_1, \mu_1) = \sum_{n=0}^{\infty} \left\{ A_n \frac{a_1^n}{r_1^{n+1}} + \sum_{k=0}^{\infty} B_k \frac{(n+k)!}{n!k!} \frac{r_1^n a_2^k}{R^{n+k+1}} \right\} P_n(\mu_1)$$

where r_1 and $\mu_1 = \cos\theta_1$ are respectively the radius and the polar angle cosine in the spherical coordinates relative to the first particle center with z-axis directed to the second particle center, R is the distance between them, $P_n(\mu_1)$ are the Legendre polynomials. Turning to the coordinate system with the pole in center of the second particle, we obtain a similar expression. Each of them must fulfil its own pair of boundary equations

$$\phi\big|_{r_1=a_1} = \phi_{1s}, \quad -\frac{a_1^2}{2}\int_0^\pi \left.\frac{\partial\phi}{\partial r}\right|_{r_1=a_1} \sin\theta_1 d\theta_1 = q_1,$$
$$\phi\big|_{r_2=a_2} = \phi_{2s}, \quad -\frac{a_2^2}{2}\int_0^\pi \left.\frac{\partial\phi}{\partial r}\right|_{r_2=a_2} \sin\theta_2 d\theta_2 = q_2,$$

resulting in the following equations for the coefficients A_n and B_n $(n \neq 0)$ [2]

$$\begin{aligned} A_n + \left(\frac{a_1}{R}\right)^{n+1}\sum_{k=1}^{\infty} B_k\frac{(n+k)!}{n!k!}\left(\frac{a_2}{R}\right)^k + q_2\left(\frac{a_1}{R}\right)^{n+1} &= 0, \\ B_n + \left(\frac{a_2}{R}\right)^{n+1}\sum_{k=1}^{\infty} A_k\frac{(n+k)!}{n!k!}\left(\frac{a_1}{R}\right)^k + q_1\left(\frac{a_2}{R}\right)^{n+1} &= 0. \end{aligned} \tag{3}$$

Considering the zeroth harmonic one finds [2]

$$\begin{aligned} \phi_{1s} &= \frac{q_1}{a_1} + \frac{q_2}{R} + \frac{1}{R}\sum_{k=1}^{\infty} B_k\left(\frac{a_2}{R}\right)^k, \\ \phi_{2s} &= \frac{q_2}{a_2} + \frac{q_1}{R} + \frac{1}{R}\sum_{k=1}^{\infty} A_k\left(\frac{a_1}{R}\right)^k. \end{aligned} \tag{4}$$

One can see from expressions (1), (3), (4) how to obtain the potential coefficients S_{11}, S_{12} and S_{22}. We should first solve system (3) providing $q_2 = 0$ to find A_n, B_n and then calculate the surface potentials ϕ_{1s}, ϕ_{2s} from (4) that yield S_{11} and S_{12} according to (1). Then, taking q_1 equal to zero, we obtain S_{22} in a similar way. This procedure can be abbreviated as

$$S_{11} = \left.\frac{\phi_{1s}}{q_1}\right|_{q_2=0}, \qquad S_{12} = \left.\frac{\phi_{2s}}{q_1}\right|_{q_2=0}, \qquad S_{22} = \left.\frac{\phi_{2s}}{q_2}\right|_{q_1=0}. \tag{5}$$

The potential coefficients can also be found analytically. We can see from (3) that the expansion coefficient A_k is expressed in terms of q_1 and q_2 as $A_k = M_k q_1 + N_k q_2$ and we can get M_k and N_k that are only determined of ratios a_1/R and a_2/R. Then one finds from (5)

$$S_{12} = \frac{1}{R} + \frac{1}{R}\sum_{k=1}^{\infty} M_k\left(\frac{a_1}{R}\right)^k, \qquad S_{22} = \frac{1}{a_2} + \frac{1}{R}\sum_{k=1}^{\infty} N_k\left(\frac{a_1}{R}\right)^k. \tag{6}$$

On the basis of symmetry, we can obtain S_{11} from S_{22} by permutation of a_1 and a_2.

Previously [2] system (3) was solved with regard for all the terms up to and including R^{-21}. In this case, in (6) we should ignore all the terms higher than eighth term. After canceling the terms and letting again to infinity we have a range of geometrical series that converges to

$$\begin{aligned} S_{a,11} &= \frac{1}{a_1} - \frac{a_2^3}{R^4}\left\{\frac{R^2}{R^2-a_2^2} - \frac{a_1 a_2^3(R^2+2a_2^2)}{R^6} + \frac{a_1 a_2^3 R^2}{(R^2-a_1^2)^3}\left[1+\frac{a_1^2}{R^2}+\frac{2a_2^2(1+2a_1^2/R^2)}{R^2-a_1^2}\right]\right\}, \\ S_{a,12} &= \frac{1}{R} + \frac{a_1 a_2}{R^3}\left(1 - \frac{R^2}{R^2-a_1^2} - \frac{R^2}{R^2-a_2^2} + \frac{R^2}{R^2-a_1^2-a_2^2}\right), \\ S_{a,22} &= \frac{1}{a_2} - \frac{a_1^3}{R^4}\left\{\frac{R^2}{R^2-a_1^2} - \frac{a_1^3 a_2(R^2+2a_1^2)}{R^6} + \frac{a_1^3 a_2 R^2}{(R^2-a_2^2)^3}\left[1+\frac{a_2^2}{R^2}+\frac{2a_1^2(1+2a_2^2/R^2)}{R^2-a_2^2}\right]\right\}. \end{aligned} \tag{7}$$

Transition from the potential coefficients to the capacitance coefficients is given by [1]

$$C_{11} = \frac{S_{22}}{\Delta}, \qquad C_{12} = -\frac{S_{12}}{\Delta}, \qquad C_{22} = \frac{S_{11}}{\Delta} \tag{8}$$

where $\Delta = S_{11}S_{22} - S_{12}^2$ is a determinant of system (1). In the present paper to obtain the capacitance coefficients C_{11}, C_{12} and C_{22} we have numerically solved system (3) putting $n_{\max} = 500$. However, when the macroparticles are far apart so that $R \gg \max(a_1, a_2)$ one can use approximate analytical expressions including monopole and dipole terms. Neglecting there a_1 and a_2 with respect to R there we have well-known asymptotics $C_{11} \to a_1$, $C_{12} \to -a_1 a_2/R$, $C_{22} \to a_2$.

When the spheres are close together the accuracy of computation in the spherical coordinates decreases significantly. In this limiting case we can use asymptotical formulae [4, 5]

$$\begin{aligned} C_{11} &= \frac{a_1 a_2}{a_1+a_2}\left[\frac{1}{2}\ln\left(\frac{2a_1 a_2}{a_1+a_2}\frac{1}{L}\right) - \psi\left(\frac{a_2}{a_1+a_2}\right) + O\left(L^2\right)\right], \\ C_{12} &= -\frac{a_1 a_2}{a_1+a_2}\left[\frac{1}{2}\ln\left(\frac{2a_1 a_2}{a_1+a_2}\frac{1}{L}\right) + \gamma + O\left(L^2\right)\right], \\ C_{22} &= \frac{a_1 a_2}{a_1+a_2}\left[\frac{1}{2}\ln\left(\frac{2a_1 a_2}{a_1+a_2}\frac{1}{L}\right) - \psi\left(\frac{a_1}{a_1+a_2}\right) + O\left(L^2\right)\right] \end{aligned}$$

where $L = R - a_1 - a_2$ is the separation distance, $\psi(z)$ is the logarithmic derivative of the gamma function [6], $\gamma = \psi(1) = 0.5772\ldots$ is the Euler's constant.

Results obtained by the described methods were compared with data obtained using the bispherical coordinate system [7, 8], as illustrated in Figure 1. One can see from the figure that the use of the bispherical coordinates provides the most accurate picture of the capacitance coefficients behaviour for all L. This is also given by asymptotical results for near approach of the macroparticles, but when $L > 0.1$ μm a discrepancy is noticeable. Numerical solving of (3) deals with $n_{\max} \times n_{\max}$ matrix and slowly convergent series that provides computational problems when L is small. However, the results obtained from there are wholly satisfactory up to $L = 10^{-3}$ μm. The approximate expressions are seen from the figure to be only possible to use for L being greater than a micrometer.

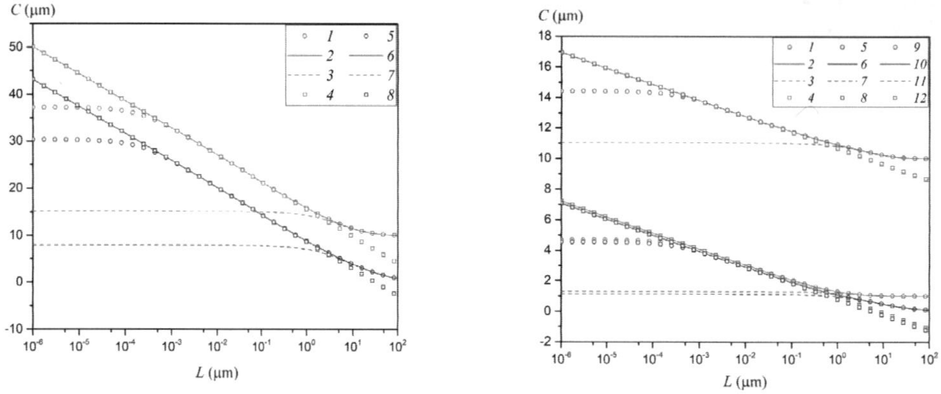

Figure 1: The capacitance coefficients C_{11} (curves 1–4), $|C_{12}|$ (curves 5–8) and C_{22} (curves 9–12) of equal ($a_1 = a_2 = 10$ μm, left) and different macroparticles ($a_1 = 1$ μm, $a_2 = 10$ μm, right) as functions of the separation distance. Circles 1, 5, 9 denote the solution of (3) for $n_{\max} = 500$, lines 2, 6, 10 give the data obtained in the bispherical coordinates [7, 8]. The approximate analytical expressions and asymptotical formulae from [4,5] are denoted with curves 3, 7, 11 and squares 4, 8, 12 respectively.

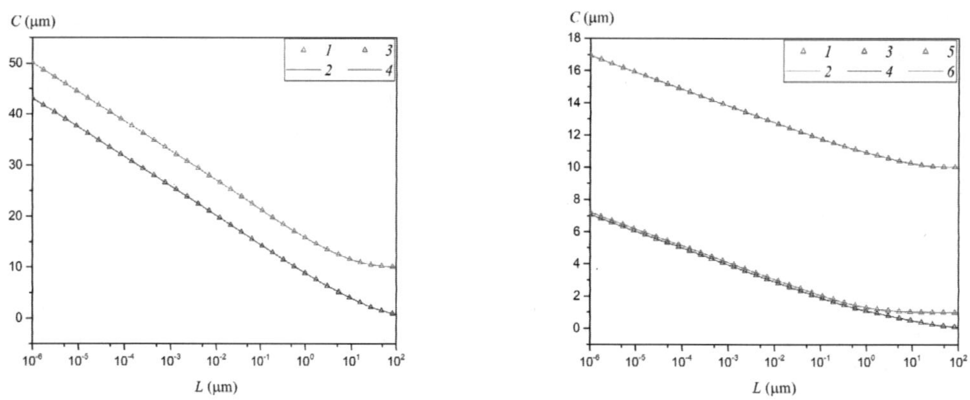

Figure 2: The capacitance coefficients C_{11} (curves 1, 2), $|C_{12}|$ (curves 3, 4) and C_{22} (lines 5, 6) of equal ($a_1 = a_2 = 10$ μm, left) and different macroparticles ($a_1 = 1$ μm, $a_2 = 10$ μm, right) versus L. Triangles 1, 3, 5 denote sewing of the results given by the approximate and asymptotical formulae. Lines 2, 4, 6 give the data obtained in the bispherical coordinates [7, 8].

3. Interaction potential and force between two macroparticles

With potential or capacitance coefficients, one can express the interaction potential of particles in terms of their charges or electric potentials respectively. As the charges are defined the interaction potential is [9]

$$U = \frac{1}{2}\left(S_{11}q_1^2 + 2S_{12}q_1q_2 + S_{22}q_2^2\right). \tag{9}$$

Differentiation of (9) with respect to R yields the interaction force with the sign reversed. Figure 3 shows the desired force F_{21} taken relative to the Coulomb force F_C describing the interaction between point charges.

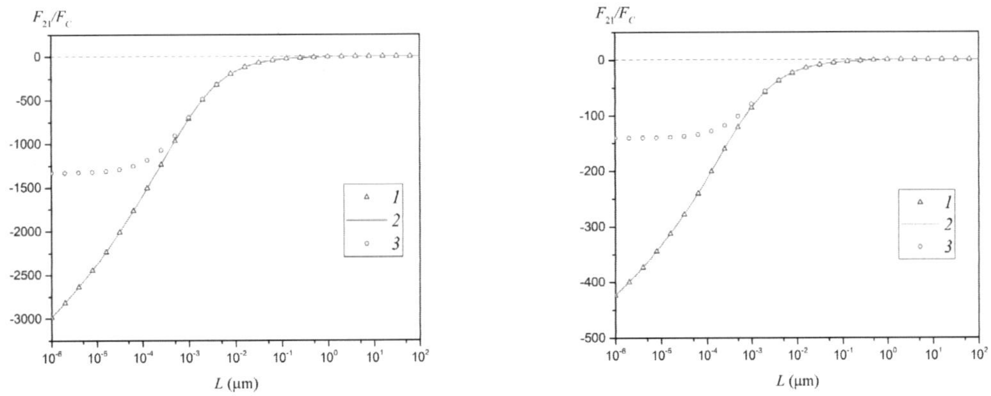

Figure 3: The reduced force between macroparticles of equal ($a_1 = a_2 = 10$ μm, left) and different radii ($a_1 = 1$ μm, $a_2 = 10$ μm, right) versus L. Symbols 1 denote numerical differentiation of (9) where the potential coefficients are obtained from the capacitance coefficients shown in Figure 2. Curves 2 and circles 3 give respectively the results obtained in the bispherical coordinates and through the use of (7) and (9). The dashed lines denote the equality $F_{21} = F_C$. The charges are $q_1 = 10^2\,e$, $q_2 = 10^3\,e$ in both cases.

4. Conclusions

Accurate and approximate calculation methods have been considered for the capacitance and potential coefficients of two conducting macroparticles. The limits of each approach applicability have been defined. Based on the data obtained the interparticle force has been calculated. It has been shown that the use of the asymptotical formulae for small interparticle distances and the analytical expressions for large ones provides good agreement with the results obtained in the bispherical coordinate system. The proposed method is not concerned with summation of infinite series and thus seems to be advantageous on the computational side.

Acknowledgments

This work was supported by the Russian Science Foundation (project no. 16-12-10424).

References

[1] Smythe W B 1988 Static and dynamic electricity
[2] Maxwell J K 1891 A treatise on electricity and magnetism
[3] Filippov A, Derbenev I 2016 *JETP* **123.6** pp 1099–109
[4] Russel A 1909 *Proc. R. Soc. Lond. A* **82** pp 524–31
[5] Lekner J 2011 Capacitance coefficients of two spheres *Journal of Electrostatics* **69** pp 11–14
[6] Davis P J 1972 Gamma function and related functions *Handbook of mathematical functions* pp 253–93
[7] Filippov A 2009 *JETP* **109.3** pp 516–29
[8] Filippov A 2009 *Contributions to Plasma Physics* **49.7–8** pp 431–45
[9] Batygin V, Toptygin I 1964 Problems in Electrodynamics

The effect of electrically conductive additives on the plasma pyrolysis of heavy hydrocarbons

K G Sadikov[1], A O Sofronitskiy[1,2] and V M Larionov[1]

[1] Kazan Federal University, Institute of Physics, 420008, Kremlevskaya St, 18, Kazan, Russian Federation
[2] Kazan National Research Technical University named by A.N.Tupolev, 420111, K.Marx str., 10, Kazan, Russian Federation

E-mail: artempic8@mail.ru

Abstract. It's shown that the electric discharge initiation of in-situ combustion can be executed by entering conductive additives to hydrocarbon raw materials. It is observed, that the most of all the soot is formed from aromatic hydrocarbons during the plasma pyrolysis. Cracking of hydrocarbons by electric discharge, with conducting additives and precursors of catalysts, leads to formation of carbon and metal nanoparticles.

1. Introduction

In the conditions of exhaustion of traditional energy resources, constantly increasing need of economy for hydrocarbons, oil production, and processing of heavy oils is an actual problem. One of the ways of extracting hardly removable oil is intra sheeted burning - a process at which the thermal energy is used received by burning oil in the formation when downloading in air layer is used. As a result of intra sheeted burning, the oil adjacent to a burning zone is diluted and gradually forced out to the second well, from which it is subsequently pumped out. The initiation of the formation of combustion occurs either forcibly or spontaneously. For acceleration of process of burning in layer initiation can be carried out using a gas burner, which is lowered into the bottom hole zone. Also the Immersed electric heaters and incendiary chemical mixtures are also used. Further maintenance of burning is carried out by downloading necessary the amount of oxygen (air). There were attempts of creation of the centers of burning by means of induction currents [1] and electric discharge of radio-frequency range.

In recent years, electrical discharges in the oil industry have began to be applied as the tool. For example, in work [2] the method of pulse plasma processing of the well was discussed with the purpose of elimination of contamination of the punched part and increasing the return of layers. Modern innovative technologies for well-drilling, especially when drilling hard rocks, suggest the use of plasma tools instead of rotating drills [3]. Various ways of influencing electric discharges on liquid hydrocarbons are discussed [4].

In this paper, we propose a new method for initiating intra sheeted burning using an electric discharge of direct or alternating current in a medium of electrically conductive liquid hydrocarbons.

2. Initiation of intra sheeted burning by electric discharge

The main idea of initiation of intra sheeted burning by electric discharge consists that the high-voltage electrode is lowered to the well, and include high voltage in an estimated zone of burning. An electrical discharge occurs between the electrode and oil treated processed by electroconductive additives. Further, when air is injected, the combustion zone extends to all depth of the oil tank. Figure 1 presents a schematic diagram of the initiation of intra sheeted burningby electric discharge.

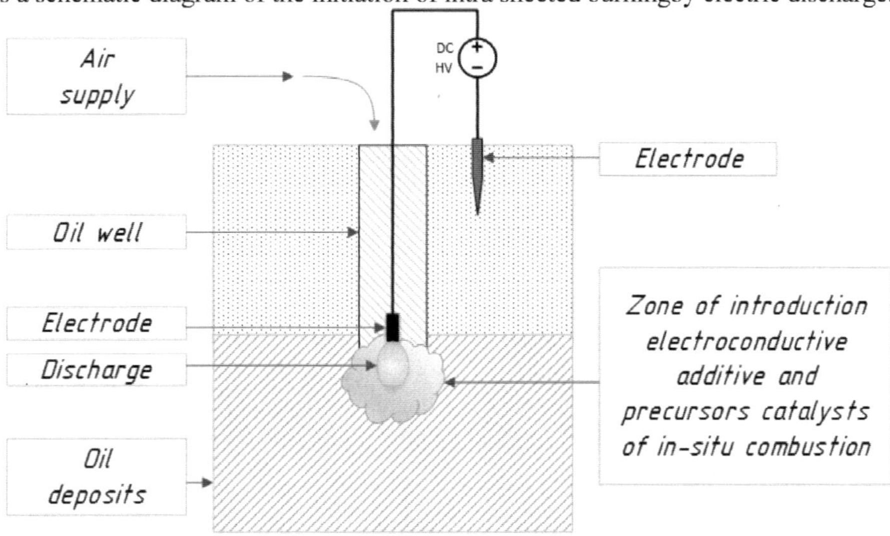

Figure 1. Schematic diagram of the initiation of intra sheeted burning by means of an electric discharge.

For the primary processing of oil in the conditions of the layer, enter precursors of catalysts which get a burning zone, are restored to catalystically active metal complexes into a zone of burning. The Catalytic complexes in the conditions of intra sheeted burning promote primary oil refining by decomposition of heavy hydrocarbon components of oil into easier fractions. Precursors of of catalysts download in layer in the form of solutions of organometallic complexes in organic solvents. Therefore, it would be advisable to pump electroconductive additives together with a solution of precursors of catalysts.

In the processes of intra sheeted burning of one of the main problems is the coking of sand, it reduces the completeness of extraction of hydrocarbon raw materials. Therefore, another task of this work was to determine the conditions under which the chemical processes of burning (oxidation) soot is the most of all formed.

Proceeding from the above, the study of the interaction of an electric discharge with hydrocarbons of oil in the presence of organometallic complexes is an actual task.

3. The experimental part

As it is well-known, fluid hydrocarbons poorly carry electric current, therefore, it is rather difficult to organize an electrical discharge between electrodes, one of which is a hydrocarbon feedstock. To solve this problem, electroconductive additives soluble in fluid hydrocarbons were used in this paper. One of the effective formulations was a solution of sodium formate in acetone. The saturated solution of sodium formate was prepared in acetone with the concentration of 0.037 mg / ml. From this initial solution, tests of hydrocarbons with different specific resistivities were prepared. At an electroconductive concentration of 20%, the resistance of the hydrocarbon test was 1 MΩ. This concentration was chosen for the experiments. For the experimental studying of the plasma pyrolysis process, individual alkane hydrocarbons and aromatic hydrocarbons were used.

The experimental setup consisted of a reactor, a gas supply device, a power source and measuring devices. The reactor was made of the transparent heat-resistant glass, in which there was a metal container for fluid hydrocarbon samples. The test of liquid hydrocarbons was a cathode. An anode was entered into the reactor from above. The anode consisted of coaxially arranged tubes, at the end of which there was an insert of silicon carbide. The tip had a central hole through which a working inert gas, helium, was carried out. The gaseous products which are formed in the discharge zone were discharged through the side openings of an external tube.

4. The results of experiments

To determine the optimum input power, the volt-ampere characteristics for various tests of hydrocarbon were taken. The volt-ampere characteristic of a discharge for a sample of oil with an electrically conductive additive is presented in Figure 2. At the current strength of 250 mA, the discharge power was 115W. Plasma pyrolysis of hydrocarbon feedstock passed efficiently both on a direct current and alternating current at a frequency of 50 Hz. During the experiments it was noticed that at increase in current intensity of plasma pyrolysis occurred more intensively.

To ignite hydrocarbons in air, at atmospheric pressure it is required to carry out the electrical breakdown of the gas which is between the electrode and hydrocarbons. During the experiment it was necessary to determine the interelectrode distance at which there will be an electrical breakdown of air __ at atmospheric pressure and voltage 2100 V. The value of the interelectrode distance for air at atmospheric pressure calculated according to Paschen's law made 0.36 mm. The experiment was carried out in the air environment, the glass reactor vessel was dismantled. During the experiment, it was demonstrated the possibility of igniting fluid hydrocarbons with electrically conductive additives by means of an electric discharge.

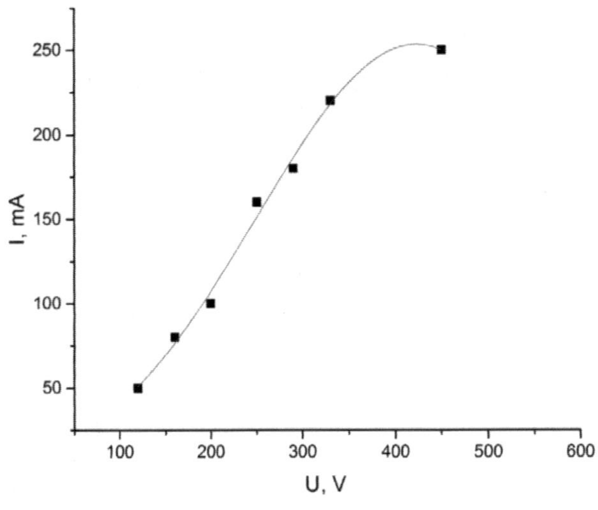

Figure 2. Voltage-current characteristic of a discharge on a direct current.

During the experiments in all cases, soot and hydrocarbon gas were formed. Experiments were carried out on individual hydrocarbons, such as, octane, toluene and mixtures thereof. The most fine-grained was the soot obtained in the plasma pyrolysis of toluene, the size of the soot particles was 10-50 nm. The amount of soot formed in the case of toluene was maximal.

The influence on the plasma pyrolysis of liquid hydrocarbons of additives - transition metal resins in the amount of 1% of the total mass of liquid hydrocarbons was studied. The addition of resins did not have a significant effect on the current-voltage characteristics of the discharge. Carbon black, obtained by the pyrolysis of liquid hydrocarbons with the addition of copper rubber, had a red metallic tint, which is probably due to the presence of copper particles in this carbon black. An analysis of the elemental composition of this soot was carried out. The measurements were carried out on an auto-emission high-resolution scanning electron microscope Merlin (Carl Zeiss). The microscope is equipped with the AZtec X-Max energy dispersion spectrometer. Elemental analysis was performed with an accelerating voltage of 20 keV and a working interval of 10 mm. The depth of sounding is about 1 micron. Elemental analysis of soot particles obtained by plasma pyrolysis of toluene with the addition of copper resinate showed that this soot contains about 1% copper. As you increase, you can see that the copper particle size is less than 90 nm, Figure 3.

Figure 3. The image of particulate matter at 20,000 times magnification.

5. Conclusions

During the experiments, it was shown that this method can be used to ignite hydrocarbons in the conditions of the formation. The amount of soot formed in the case of aromatic hydrocarbons was maximal, i.e. The main source of coke formation is aromatic compounds. The dispersion of different soot samples was determined. The most fine-grained was the carbon black, obtained by plasma pyrolysis of aromatic hydrocarbons. Copper nanoparticles were also produced, which were formed from copper.

6. Acknowledgements

The work is performed according to the Russian Government Program of Competitive Growth of Kazan Federal University.

References

[1] Patent RU 2477788 Method of underground gasification.
[2] Pashchenko A. F., Ageev N. P., 2016. Increased Oil Recovery by Application of Plasma Pulse Treatment , *Society of Petroleum Engineers* SPE-183057-MS
[3] Kocis I, Kristofic T, Gajdos M, Horvath G, and Jankovic S. 2015. Utilization of Electrical Plasma for Hard Rock Drilling and Casing Milling. *SPE/IADC Drilling Conference and Exhibition*
[4] Timerkaev B A, Sofronitskiy A O and Andreeva A A. 2016 *Journal of Physics: Conference Series,* Volume **669**, conference 1, 012062. doi:10.1088/1742-6596/669/1/012062

IOP Conf. Series: Journal of Physics: Conf. Series **927** (2017) 012047 doi:10.1088/1742-6596/927/1/012047

Some numerical simulation results of the dynamic temperature distribution in dc plasma torch «Thermoplasma 50-01»

A Saifutdinov[1,2], S Fadeev[1] , I Fayrushin[2]
[1] Kazan Federal University, Kazan, Russia
[2] Saint-Petersburg State University, Saint-Petersburg, Russia

E-mail: as.uav@bk.ru

Abstract. A DC plasma torch "Thermoplasma 50-01" has been modeled and simulated by developing a 2D axisymmetric model of laminar flow and heat transfer coupled to electromagnetic fields. As a result of the numerical solution, the dynamics of the formation of the temperature field and the velocity field in the plasma torch channel and at its exit is presented. The numerical results of the gas temperature and axial velocity result to be quite satisfactory.

1. Introduction

DC arc plasma torches have been known, studied and used for many decades. They used to generate thermal plasma jets in different industrial applications, such as thermal plasma waste treatment, atmospheric or low-pressure plasma spraying, plasma-assisted chemical vapor disposition, plasma preparation of ultra finepowder, etc [1-5].

At present, new plasma torches are being developed and created all over the world. In particular, modern developments is the «Thermoplasma 50-01» plasma torch, developed in Khristianovich Institute of Theoretical and Applied Mechanics of the Siberian Branch of the RAS [7-10].

One of the important tools for understanding physical processes in plasma torches is numerical modeling. In the last 10-20 years, much research has been undertaken to improve modeling of electric arcs, with the ambitious goal to be able to simulate numerically a working device, like a plasma torch [11, 12].

More recently, in [13] and [14] the complete arc and jet dynamics are simulated with a thermodynamic-equilibrium-based model.

The purpose of this work is numerical simulation of the plasma jet dynamics of the «Thermoplasma 50-01» plasma torch.

2. Description of the Mathematical Model

The continuum assumption is valid and the plasma is considered as a compressible, perfect gas in Local Thermodynamic Equilibrium (LTE), hence characterized by a single temperature T for all its species (atoms, ions, electrons, molecules); the quasi-neutrality condition holds; the plasma is optically thin; Hall currents, gravitational effects, and viscous dissipation are considered negligible.

As the plasma is a conducting fluid, its description requires the solution of the fluid conservation and electromagnetic equations; which, according to the assumptions stated above, are given by:

$$\frac{\partial \rho}{\partial t} + \nabla\left(\rho \vec{V}\right) = 0, \tag{1}$$

$$\rho\left(\frac{\partial \vec{V}}{\partial t} + \vec{V} \cdot \nabla \vec{V}\right) = \vec{j} \times \vec{B} - \nabla\left[P + \frac{2}{3}\mu\left(\nabla \cdot \vec{V}\right)\right] + 2\nabla \cdot (\mu S), \tag{2}$$

$$\rho c_p\left(\frac{\partial T}{\partial t} + \vec{V} \cdot \nabla T\right) - \frac{dP}{dt} = \nabla \cdot (\lambda \cdot \nabla T) + \vec{j} \times \vec{E} + \frac{5k_B}{2e}\vec{j} \cdot \nabla T - Q_{rad}, \tag{3}$$

$$\nabla \cdot (\sigma \cdot \nabla \varphi) = 0, \quad \vec{E} = -\nabla \varphi, \quad \vec{j} = \sigma \vec{E}, \quad \nabla^2 \vec{A} = -\mu_0 \vec{j}, \quad \vec{B} = \nabla \times \vec{A}, \tag{4}$$

where ρ is the fluid density; t is time; $\vec{V}$ velocity; $\vec{j} \times \vec{B}$ represents the Lorentz force, with $\vec{j}$ as the current density and $\vec{B}$ as the magnetic field; P is pressure; μ is the dynamic viscosity; c_p is the gas specific heat at constant pressure; T is the temperature; $\vec{j} \times \vec{E}$ is the Joule heating term, with $\vec{E}$ as the electric field; λ is the thermal conductivity; σ is the electrical conductivity; φ is the electric potential; $\vec{A}$ is the magnetic vector potential; μ_0 is the permeability of free space; the term proportional to $\vec{j} \cdot \nabla T$ represents the diffusion of electron enthalpy with k_B as Boltzmann's constant and e as the elementary charge; Q_{rad} is specific radiation power, depending on temperature.

The computational domain consisted of two parts: a plasma channel inside the plasma torch and a plasma jet at the output of the plasma torch. Table 1 shows the boundary conditions used in the simulation, where p_{out} represents the outlet pressure equal to 760 Torr, h_w is the convective heat transfer coefficient at the anode wall equal to 3500 $W/(m^2 \cdot K)$, T_0 a reference cooling water temperature of 277 K. $T_2 = 3600K$, $T_3 = 500K$ are temperature of cathode and anode. The arc current is current I =200 A and air mass flow rate $\dot{m} = 1.5\, g/s$.

Table 1. Boundary conditions.

	Boundary conditions
Inlet	$T = T_0$, $-\int_{\partial\Omega} \rho(\vec{v} \cdot \vec{n}) d_{bc} dS = \dot{m}$, $\vec{j} \cdot \vec{n} = 0$, $\vec{A} \times \vec{n} = 0$;
Cathode	$T = T_2$, $\vec{v} = 0$, $\vec{j} \cdot \vec{n} = I/(\pi R_c^2)$ $\vec{A} \times \vec{n} = 0$;
Anode	$T = T_3$, $\vec{v} = 0$, $\varphi = 0$, $\vec{A} \times \vec{n} = 0$;
Outlet	$\nabla T = 0$, $p_{out} = p_0$, $\vec{j} \cdot \vec{n} = 0$, $\vec{A} \times \vec{n} = 0$;
Interelectrode insert	$-\vec{n} \cdot \lambda \vec{\nabla} T = h(T_0 - T)$, $\vec{v} = 0$, $\vec{j} \cdot \vec{n} = 0$, $\vec{A} \times \vec{n} = 0$.

3. Numerical simulation and results

In Fig. 1 shows the dynamics of the formation of the temperature field both the plasma torch channel and its output. It is seen that the temperature in the near-cathode region reaches 30000K, and in the near-anode region and at the exit of the plasma torch it is 10000-15000 K. The velocity in the near-cathode region reaches 2500 m/s and in the near-anode region and at the exit of the plasma torch it is 1000 m/s. The plasma jet is set in a time equal to 1.3-1.5 ms.

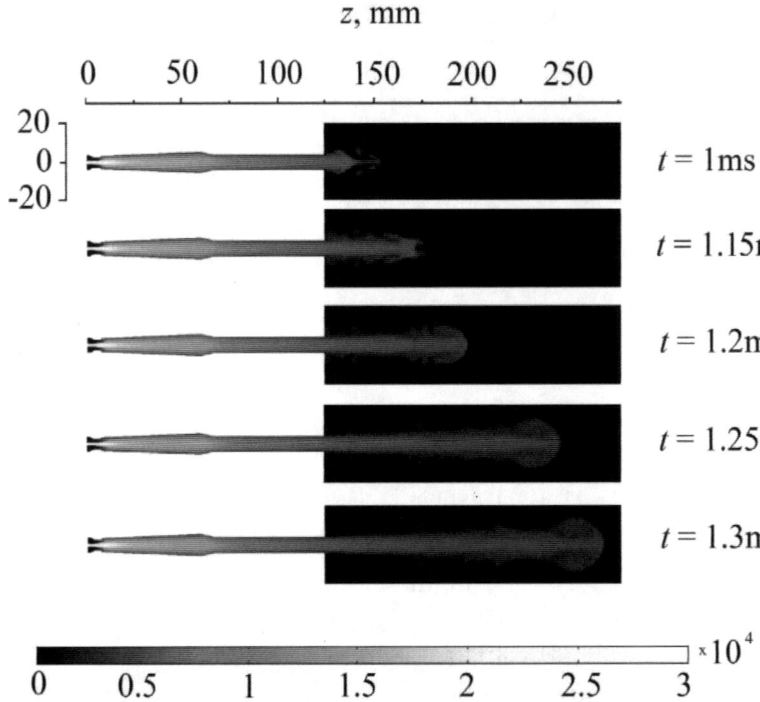

Figure 1. Dynamics of the temperature field in the plasma torch "Thermoplasma 50-01" channel and its output.

4. Conclusions

A DC plasma torch "Thermoplasma 50-01" has been modeled and simulated by developing a 2D axisymmetric model of laminar flow and heat transfer coupled to electromagnetic fields. In order to solve the partial differential equations of electric currents and magnetic fields, both in the gas than in the anode region, we have contemplated appropriate boundary conditions in the modeling work. Lorentz forces and Joule heating effects have been modeled, coupled to the physical model of the plasma torch and finally computed. The dynamics of the formation of the temperature field and the velocity field in the plasma torch channel and at its outlet is presented.

The numerical results of the gas temperature and axial velocity result to be quite satisfactory. We foresee to develop a more complete reproduction of the thermal and fluid phenomena in a future three dimensional model, which might include the modeling of other issues. In particular, in the future work, it is necessary to consider the effect of gas supply for the anode curtain.

Acknowledgments

The reported study was funded by RFBR, according to the research project No. 16-38-60187 mol_a_dk.

References

[1] Pfender E 1999 *Plasma Chem. Plasma Proc.* **19** 1–31
[2] Murphy A B 1996 *J. Phys. D: Appl. Phys.* **29** 1922–1932
[3] Pan Wenxia, Meng Xian, Chen Xi et al 2006 *Plasma Chem. Plasma Process.* **26** 335–345
[4] Robert C and Tucker Jr 2013 *ASM Handbook, Volume 5A: Thermal Spray Technology* (ASM International) p 412

[5] Ilyushchenko A F, Shevtsov A I, Okovity V A et al 2011 *The formation of thermal coatings and their modeling* (Minsk: Belarus, Nauka) p 357

[6] Zasypkin I V, Zhukov M F 2006 *Thermal Plasma Torches: Design, Characteristics, Application* (Cambridge International Sciences Publishing, Cambridge) p 720

[7] Kuzmin V I, Mikhal'chenko A A, Kovalev O B, Kartaev E V and Rudenskaya N A 2012 *J. of Thermal Spray Technology* **21** 159–168

[8] Kuzmin V, Cartan E, Kornienko E et al. 2014 Plasma spraying of powder coatings with gas dynamic focusing of the dispersed phase. Current problems in engineering = Actual problems in machine building: 1 Materials Intern. Scientific And Practical. Conf., Novosibirsk, March 26, 2014 – Novosibirsk: Publishing House of the NSTU 482–488.

[9] Kuzmin V I, Grigoriev S N, Kovalev O B et al 2013 *Friction and Wear* **34** 221–226

[10] Kuzmin V I, Mikhal'chenko A A, Kartaev E V et al 2012 *Journal of Thermal Spray Technology* **21** 159–168

[11] Trelles J, Chazelas C, Vardelle A and Heberlein J 2009 *Journal of Thermal Spray Technology* **18** 728–752

[12] Vardelle A, Moreau C, Themelis N and Chazelas C 2015 *Plasma Chem Plasma Process* **35** 491–509.

[13] Colombo V. and Ghedini E 2005 "Time dependent 3-d simulations of a dc non-transferred arc plasma torch: anode attachment and downstream region effects", in Proc. ISPC, pp. 169-170

[14] Marchand C, Chazelas C, Mariaux G and Vardelle A 2007 *J. Thermal Spray Tech.* **16** 705–712

Periodical plasma structures controlled by external magnetic field

I V Schweigert[1] and M Keidar[2]

[1]Institute of Theoretical and Applied Mechanics, Novosibirsk, 630090, Russia
[2]George Washington University, Washington D.C. 20052, USA

ischweig@itam.nsc.ru

Abstract. The plasma of Hall thruster type in external magnetic field is studied in 2D3V kinetic simulations using PIC MCC method. The periodical structure with maxima of electron and ion densities is formed and becomes more pronounced with increase of magnetic field incidence angle in the plasma. These ridges of electron and ion densities are aligned with the magnetic field vector and shifted relative each other. This leads to formation of two-dimensional double-layers structure in cylindrical plasma chamber. Depending on Larmor radius and Debye length up to nineteen potential steps appear across the oblique magnetic field. The electrical current gathered on the wall is associated with the electron and ion density ridges.

1. Introduction

Recently some methods to control the Hall thruster characteristics with applying the oblique magnetic field with respect to the channel walls are widely discussed [1,2]. Nevertheless with increasing the inclination of the magnetic field, discharge plasma properties can essentially change. For example, a several stationary, magnetized, two-dimensional weak double-layers were observed in a laboratory experiment for this type of plasma [3]. The double-layer potential drops were found to be followed across any given potential profile. In [4] it was shown in PIC simulations that weak magnetization results in the double-layer electric-field alignment of particles accelerated by these potential structures and that strong magnetization results in their magnetic-field alignment. A morphological invariance in two-dimensional double-layers with respect to the degree of magnetization observed in [4] implied that the potential structures scale with Debye lengths rather than with gyroradii. In this work, in kinetic simulations we consider the direct current discharge plasma in the external oblique magnetic field at low gas pressure, $P=10^{-4}$ Torr. The purpose is to study the plasma structure modification with changing the electron temperature, magnetic field strength and incidence angle for the conditions similar to the Hall thruster.

2. The physical model and simulation algorithm

In our simulations, the plasma is embedded in a cylindrical metallic chamber with the radius of 4 cm and the height H=10 cm. The calculation domain is shown in Figure 1. The cathode made from metal material with the radius of 3 cm is placed 0.3 cm apart from the chamber bottom. All walls of the chamber made from stainless steel are grounded and the cathode is biased with -90 V. A plate made from Al_2O_3 is placed near the side wall. In some variants we considered the secondary electron emission (SEE) effect on the sheath near the wall.

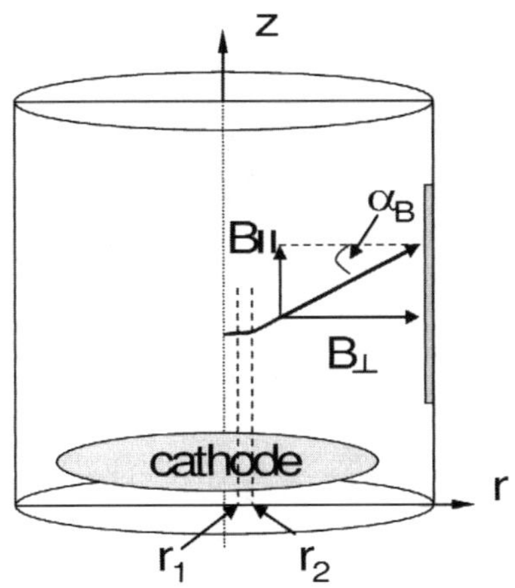

Figure 1. Scheme of calculation domain. **B is the magnetic field vector, α_B is angle between B and the normal to the side wall.**

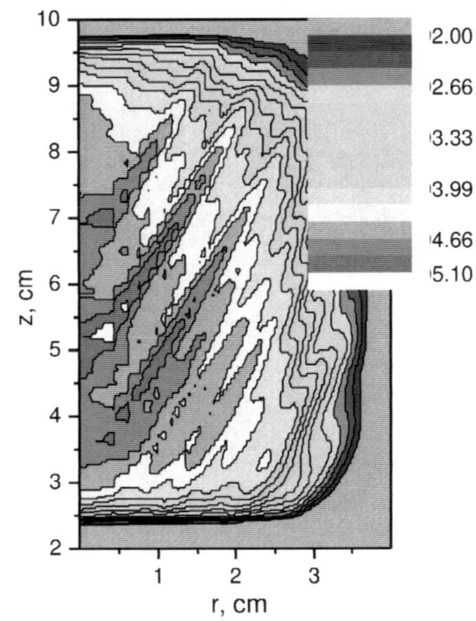

Figure 2. Distribution of potential φ (V) for B=50 G, α_B =65° and Te=5 eV.

The external magnetic field B is assumed to be constant over the plasma volume. The magnetic field is axially symmetrical. To avoid the singularity at r=0, we took the magnetic field angle α_B in the following form: α_B =0 at $r<r_1$, $\alpha_B = \alpha_0$ at $r>r_2$, α_B is approximated by a quadratic spline function at $r_1<r<r_2$. In simulations, we took r_1=0.3 cm and r_2=0.6 cm (see Figure 1). The α_0 ranges from 0 to 77° in different variants, the strength of magnetic field B is from 25 to 100 G.

The magnetized plasma is described with the system of equations including the Boltzmann equations for electrons and ions and fast neutral atoms. The distribution functions for electrons $f_e(t,x,\mathbf{v})$ and ions $f_i(t,x,\mathbf{v})$ are calculated by solving the Boltzmann equations

$$\frac{\partial f_e}{\partial t} + \mathbf{v}_e \frac{\partial f_e}{\partial \mathbf{r}} + \frac{e(\mathbf{E} + \mathbf{v}_e \times \mathbf{B})}{m_e} \frac{\partial f_e}{\partial \mathbf{v}_e} = J_e, \quad n_e = \int f_e d\mathbf{v}_e \, ,$$

$$\frac{\partial f_i}{\partial t} + \mathbf{v}_i \frac{\partial f_i}{\partial \mathbf{r}} + \frac{e(\mathbf{E} + \mathbf{v}_i \times \mathbf{B})}{m_i} \frac{\partial f_i}{\partial \mathbf{v}_i} = J_i, \quad n_i = \int f_i d\mathbf{v}_i \, ,$$

and the Poisson equation for the electric potential φ is also solved:

$$\Delta\varphi = 4\pi e\left(n_e - n_i\right), \quad \mathbf{E} = -\frac{\partial \varphi}{\partial \mathbf{r}} .$$

where $\mathbf{v}_e$, $\mathbf{v}_i$, n_e, n_i, m_e, m_i are the velocity vectors, densities and masses of electrons and ions respectively, **E, B** are the electrical and magnetic field vectors, J_e, J_i are the collisional integrals for electrons and ions. The Particle in Cell Monte Carlo Collision (PIC MCC) method [5] was applied to solve the Boltzmann equations, which are two-dimensional in space and three-dimensional in velocity space. The system of equations are solved self-consistently with PlasmaNOV code [6]. For the electrons, the model includes elastic scattering, excitation and ionization collisions are accounted; for ions, the resonant charge exchage collisions with background argon gas are included. Additionally, the external ionization is included as electron-ion pairs generation with the Maxwellian distributions with the electron temperature T_e from simulation and the ion temperature T_i = 0.05 eV. The calculation grid is uniform over z-direction and nonuniform over radius condensing with increasing r.

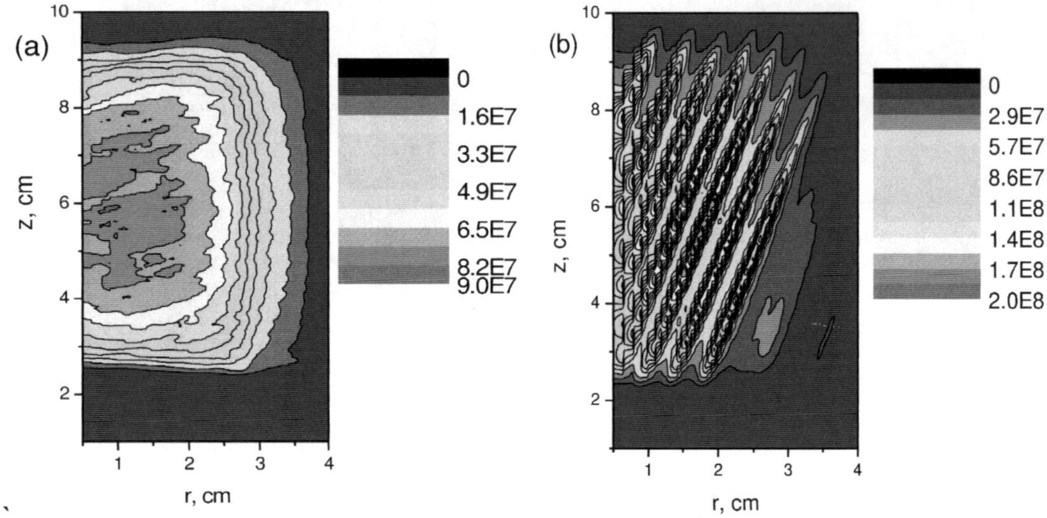

Figure 3. Electron density distribution n_e, cm^{-3}, for α_B=27° (a) and 77° (b), B=50 G, T_e=2.5 eV.

3. Results

The periodical structure with ridges of ion and electron densities is observed for larger inclination of magnetic field. For α_B smaller than 27° shown in Figure 3(a), the electron density is almost uniform in the central part of the chamber. In general, the plasma looks very similar to the case of α_B =0. The developed sheath forms near the cathode, with the potential drop of approximately 90 V. A weaker sheath screens plasma from walls of the chamber with the potential drop of ≈ 3 V. The sheaths can be seen in Fig.3 as areas with the depleted electron density.With increasing α_B the periodical plasma structure becomes clearly visible (see Figure 2 and 3). The electron and ion ridges are shifted with respect to each other and double–layer structure appears across the magnetic field and along the potential rise. The double-layers structure forms due to a distortion of local quasineutrality in the presence of inclined magnetic field. The electron is shifted from the ion in the direction normal to the magnetic field and a local charge appears. The structure occupies the quasineutral part of plasma in which the electrical field is small. Within the cathode sheath it does not appear even for large α_B. More complex situation takes place in the wall sheath. As the potential drop over the wall is small (3 - 8 V), the Lorentz and electrical forces are comparable there. Therefore the wall sheath becomes modulated by B-field with increasing α_B.

In Figure 4(a), the electron current j_e near the wall is shown for two values of α_B. It is seen that the currents approaching the wall are affected by a variation of α_B. The j_e profile over z taken at r =3 cm is almost uniform for α_B=10° and has peaks for α_B=65°. Each electron current peak is splitted with a scale of $2r_L$, where r_L is Larmor radius. The ion current near the wall also exhibits peaks for larger α_B. This effect can lead to an additional local erosion of wall material.

Figure 4(b) shows the potential distribution across B-field. The charge distribution has negative and positive ridges, since the sheets of large negative and positive charges appear in quasineutral plasma due to relative shift of n_e and n_i ridges in the direction of increasing potential and across B-field. This structure is called as double-layers and characterized with the non-monotonic potential distribution. Cross sections of the potential distribution in Figure 4 are taken along the dashed lines which start at r=0 and z= 7cm, 8 cm and 9 cm. For this case the potential bumps across magnetic field is about 0.5 V. The presence of magnetic field enhances the charge separation. For this case we can distinguish seven double-layers. The double-layers form due to a distortion of local quasineutrality in the presence of oblique magnetic field. When electron-ion pair appears after an ionization event an electron begins

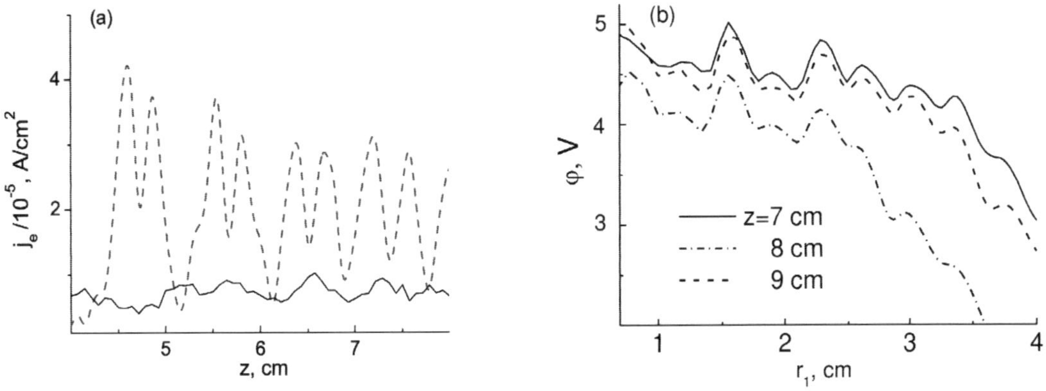

Figure 4. (a) Electron current density j_e (A/cm^2) plotted over z near side wall, for α_B=10° (solid) and 65° (dashed line); (b) Potential profile for α_B =65°, plotted along axis r_p which is normal to B and starts from r=0 and z=7, 8 and 9 cm. B=50 G, T_e= 5 eV.

Larmor gyramotion. The electron is shifted from the ion in the direction normal to B-field and a local charge appears.

The characteristics of plasma structure such as the number of peaks, gap between them, their broadening depend on the Larmor radius $r_L \sim T_e^{0.5}/B$, Debye length and the size of quasineutral plasma. The structure exists within some ranged of r_L and n_e. With increasing r_L and decreasing n_e the density peaks begin to overlap due to increasing broadening and the plasma loses the periodical structure.

In conclusion, the 2D3V PIC MCC simulations of dc discharge plasma in the cylindrical chamber at low gas pressure have been performed. The plasma was maintained by external ionization and confined by inclined magnetic field. The periodical structures with ridges of ion and electron densities have been found for larger inclination of magnetic field. The electron and ion ridges were shifted with respect to each other and periodical double-layer structure appears across magnetic field vector and along the potential rise. The flow channels of current to side wall are associated with ridges of electron and ion densities.

4. Acknowlegements
We are grateful to the Russian Foundation of Basic Research, project No. 15-02-02536, for supporting the development of software.

References
[1] Miedzik J, Barral S and Danilko D 2015 *Phys. Plasmas* **22** 043511
[2] Xu K G, Dao H and Walker M L R 2012 *Phys. Plasmas* **19** 103502
[3] Intrator T, Menard J and Hershkowitz N 1993 *Physics of Fluids* B **5** 806
[4] Borovsky J E and Joyce G 1983 *J. Plasma Phys.* **29** 45
[5] Birdsall C K and Langdon A B 1985 *Plasma Physics Via Computer Simulation* (New York: McGraw-Hill)
[6] Schweigert I V, Kaganovich I D and Demidov V I 2013 *Phys. Plasmas* **20** 101606
Schweigert I V, Langendorf S J, Walker M L R and Keidar M 2015 *Plasma Sources Sci. Technol.* **24** 025012

Transition in discharge plasma of Hall thruster type in presence of secondary electron emissive surface

I V Schweigert, M A Yadrenkin and V P Fomichev

Institute of Theoretical and Applied Mechanics, Novosibirsk, 630090, Russia

ischweig@itam.nsc.ru

Abstract. Modification of the sheath structure near the emissive plate placed in magnetized DC discharge plasma of Hall thruster type was studied in the experiment and in kinetic simulations. The plate is made from Al_2O_3 which has enhanced secondary electron emission yield. The energetic electrons emitted by heated cathode provide the volume ionization and the secondary electron emission from the plate. An increase of the electron beam energy leads to an increase of the secondary electron generation, which initiates the transition in sheath structure over the emissive plate.

1. Introduction

Plasma properties are strongly dependent on the processes of plasma-wall interaction. The secondary electron emission (SEE) from the chamber wall strongly affects the properties of Debye sheath on the wall, which screens plasma from the surface. Using materials with enhanced SEE for chamber walls and electrodes can change the classical structure of wall Debye sheath (see, for example, [1-4]). Rearrangement of Debye sheath for the materials with large SEE depends on plasma electrons temperature T_e and energy of emitted electrons in the collisionless plasma [5,6]. The variations of Debye sheath structure leads to change of plasma parameters. For example, in experiments investigating Hall thruster plasma, the wall material with enhanced SEE yield led to saturation of the T_e growth with increasing of applied voltage, but for material with small SEE yield the T_e demonstrated linear growth with voltage [7].

In this work, in the experiment and kinetic simulations we study the DC discharge plasma of Hall thruster type, controlled by electron beam from heated cathode. The aim of work was to investigate the effect of variation of the electron beam energy ε_e, on the structure of wall sheath near a Al_2O_3 plate, having enhanced SEE yield, for the conditions similar to the plasma of Hall thruster.

2. The experimental setup

In the experiment, the direct current discharge glows in a cylindrical chamber with radius of 8 cm and H=15 cm at low gas pressure, $P=4\times10^{-4}$ Torr. The walls of the chamber are grounded. The external magnetic field (B=24 G) is generated by coils of wire (solenoid) and directed along the axis of symmetry. The cathode is the system of tungsten wires which are heated and emit electrons. This cathode biased with the voltage of 0 - 300 V. The floating plate made from Al_2O_3 is placed 9 cm apart from the cathode. An additional mesh electrode is placed 0.5 cm apart from the plate surface. This electrode can be used to measure the plasma potential and to control the secondary electron emission current from the surface.

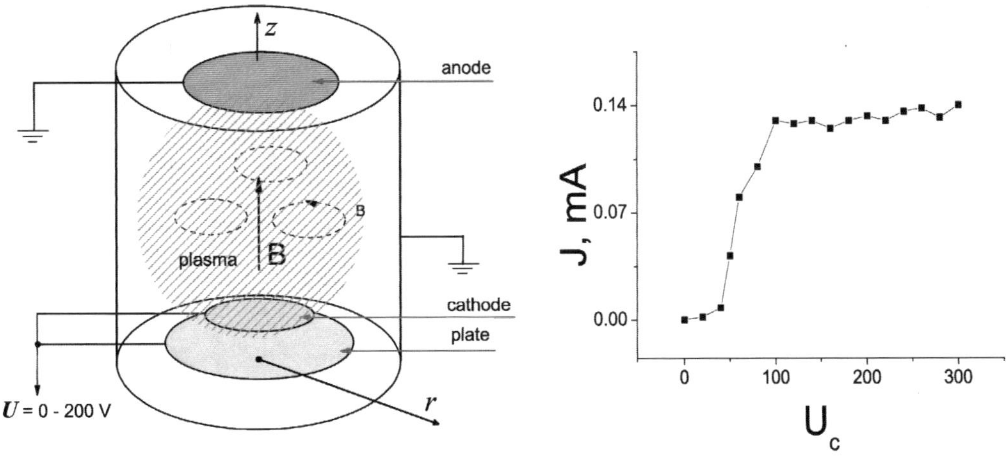

Figure 1. Calculation domain (left) and the discharge current as function of applied voltage (right), observed in the experiment.

3. Theoretical model

The magnetized plasma is described with the system of equations including the Boltzmann equations for electrons and ions and fast neutral atoms. The distribution functions for electrons $f_e(t,x,\mathbf{v})$ and ions $f_i(t,x,\mathbf{v})$ are calculated by solving the Boltzmann equations

$$\frac{\partial f_e}{\partial t} + \mathbf{v}_e \frac{\partial f_e}{\partial \mathbf{r}} + \frac{e(\mathbf{E} + \mathbf{v}_e \times \mathbf{B})}{m_e} \frac{\partial f_e}{\partial \mathbf{v}_e} = J_e, \quad n_e = \int f_e d\mathbf{v}_e ,$$

$$\frac{\partial f_i}{\partial t} + \mathbf{v}_i \frac{\partial f_i}{\partial \mathbf{r}} + \frac{e(\mathbf{E} + \mathbf{v}_i \times \mathbf{B})}{m_i} \frac{\partial f_i}{\partial \mathbf{v}_i} = J_i, \quad n_i = \int f_i d\mathbf{v}_i ,$$

and the Poisson equation for the electric potential φ is also solved:

$$\Delta\varphi = 4\pi e(n_e - n_i), \quad \mathbf{E} = -\frac{\partial \varphi}{\partial \mathbf{r}} .$$

where $\mathbf{v}_e$, $\mathbf{v}_i$, n_e, n_i, m_e, m_i are the velocity vectors, densities and masses of electrons and ions respectively, **E, B** are the electrical and magnetic field vectors, J_e, J_i are the collisional integrals for electrons and ions. The Particle in Cell Monte Carlo Collision (PIC MCC) method [8] was applied to solve the Boltzmann equations, which are two-dimensional in space and three-dimensional in velocity space. The system of equations are solved self-consistently with PlasmaNOV code [9]. For the electrons, the model includes elastic scattering, excitation and ionization collisions are accounted; for ions, the resonant charge exchage collisions with background argon gas are included.

In simulations, the applied voltage U_c is ranged from 0 to 300 V. For these plasma parameters the electron Larmor radius r_L is comparable to the Debye length λ_D, $r_L < \lambda_D \approx 0.1\div0.5$ cm. The plasma frequency $\omega_p < \Omega_e \approx (5\div50)\times10^8$ s^{-1}, where Ω_e is the electron gyrofrequency.

In calculations, the electron time step is $(2\div5)\times10^{-12}$ s. The grid is uniform in z-direction and nonuniform over radius condensing with decreasing r. The total number of pseudoparticles for PCC MCC algorythm is chosen so that there is an average of approximately 100 positive and negative particles per Debye sphere.

In Figure 1 (left), the calculation domain is shown. The plasma is embedded in a cylindrical chamber with the radius of 8 cm and the height of 10 cm. The cathode radius is 2 cm.

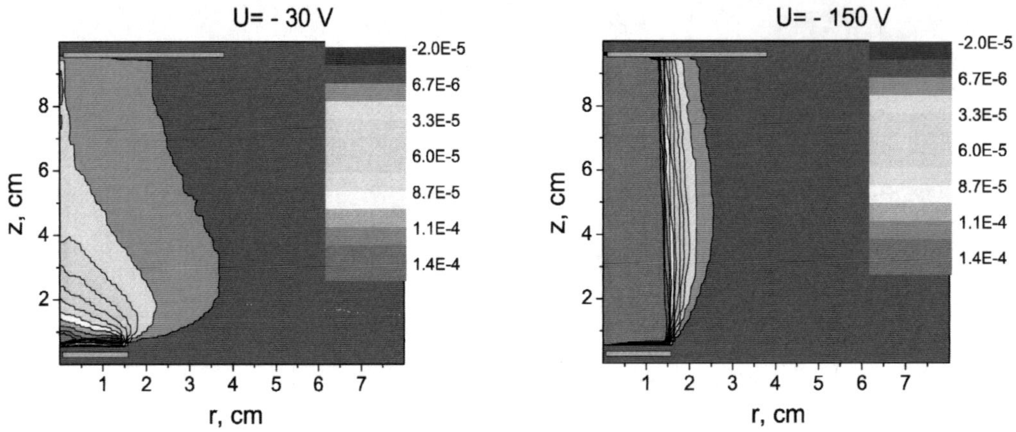

Figure 2. Electron current density for U_c=30 V (left) and U_c=150 V (right).

4. Results

In the experiment, the heating of tungsten wires of the cathode with 12 A current gives us the maximum of the emissive electron current of 100 - 140 μA. The emissive electrons crossing the cathode sheath gain the energy of $e\varphi_c$, where φ_c is the potential drop over the cathode sheath. Thus the energy of beam electrons rises with increase of applied voltage U_c. In Figure 1 (right), the measured current to the anode plate as function of U_c is shown for the grounded anode and wall of the chamber. It is seen that for smaller U_c <50 V, only a minor part of electrons reaches the anode. The computed electron current distribution is shown in Figure 2 for U_c =30 V and 150V.

To study the sheath modification with an increase of the electron beam energy, in the experiment, the measurements of floating potential on the Al_2O_3 plate surface and 0.5 cm apart from the surface were done for different applied voltages. The results are given in Figure 3 (left). Note that the difference in the potentials is essential for U_c < 100 V. For U_c > 100 V, the potentials on the plate surface and 0.5 cm apart for the surface practically coincide.

It is seen that the floating potential of the plate surface does not change with U_c. In contrary, the plasma potential at z=0.5 cm apart from the surface varies essentially. It manifests the collapse of sheath near the emissive surface for the case when secondary emission becomes pronounced.

As simulations show, for U < 100 V, the potential drop near the floating plate surface $\Delta\varphi$ is set by currents of the beam electrons j_{be} and secondary electrons, $j_{be} = j_{se}$. The ion current is negligible. The transition happens at U ≈ 100 V. It is clearly indicated in Figure 3 (right) by a decrease of difference in potentials of plate surface and 0.5 cm apart.

The mean temperature of the plasma electrons also decreases at the point of the transition. In this new regime (collapsed sheath), the cold plasma electrons start to contribute to the current balance on the plate. Now the currents of the cold plasma electrons j_{pe}, beam electrons j_{be} and secondary electrons, $j_{be} + j_{pe} = j_{se}$ set the potential drop near the floating plate. Since the density of the slow plasma electrons is much larger than that of the beam electrons, a small decrease of $\Delta\varphi$ leads to considerable increase of the plasma electron current. With increasing U_c the potential drop $\Delta\varphi$ tends to value of T_e.

In conclusion, we have studied the sheath formation in DC discharge plasma in front of floating emissive plate made of Al_2O_3 with enhanced secondary emission coefficient. The discharge operation is maintained by beam of energetic electrons from the heated cathode. The voltage is applied to the cathode, and the energy of electron beam is set by this applied voltage. We have found the collapse of sheath near the floating plate with increasing applied voltage.

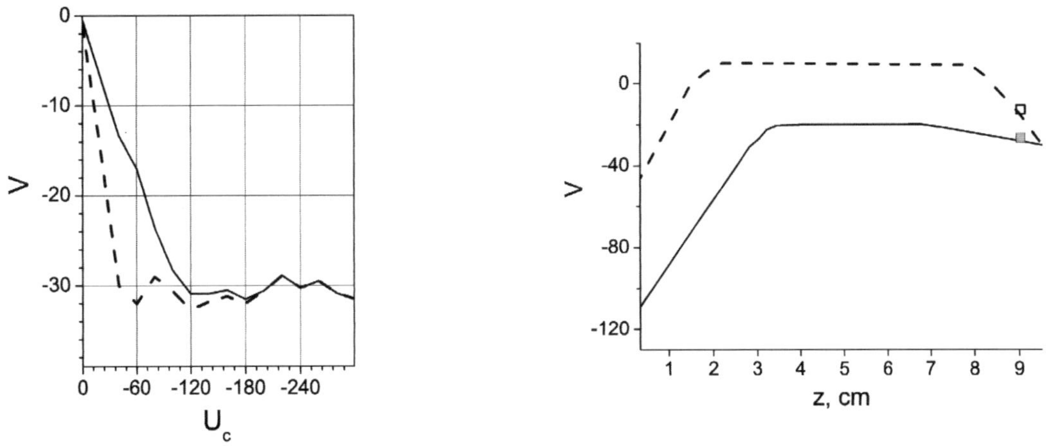

Figure 3. Left: the floating potential, measured in the experiment, on the plate surface (solid line) and 0.5 cm apart from the surface (dashed line) for different applied voltages. Right: calculated potential distribution over axis of symmetry for U_c = 60 V (dashed line) and 110 V (solid line).

5. Acknowlegements

We are grateful to the Russian Science Foundation for supporting this work (project 17-19-01375), and also to the Russian Foundation of Basic Research (project 15-02-02536), for supporting the development of software.

References

[1] Campanell M D, Khrabrov A V and Kaganovich I D 2012 *Phys. Rev. Lett.* **108** 235001
Campanell M D, Khrabrov A V and Kaganovich I D 2012 *Phys. Rev. Lett.* **108** 255001
[2] Morozov A I and Savel'ev V V 2007 *Plazma Phys. Reports* **33** 20
[3] Keidar M, Boyd I D and Beilis I I 2001 *Physics of Plasmas* **8** 5315
[4] Schweigert I V, Langendorf S J, Walker M L R and Keidar M 2015 *Plasma Sources Sci. Technol.* **24** 025012
[5] Hobbs G D and Wesson J A 1967 *Plasma Phys.* **9** 85
[6] Sheehan J P, N. Hershkowitz N, Kaganovich I D et al 2013 *Phys. Rev. Lett.* **111** 075002
[7] Raitses Y, Smirnov A, Staack D and Fisch N 2006 *Phys. Plasmas* **13** 014502
[8] Birdsall C K and Langdon A B 1985 *Plasma Physics Via Computer Simulation* (New York: McGraw-Hill)
[9] Schweigert I V, Kaganovich I D and Demidov V I 2013 Phys. Plasmas, **20**, 101606

Effect of secondary electron emission on subnanosecond breakdown in high-voltage pulse discharge

I V Schweigert[1], A L Alexandrov[1], P Gugin[2], M Lavrukhin[2], P A Bokhan[2], Dm E Zakrevsky[2]

[1] Institute of Theoretical and Applied Mechanics, Novosibirsk, 630090, Russia
[2] Institute of Semiconductor Physics, Novosibirsk 630090, Russia

ischweig@itam.nsc.ru

Abstract. The subnanosecond breakdown in open discharge may be applied for producing superfast high power switches. Such fast breakdown in high-voltage pulse discharge in helium was explored both in experiment and in kinetic simulations. The kinetic model of electron avalanche development was developed using PIC-MCC technique. The model simulates motion of electrons, ions and fast helium atoms, appearing due to ions scattering. It was shown that the mechanism responsible for ultra-fast breakdown development is the electron emission from cathode. The photoemission and emission by ions or fast atoms impact is the main reason of current growth at the early stage of breakdown, but at the final stage, when the voltage on discharge gap drops, the secondary electron emission (SEE) is responsible for subnanosecond time scale of current growth. It was also found that the characteristic time of the current growth τ_s depends on the SEE yield of the cathode material. Three types of cathode material (titanium, SiC, and CuAlMg-alloy) were tested. It is shown that in discharge with SiC and CuAlMg-alloy cathodes (which have enhanced SEE) the current can increase with a subnanosecond characteristic time as small as τ_s =0.4 ns, for the pulse voltage amplitude of 5- 12 kV..

1. Introduction

In the experiments with high-voltage open discharge in helium [1,2] a new method was proposed for switching of power pulses with time less than a nanosecond. The high-voltage open discharge proved to be an useful tool for producing high-energy electron beams or subnanosecond high-power electrical switch. But the mechanism of this fast current rise in this type of discharge was not completely clear.

Recently the current exponential growth within a subnanosecond was registrated [1,2] during breakdown in high voltage open (HVO) discharges in helium. The discharge operates between two parallel plane cathodes and a mesh anode with 90% transparency between them. Applying high voltage leads to the generation current with density of 200 - 400 A/cm^2 and current increase rate of 500 A/(cm^{2}ns). Such type of discharges with unique parameters can be used as a source of electron beams or as subnanosecond high-power switches.

The earlier developed models of breakdown in helium [3] are not able to explain such ultra-fast development of current. The electron emission from the cathode due to ions or fast atoms bombardment is usually considered as a main process of breakdown development. However the subnanosecond times of breakdown in HVO discharge can not be supported by emission with ions and fast atoms bombardment due to their inertia. The photoemission by resonant radiation from atoms excited by electron impacts is also unable to maintain the observed current growth rate since for these

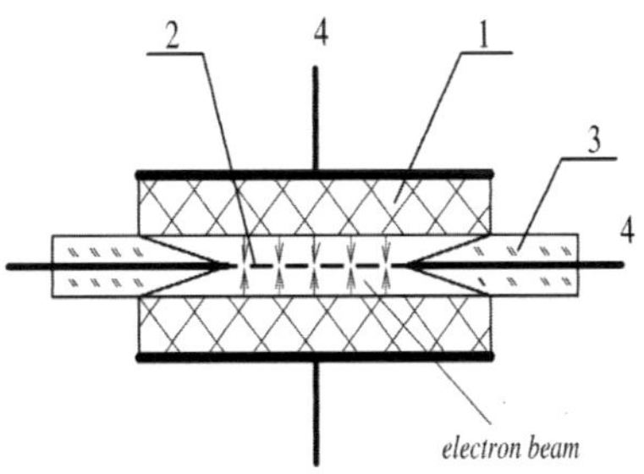

Figure1. Scheme of planar discharge cell with transparent anode between symmetrical cathodes: 1) – cathodes, 2) – mesh anode, 3) – insulator, 4) – lead-in.

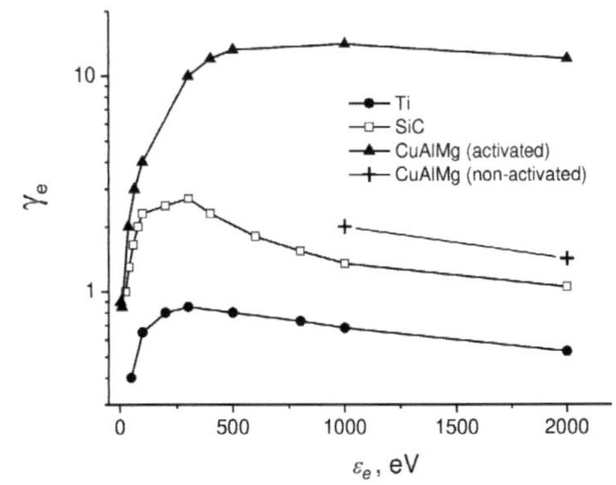

Figure 2. Secondary electron emission coefficient as function of the electron energy for titanium, silicon carbide and CuAlMg-alloy cathodes.

times the resonant photons are trapped in plasma due to reabsorption. In works [1,2] it was proposed that the key role in photoemission play the photons of resonant radiation, having Doppler shift. The fast atoms, appearing after elastic scattering of fast ions, have large energy (1-500 eV) and due to their large velocity, radiate resonant photons with a Doppler shift. The same occurs, due to momentum transfer, when atom of buffer gas is excited by ion impact or fast atom impact. Another significant source of Doppler-shifted (DS) photons is reaction of resonant excitation transfer from excited buffer gas atoms to fast atoms. The DS photons are not reabsorbed in buffer gas and reach the cathode surface immediately, producing intensive photoemission. In first attempts to simulate ultra-fast breakdown [4] the account of photoemission by DS photons allowed to explain the nanosecond times of plasma accumulation before the final stage of breakdown. But, as was found next, the real explanation of subnanosecond current front rise belongs to secondary electron emission (SEE) from cathode surface at the final breakdown stage. When the discharge current reaches magnitudes of tens A/cm^2, the voltage on the discharge gap begins to diminish due to properties of electric circuit. The hot electrons, accumulated in bulk plasma, become able to overcome the cathode potential fall and begin to bombard the cathode. This leads to enlarging of current and thus to further decrease of cathode voltage. So the SEE grows exponentially and becomes the dominant process in subnanosecond current growth at the final stage of breakdown. In this work, the aim was to study the influence of cathode SEE yield on the minimal achievable time (or maximal rate) of current growth. The experimental investigations and simulations were performed for three different types of cathode materials, having different SEE yield, namely: titanium, silicon carbide (SiC) and CuAlMg alloy.

2. Experimental technique

In experiment, the breakdown in helium is studied in a planar pulse discharge with generation of two oppositely directed electron beams. The scheme of experimental cell is shown in Figure 1. It has two round cathodes with area of 16 cm^2 and a mesh-anode with the geometric transparency of 0.9 is placed between them. The inter-cathode distance is 6 mm. The cathodes and anode are isolated with a set of plates from glass. The pulse voltage is simultaneously applied to the cathodes and its amplitude ranges from 4 kV to 12 kV in different experiments. The gas pressure varies from 10 Torr to 35 Torr.

The voltage shape is recorded using oscilloscope Tektronix DPO 70804C with a bandwidth of 8 GHz. The registration circuit and other experimental conditions were described in details in [1].

In the experiments three types of cathodes made from different materials were tested. All these materials, titanium (Ti), silicon carbide (SiC), and CuAlMg alloy have enhanced secondary electron yield, but the dependences of secondary electron emission coefficient (SEE) γ_e from the electron

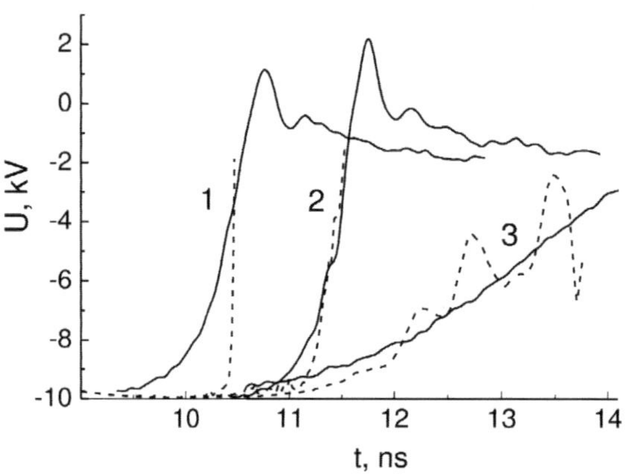

Figure 3. Voltage waveforms measured (solid lines) and calculated (lines with symbols) for the cases with cathodes made from CuAlMg-alloy (1), silicon carbide (2) and titanium (3) for U_a=10 kV and P=25 Torr.

energy ε_e are very different. For Ti the SEE coefficient γ_e varies from 0.4 to 0.9 in the range from 50 eV to 5 keV, with maximum at 300 eV [5]. For SiC the γ_e has similar behavior as for Ti for energy range 20~eV - 400~eV [6], but it is nearly three times larger, γ_e= 0.9 - 2.5. For CuAlMg alloy γ_e can be very different depending on the state non-activated and activated by annealing [7]. It should be noted that for non-activated CuAlMg alloy the γ_e is close to observed for SiC. The activation occurs due to diffusion of Mg to the surface and its oxidation. So, the annealed CuAlMg alloy demonstrates γ_e up to 12.5 in the electron energy range 500 eV - 2 keV. The energy dependence of γ_e for all three types of cathode material is shown in Figure 2. The values of γ_e for inactivated CuAlMg alloy from [7] are also shown in Figure 2 by crosses. There is lack of empirical data for inactivated CuAlMg at low-energies. The comparison of the measured current growth times in the discharge with different cathodes allows us to estimate the contribution of SEE in the electron avalanche development.

3. Physical model

The discharge plasma is described with the system of equations including the Boltzmann equations for electrons, ions and fast neutral atoms. The distribution functions for electrons $f_e(t,x,\mathbf{v})$, ions $f_i(t,x,\mathbf{v})$ and fast atoms $f_a(t,x,\mathbf{v})$ are calculated from the Boltzmann equations

$$\frac{\partial f_e}{\partial t} + \mathbf{v}_e \frac{\partial f_e}{\partial \mathbf{r}} + \frac{e\mathbf{E}}{m_e} \frac{\partial f_e}{\partial \mathbf{v}_e} = J_e, \quad n_e = \int f_e d\mathbf{v}_e ,$$

$$\frac{\partial f_i}{\partial t} + \mathbf{v}_i \frac{\partial f_i}{\partial \mathbf{r}} + \frac{e\mathbf{E}}{m_i} \frac{\partial f_i}{\partial \mathbf{v}_i} = J_i, \quad n_i = \int f_i d\mathbf{v}_i ,$$

$$\frac{\partial f_a}{\partial t} + \mathbf{v}_a \frac{\partial f_a}{\partial \mathbf{r}} = J_a, \quad n_a = \int f_a d\mathbf{v}_a ,$$

and the Poisson equation for the electric potential φ :

$$\Delta\varphi = 4\pi e(n_e - n_i), \quad \mathbf{E} = -\frac{\partial \varphi}{\partial \mathbf{r}} .$$

where $\mathbf{v}_e$, $\mathbf{v}_i$, $\mathbf{v}_a$, n_e, n_i, n_a, m_e, m_i are the velocities, densities and masses of electrons, ions and fast atoms respectively, $\mathbf{E}$ is the electrical field, J_e, J_i, J_a are the collisional integrals for electrons, ions and fast atoms, which include elastic and inelastic scatterings with He atoms. The PIC MCC method was applied to solve the Boltzmann equations. The system of equations are solved self-consistently. For the electrons, the model includes the following collisions: 1) elastic, or momentum transfer; 2)

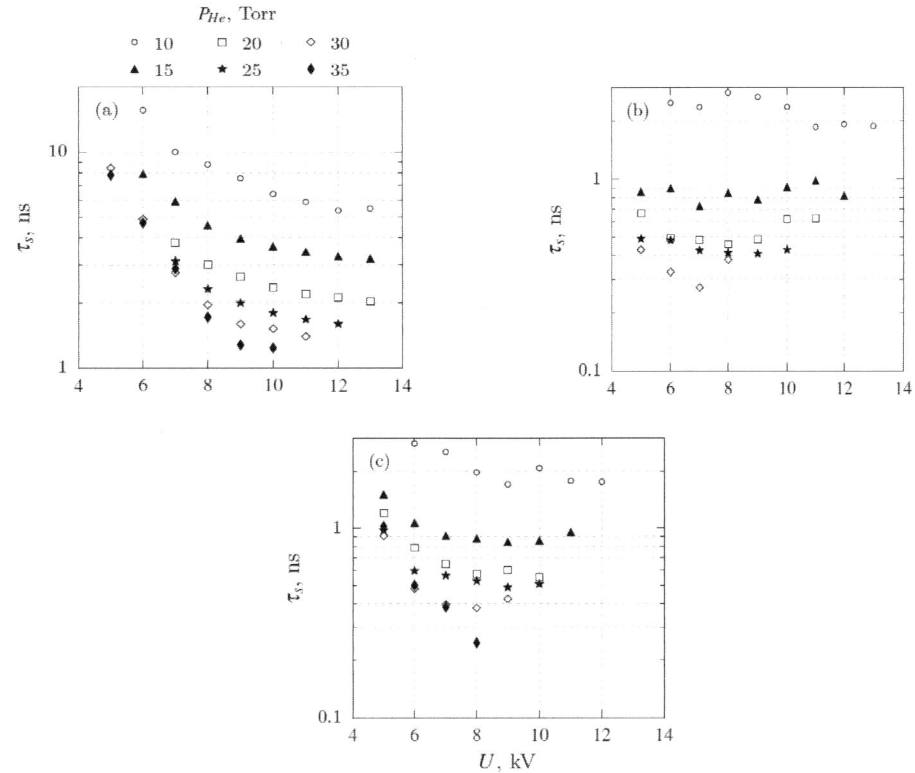

Figure 4. Characteristic time of current front rise τ_s from pulse voltage amplitude for cathodes made from Ti (a), SiC (b) and for CuAlMg-alloy (c) for different gas pressures.

electron impact excitation of He atom; and 3) electron impact ionization of He. For the He^+ ions, four types of collisions with the background atoms are taken into account: 1) elastic collision; 2) resonant charge exchange collision, or backward elastic scattering; 3) ion impact excitation of He atom; 4) ion impact ionization of He atom. The fast neutral atoms He_f appear during ions scattering on the background gas. We include into consideration only the fast atoms with the energy $\varepsilon_a > 1$ eV.

The collisions of the fast atoms include also four types of reactions: 1) elastic scattering; 2) atom impact excitation; and 3) atom impact ionization. The scattering of heavy particles are assumed to be isotropic in the center of mass system. Also, the 4-th reaction of collisional excitation transfer (CET) is taken into consideration, as a possible source of excited fast atoms.The details of physical model and the crossections of described above reactions can be found in publications [4,8,9].

The most important process, defining the development of breakdown, is the electron emission from the cathode. As was mentioned above, the resonant photons with a Doppler-shifted frequency can reach the cathode instantly without reabsorption. Also elecrtons are emitted by incndent ions or fast atoms. As was shown in [8,9], the secondary electron emission (SEE) becomes a dominant process in current growth at the final stage of breakdown.

At the stage of preliminary plasma accumulation, the high energy electrons are trapped between two cathodes by the kV-potential drop over the cathode sheaths. While current rise, the discharge voltage begins to decrease and cathode potential drop diminishes. The high energy electrons accumulated in the discharge volume start to bombard the cathodes and the breakdown current grows exponentially with subnanosecond characteristic time. With increasing gas pressure the contributions from the photoemission and secondary electron emission enlarge and the characteristic time of - breakdown decreases. The purpose of this work was to study the effect of SEE yield emission on the breakdown time for three types of cathode materials with enhanced SEE emission yield, and to find minimal breakdown time with a combined effects of cathode material with high SEE and variation of the gas pressure.

4. Results

The typical waveform of voltage pulses on the discharge gap are shown in Figure 3 for the discharge cells with Ti, SiC and CuAlMg-alloy cathodes for the voltage amplitude U_a= 10 kV and for helium gas pressure P = 25 Torr. For the case with CuAlMg-alloy cathode the simulated dU/dt is larger than observed in the experiment, likely because of the specific state of the CuAlMg- alloy surface in the experiment. The value of γ_e for CuAlMg-alloy cathode is very sensitive to a degree of alloy activation due to magnesium diffusion to the surface and its oxidation by heating. For the case with SiC-cathode $\tau_s \approx 0.4$ ns and computed and measured voltage profiles are in good agreement. For Ti-cathode $\tau_s \approx 2$ ns. Calculated profile demonstrates the pronounced oscillations likely due to the simplified external circuits taken in simulations.

As seen in Figure 4, the switching time is very sensitive both to the gas pressure and the cathode material (SEE yield). An increase of gas pressure diminishes the breakdown time. However if the secondary electron emission coefficient is small like for titanium, τ_s remains larger than 1 ns even with a variation of gas pressure between 20 Torr and 40 Torr (see Figure 4(a)). For the materials with enhanced SEE like SiC and alloy, an decrease of gas pressure from 10 Torr to 30 Torr allows us to reduce τ_s from 2 ns to 0.4 ns (see Figure 4 (b),(c)).

In conclusion, the experimental study and simulations of SEE yield, voltage magnitude and gas pressure confirmed that the SEE makes main contribution in current growth on the final stage of breakdown. The $\tau_s \approx 0.4$ ns were observed for SiC and CuAlMg cathodes, but for titanium cathode only $\tau_s \approx 2$ ns was obtained. The results of experiment and simulations are in good agreement. We also found out that the increase of gas pressure from 10 Torr to 30 Torr helps to decrease the τ_s. Altering pulse voltage amplitude within 5 kV - 10 kV weakly changes τ_s.

5. Acknowlegements

We are grateful to the Russian Science Foundation for supporting this work, project 14-19-00339. The development of software was supported by grant of Russian Foundation of Basic Research No. 15-02-02536.

References

[1] Bokhan P A, Gugin P P, Zakrevsky Dm E and Lavrukhin M A 2013 *Phys. Plasmas* **20** 033507
[2] Bokhan P A, Gugin P P, Zakrevsky Dm E and Lavrukhin M A 2013 *Tech. Phys. Lett.* **39** 775
[3] Hartmann P, Donko Z, Bano G et al 2000 *Plasma Sources Sci. Technol.* **9** 183
[4] Schweigert I V, Alexandrov A L, Zakrevsky Dm E and Bokhan P A 2014 *Phys. Rev. E* **90** 051101(R)
[5] Walker C G H, EL-Gomati M M, Assa'd A M D, and Zadrazil M 2008 *Scanning* **30** 365
[6] Viel-Inguimbert V 2003 *Proc. of 28th International Electric Propulsion Conf.* (Toulouse, France) IEPC-2003-258
[7] Lepeshinskaya V N and Stuchinskiy G B 1960 *Fizika Tverdogo Tela (Solid State Physics)* **2** 1328 (in Russian).
[8] Schweigert I V, Alexandrov A L, Zakrevsky Dm E and Bokhan P A 2015 *Plasma Sources Sci. Technol.* **24** 044005
[9] Schweigert I V, Alexandrov A L, Zakrevsky Dm E and Bokhan P A 2016 *Plasma Phys. Reports* **42** 666

IOP Conf. Series: Journal of Physics: Conf. Series **927** (2017) 012051 doi:10.1088/1742-6596/927/1/012051

Influence of DC and AC external electric field on the propagation of "plasma bullets" along DBD helium plasma jet

Yu S Akishev[1,2], V B Karalnik[1], M A Medvedev[1], A V Petryakov[1], N I Trushkin[1], A G Shafikov[1]

[1] SRC RF TRINITI, 108840, Moscow, Troitsk, Pushkovykh street, 12, Russia
[2] NRNU MEPhI, 115409, Moscow, Kashirskoe shosse, 31, Russia

Abstract. Plasma jet generated by sinusoidal DBD in a tube blown with He is not uniform along its length because of existence of so called "plasma bullets" (ionization waves) travelling in jet each half-cycle. The bullet velocity and the visual shape of plasma jet will change, if to vary the ambient conditions for plasma jet. This work is devoted to the detailed experimental study on influence of the external DC and AC electric field with different amplitude and polarity on the bullet propagation in plasma jet. It has been established that one can change appreciably the plasma bullet velocity and even wholly suppress the bullets in the plasma jet.

1. Introduction

Non-thermal plasma jet can be generated by different discharges in narrow tube blown with various gas mixtures. The most popular source of non-thermal plasma jet is a sinusoidal dielectric barrier discharge (DBD) in He flow [1,2]. The detailed review of experimental and theoretical investigations devoted to different properties of such plasma jet was done in paper [3]. One of the interesting phenomena observed in DBD plasma jet is so called "plasma bullets" travelling along plasma jet. These "bullets" have been for the first time revealed by M. Teschke et al [4]. The bullets are characterized by higher electric field and higher light emission. In other words, the bullets flash the trajectory of plasma jet. This is a reason why the visual length of plasma jet is determined by a maximum distance that can be passed by a bullet after its run-out from the DBD zone. Note plasma bullet, in fact, is the ionization wave, the propagation mechanism of which is similar to the propagation mechanism of the head of normal streamer. According to this mechanism, the electric field strength in/around streamer head plays a crucial role in the propagation of the streamer. Based on this similarity, one may state that the bullet propagation velocity will depend not only on the parameters of the applied voltage (amplitude and frequency of sinusoidal voltage) influencing the electric field in the bullet but the polarity and amplitude of DC and AC external electric field formed artificially around the plasma jet as well. Moreover, in principle it is possible to arrange the configuration of the electric field around plasma jet which suppresses wholly the bullets. This work is devoted to the detailed experimental study of influence of the DC and AC external electric field formed artificially around the plasma jet on both the bullet propagation velocity in He plasma jet and possibility for existence of the bullets in plasma jet.

2. Experimental setup

Non-thermal plasma jet in helium flow was generated by sinusoidal dielectric barrier discharge (DBD) operating at atmospheric pressure. Gas discharge was excited in the quartz tube with inner and outer diameters of 2.5mm and 4mm, respectively. The high-voltage electrode was a stainless steel wire with a diameter of 1 mm. The frequency of the applied voltage is equal to 100 kHz; the amplitude of the applied voltage was varied up to 4 kV; gas flow velocity at the DBD exit is equal to 40 m/s. Plasma jet enters the ambient air at room temperature and is directed towards thin metallic disc of 120 mm in diameter located at a distance of 36 mm from the tip of HV-electrode (or 28 mm from the discharge tube exit). In some experiments, we used the disc with a circular hole of 4 mm in diameter at his center. Metallic disc can be supported either at floating potential or stressed by DC or AC voltage. Sketch of the experimental setup is shown in figure 1.

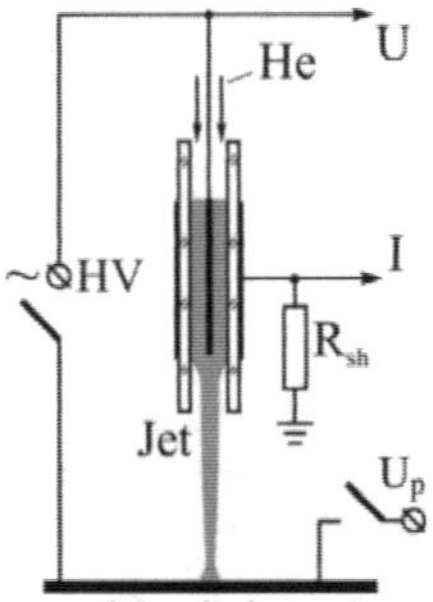

Figure 1. Sketch of the experimental setup used for generation of a plasma jet by DBD in He flow

3. Experimental results and discussions

The images of plasma jet striking the metallic disc stressed by different DC potentials are shown in figure 2. All photos were taken at the same exposure time equal to 250 μs that is much longer compared to the period T of sinusoidal voltage (T = 10 μs).

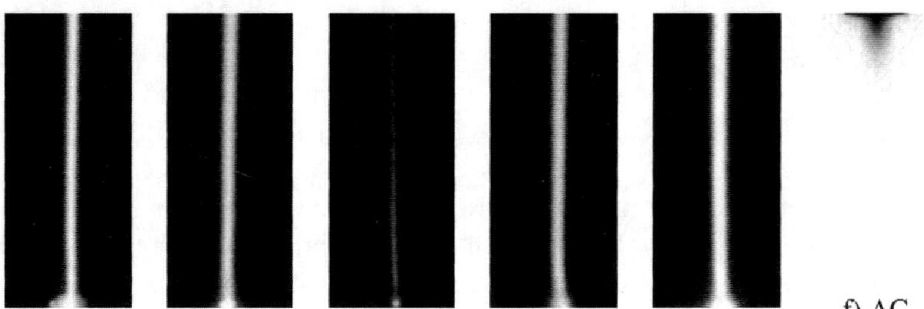

Figure 2. The images of plasma jet colliding with metallic disc. The top of each jet corresponds to the exit from a discharge tube. The disc is stressed by different DC potentials (a)-(e); the enlarged negative image of plasma jet with visual length less of 2 mm, the disc is connected to the HV electrode (f). Amplitude of sinusoidal voltage powering the DBD is 3 kV. Exposure time for shots (a)-(e) is equal to 250 μs and 0.5 s for shot (f).

Note that in contrast to the dielectric substrate, plasma jet practically does not spread over the metallic disc. A reason is an absence on the metallic surface of the tangential component of electric field which is necessary for plasma spreading [5]. However, the diameter of area activated by plasma jet on the metallic disc is much larger compared to the size of visual plasma spot on the surface. This statement was proved by measurement of a radial distribution of contact angle of the water droplets deposited on the metallic surface treated by plasma jet with the small visual spot. It means that plasma jet contains a lot of reactive species which do not emit the light but they are transported by gaseous jet and spread (together with the spreading jet) over the metallic surface in the radial direction.

Another feature of the experimental results presented in figure 2 is the strong dependence of the brightness of plasma jet on DC potential applied to the metallic disc - the higher DC potential, the stronger brightness of the jet. Such dependence is more pronounced for the positive polarity. It is clear that an existence of DC potential on the disc influences the voltage drop along plasma jet owing to changing the potential difference between HV electrode and disc in every half-cycle. For instance, DC positive potential increases the voltage drop along plasma jet in the negative half-cycle and diminishes it in the positive half-cycle. For DC negative potential, everything will be on the contrary.

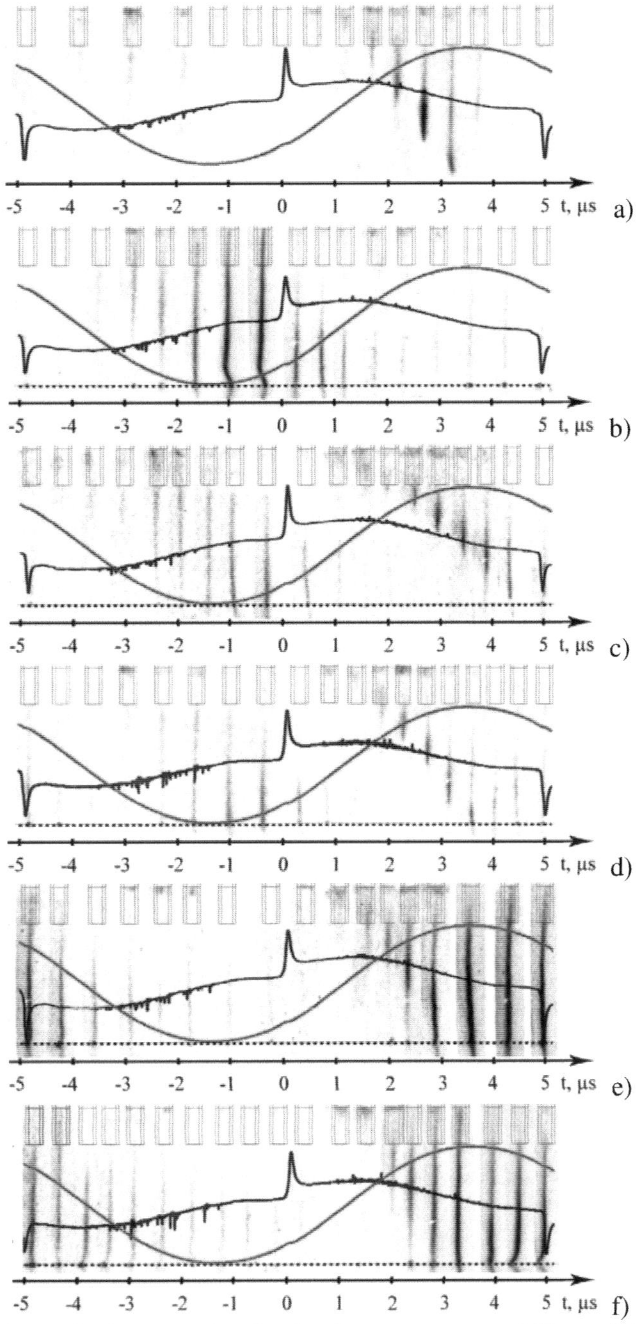

Figure 3. The set of plasma jet images taken by fast frame camera at different DC potentials Up on the disc. The shots are distributed over a single cycle of the current and voltage of DBD. Each isolated shot correlates with the image of the discharge tube shown in the top of every shot. The exposure time of every shot is 200 ns. a) Free jet; Up = +2 kV (b); +1 kV (c); 0 kV (d); -1 kV (e); -2 kV (f).

In the case of the disc connected to the HV electrode (figure 2.f), the electric field strength in the space between HV electrode and disc is equal to zero at any moment. It leads to that the luminosity of plasma jet disappears in this space - only the dim light has been emitted from very short area located in vicinity of the discharge tube exit. This is a reason why we used a very long exposure time of 0.5 s in order to get the image of the dim plasma jet. Based on this experiment we can conclude that plasma bullets cannot develop in plasma jet with zero electric field. However, the absence of plasma bullets in plasma jet does not mean that dark plasma jet is not enriched with the reactive species. This statement was proved again by the measurement of a radial distribution of contact angle of the water droplets deposited on the metallic surface treated by dark plasma jet. As it turned out, the dark plasma jet treats the substrate with the efficiency close to that for bright plasma jet with bullets.

The images presented in figure 2 give only general impression about influence on plasma jet of DC potential applied to the disc. To clarify in more detail what happens in plasma jet during each half-cycle of sinusoidal voltage, we have taken the shots of jet with short exposure time and synchronized with the voltage and current of DBD. The obtained results for different DC potentials on the disc are presented in figure 3. For the reference, the set of images of free jet is presented in figure 3a. In free plasma jet, the bullets can be seen clearly only in positive half-cycle.

The disk under floating potential Up = 0 kV leads in itself to change of behavior of plasma bullets (see figure 3d). A reason of that is an existence of the stray capacitance of the disc regarding the ground. This capacitance increases the current flowing through plasma jet to the ground. In this case plasma bullet appears a bit later and propagates more slowly compared to that in free jet. In the negative half-cycle, plasma bullet again does not happen but the brightness of plasma jet increases for short time simultaneously throughout its length. Positive potential Up = +1 kV on the disc leads to still more lag in arising of plasma bullet and its deceleration (see figure 3c). As for negative half-cycle, the increase of brightness of plasma jet along its full length lasts a bit longer. Further growth of positive potential (Up = +2 kV) results in a full cessation of the bullet appearance in positive half-cycle (see figure 3b) but in negative half-cycle both the brightness of full plasma jet and the duration of stage with bright plasma grow up still more appreciably.

The applying of DC negative potential to the disc leads to the acceleration of plasma bullet movement in positive half-cycle - the higher negative potential, the faster movement – (see figures 3e and 3f). The arrival of the bullet to the disc is followed by formation of plasma filament when plasma jet increases appreciably its brightness simultaneously throughout its full length. The duration of this bright stage increases with the increase of negative potential. In negative half-cycle, the length of bright plasma filament formed in previous positive half-cycle diminishes due to its reducing towards a disk.

Close examination of the images presented in figure 3 allowed us to determine the distribution of plasma bullet velocity along the jet for different experimental conditions. The summary of the obtained results is shown in figure 4.

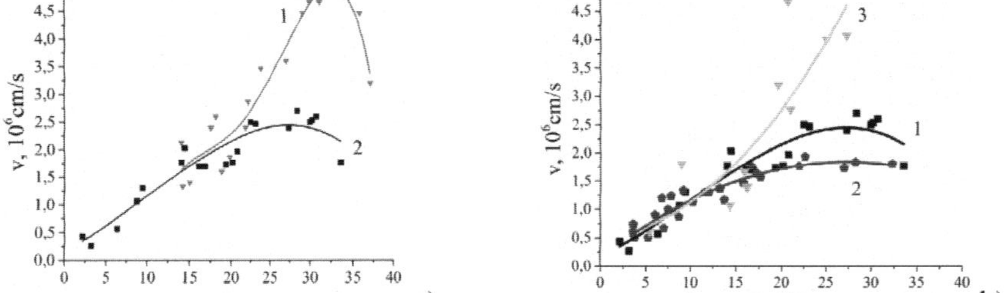

Figure 4. The distribution of plasma bullet velocity along He plasma jet. a) free plasma jet (1); plasma jet colliding with the disc under floating potential (2). b) plasma jet colliding with the disc: Up = 0 kV

(1), Up = +1 kV (2), Up = -1 kV (3). x is the coordinate along the plasma jet; x = 0 corresponds to the exit of the discharge tube.

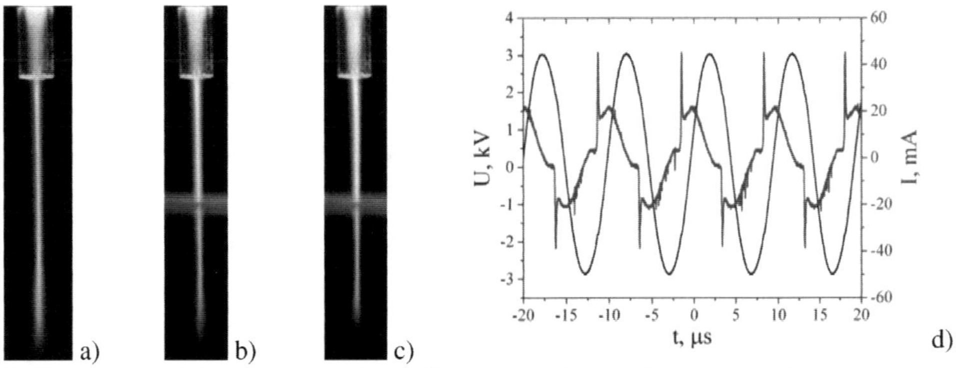

Figure 5. The images of the plasma jets for different conditions of their propagation.
a) plasma jet in free space; total plasma jet length is 43 mm; DBD power is 2.9 W.
b) plasma jet meets the metallic disc with a hole of 4 mm in diameter and passes through this hole; the disc is non-grounded (floating potential); total plasma jet length is 41 mm; DBD power is 3.0 W.
c) plasma jet meets the metallic disc with a hole of 4 mm in diameter and passes through this hole; the disc is grounded; total plasma jet length is 39 mm; DBD power is 4.1 W.
d) the current and voltage waveforms of DBD which practically the same for a), b) and c) cases.

We studied the possibility of plasma jet to pass through the metallic substrate (disc) having a hole. The obtained results are presented in figure 5. As it turned out, plasma jet can pass through the hole if its diameter exceeds 3 mm whether the disk is grounded or not. Besides, the disc increases the brightness of the jet between the discharge exit and disc. This effect is more pronounced for the case of the grounded disc. We attribute this effect to an increase the stray capacitance provided by the disc that leads to increasing of both the displacement current between the lateral area of plasma jet and disc and conductivity current flowing through plasma jet to HV electrode.

4. Conclusion

It was shown that interaction of He plasma jet with the metallic substrate placed transverse to the jet depends on the polarity and amplitude of potential applied to the substrate. It has been established that one can change appreciably the plasma bullet velocity and even wholly suppress the bullets in the plasma jet by changing the potential on the substrate. It is found out that the conductivity current flowing through plasma jet depends on the stray capacitance formed around the jet - the higher stray capacitance, the higher conductivity current flowing through plasma jet and higher its brightness.

5. References

[1] Akishev Yu, Balakirev A, Grushin M, Karalnik V, Kochetov I, Napartovich A, Petryakov A and Trushkin N 2015 *IEEE Trans. Plasma Sci.* **43** 745
[2] Korolev Yu 2015 *Russian Journal of General Chemistry* **85** 1311
[3] Lu X, Naidis G, Laroussi M and Ostrikov K 2014 *Phys. Rep.* **540** 123
[4] Teschke M, Kedzierski J, Finantu-Dinu E, Korzec D and Engemann J 2005 *Trans. Plasma Sci.* **33** 310
[5] Akishev Yu, Aponin G, Balakirev A, Grushin M, Karalnik V, Petryakov A and Trushkin N 2013 *J. Phys. D: Appl. Phys.* **46** 464014

Acknowledgments

This work was supported by Russian Foundation for Basic Research (grant № 17-02-00234).

IOP Conf. Series: Journal of Physics: Conf. Series **927** (2017) 012052 doi:10.1088/1742-6596/927/1/012052

The role of secondary processes in kinetics of triplet states of a hydrogen molecule in an ECR discharge

V A Shakhatov[1], Yu A Lebedev[1], A Lacoste[2] and S Bechu[2]

[1]Topchiev Institute of Petrochemical Synthesis of the Russian Academy of Sciences (TIPS RAS), Leninsky Prospect, 29, Moscow, 119991, Russia
[2]LPSC, Universite Joseph Fourier Grenoble 1, CNRS/IN2P3, Grenoble INP, 53, Avenue des Martyrs, 38026 Grenoble, France

E-mail: shakhatov@ips.ac.ru

Abstract. An electronic state sensitive semiempirical collision-radiative model of hydrogen plasma of ECR-discharge is used to analyze the applicability of emission of triplet states of molecular hydrogen for plasma diagnostics. It is shown that secondary processes make the greatest contribution to the kinetics of population-depopulation of triplet states $a^3\Sigma_g^+$, $c^3\Pi_u$, $d^3\Pi_u$, $e^3\Sigma_u^+$, $g^3\Sigma_g^+$, $h^3\Sigma_g^+$, $i^3\Pi_g$ and $r^3\Pi_g$. The secondary processes give the smallest contribution to the excitation and deactivation of triplet states $f^3\Sigma_u^+$ и $k^3\Pi_u$. Thus a simplified coronal model (electron impact excitation followed by radiative decay) can be used to process the intensities of the dipole allowed $f^3\Sigma_u^+ \to a^3\Sigma_g^+, g^3\Sigma_g^+$ and $k^3\Pi_u \to a^3\Sigma_g^+$ transitions. The complicated collision-radiative model should be used for other transitions.

1. Introduction

Interest in the hydrogen-containing low-temperature plasma of the ECR-discharge is associated with the problem of creating the gas-discharge sources of negative atomic hydrogen ions H^- [1]. Studies [2,3] show that the dominant role in the formation of H^- in the ECR-discharge is played by processes involving vibrationally excited hydrogen molecules in the ground electronic state $H_2(X^1\Sigma_g^+,v)$. An important stage in creating and optimizing the parameters of gas-discharge sources H^- is the development of non-perturbing and rapid methods for the diagnosis of $H_2(X^1\Sigma_g^+,v)$ (or vibrational temperature $T_V\left(X^1\Sigma_g^+\right)$ of the first level) in plasma.

Increasingly widespread in the diagnostics of the vibrational temperature of the ground state in hydrogen plasma are experimental-calculation techniques that are based on a combination of emission spectroscopy and collision-radiative plasma models [4-7]. A simplified coronal model is often used to process results of emission spectroscopy [8]. For example, in processing the emission spectra of the Fulcher system of the hydrogen molecule $H_2(d^3\Pi_u \to a^3\Sigma^+_g)$ it is assumed that the radiating state of the $d^3\Pi_u$ state is excited by the electron impact, and deactivated by radiative decay. The presence of additional (secondary) collision-radiative processes may violate this assumption. Neglecting of secondary processes in the simplified models can lead to overestimated values of $T_V\left(X^1\Sigma_g^+\right)$. The question on the role of secondary processes in the mechanisms of excitation of the triplet states of molecules is still not fully understood.

In the present paper this problem is investigated by modeling the processes of population and deexcitation of the triplet states of a hydrogen molecule in an ECR-discharge. It is a continuation of [8], in which studies of the distribution function for the rotational and vibrational levels of a hydrogen molecule in triplet state $d^3\Pi_u$ in the ECR-discharge in hydrogen were performed by emission spectroscopy.

Results of study the problem for nonequilibrium DC and microwave discharges are presented in [7, 9]. From the point of view of the problem under consideration, the ECR-discharge is of particular interest, since the electron energy distribution function (EEDF) in this discharge differs significantly from the EEDF in other discharges in that it is enriched by electrons in the high-energy region compared to other discharges. This can lead to a change in the ratio of the channels in the triplet states kinetics.

2. Experimental conditions

The study of the processes of population and deexcitation of the triplet states of a hydrogen molecule is performed for the conditions described in [1]. ECR discharge was created in a chamber of cylindrical geometry (diameter D = 100 mm) made of stainless steel in a stream of hydrogen in the pressure range 0.001-0.01 Torr. The plasma was isolated from the chamber walls by a Pyrex cylinder. Microwave power in the range 200 - 1200 W at 2.45 GHz in continuous generation mode was supplied to four dipolar sources. The wall temperature of the T_W chamber was kept at room temperature. The value of the translational temperature T_g measured on the discharge axis in the region of dipolar sources measured by the emission spectroscopy method was 600 ± 650 K and decreased to 430 ± 50 K at a distance of 90 mm from them [8]. The EEDF measured by the probe method is bimaxwellian and enriched by electrons in the high-energy region [1]. It is determined by the temperatures (T_e^c, T_e^h) and the concentrations (N_e^c, N_e^h) of the "cold" (T_e^c, N_e^c) and "hot" (T_e^h, N_e^h) electrons corresponding to the low- and high-energy part of the EEDF. When the gas pressure p varied from 0.25×10^{-2} to 0.96×10^{-2} Torr, the concentration and temperature of the "cold" and "hot" electrons measured by the probe method lay in the ranges: N_e^c=(0.13-2.3)×10^{10} cm^{-3} and N_e^h=(0.3-0.8)×10^{10} cm^{-3}; T_e^c=0.6-2.0 eV and T_e^h=2.0-6.0 eV. The ratio of the concentration of the negative ion of the hydrogen atom to the electron concentration measured by the laser-induced photo-detachment in the ECR-discharge is 0.01 – 0.1. The content of atomic hydrogen P_H in the plasma does not exceed 4%. Since the used collision-radiative model is semiempirical, the parameters T_e^c, T_e^h, N_e^c, N_e^h T_g, p (or the total concentration of particles N), D, P_H and T_W are the initial data in the model. kinetics.

3. Semiempirical level-to-level collision-radiative model of hydrogen plasma in ECR-discharge

In the model developed in [7] and used in this paper, the number of electronically excited states of a hydrogen molecule is significantly increased, compared with those described in [4-6], and the kinetic scheme is extended in the part of the collisional-radiative processes of interest for emission spectroscopy of hydrogen plasma. It additionally takes into account processes involving hydrogen molecules in singlet ($N^1\Lambda_\sigma$=$2B^1\Sigma_u^+$, $2C^1\Pi_u$, $3B'^1\Sigma_u^+$, $3D^1\Pi_u$, $4B''^1\Sigma_u^+$, $4D'^1\Pi_u$, $2EF^1\Sigma_g^+$, $3I^1\Pi_g$, $3H\overline{H}^1\Sigma_g^+$, $3GK^1\Sigma_g^+$, $3J^1\Delta_g$, $4P^1\Sigma_g^+$, $4O^1\Sigma_g^+$, $4R^1\Pi_g$, $4S^1\Delta_g$, $O^1\Sigma_g^+$) and triplet ($N^3\Lambda_\sigma$ = $2b^3\Sigma_u^+$, $2c^3\Pi_u$, $2a^3\Sigma_g^+$, $3e^3\Sigma_u^+$, $3d^3\Pi_u$, $3h^3\Sigma_g^+$, $3g^3\Sigma_g^+$, $3i^3\Pi_g$, $4f^3\Sigma_u^+$, $4k^3\Pi_u$, $4r^3\Pi_g$, $4s^3\Delta_g$) states. To determine the level-rate coefficients of the processes describing the collisions of electrons with heavy particles, we used the EEDF, obtained as a result of probe measurements for the experimental conditions from [1]. A list of collision-radiative processes with a corresponding database (or the total concentration of particles N), D, P_H and T_W are the initial data in the model. kinetics.

4. Results and discussion

As a result of calculations it was shown that the value of the electronic relaxation time of the populations of triplet states of the hydrogen molecule does not exceed a tenth of a second. The values of the stationary concentrations of the excited molecules $H_2\left(N^3\Lambda_\sigma\right)$ lie in the range 1.7×10^4 - 8.2×10^7 cm^{-3}. The stationary distributions of the populations of the triplet states $N^3\Lambda_\sigma$ and the vibrational levels of the ground state of hydrogen molecule $X^1\Sigma_g^+$, differ from the Boltzmann distributions. The calculated values of the temperature of the first vibrational level $T_V\left(X^1\Sigma_g^+\right)$ lie in the range from 1700 K to 3000 K. In the plasma under optimal conditions for the formation of vibrationally excited molecules $H_2(X^1\Sigma_g^+, v)$, processes involving $H_2(X^1\Sigma_g^+, v = 1-4)$ can contribute to the excitation kinetics of triplet states $c^3\Pi_u$, $d^3\Pi_u$, $e^3\Sigma_u^+$, $h^3\Sigma_g^+$, $i^3\Pi_g$ and $r^3\Pi_g$.

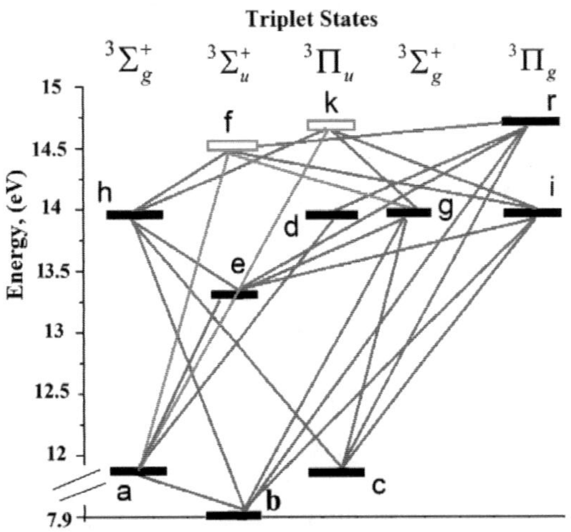

Figure 1. Diagram of radiative transitions between triplet states. Blue lines denote the transitions $f^3\Sigma_u^+ \rightarrow a^3\Sigma_g^+, g^3\Sigma_g^+$ and $k^3\Pi_u \rightarrow a^3\Sigma_g^+$, for processing the intensities of which a simplified coronal model can be applied. Red lines denote the transitions, for processing the intensities of which the semiempirical level-to-level collision-radiative model should be applied.

Calculations show that the mechanisms of excitation and deexcitation differ for different triplet states of the hydrogen molecule and they are determined by volumetric collisional-radiative processes. The composition and number of secondary processes and their hierarchy vary with the gas residence time in the discharge zone τ_D. Secondary processes make the greatest contribution to the kinetics of triplet states $a^3\Sigma_g^+$, $c^3\Pi_u$, $d^3\Pi_u$, $e^3\Sigma_u^+$, $g^3\Sigma_g^+$, $h^3\Sigma_g^+$, $i^3\Pi_g$ and $r^3\Pi_g$. The smallest contribution is made by secondary processes to the kinetics of triplet states $f^3\Sigma_u^+$ and $k^3\Pi_u$. Dipole allowed transitions $f^3\Sigma_u^+ \rightarrow a^3\Sigma_g^+, g^3\Sigma_g^+$ and $k^3\Pi_u \rightarrow a^3\Sigma_g^+$ can be used for spectral diagnostics of hydrogen plasma of ECR-discharge on the base of simplified coronal model. This is reflected in figure 1, which shows the radiative transitions, which can and cannot be used within the framework of the simplified coronal model.

The results of analysis of population-depopulation mechanism of the $H_2(d^3\Pi_u)$ state are of particular interest, since this state is often used in the spectral diagnostics of gas discharges. At $\tau_D \geq 10^{-8}$ s, along with the process (1)

$$H_2(X^1\Sigma_g^+, v = 0) + e \rightarrow H_2(N^3\Lambda_\sigma) + e \text{ , (1)}$$

the processes (2) caused by collisions of the first and second kinds of electrons with excited hydrogen molecules $H_2(d^3\Pi_u)$ and $H_2(g^3\Sigma_g^+)$ begin to play an appreciable role in the excitation-deexcitation mechanism of the $d^3\Pi_u$ state of the hydrogen molecule

$$H_2(d^3\Pi_u) + e \Leftrightarrow H_2(g^3\Sigma_g^+) + e. \quad (2)$$

At $\tau_D \geq 10^{-7}$ s and p=0.25×10^{-2} Torr, N_e^c=0.13×10^{10} cm^{-3}, N_e^h=0.46×10^{10} cm^{-3} и T_e^c=0.7 eV, T_e^h=2.7 eV processes (1, 2) and

$$H_2(d^3\Pi_u) \rightarrow H_2(a^3\Sigma_g^+) + h\nu \quad (3)$$

are responsible for the stationary concentration of hydrogen molecules in $d^3\Pi_u$ state (1.9×10^5 cm^{-3}). The electronic relaxation time of the $d^3\Pi_u$ state population is equal to τ_R^e = 10^{-3} s. With increasing the pressure up to p =0.96×10^{-2} Torr at N_e^c=2.3×10^{10} cm^{-3}, N_e^h=0.8×10^{10} cm^{-3}, T_e^c=2.0 eV и T_e^h=6.0 eV an additional contribution to the kinetics of the production of the $d^3\Pi_u$ state can give a process involving hydrogen molecules $H_2\left(X^1\Sigma_g^+, v=1\right)$

$$H_2\left(X^1\Sigma_g^+, v=1\right)+e \rightarrow H_2\left(d^3\Pi_u\right)+e. \quad (4)$$

Processes (1-4) determine the stationary concentration (~ 9×10^6 см$^{-3}$) of molecules in $d^3\Pi_u$ state. The electronic relaxation time τ_R^e of the $d^3\Pi_u$ state population is of 2×10^{-2} s.

Prevailing of secondary processes (2) in the kinetics of the $d^3\Pi_u$ state of a hydrogen molecule in a wide range of plasma conditions limits the use of the Fulcher system $H_2\left(d^3\Pi_u^- \rightarrow a^3\Sigma_g^+\right)$ for spectral diagnostics of a stationary ECR-discharge in the approximation of simplified coronal model.

This conclusion differs from that obtained earlier for a microwave discharge [7] and a DC discharge [9]. This is due to the fact that the energy distribution of electrons in the ECR discharge is enriched by fast electrons in comparison with the distribution in these discharges. This difference leads to a higher rate of excitation of triplet states, an increase in their concentrations and, correspondingly, to an increase in the role of secondary processes in their population.

5. References

[1] Bechu S, Soum-Glaude A, Bes A, Lacoste A, Svarnas P, Aleiferis S, Ivanov A A and Bacal Jr M 2013 *Physics of Plasmas* **20** 101601
[2] Bacal M and Wada M 2015 *Appl. Phys. Rev.* **2** 021305
[3] Nicolopoulou E, Bacal M and Doucet H J 1977 *J. Physique* **38** 1399
[4] Shakhatov V A and Lebedev Yu A 2011 *High Temperature* **49** 257
[5] Shakhatov V A, Lebedev Yu A, Lacoste A and Bechu S 2015 *High Temperature* **53** 569
[6] Shakhatov V A, Lebedev Yu A, Lacoste A and Bechu S 2016 *High Temperature* **54** 120
[7] Shakhatov V A and Lebedev Yu A 2017 *High Temparature* **55** 3
[8] Shakhatov V A, Lebedev Yu A, Lacoste A and Bechu S 2016 *High Temperature* **54** 467
[9] Shakhatov V A and Lebedev Yu A 2017 *Plasma Phys. Rep.* **43** (to be published)

Antimicrobial properties of nanostructured plasma condensates from medical implants

I I Shamsutdinov[1], L T Bayazitova[2,3], M M Grebenshchikova[1], M M Mironov[1]

[1]Department of Plasma Technology and Nanotechnology of High Molecular Weight Materials, Kazan National Research Technological University, Kazan, Russia
[2]The federal budgetary research institution "Kazan Research Institute of Epidemiology and Microbiology" Kazan, Russia
[3]Federal State Budget Educational Institution of Higher Education "Kazan State Medical University" Kazan, Russia

E-mail: mironov.medinstrument@yandex.ru

Abstract. Thin film coatings of titanium nitrides and hafnium for implants were investigated. The HfN + TiN coatings were condensed from the plasma phase. Analysis of the structure and composition revealed multilayers of the coating, the composition of the layers, and the presence of columnar crystals. The coating thickness is 2-5 μm, with a characteristic size of grains 20-50 nm. Antimicrobial properties for *staphylococcus* and *pseudomonas aeruginosa* have been established, a technique for determining the antimicrobial properties of coatings has been demonstrated, and medical approbation of implants with a coating has been carried out.

Superhard thin-film coatings made of nitrides Ti, Al, Cr, Zr, Mo are widely used in the production of metal-cutting tools made of heat-resistant steels and sintered metal-ceramic alloys. The world leaders in the development of technology and equipment in this area are Oerlikon Balzers AG and Diener Elektronik gmbh, Ion Bond gmbh. Condensation of coatings is performed on magnetron installations and installations with electric arc evaporation of metal.

The use of coatings for medical devices is limited, due to the diversity and specificity of medical and technical requirements [1], including the assessment of biological effects in accordance with GOST (ISO) 10993-2011. The technology of plasma condensation with ion bombardment (CIB) and surface cleaning with high-energy metal ions Ti, Cu, Cr allows to significantly increase the adhesion of the coating to the substrate. Prospective is the use of thin-film coatings to protect the human body from the migration of toxic elements Cr, Mo, Co from steel V and Mo from titanium implant alloys, which can cause negative reactions of the organism with the development of metallurgy, allergic conditions and inflammation [2].

The condensation of the investigated coatings [3] was carried out in a nitrogen atmosphere at a pressure of about 10^{-1} Pa in an installation with three electric arc evaporators. Surface cleaning was carried out with titanium ions to a temperature of titanium implants not exceeding 600 ° C, i.e. the beginning of the phase changes leading to the growth of the grain of the β-structure. For single-use skeletal traction locks of Cr18N9 steel, the temperature during cleaning was reduced to 300-400 ° C.

Measurement of the microhardness of the coating showed a value of 14-30 GPa, with a thickness of 2-5 μm. The thickness was measured both on the samples from the polycorundum, and on the bends of the spokes, screws, osteosynthesis plates, and the cleavage of the coating, fixed by an electron microscope. Estimation of porosity showed that the pore volume in the coating does not exceed 1-2% at a thickness of 2 μm. The pores were fixed on the scaffold by microscopic survey and are seen in Figure 1. It should be noted that the thickness of the coating on the inner surface of the hole in the osteosynthesis plates was understated and amounted to about 1 μm. Multilayer nanostructured coatings consist of columnar crystals of titanium and hafnium nitrides with hetero-thermal chemical bonds and have characteristic dimensions of 5-20 nm.

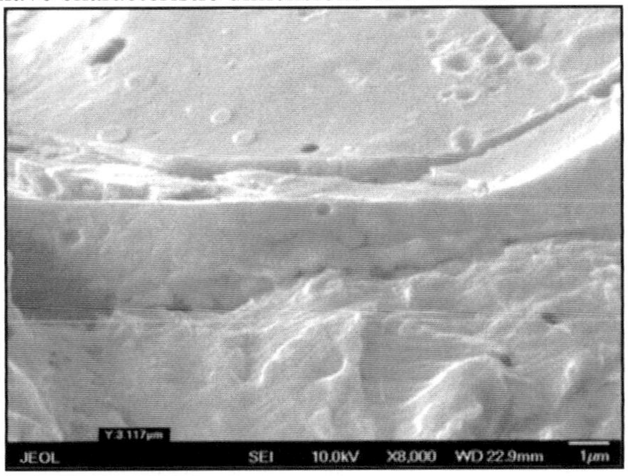

Figure 1. SEM image of fracture of the coating and the type of closed pores.

Investigations of the structure of the coating, both on the surface and on the thickness, revealed the characteristic dimensions of the HfN + TiN crystals of 20-50 nm and the height in the thickness of the coating (Figures 2 and 3).

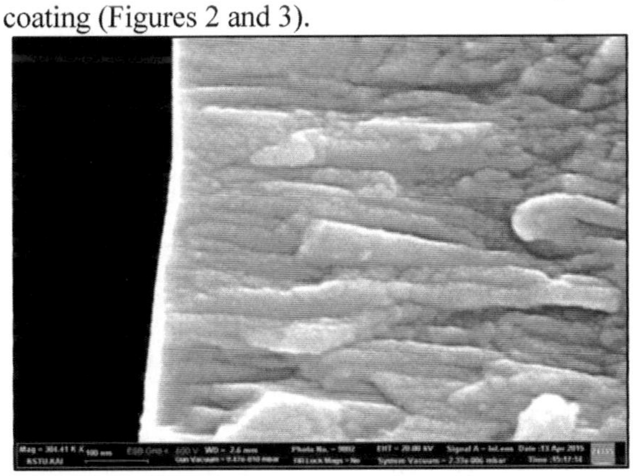

Figure 2. SEM image of column crystals of nitrides.

The layer composition in the electron spectrum was analyzed by a high-resolution energy dispersion spectrometer and is shown in Figures 4 and 5. The coating cross-section in the thickness of the Ga ion flux established a multi-layer coating consisting of nitride layers with a different layer composition and a layer thickness of 20-30 nm. At the same time, the integrity of the crystal is not disturbed, but a transition is visible, a thin layer that is transitional in composition. X-ray diffraction analysis recorded the phases of HfN, $TiN_{0,9}$, and $TiHfN_2$ with crystal growth by the crystallographic plane 111 parallel to the base. The composition analyzer established a molar ratio of titanium to hafnium in equal parts with a deviation of 10%.

The guarantee of the absence of toxicity is the study of the migration of hafnium and titanium ions from the coating surface and toxic vanadium from the titanium alloy through the coating. The concentration of hafnium ions from the 20-day stretch coating was less than 10^{-9} mol / liter, and the vanadium concentration decreased by 5 times in comparison with the uncoated sample [4]. This characterizes the barrier effect of the coating, and a small migration of ions - biocompatibility.

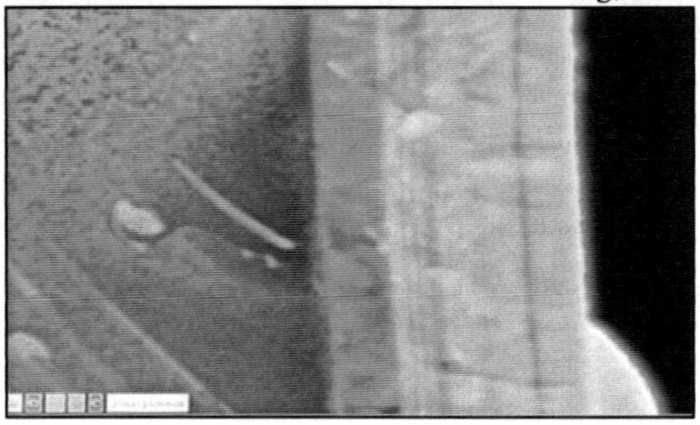

Figure 3. Coating section and composition analysis point.

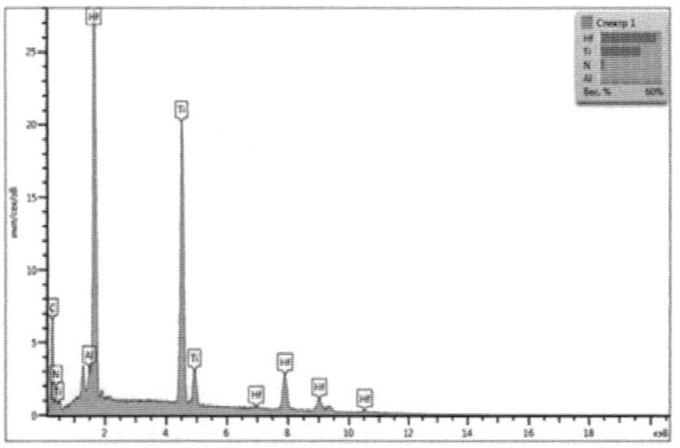

Figure 4. Spectrum of the energy of electrons at the point of analysis.

Preclinical tests on animals showed the main effects and advantages of implants coated with titan-hafnium nitrides:

1) Accelerated wound healing;
2) Absence of allergic reactions;
3) Decreased reactive protein values that do not exceed the normative values in comparison with the increased index from the materials of other implants.

The absence of toxic effects of coatings containing titanium nitride and hafnium on the hepatobiliary system, biochemical indicators of blood give rise to further introduction of nitride coatings, expansion of their application spectrum and types of osteosynthesis with intramedullary, bone and extra-focus fixation [7]. Positive biological activity of nitride titanium-garnet coatings is related to their antimicrobial properties.

At present, there are no standardized methods for reliable determination of the degree of antibacterial action of metal-like antimicrobial surfaces. Testing the degree of sensitivity of microbial culture to antimicrobial drugs is carried out in accordance with MUK 4.2.1890-04 "Determination of the sensitivity of microorganisms to antibacterial drugs." The degree of sensitivity of bacteria to antimicrobial drugs is determined by the diameter of the inhibition zone (absence) of growth of test strains on the surface of the Muller-Hinton culture medium. When measuring the zones, growth retardations are oriented towards the zone of complete suppression of visible growth. The initial suspension of the inoculum contains 1.5 x 10^{8} colony forming units - CFU / ml, which corresponds to 0.5 units according to the McFarland standard. Applying student criterion 0.95, you can compare the results of studies and obtain reliable indicators.

The essence of the technique developed for metal and metal-like coatings is the direct contact of the coating with the culture suspension of the microorganism throughout the experiment, i.e. the coating is immersed in a suspension of the culture suspension of the test microorganism. From special conditions it is necessary to allocate: 1) the concentration of microflora is reduced to 10^{5} CFU / ml; 2)

The metal of the substrate for coatings is limited by the titanium alloy VT6 and steel Cr18N9; 3) in the experiment, tenfold dilutions of the tested microbial culture are used; After incubation, a seeding is carried out on a solid nutrient medium followed by counting the colonies. The effect is on the multiplying microflora in the growth stage.

In the experiment, titanium substrates of VT6 alloy, 20X20 mm in size, were used with a two-sided coating of nitrides of titanium and hafnium with their ratio in the coating 50% to 50% by weight of metals and a coating thickness of 3-4 μm. In control - in the absence of coatings on the plates, a normal growth of culture takes place in the nutrient medium with a sharp increase in its concentration exceeding 10^8 CFU / ml. The results are illustrated in Figure 6. Growth starts quickly within 4-6 hours and lasts several days until the nutrient medium is completely depleted, after which the death of microorganisms begins (decrease in CFU). The last increase in microflora growth, necrotic growth, is caused by reproduction of the culture on the products of dead bacteria. A completely different picture is observed in the presence of nitride coated plates in a medium with a microflora. In the first hours there is an increase in the number of microflora by about an order of magnitude, but then there is an intensive dying out of microorganisms up to the complete destruction of the culture. This is the case with staphylococcus and, to a lesser extent, Pseudomonas aeruginosa, which is due to the difference in the structure of the cell wall.

Hospital strains of microflora showed similar sensitivity to the coating material in the experiment. The antimicrobial effect of the coating was confirmed, despite the known resistance of hospital strains to antibiotics. Antimicrobial effect from coatings from titanium nitride is not revealed.

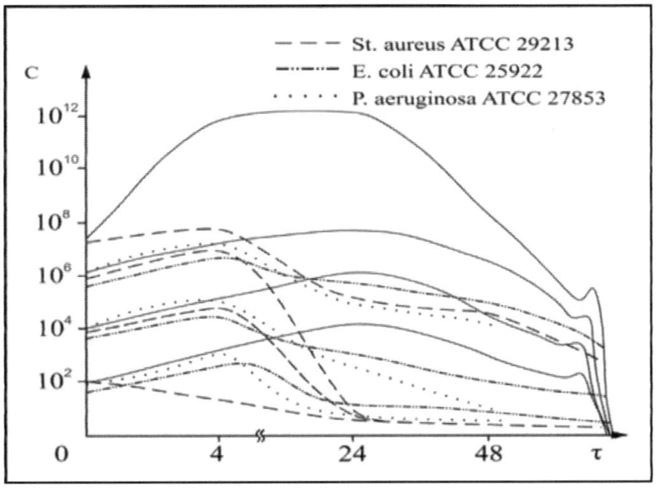

Figure 5. Dependence of the C-concentration of CFU on the growth time (τ)

The fungicidal effect of the titanium naphtha coating [5] was evaluated on samples of genuine leather with this coating. The coating had a scaly, rather than a continuous structure and the dimensions of its structural elements did not exceed 2 μm [6]. Peculiarities of experiments with pathogenic cultures of fungi consist in a low rate of their growth and resistance to external factors. In the experiment, a pathogenic mycotic culture from the trichophyton family was used. The evaluation results of the experiment showed the presence of fungicidal properties in the nitride titanium-garnet coating.

According to the experimental data, the estimated antimicrobial efficacy of the nitride titanate coating is approximately 40 times lower than the antimicrobial activity of benzylpenicillin.

References

[1] I.Sh.Abdullin, M.M.Mironov, G.I. Garipova Antibacterial and biologically resistant coatings for medical implants and instruments. Medical Technik 2004, № 4. pp. 20-22.

[2] I.F.Ahtamov, P.S. Andreev, E.B. Gatina, E.I. Aliev, The first experience of testing of implants coated with nitrides of titanium and hafnium. Practical medicine, 2015, № 4. pp. 21-24.
[3] Material of bactericidal coating. The patent for the invention № RU 2554773 from 27.06.2015. Authors: Mironov M.M. and others
[4] M.M. Grebenshchikova, M.M. Mironov, E.V. Starodumova , Study of the metal ions migration from protective nanostructured coatings for implants. Bulletin of the Technological University 2016, vol.16, № 17, pp. 49-50.
[5] V.P. Veinov, L.T. Bayasitova, E.B. Haldeeva, M.M. Mironov, M.M. Grebenshchikova, Antimicrobial activity of the coatings with the hafnium. Scientific session KNRTU 2016, pp. 486
[6] M.M. Mironov and M.M. Grebenshchikova. Structure and properties of the surface of collagen. Journal of Physics: Conference Series, Volume 789 (2017), conference 1.
[7] I.F. Ahtamov, M.M. Mironov, A.I.Uocef Evaluation of blood markers on osteosynthesis by implants coated with superhard metals. Practical medicine 2016, № 3. . pp. 121-124.

IOP Conf. Series: Journal of Physics: Conf. Series **927** (2017) 012054 doi:10.1088/1742-6596/927/1/012054

One-step synthesis of N-doped graphene in a plasma jet reactor

M B Shavelkina, R Kh Amirov, S L Kanashenko, A V Naumkin, I R Vahitov , A I Gumarov, I V Yanilkin and L R Tagirov

[1]Joint Institute for High Temperatures of Russian Academy of Sciences, Izhorskaya Street 13/2, 125412 Moscow, Russia
[2] Research Institute of Biomedical Chemistry, Pogodinskaya Street 10/8, 119121, Moscow, Russia
[3]A.N.Nesmeyanov Institute of Organoelement Compounds of Russian Academy of Sciences, Vavilova Street 28, 119991, Moscow, Russia
[4]Institute of Physics of Kazan Federal University, Kremlyovskaya Street 16a, Kazan, Russia

E-mail: mshavelkina@gmail.com

Abstract. The possibility of doping graphene during its synthesis in a plasma jet of nitrogen has been studied. Direct current plasma torch with power of up to 40 kW was used as plasma jet generator. The source of carbon was propane-butane mixture, acetylene or methane. Synthesized materials are characterized by scanning electron microscopy, X-ray photoelectron spectroscopy and thermogravimetric analysis. It has been shown that XPS spectra of graphene flakes produced in nitrogen plasma differ in atomic nitrogen content. The maximum degree of nitrogen doping of graphene was obtained at decomposition of acetylene at 77 Torr.

1. Introduction

Graphene is a single graphite plane in which sp^2-hybridized carbon atoms form a hexagonal lattice. For the use of graphene in the majority of the applications, a purposeful modification of its electronic structure is required. This has led to an active search for methods of graphene chemical modification by various atoms and functional groups. Obtaining functionalized single-layer and multilayer graphene using efficient cost-effective methods has opened up new possibilities in the production of multifunctional nanostructured carbon materials, dispersions and hybrids of carbon and polymers that find their use in machinery building, catalysis, energy engineering and biomedicine, production of protective coatings, intellectual materials and systems [1].

In this paper, the possibility of graphene materials doping with nitrogen directly in the course of synthesis in a plasma jet is investigated. In [2], the efficiency of graphene synthesis was demonstrated using helium and argon as plasma-forming gas. In contrast to [2], in the present work nitrogen is used as a plasma-forming gas.

2. Methods

To synthesize N-graphene, a plasma jet reactor was used on the basis of a powerful dc plasma torch with an expanding anode channel [2]. The precursor is introduced into the plasma torch simultaneously with a plasma-forming gas. Hydrocarbon decomposes both in the arc discharge region and in the plasma jet to form a vapor-gas mixture, which, as it cools condenses in the form of flakes in the collector. Pressure in the reactor was 77-710 Torr and was maintained under dynamic pumping by a water pump. The arc current was kept constant and corresponded to 300-350 A. Power of the plasma torch reached 35 kW. Methane, mixture of propane-butane and acetylene were used as a source of carbon. Nitrogen was a plasma-forming gas. The rate of hydrocarbons and nitrogen varied independently of each other. The synthesis parameters are presented in table 1. For each experiment, the hydrocarbons were introduced after the temperatures in the collector would reach 350-400 °C. Monitoring the collector temperature was carried out using the chromel-alumel and chromel-copel thermocouples.

Table 1. Technological conditions.

Power (kW)	Current (A)	Voltage (V)	Gas pressure (Torr)	Nitrogen flow rate (g/s)	Hydrocarbons flow rate (g/s)
30-40	150 -450	60-110	77-740	1.5-2.0	0.05-0.4

The synthesis products were examined by traditional methods for studying doped graphene [11]. To directly register the dimensional parameters and images of plasma torch pyrolysis products, a scanning electron microscope (SEM) - Hitachi S5500 with in-lens technology was used to obtain both ultra-high resolution and high sensitivity of EDS analysis. The survey was conducted on 3 detectors (SE-mode, BF-STEM (light-field mode), DF-STEM (dark-field mode).

To evaluate the efficiency of the synthesis, phase composition and thermal stability of carbon products, thermal analysis methods were used: DSC and TG (STA 449 F3 Jupiter platform from Netzsch) with linear heating in air at 10 K/min in temperature range of 20-1000 °C.

X-ray photoelectron spectroscopy (XPS) was used for quantitative study of the elemental composition and chemical state of atoms. XPS measurements were taken in an ultrahigh vacuum chamber ($\sim 5 \times 10^{-10}$ mbar) equipped with an X-ray source (Mg Kα, 12.5 kV, 250 W) and a hemispherical energy analyzer Phoibos 150 SPECS Gmb Germany).

Nitrogen plasma spectra were taken at a distance of 2 mm from the cathode on a three-channel AvaSpec2048 optical fiber spectrometer with a spectral resolution of 0.2 nm in the spectral range of $200 \div 1100$ nm. Using a quartz condenser f = 150 mm, a sharp image of the plasma pinch was formed on a scale of 1.0 : 1.0. The input end of AvaSpec 2048 light guide perpendicular to the plane of this image could move in this plane, cutting out the necessary glow zone with a 1 mm diameter collimator aperture.

To determine the electron temperature, we used the relative intensities approach to the lines of particles with an equal multiplicity of ionization (method of Boltzmann exhibitors).

3. Results and discussions

The design feature of the plasma torch is the expanding anode channel, it provides an increasing volt-ampere characteristic (VAC) of the discharge in the region of high currents. Figure 1 shows the VAC at two pressures - 350 and 710 Torr. As it is seen from the figure, at a pressure of 710 Torr the dependence has a descending character in the region of small currents, while at currents exceeding 350 A the VAC is increasing. At a pressure of 350 Torr, the increasing character of VAC is observed starting from 200 A. The increasing character of VAC provides for a stable operation of the plasma torch. The use of the expanding channel of the output electrode in the plasma torch makes it possible to substantially increase the velocity of the cold gas at the inlet to the channel and intensify the heat exchange between the arc and the plasma-forming gas, which contributes to a decrease in the arc length, more even distribution of the heat fluxes into the electrode walls, and, in general, ensures the stability of the arc burning. Figure 2 illustrates the stability of the electrical parameters of the plasma torch. Voltage fluctuations do not exceed 5%.

Figure 3 shows an example of the emission spectrum of nitrogen plasma and plasma with addition of methane. Lines H are clearly visible in the spectrum, and when the carbon precursor is added, the lines CN (380 nm) and C appear.

Based on the recorded spectra, the plasma temperatures and electron concentrations were determined. For nitrogen plasma without addition of the hydrocarbons , the electron temperature was 12,000 K (pressure 740 Torr). Electron concentration was 1.5 x 10^{17} cm^{-1}. When methane was added at a rate of 0.3 g/s, the temperature dropped to 9500 K, and the electron concentration was 5.8 x 10^{16} cm^{-1}.

Figure 4 presents the morphology of the carbon deposit collected on the collector. It is seen that the precipitate produced at propane-butane decomposition at 350 Torr is presented by flakes with the size of 200 to 500 nm. The observed morphology is typical for the synthesis of graphene using plasma jets. This confirms the possibility of plasma-jet synthesis without the use of catalysts and which does not require an additional purification stage.

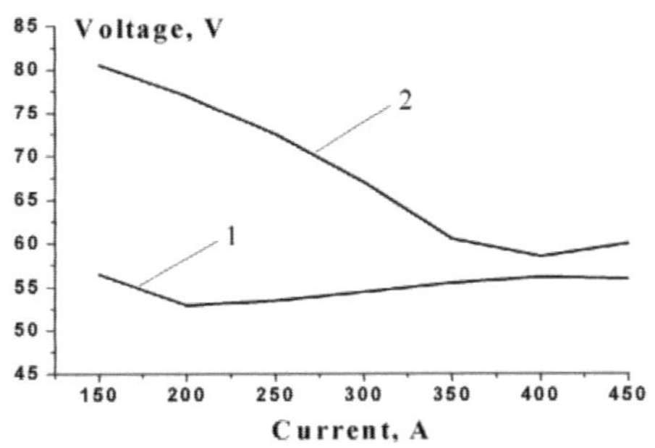

Figure 1. Dependence of the arc burning voltage on the current at pressure change and constant flow rate of nitrogen (2 g / s).

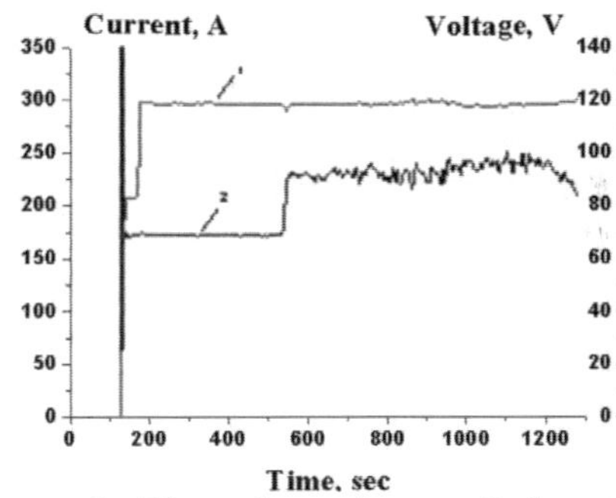

Figure 2. Time dependence of the current intensity and the arc burning voltage at a pressure of 710 Torr.

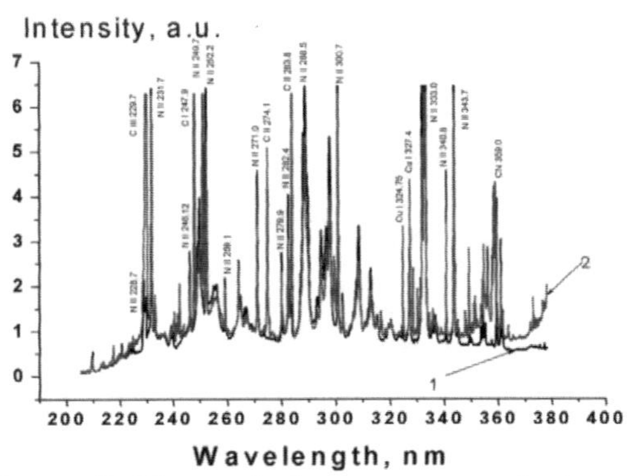

Figure 3. The emission spectrum of nitrogen plasma with addition of methane (740 Torr, 300A, 0.3 g/s).

Figure 4. STEM image of graphene sheets produced at propane-butane decomposition at 350 Torr.

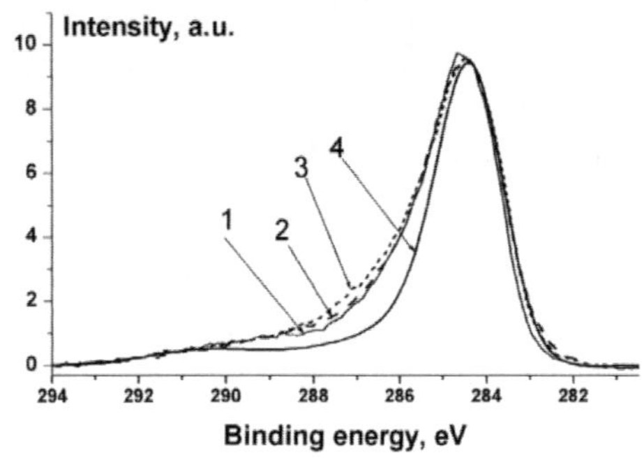

Figure 5. Photoelectron C1s spectra of synthesized samples: 1- sample 1; 2 - sample 2; 3 - sample 3; 4 – pure HOPG.

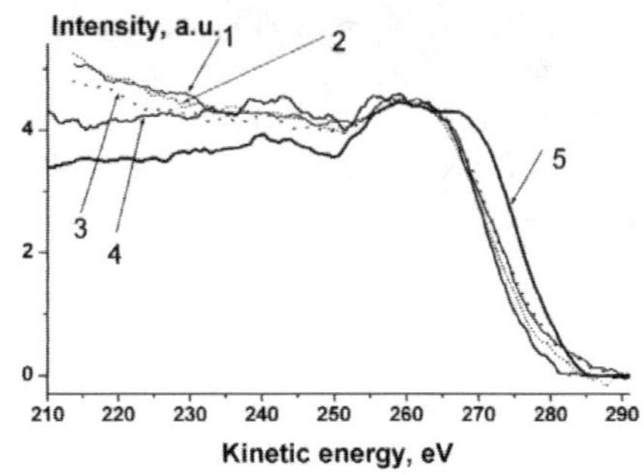

Figure 6. KVV Auger spectra: 1- sample 1; 2 - sample 2; 3 - sample 3; 4 – HOPG, 21.875°; 5 – HOPG, 78.125°.

Figures 5-6 present photoelectron C 1s spectra and LVV Auger spectra of samples synthesized by decomposition of propane-butane mixture and acetylene at 77-710 Torr (table 2) and highly oriented pyrolytic graphite (HOPG). The C1s spectra of HOPG were used to determine the fraction of sp^2 hybridization and compensation for surface charging. The Auger spectra were used to determine the interaction between graphene layers.

Table 2. The proportion of the sp^2-state and relative atomic concentrations of the elements.

Sample number	Synthesis condition	sp^2, rel. unit	C	O	N	Cu	Si
1	77 Torr, C_2H_2	0.79	85.3	5.6	8.0	1.1	
2	150 Torr, C_2H_2	0.83	75.9	11.2	3.8		9.1
3	350 Torr, C_3H_8-C_4H_{10}	0.79	87.4	7.2	3.6	1.8	

The sp^2 hybridization degree for samples 1, 2 and 3 was 0.79, 0.83 and 0.79, respectively (table 2). The differences in the C 1s spectra are due to the difference in the relative concentrations of carbon groups containing oxygen (figure 5). Figure 6 shows fitting the C 1s spectra of sample 3 into components that correspond to the C-N, C-O-C/C-OH, O-C-O/C=O and C(O)O groups [3]. The presence of oxygen relates to the leakage of air during dynamic evacuation.

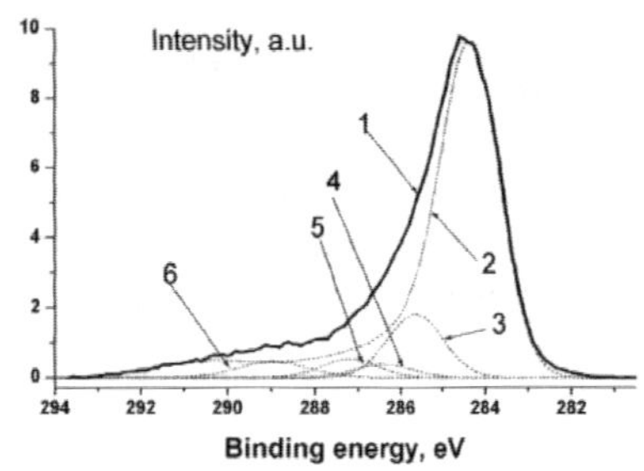

Figure 7. The C 1s spectrum decomposition into components; 1- sample 3, 2 – sp^2-state, 3 - C-N, 4 - C-O-C/C-OH , 5 - O-C-O/C=O, 6 – C(O)O.

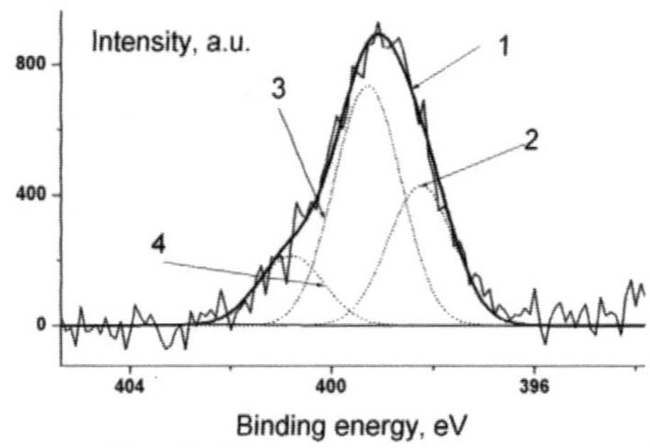

Figure 8. The N 1s spectrum decomposition into components; 1 - sample 3, 2 – pyridinic, 3 – C-NH, 4 – graphitic.

According to the C 1s spectra, the atomic percentage of nitrogen in N-graphene consisted is 8%.

As we have seen from figure 6 the C KVV Auger spectra of the investigated samples are close to that of HOPG, measured at an angle of 21.875° corresponding to the information depth close to one layer of graphite; it indicates a weak interaction of π-systems of layers of the investigated samples. This interaction, on one hand, can be caused by a small number of graphene layers in packages and/or, on the other hand, by the presence of oxygen and nitrogen, which interfere with their interaction. For samples 2 and 3, the binding energies of the C 1s spectrum of HOPG inscribed in the original photoelectron spectra at the best coincidence of the low-energy edges and the satellite-normalized intensity are 284.40 and 284.37 eV, respectively, which indicate their good electrical conductivity. For sample 1, the corresponding value is 284.57 eV. According to [3], the shift of the peaks toward higher binding energies indicates a decrease in the conductivity of the samples under the action of photons and secondary electrons.

It should be noted that, despite the same relative concentrations of the state with sp^2 hybridization in the spectra of C 1s of samples 1 and 3, they differ in relative concentrations of oxygen- and nitrogen-containing carbon groups, which follows from the results of the decomposition of spectra C 1s into components.

When analyzing the XPS spectra of N 1s (figure 8), several states were identified. Peaks located at 398.2, 399.5 and 400.8 eV refer to pyridinic, amino and graphitic nitrogen, respectively. The binding energies are in accordance with the published values [4 - 6].

4. Conclusion

The research study for direct synthesis of N-graphene during decomposition of hydrocarbons in a plasma jet reactor based on a dc plasma torch with an expanding anode channel with power of up to 40 kW was carried out. The electric characteristics of the plasma torch were studied depending on the current, pressure, nitrogen flow rate and carbon source (ethanol, propane-butane, methane, acetylene). Spectral measurements of plasma arc radiation in the plasma torch in the visible spectral range were performed depending on the carbon source (ethanol, methane). The resulting deposit in the plasma jet was presented by flakes containing from 2 to 5 layers of graphite. Study of the thermal stability of N-graphene system shows that the nitrogen impurity significantly increases the temperature stability limit compare to that of pure graphene. Photoelectron spectroscopy has shown that in the synthesized N-graphene, pyridine nitrogen, which is not an electron donor, predominates. The dependence of the synthesis parameters on the electronic characteristics of the obtained N-graphene has been established.

5. Acknowledgements

The authors gratefully acknowledge Russian Foundation for Basic Research for the support by grants No 15-08-00165, No 16-08-00145.

6. References

[1] Wang H, Yuan X, Wu Y, Huang H, Peng X, Zeng G, Zhong H, Liang J and Ren M M 2013 *Adv. Colloid Interfac.* **195–196** 19
[2] Amirov R, Isakaev E, Shavelkina M and Shatalova T 2014 *J. of Phys.: Conf. Ser.* **550** 0120
[3] Wei D C, Liu Y Q, Wang Y, Zhang H L, Huang L P and Yu G 2009 *Nano Lett.* **9** 1752.
[4] Li X L, Wang H L, Robinson J T, Sanchez H, Diankov G and Dai H J 2009 *J. Am. Chem. Soc.* **131** 15939
[5] Zhang, L S, Liang X Q, Song W G and Wu Z Y 2010 *Physical Chemistry Chemical Physics* **12** 12055
[6] Artyushkova K, Kiefer B, Halevi B, Knop-Gericke A, Schlogl R and Atanassov P 2013 *Chemical Communications* **49** 2539

Mathematical modelling of RF plasma flow at low pressure with electrodynamics

A Yu Shemakhin[1], V S Zheltukhin[2]

[1]Kazan Federal University, 18 Kremlevskaya Street, Kazan, 420008, Russia
[2]Kazan State Technological University, 68 Karl Marx Street, Kazan, 420015, Russia

E-mail: shemakhin@gmail.com

Abstract. The mathematical model of the RF plasma at low pressure in both free-molecule and transition flow at Knudsen $0.03 \leq Kn \leq 3$ is described. The model is based on the statistical approach for the neutral component of the plasma together with the continuum model for electron, electromagnetic field and metastable components. Results of plasma flow parameters calculations and testing results of electric field calculations are presented.

1. Introduction

RF plasma discharges at low pressures (p = 0.15 − 150 Pa) is widely used for the modification of various materials: dielectrics, conductors, and semiconductors. The plasma has the following characteristics: ionization degree is 10^{-7}–10^{-5}, electron density $n_e=10^{15}$–10^{19}m^{-3}, electron temperature T_e=1–4 eV, temperature of atoms and ions in the plasma bunch $T_a=(3\text{-}4)\cdot 10^3$ K, in the plasma jet $T_a=(3.2\text{-}10)\cdot 10^2$ K [1-6]. One of important aspect is to investigate how the electromagnetic field influence on the plasma flow. For this purpose, equations for electric field was included with the system of equations of the plasma flow and it was made testing calculations ones.

.

2. Mathematical model

Hybrid model of the RF plasma is constructed, which includes kinetic model for carrier gas and continuous model for electron gas, metastable atoms and electric field. The model includes:

1) The Boltzmann's transport equation for neutral atoms

$$\frac{\partial f}{\partial t} + \mathbf{c} \cdot \frac{\partial f}{\partial \mathbf{r}} + \tilde{\mathbf{F}} \cdot \frac{\partial f}{\partial \mathbf{c}} = S(f),$$
$$f(\mathbf{c}, \mathbf{r}, 0) = f_0(\mathbf{c}, \mathbf{r}), \quad \tilde{\mathbf{F}} = -(1/m_a) \text{ grad } W_T \tag{1}$$

2) The equation of the electron gas continuity:

$$\frac{\partial n_e}{\partial t} - \text{div}(D_a \text{grad}\, n_e - \mathbf{v}_a n_e) = \nu_i n_e + R_1 n_e n_a + R_2 n_m^2 + R_3 n_m n_e - R_4 n_e^2 - R_5 n_e^3,$$
$$n_e\big|_{inlet} = n_{e,inlet}, \; n_e\big|_{outlet} = 0, \; n_e\big|_{walls} = 0, \quad n_e\big|_{body} = 0, \; n_e\big|_{t=0} = n_{e0} \tag{2}$$

3) The equation of the electron heating:

$$c_p \rho_e \frac{\partial T_e}{\partial t} - \text{div}(\lambda_e \text{ grad}\, T_e - \frac{5}{2} k_B n_e T_e \mathbf{v}_e) + \frac{3}{2} k_B \delta \nu_c n_e (T_e - T_a) = \sigma E^2 - \nu_i n_e E_I - I_1 R_3 n_m n_e, \tag{3}$$

$$T_e\big|_{inlet} = T_{e,inlet},\ T_e\big|_{outlet} = T_{e,room},\ T_e\big|_{walls} = T_{e,room},\ \frac{\partial T_e}{\partial \mathbf{n}}\bigg|_{body} = 0,\ T_e\big|_{t=0} = T_{e0},$$

4) The equation of the metastables continuity:

$$\frac{\partial n_m}{\partial t} - \mathrm{div}\left(D_m \mathrm{grad}\,(n_m)\right) = R_6 n_e n_a - R_2 n_m^2 + R_3 n_m n_e - R_7 n_m - R_8 n_m n_a - R_9 n_m \tag{4}$$

$$n_m\Big|_{inlet,\ outlet,\ sides} = 0,$$

5) The equation of electric fields described by the inhomogeneous wave equation known as Telegraphers equation[7].

$$\left(\Delta - \mu_0\varepsilon_0 \frac{\partial^2}{\partial t^2}\right)\mathbf{E}(\mathbf{r},t) = \mu_0 \frac{\partial}{\partial t}\mathbf{j}(\mathbf{r},t), \tag{5}$$

System of equation (1)-(5) is closed by the following formulae

$$p_a = n_a k_B T_a,\quad \mathbf{v}_a(\mathbf{r},t) = \int_{-\infty}^{\infty} \mathbf{c} f(\mathbf{c},\mathbf{r},t) d\mathbf{c},\quad \mathbf{v}_e = \mathbf{v}_a - (D_a / n_e)\mathrm{grad}\, n_e. \tag{6}$$

Here $\mathbf{E}(\mathbf{r},t)$ is the electric field and $\mathbf{j}(\mathbf{r},t)$ is the current, **c** and **r** are vectors of velocity and coordinates of atoms, $f(\mathbf{c}, \mathbf{r}, t)$are function of distribution of neutral atoms by velocities, f_0 is Maxwell function of distribution by velocities, $S(f)$is collisions integral, $\tilde{\mathbf{F}}$ is force, which influence on neutral atoms in elastic collision process, W_T is energy, which transferred to atoms in elastic collision process, n_e is density of electrons, D_a is ambipolar diffusion coefficient, λ_e is coefficient of thermal conductivity of electron gas, ν_i is ionization frequency, c_p is heat capacity of the electron gas, ν_c is the frequency of elastic collisions of electrons with atoms and ions, σ is plasma conductivity, **E** is the electric field vector, $E=|\mathbf{E}|$, E_i is ionization potential, k_B is Boltzmann constant, $\delta = m_e / 2m_a$, m_a is atom mass, m_e is electron mass, R_{vk} is radius of the cylindrical vacuum chamber, L_{vk} is length, R_{rk} is the radius of the plasma torch outlet, subscripts *inlet*, *outlet*, *body*, *walls* devoted parameter value on inlet and outlet of the chamber, on the walls of the sample, and the vacuum chamber, respectively, $W_T = \int E_c dV dt$, dV is volume element, $E_c = \frac{3}{2} k_B \delta \nu_c n_e (T_e - T_a)$ is energy of elastic collisions transport. Coefficients D_a, ν_i, λ_e are functions of electron temperature T_e, I_1 =11.56 eV is excitation energy, R_1 is impact ionization rate coefficient, R_2 is Penning ionization rate coefficient, R_3 is step ionization rate coefficient, R_4 is photo recombination rate coefficient, R_5 is triple recombination rate coefficient, R_6 is excitation of metastables rate coefficient, R_7 is radiative recombination rate coefficient, R_8 is collisional quenching rate coefficient, R_9 is electron deexcitation rate coefficient.

The plasma is quasi-neutral and therefore approach (5) is applicable.We assume that **E** field and **j** current are changing harmonically in time, and can be used complex amplitude method. Given that

$$\mathbf{E}(\mathbf{r}) = \mathbf{E}_{\mathrm{Re}}(\mathbf{r}) + i\cdot\mathbf{E}_{\mathrm{Im}}(\mathbf{r}),\ \mathbf{j} = \sigma \cdot \mathbf{E}(\mathbf{r}) \tag{7}$$

we have the equations for real and the imaginary part of **E** field:

$$\Delta\,\mathbf{E}_{\mathrm{Re}}(\mathbf{r}) = \mu_0\varepsilon_0\varepsilon\omega^2 \cdot \mathbf{E}_{\mathrm{Re}}(\mathbf{r}) + \sigma\mu_0\omega \cdot \mathbf{E}_{\mathrm{Im}}(\mathbf{r}),$$
$$\Delta\,\mathbf{E}_{\mathrm{Im}}(\mathbf{r}) = \mu_0\varepsilon_0\varepsilon\omega^2 \cdot \mathbf{E}_{\mathrm{Im}}(\mathbf{r}) - \sigma\mu_0\omega \cdot \mathbf{E}_{\mathrm{Re}}(\mathbf{r}). \tag{8}$$

Values of the electric fields on the boundarycan be calculated using Biot-Savarts law for the electric field's vector potential:

$$\mathbf{E}(\mathbf{r}) = -i\omega \frac{\mu_0}{4\pi} \int \frac{\mathbf{j}(\mathrm{s})_{coil} + \mathbf{j}(\mathrm{s})_{el}}{|\mathbf{r} - \mathrm{s}|} d^3\mathrm{s}, \tag{9}$$

where $\mathbf{j}(\mathrm{s})_{coil}$ is current on the coil and $\mathbf{j}(\mathrm{s})_{el}$ is current of the electrons.

3. Metod of calculation and results

Bird's method [8] is modified by introducing the spatial heat source power W_T. A four-step iterative process is constructed for solving the problem (1) - (6). Equation (1) is solved by modified Bird's method. Equations (2)-(5) is solved by finite volume method (FVM) with specified numerical scheme. Software package for finding the spatial distribution of the main plasma characteristics is developed. The package use OpenFOAM [9] library environment and works under the Linux operating system.

The gas-dynamic characteristics of low pressure RF plasma both an undisturbed jet and the jet with overflowing sample in the vacuum chamber of R_{vk}=0,2 m, L_{vk}=0,5 m and R_{gk}=0,012 m, on the center of the base plate is carried out. A cylindrical sample of radius R_b=0,03 m and a height L_b=0.02 m located in the plasma jet at a distance L_{tb}=0,02 m. Flow input parameters are the following: the plasma forming gas is argon, gas flow rate G=0,02–0,24 g/s, pressure P_{inlet}= 3,5–135 Pa, the temperature T_{inlet}= 400–600 K, the degree of ionization $\delta_i=10^{-4}$. The initial pressure in the vacuum chamber P_0 =0,35–13,5 Pa. Distributions of temperature, velocity, and pressure of carrier gas as well as electron temperature and density are obtained and described in previous work [10].

A model of the two-spiral coil around the tube of 0.15 m by length on position 0 cm and on 5 cm from the beginning of tube is used for testing of calculation of the equations(8). The distribution of current **j** on each spiral on coil was set constantly 30 A/m^2. For calculating of **E** field was written the program on C++ language, which uses OpenFOAM [9] library. System (8) was solved iteratively till establishing. On each iteration produced recalculating of boundary conditions on tube by relation (9) for $\mathbf{j}(\mathrm{s})_{coil}$ and for electron volume charge current $\mathbf{j}(\mathrm{s})_{el}$ separately. The boundary condition on the tube surface setups after summation of currents by relation (9) and iterative process continue to next step.

Figure 1 illustrates distributions of the $\mathbf{E}_{\mathrm{Im}}$ along axis (left side) and in slice (right side). It is shown that maximum of module $\mathbf{E}_{\mathrm{Im}}$ is on the tube boundary where it equals to 6440 V·m^{-1}, and $\mathbf{E}_{\mathrm{Im}}$=0 in the centre of tube. It is shown that the vector of $\mathbf{E}_{\mathrm{Im}}$ has tangential direction. The results of calculations are quality in agree with known experimental data.

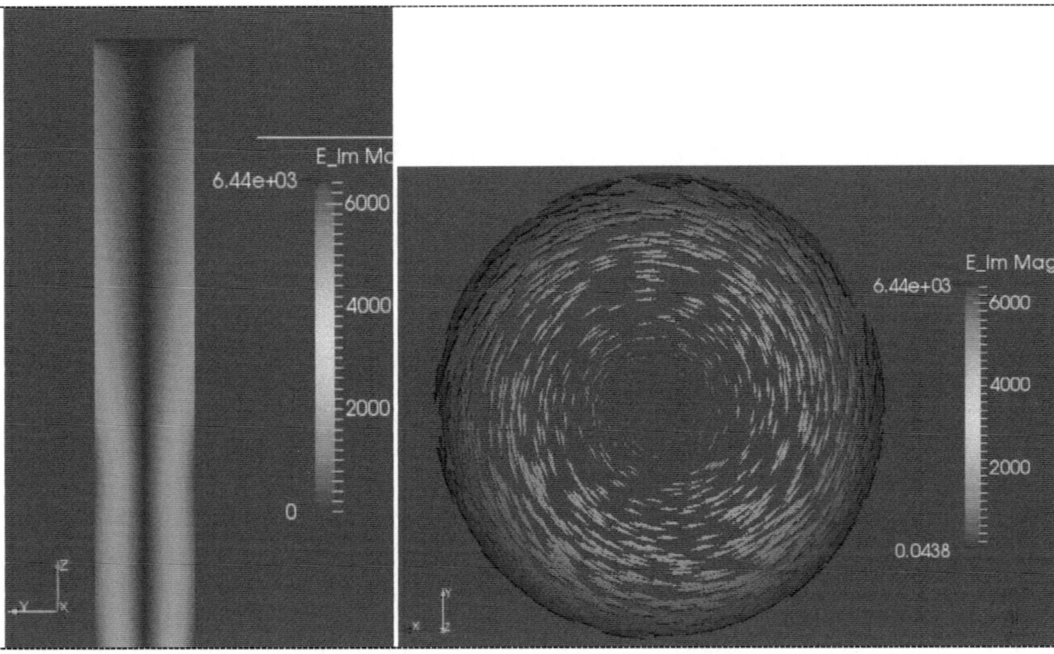

Figure 1. $\mathbf{E}_{Im}$in OZ-direction (left side) and $\mathbf{E}_{Im}$ in slice (right side) distributions.

Acknowledgments
The reported study was funded by RFBR according to the research project № 16-31-60081.

References

[1] Abdullin I S, Zheltukhin V S, Sagbiyev I R and Shayekhov M F 2007 *Modifikaciya Nanosloev vysokochastotnoj Plazme Ponizhennogo Davleniya* (Kazan: Izdatel'stvo Kazanskogo Tekhnologicheskogo Universiteta)
[2] Khubatkhuzin A A, Abdullin I S, Gatina E B, Zheltukhin V S and Shemakhin A Y 2012 *Vestnik Kazanskogo tekhnologicheskogo universiteta* **15** 72
[3] Khubatkhuzin A A, Abdullin I S, Bashkirtsev A A and Gatina E B 2012 *Vestnik Kazanskogo tekhnologicheskogo universiteta* **15** 71
[4] Abdullin I S, Zheltukhin V S, Khubatkhuzin A A and Shemakhin A Y 2014 *Matematicheskoe Modelirovanie Gazodinamiki Strujnykh Techenij Vysokochastotnoj Plazmy Ponizhennogo Davleniya* (Kazan: Izdatel'stvoKazanskogo Tekhnologicheskogo Universiteta)
[5] Zheltukhin V S and Shemakhin A Y 2011 *Uchenye Zapiski Kazanskogo Universiteta. Seriya Fiziko-Matematicheskie Nauki* **4** 135
[6] Zheltukhin V S and Shemakhin A Y 2014 *Mathematical models and computer simulations* **6** 101
[7] John D. Kraus. *Electromagnetics. Third.* (New York, NY: McGraw-Hill, 1984), pp. 380—419
[8] Bird G A 1976 *Molecular Gas Dynamics and the Direct Simulation of Gas Flows* (Clarendon Press: Oxford)
[9] OpenFOAM *OpenFOAM Foundation. Free Open Source CFD* 2011–2016 http://www.openfoam.org
[10] A.Yu. Shemakhin and V.S. Zheltukhin 2017 *Mathematical modeling of rf plasma flow with metastable atoms at low pressure* (Mathematica Montisnigri XXXIX), p.126-132

Molecular dynamics simulation of ion sputtering of the solutions of sodium and calcium chlorides

N A Sirotkin, D L Gurina and V A Titov

G.A. Krestov Institute of Solution Chemistry RAS, Ivanovo, 153045, Russia

E-mail: alexsad8@yandex.ru

Abstract. The ionic sputtering of sodium chloride and calcium chloride solutions was studied by classical molecular dynamics method. It is shown that the ions of the solute transferred into the gas phase both in the form of hydrated ions and in the form of ion pairs in water clusters. The threshold character of the ions transfer process is established. The calculated transfer coefficients of water and anions are in a good agreement with the experimental data. The fraction of water molecules sputtered in the form of clusters increases with the energy inputted in the solution. The fraction of water clusters in the gas phase is 40% at the inputted energy is 45 kJ mol^{-1}.

1. Introduction

Investigation of gas discharge plasma in contact or in liquids is one of the priority areas in plasma physics [1]. Atmospheric pressure discharges with liquid electrolyte electrodes are of interest as sources of active species for a variety of plasma-chemical applications [2-5]. Furthermore, the plasma-solution system is non-equilibrium in which the processes of formation of active particles in the liquid phase and in the plasma are closely interrelated. The properties of discharges with liquid cathode depend on the transfer of solvent and dissolved substances from the cathode to the gas phase. In discharges with an electrolyte cathode, the surface of the cathode is bombarded by positive ions, which leads to the transfer to the gas phase of the molecules of the solvent and of the components of the solute [6].

There are several possible mechanisms of particles transfer to the gas phase. First, the transfer of particles to the gas phase is similar to cathode sputtering [7]. In this case, the positive ions accelerated by the near-electrode potential drop collide with the surface of the liquid cathode and sputter out drops of liquid from its surface. Cserfalvi and Mezei [8] assumed that the sputtered drop passes through four stages: desolvation, charge stripping, bond breaking, and atomization, sometimes ending in the generation of a free atom of metal. Maksimov et al. [9] suggested that these solvated ions are transferred into the discharge by sputtering of water clusters from the solution cathode. In the discharge, the ions are excited to high vibrational energy levels. The vibrationally excited ionic salts transition from ionic states to covalent states at these high vibrational levels and dissociate as neutral atoms in the discharge. Another possible transfer mechanism is the thermal evaporation of droplets from the surface of the electrolyte cathode, desolvation and further thermal excitation of metal atoms because there is a plasma zone above the surface of the electrolyte cathode where the gas temperature is 1500-3000 K [10]. Finally, electrospray liquid free-jet formation on the solution surface and the formation of a Taylor cone have also been put forth as possible mechanisms [11].

The transfer process can be characterized by a transfer coefficient – the number of particles transferred from liquid to gas phase per one ion bombarding solution surface. In [6, 12] the authors have shown that the transfer coefficients for water are about 200 – 500 molecule ion^{-1} and for solute these values are 0.01-1 particle ion^{-1}. However, a detailed study of the transfer processes is difficult due to the complexity of experimental measurements of transfer coefficients. Nevertheless, it is possible to study the ion sputtering processes of liquid media in a glow discharge at atmospheric pressure by means of numerical simulation methods such as molecular dynamics. For example, the authors [13] investigated the water molecules sputtering from the liquid phase under the falling ion with an initial energy of 50 – 500 eV. According to the results of the simulation, the authors showed that the value of transfer coefficient of water molecules in the gas phase is about 100 – 500 molecule ion^{-1}. The number of sputtered water molecules, liquid temperature, and ion depth penetration after ions bombardment of O^+ with kinetic energies of 10 eV and 100 eV on the liquid surface were investigated in [14]. The average number of sputtered water molecules by ion was 0.5 at 10 eV and 7.0 at 100 eV.

In this context, the aim of our work was to study the process of plasma sputtering more complex systems such as water solution of sodium and calcium chloride by means of the molecular dynamics method. Furthermore, the flow of bombarding positive ions was increased (up to 20 ions) and the transfer coefficients of both solvent and solute were evaluated.

2. Experimental part

The plasma sputtering of sodium and calcium chlorides solution is studied by classical molecular dynamics method, implemented in a software package Gromacs-4.5.4, according to the algorithms described in [15]. The cell with periodic boundary conditions contained 8112 water molecules (SPC/E model), 80 Na^+ ions or 40 Ca^{2+} and 80 Cl- ions, which corresponds to a concentration is 0,5 mol l^{-1}. Cell size 4.9596×4.9596×9.9192 nm corresponds to the density of solution 1026.8 kg m^{-3} at T = 298 K. Equilibration of the system was carried out in NVT-ensemble for 500 ps with a time step 1 fs. Equilibration achievement determined by analysing of the radial distribution functions, the value of the potential energy and temperature and energy fluctuations during the simulation. The cell obtained in such way then was used to create box consisting of a layer of the solution with thickness 9.9192 nm, and the vacuum layer with thickness 80.0808 nm. The ion with atomic mass equals 23 a.u. and charge +1 was selected as the bombarding ion. The solution was bombarded by one, five, eight, nine, thirteen and twenty ions with energies from 60 eV to 500 eV. Simulation was performed in NVE-ensemble, bond lengths in the molecules were restricted by the LINCS algorithm. Five simulations of each system were carried out to average statistical error.

3. Results and discussions

Sputtering both the solvent (water) and the solute is observed under the bombardment of the solution by the incident ions. It should be noted that the process of sputtering of the solution surface depends on a number of the bombardment ions and energy of ions (Figure 1). When incident ion collides with the solution surface its energy is transferred to nearest water molecules. Then these molecules interact with neighbouring ones and it results in increasing the kinetic energy of the particles in the region of impact and breaking of hydrogen bonds between water molecules. Having obtained enough kinetic energy some of the water molecules transfer into the gas phase. Thus in the solution cavity is formed due to the intensive destruction of hydrogen bonds and the sputtering of water molecules. Cavity size varies from 0.31 nm^3 at the energy of the incident ion 250 eV to 1.97 nm^3 at the energy of the incident ion 500 eV. At the same time, the incident ion overcomes a distance of 2 to 3.5 nm in the solution. Simulation results show that the water molecules sputtered due to impact the incident ion with the energy of 250 eV exist as monomers in the gas phase, whereas dimers, trimers and larger clusters of water molecules can be observed in the gas phase in the case when the initial energy of the incident ion is 500 eV. When the solution is bombarded only one ion the transition of the solute ions in the gas phase is absent. Mainly monomers of water are transferred into the gas phase at a relatively small amount of energy inputted in the solution (0.14 – 0.25 eV particle^{-1}, 13.5 – 24 kJ mol^{-1}). The dimers, trimers and large clusters of water appear in the gas phase at the increasing energy of bombarding ions.

In contrast, the bombardment of the solution surface by several ions results in altering of the sputtering process. The components of the dissolved substance are detected in the gas phase when the threshold value of the input energy (0.3 eV particle^{-1}, 30 kJ mol^{-1}) is reached. Sputtered of solvent molecules is occurred already along the entire cell surface without cavity formation. Moreover, the sodium and chloride ions exist mainly as ion pairs in a gas phase, and about 30% of all ions are presented separately each other in hydrated complexes. Also, the increasing of the incident ions number leads to increased bombarding ion track in the liquid phase - up to 7.5 nm. On the basis of the results obtained by molecular dynamics, we were able to estimate the amount of ejected into the gas phase ions per one bombarding ion (transfer coefficients). Obtained data are summarized in Table 1 in comparison with experimental ones. The difference between the transfer coefficients for cations and anions is possibly due to the electric field that exists in real plasma cell and which inhibits transferring of positive ions to the condensate. Computer simulation results show that two of the eight solute ions transfer separately into the gas phase under the action of several bombarding ions with a low energy (200 eV). It should be noted that under experimental conditions the transfer of the cluster containing only the cation to the condensate is very difficult because of the presence of interfering electric field. Perhaps such phenomenon can be a reasonable explanation of the stoichiometry violation when the solute transfers to the gas phase. However, in computer simulations the fraction of ions sputtered separately is only 30%. Thus, molecular dynamics simulation results allow concluding that the plasma sputtering of the solution by ion bombardment may cause transfer of the solute ions into the gas phase preferably in the form of ion pairs surrounded by water clusters.

The number of sputtered molecules and the fraction of molecules sprayed into clusters increases with increasing inputted energy. In addition, the water clusters become larger. The geometric criterion of the hydrogen bond was used at the determining the number of molecules in the water cluster. As shown in Figure 2, the water clusters of 5 or more molecules are starting to prevail in the gas phase at the energy reaches ~ 30 kJ mol^{-1}. The maximum fraction of the transferred clusters reaches 40%, and half of them are large clusters with 7 – 10 or more water molecules. Thus, the results of computer simulation by the method of classical molecular dynamics confirm the experimentally observed threshold character of the transfer of cations from the liquid cathode. It has been established that the components of the solute were transferred into the gas phase both in the form of hydrated ions and in the form of ion pairs in water clusters.

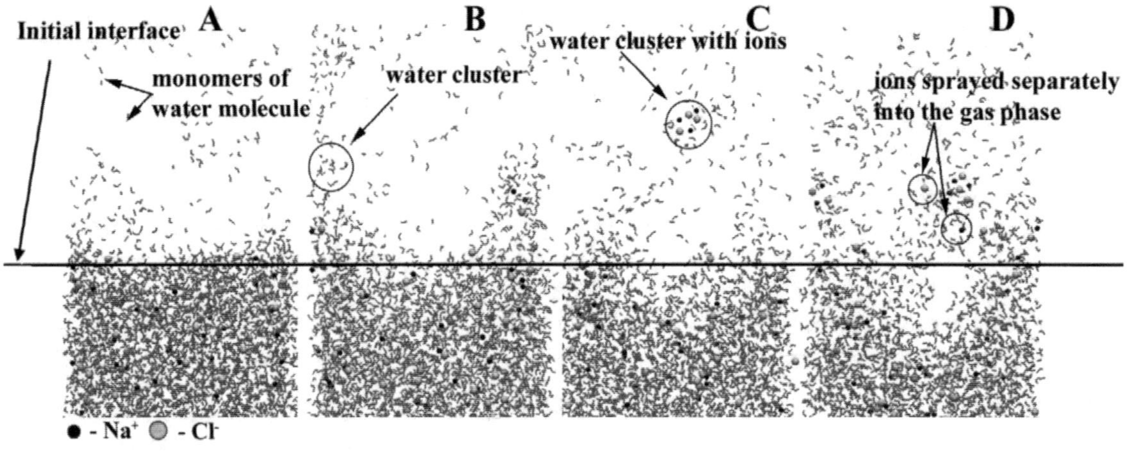

Figure 1. Configuration of the cell at 10 ps after the impingement of the: A) 1 ion with E_i =500 eV B) 5 ions with E_i = 300 eV C) 9 ions E_i = 300 eV D) 13 ions with E_i = 200 eV.

Table 1. The transfer coefficients (particle ion^{-1}) for water molecules $s(H_2O)$ and ions (s) at different numbers of bombarding ions (N_i) and energy of ions (E_i).

N_i	E_i (eV)	$s(H_2O)$	s(Cl)	s(Na)	s(Ca)	Experiment[a] [16] $s(H_2O)$	Experiment[a] [16] s(Cl)	Experiment[a] [16] s(Na)	MD[b] [13] $s(H_2O)$
1	250	50±5				300±50	0.213±0.018	0.005±0.001	220±10
1	500	450±15							
5	300	270±10				350±50	0.258±0.025	0.019±0.005	
8	300	310±15	0.255	0,255	0.128				
9	300	405±20	0.333	0.333	0.167	440±50	0.334±0.027	0.029±0.007	510±20
13	200	280±25	0.307	0.307	0.154				
20	60	170±25	0.305	0.305	0.152	480±50	0.395±0.032	0.038±0.006	

[a] C(NaCl) = 0.5 mol l^{-1}, i = 35 mA, 40 mA, 45 mA, 50 mA,
[b] the energies of bombarding ions are E_i = 300 eV and E_i = 500 eV.

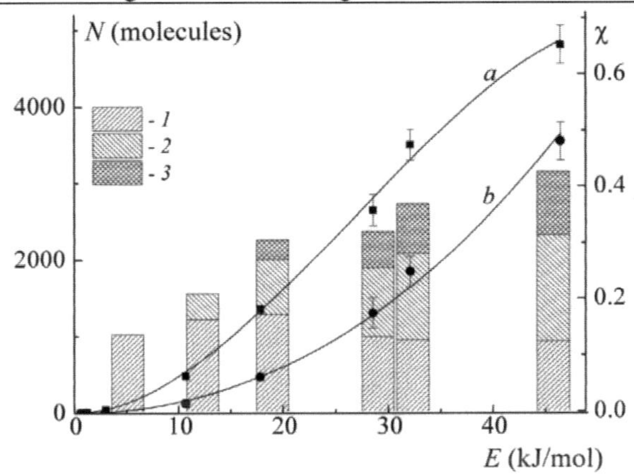

Figure 2. The number of sprayed water molecules to the gas phase (a), the number of hydrogen-bonded water molecules in the gas phase (b) and the molar fraction (χ) of the water molecules in clusters:
1 – 3-4 molecules in cluster,
2 – 5-7 molecules in cluster,
3 – more than 7 molecules in cluster.

Acknowledgments
The study was supported by Russian Foundation for Basic Research according to the research project 16–33–60061 mol_a_dk

References

[1] Bruggeman P J, Kushner M J, Locke B R, et.al. 2016 *Plasma Sources Sci. Technol.* **25** 053002
[2] Yang Y, Cho Y I, Fridman A. 2012 *Plasma Discharge in Liquid: Water Treatment and Applications* (New York: CRC Press)
[3] Yang C, Wang L, Zhu Z, Jin L, Zheng H, Belshaw N S and Hu S 2016 *Talanta* **155** 314
[4] Du C M, Wang J, Zhang L, et al. 2012 *New J. Phys.* **14** 013010
[5] Titov V A. Shikova T G, Rybkin V V, et al. 2006 *High Temp. Mater. Process.* **10** 467
[6] Khlyustova A V, Sirotkin N A, Maximov A I 2010 *High Energy Chem.* **44** 75
[7] Ecker G, Emeleus K G 1954 *Proc. P. Soc. Sec. B.* **67** 546
[8] Cserfalvi T, Mezei P 2005 *J. Anal. At. Spectrom.* **20** 939
[9] Maksimov A I, Titov V A, Khlyustova A V 2004 *High Energy Chem.* **38** 196
[10] Webb M R, Andrade F A, Gamez G, et al. 2005 *J. Anal. At. Spectrom.* **20** 1218
[11] Webb M R, Andrade F J, Hieftje G M 2007 *Anal. Chem.* **79** 7807
[12] Kutepov A M, Zakharov A G, Maksimov A I, Titov V A 2003 *High Energy Chem.* **37** 362
[13] Nikiforov A Yu 2008 *High Energy Chem.* **43** 235
[14] Minagawa Y, Shirai N, Uchida S, Tochikubo F 2014 *Jpn. J. Appl. Phys.* **53** 010210
[15] Titov V A, Sirotkin N A 2014 *Proc. VII International Symposium on Theoretical and Applied Plasma Chemistry* (Ivanovo: ISUCT) p 160
[16] Sirotkin N A, Titov V A 2016 *Plasma Phys. Tech.* **3** 126

Kinetics of the synthesis of ultradisperse cadmium and zinc oxide-like powders in a plasma-solution system

K V Smirnova, D A Shutov, V V Rybkin, A N Ivanov, M V Gromov

Ivanovo State University of Chemistry and Technology, Ivanovo, 153000 Russia

e-mail: shutov@isuct.ru

Abstract. The action of the DC glow discharge on the cadmium and zinc nitrates water solution in the absence of additional reagents and without electrodes-solution contact was shown to result in the production of the solids in the liquid phase. The atmospheric pressure direct current discharge was excited in the ambient air by applying a high direct voltage to two pointed titanium electrodes placed above liquid anode and liquid cathode in the H-shaped cell. The discharge current was varied in the range of 30-70 mA and the initional solutions concentrations were 5 mM. The kinetics of particles formation was studied using turbidimetry and nephelometry methods.

1. Introduction

In recent decades, interest in obtaining of nanomaterials has increased, which have unique mechanical, electrical, optical, magnetic and catalytic properties. Nanoparticles of metals, alloys, oxides, etc. have found wide application in microelectronics, powder metallurgy, in medicine for the precise delivery of medicines, as well as in the production of catalysts, dyes, lubricants and other fields.

There are many different methods for synthesizing micro / nano (ultradispersed) structures. Enough can be conditionally divided into three groups: first - the synthesis of nanoparticles in gaseous (vapor) phase; the second - mechanical synthesis (mechanosynthesis); the third is the synthesis of nanocharts in liquid (colloidal) systems. The latter include methods based on the action of gas discharges. In the last 10-15 years, a large number of works have appeared in the world, in which the possibilities of obtaining nanoparticles of metals, oxides and some other compounds using plasma-solution methods are shown. Various methods of excitation of discharges and plasma-chemical cells, used precisely for the purposes of synthesis, are considered in detail in the review papers [1, 2].

The most popular among researchers are nanoparticles of precious metals: Au, Pt, Ag, Pd. There are works devoted to the synthesis of Al, Ti, Fe, Co, Ni, Cu, Zn, Ge, Zr, Sn. From the composite materials, the structures of Au-Ag, Pt-Au, Ag-Pt, Co-Pt, Fe-Pt, Sm-Co, Ni-Cr, Sn-Ag, Sn-Pb, Co-B, Mo-S can be cited. Among the oxides synthesized with the help of plasma-solution systems, we can name oxides Al, Ti, Cu, Fe, Co, Zr, Ru, Sn, Ta, W, In and Zn. As far as we know there are no any papers devoted to plasma-solution synthesis of CdO. Also, there are no any data on the investigations of the kinetics of the solids formation in the liquid phase during solution treatment by discharge.

The main goal of this work was the investigation of the kinetics behaviors of the solid formation during atmospheric pressure direct current discharge treatment of the water solutions of the $Cd(NO_3)_2{\cdot}4H_2O$ and $Zn(NO_3)_2{\cdot}6H_2O$.

2. Experimental

Aqueous solutions of analytical grade $Cd(NO_3)_2{\cdot}4H_2O$ or $Zn(NO_3)_2{\cdot}6H_2O$ (JSC "Himreactiv", Russia) with a concentration of 0.005 mol/L, a volume of 200 ml, were placed in an H-shaped glass cell. The discharge was excited in the ambient air by applying a high direct voltage to two pointed titanium electrodes. So the solution surface of one of the cell is considered as liquid anode (A-cell) and of another cell as liquid cathode (C-cell). The electrode-solution surface distance was adjusted to 5 mm. The discharge current was varied in the range of 30-70 mA. The duration of solution treatment was 10 min. The experimental setup was described in detail in [3].

By the action of a discharge a colloidal solution was formed in the near-surface layer of A-cell and precipitate was formed at the bottom of the cell.

To characterize the kinetics of the particles formation process the turbidimetry and nephelometry methods were used. The intensity of light transmitted through the near-surface layer (1 mm below surface) of solution in first case, or the intensity of the scattered light in second case was measured using AvaSpec-2048FT-2 spectrometer (Avantes, Netherlands). Optical length was 45 mm in both cases. Scattering angle was 90°. The He-Ne laser (λ=632.8 nm) was used as a light source (Figure 1).

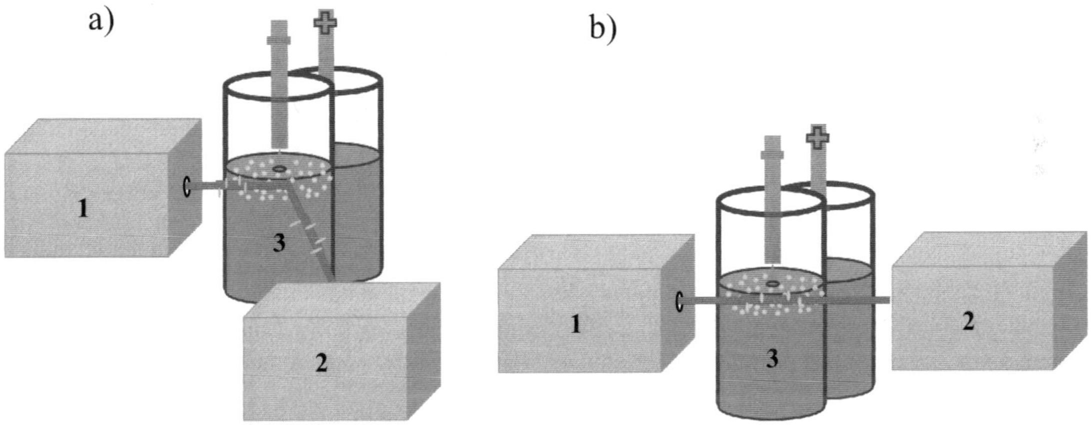

Figure 1. Schematic view of nephelometry (a) and turbidimetry (b) measurements: 1 – He-Ne laser, 2 – spectrometer, 3 – anodic cell.

3. Results and discussion

After the discharge ignition there are no any visible changes in both A-cell and C-cell at first tens seconds of solution treatment. Turbodimetry curve (Figure 2) demonstrates induction period of the process. After induction period a colloidal solution was formed in the near-surface layer of A-cell and precipitate formation started at the bottom of this cell. No colloid solution or precipitate was observed in the C-cell during treatment but a gas bubbles near the liquid phase surface always appeared.

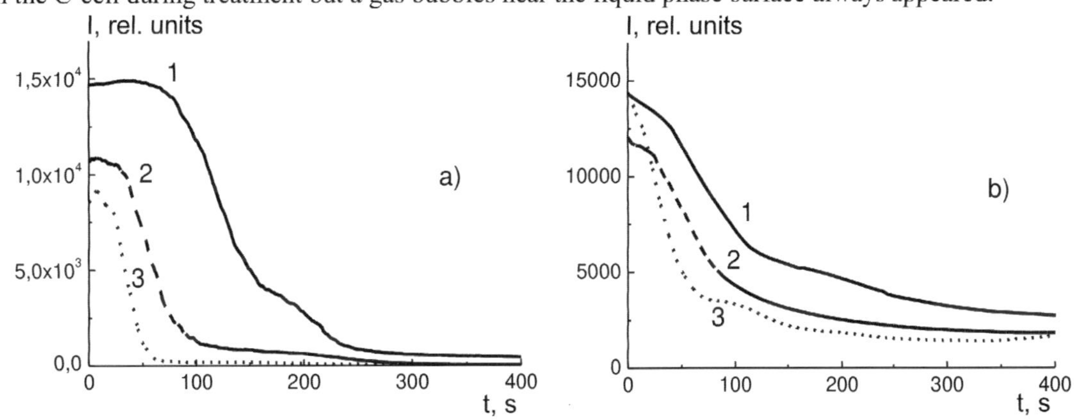

Figure 2. Turbidimetric kinetic curves of the $Zn(NO_3)_2{\cdot}6H_2O$ (a) and $Cd(NO_3)_2{\cdot}4H_2O$ (b) solutions: 1 – 30 mA, 2 – 60 mA, 3 – 70 mA.

The increase of the discharge current lead to the decrease of the induction period (figure 3a). At low input power the induction period of the Zn-containing powder formation is about two times higher than Cd-containing powder formation. But as discharge current increase the induction times becomes comparable.

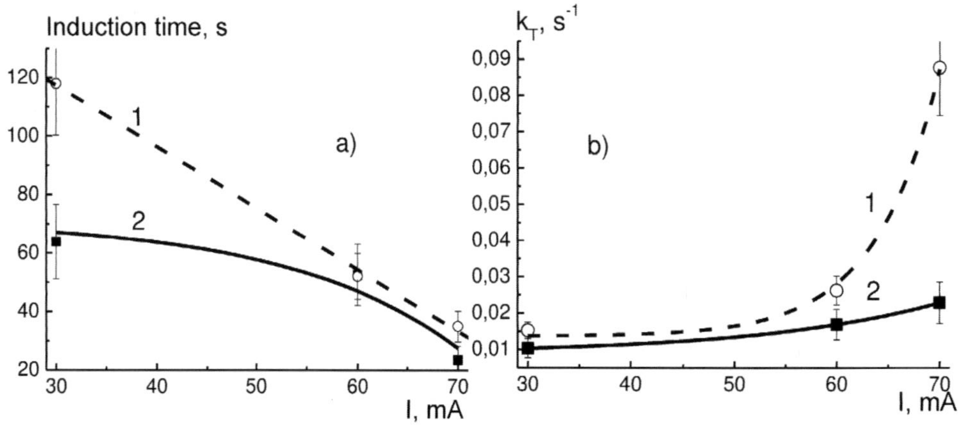

Figure 3. Induction period (a) and effective constant of the powder formation process (b): 1 – $Zn(NO_3)_2{\cdot}6H_2O$, 2 - $Cd(NO_3)_2{\cdot}4H_2O$

In turbidimetry the Beer's-low like expression for the intensity is applicable [4]:

$$I_T = I_0 e^{-\alpha_T C} \qquad (1),$$

where I_T – transmitted light intensity due to the scattering effect of particles suspended, I_0 - initial light intensity, α_T – coefficient of proportionality that included turbidity constant and light path length, and C – the concentration of scattering species.

In our case the concentration is a function of discharge treatment time, $C=C(t)$. But we don't exactly know the rate law for the particles formation process. The simplest assumption that kinetics of the species formation followed by zero-order law gives to us:

$$C = kC_0 t \qquad (2),$$

where k – effective solids formation rate constant, C_0 – initial concentration of nitrate. And the combination of (1) and (2) is:

$$I_T = I_0 e^{-\alpha_T k C_0 t} = I_0 e^{-K_T t} \qquad (3)$$

Really, the kinetics of the passed light intensity is described well with a formal first-order rate law (3) with an effective constant K_T of about 10^{-2} s^{-1} (pairwise correlation coefficient is 0.98) (figure 3b). It's should be noted that this coefficient characterizes not only solid species formation rate but includes flocculation and sedimentation processes. Also, we find out that the increase of the discharge current lead to the growth of the effective rate constant.

To check our assumption of zero-order law species formation (2) we have measured scattered light intensity during the discharge.

In the first approximation the intensity of scattered light is given by expression:

$$I_S = \alpha_S I_0 C \qquad (4),$$

where I_S –scattered light intensity, I_0 - initial light intensity, α_S – some empirical coefficient of proportionality, and C – the concentration of scattering species.

Theoretically combination of (2) and (4):

$$I_S = \alpha_S I_0 k C_0 t = I_0 K_S t \qquad (5)$$

gives the linear time's behavior of the scattered light intensity where the K_S is an effective constant. As it can be seen in figure 4a, nephelometry curve shows two near-to-linear parts. The curve consists of four accentuated region (figure 4a). There is no noticeable increase in the scattering light intensity at initial time after discharge excitation (section I) and the induction period followed by the two sections of linear growth (sections II and III). It's should be noted that curve look's like typical topochemical kinetic curve where reactions are strongly localized, in our case in the area of the discharge-liquid contact point. After the growth sections the near-to-exponentional decay of measured scattered light intensity is obtained (section IV). We assume that this interval deals with reabsorption and/or rescattering of the light in the high concentrated colloidal solution near the liquid surface.

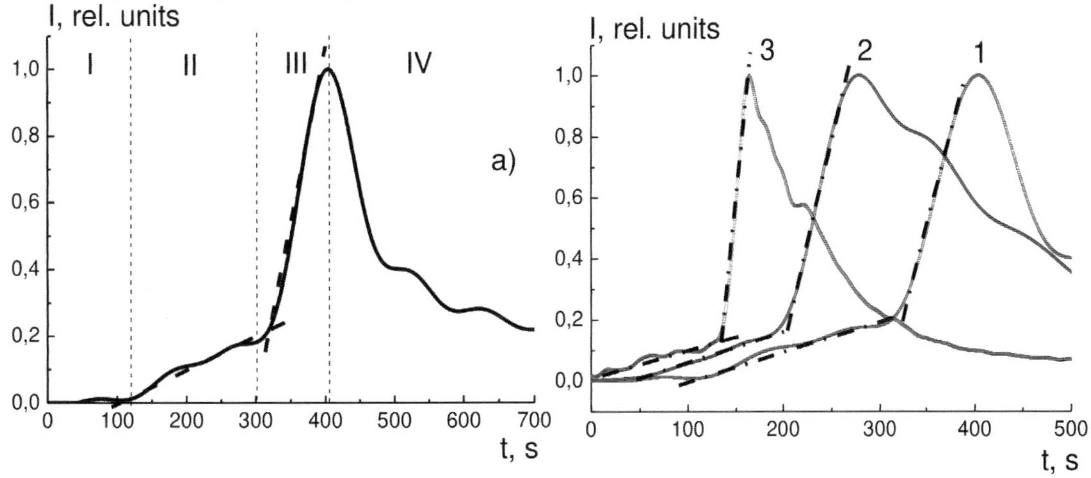

Figure 4. Different segments of nephelometry kinetic curves of the $Zn(NO_3)_2 \cdot 6H_2O$ at discharge current 30 mA (a) and nephelometry kinetic curves of the $Zn(NO_3)_2 \cdot 6H_2O$ (b) at currents of: 1 – 30 mA, 2 – 60 mA, 3 – 70 mA.

The calculated effective constants K_S are shown in table 1. The character of the constants changes with discharge current increase identical to the changes of K_T. Unfortunately, both K_S and K_T includes unknown coefficients α_S and α_T, respectively. So it is impossible to calculate effective solids formation rate constant k. Moreover, changes of the slop in nephelometry kinetic curves can be assigned with changes of the coefficient α_S, that is turbidity in fact.

Table 1. Effective constants K_S obtained from sections II and III of the figure 4b.

Discharge current (mA)	K_S^{II} (s^{-1})	K_S^{III} (s^{-1})
30	$0.94 \cdot 10^{-3}$	$1.22 \cdot 10^{-2}$
60	$1.08 \cdot 10^{-3}$	$1.46 \cdot 10^{-2}$
70	$1.3 \cdot 10^{-3}$	$3.48 \cdot 10^{-2}$

Anyway, the kinetics of the particle formation under the discharge action is complicated process and needs additional investigations.

4. Conclusion

The action of the glow discharge onto the water solution of $Cd(NO_3)_2 \cdot 4H_2O$ and $Zn(NO_3)_2 \cdot 6H_2O$ lead to the formation of the powders in near-to-surface layer of the liquid anode. Increase of the discharge current lead to the increase of the powder formation rate. The powders formation kinetics in first approximation describes by the zero-order kinetics law.

Aknowledgments
This study was supported by Ministry of Education and Science of the Russian Federation, project 3.1371.2017/.4.6

References

[1] Saito G 2015 *Journal of Nanomaterials* **2015** 21
[2] Tochikubo F 2014 *Japan. J. Appl. Phys.* **53** 126201
[3] Shutov D A, Rybkin V V, Ivanov A N, Smirnova K V 2017 *High Energy Chemistry* **51(1)** 65
[4] Lawler D M 2005 Spectrophotometry: Turbidimetry and Nephelometry *Encyclopedia of Analytical Science* (Elsevier, 2nd ed.) pp 343-351

Structure pattern peculiarities and electrical characteristics of surface dielectric barrier discharge in air driven by impulse and sine high voltage

M Sokolova[1*], V Voevodin[1], M Malashin[2], I Rebrov[2] and N Kiselev[3]

[1] Moscow Power Engineering Institute, Moscow, Russia
[2] Institute for Electrophysics and Electric Power of Russian Academy of Sciences, St.-Petersburg, Russia
[3] "Sirius" Ltd, Moscow, Russia

*E-mail: mvsokolova@mail.ru

Abstract. Structure patterns and electric characteristics of surface dielectric barrier discharges in air driven by periodic unipolar impulse voltage with ns front duration and sine voltage of the same frequency f = 15 kHz are presented. An influence of voltage impulse polarity, amplitude and impulse duration on discharge characteristics is shown to exist. The currents extracted from the surface discharge plasma layer by d.c. voltage of additional third electrode are found to depend on the waveform of the applied high voltage impulses.

1.Introduction

It has been shown in a range of works that waveform of the supply voltage is a crucial parameter for non-thermal atmospheric pressure barrier discharges which are relevant for a variety of applications [1-2]. Another prominent example of the generation of non-thermal plasmas are surface dielectric barrier discharges (SDBDs) [3]. Although SDBDs are commonly sine-driven it has been shown experimentally [4] that pulsed operation enables in a range of applications a higher yield of different species, lower power consumption and allows the study of variable HV parameters that define the discharge characteristics. The results of experimental investigation of pulsed driven SDBD can be used in modeling of the surface discharges for different conditions.

The aim of the present work was to find the main factors that define the SDBD structure for different characteristics of the applied voltage: different impulse amplitude, different polarity and impulse duration and sine voltage of the same frequency – and connect the structure peculiarities with electrical characteristics of SDBD and the yield of produced charged species.

2. Experimental set-up

The experimental set-up with apparatus for diagnostics is schematically presented in Fig. 1. The single strip arrangement featured a double-sided dielectric barrier configuration. An Alumina plate 1 mm of thickness and permittivity ε_b = 9,5 was used as a barrier, the HV electrode (1) placed on one side of the barrier was a Nickel strip (14x80 mm) 30 μm thick all edges of which were covered with an epoxy layer to prevent discharge appearance. The main electrode with the discharge (2) was a Nickel strip 1 mm wide, 80 mm long and 30 μm thick placed on the reverse side of the barrier and grounded through a specially made fast current active low inductance probe (R_{pr} = 7.8 Ohm, rise time

< 2 ns). High voltage measurements were performed by voltage probe (Tektronix P6015, rise time 4.67 ns). A third flat electrode (3) (40x60 mm) which was used in a three-electrode arrangement was placed at a distance 1 cm from the plasma layer of the discharge. This electrode having DC high potential (up to 10 kV) was used to measure the current of charged species produced by SDBDs. The three-electrode arrangement was housed in a dielectric sell with Plexiglas windows. A flow of dried air (1.7 l/min, humidity $\lambda < 1\%$) was maintained through the sell along the strip with the discharge. The discharge in the two-electrode arrangement exposed to the high-speed camera measurements was maintained in a weak blowing of open atmospheric air. The blowing was needed to carry away the ozone from the discharge. The pulsed surface discharge was driven by unipolar square wave periodic pulses of different polarity with U_m = 7 and 8 kV and f =15 kHz (T = 67 μs) using a specially developed generator [5]. The pulse width was varied from 0.5 μs to 45 μs. Most measurements were carried out with a HV slope steepness of the voltage impulse equal to 70-80 V ns^{-1} which supported a rise/fall time of 130 ns. The sinusoidal surface discharge was driven by high voltage using a special sine generator with fixed frequency f = 15 kHz and voltage amplitude $2U_m = U_{p/p}$ = 6 kV.

The optical diagnostic consisted of ultrahigh-speed camera Phantom v2511 and a photo-camera Nikon D5200.

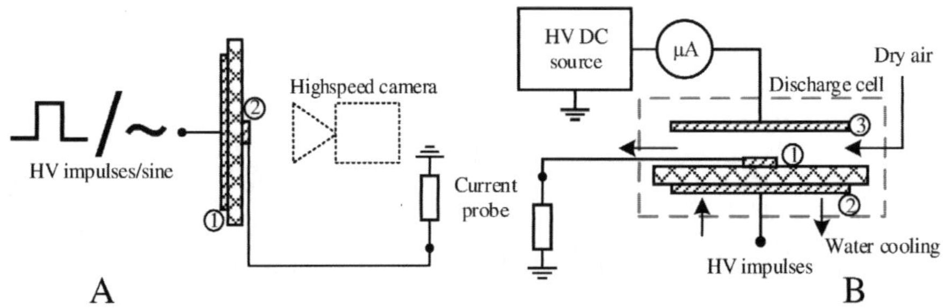

Fig. 1. The experimental set-up. A – scheme for registration of discharge structure, B – scheme for registration of extracted current

3. Experimental results and discussion

The pictures of the discharge structure made by speed camera (time exposure 17 μs) are shown in Fig.2 for rising (A and C) and for falling (B and D) slope of positive and negative impulse voltage. Fig. 2 also presents corresponding oscillograms of voltage impulse and electrical discharge current $I_{dis} = I_{tot} - I_{displ}$.

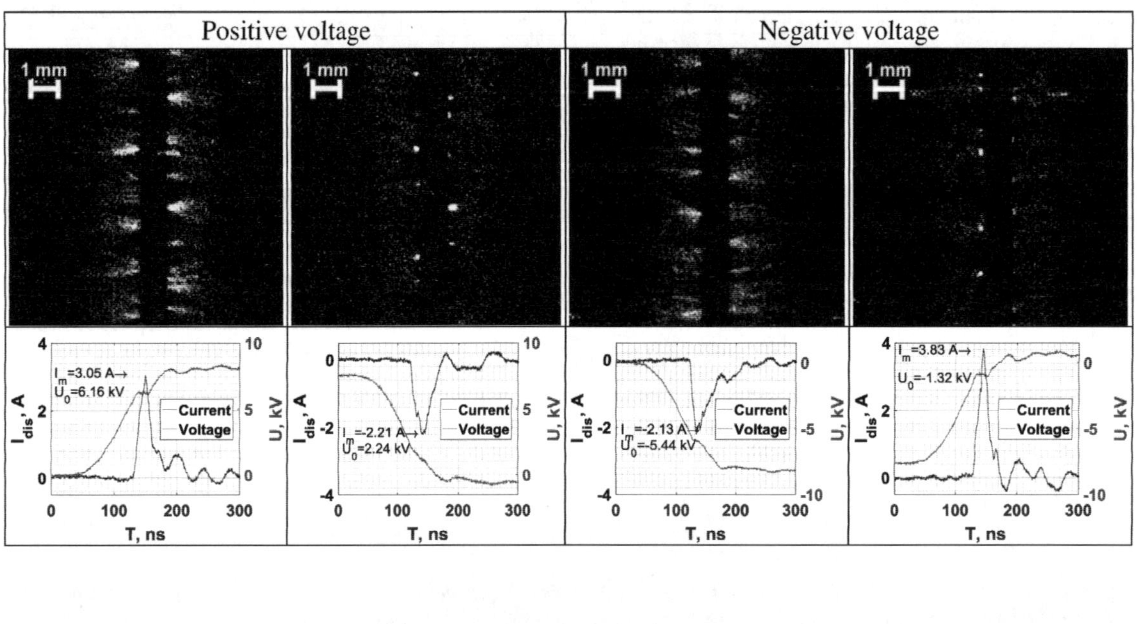

A	B	C	D

Fig. 2. Discharge structure and corresponding voltage and current curves for different voltage impulse polarity, $U_m = 8$ kV, $\tau_{imp} = 30$ μs. A, C – rising slope, B, D – falling slope

I_{tot} is the total current measured by the current probe and I_{displ} is the displacement current, i.e. the current proportional to the capacity C_{el} of the electrode system without the discharge and to the steepness of the voltage impulse front dU/dt. To measure the displacement current the electrode arrangement was replaced by a unit capacitor of the same capacitance as the discharge gap. The main measured characteristic values (maximal discharge current I_m and voltage U_0 of discharge ignition) of the discharge are given in fig. 2. The pictures show different structure of the discharge for rising and for falling slopes both for positive and for negative HV impulses. The channel structure of individual microdischarges is more pronounced for rising slope of positive HV while the discharge structure is practically the same for the impulse falling slope of both polarities and consists of short bright spots with very weak diffusive part connected to the spot. The visible length of microdischarges seen in the pictures done with speed camera is about 1.2 mm for rising impulse slopes of both polarities and U_m = 8 kV.

The discharge current in oscillograms is presented in most cases (i.e. for all parts of the HV impulse and both polarities) by a single impulse. Nevertheless there are rare cases when the current curve shows a small number of discrete impulses (not more than 4-5) although there are seen more than 30 channels in the picture of the corresponding discharge.

Analysis of packets of discharge currents (20 discharge events) for U_m= 8 kV and 30 μs pulse width shows that maximal values of the discharge current pulses at the rising part of **positive** voltage impulse can be 2-3 times higher than the same for the discharge at the falling impulse part. Contrary to it the discharge current maximal value for falling part of **the negative** voltage impulse can be 1.5-2 times higher than for the rising voltage impulse front and have longer width (35 ns). According to [6] this difference can be explained by different pre-ionization concentration of charged species in the gap left after previous discharges.

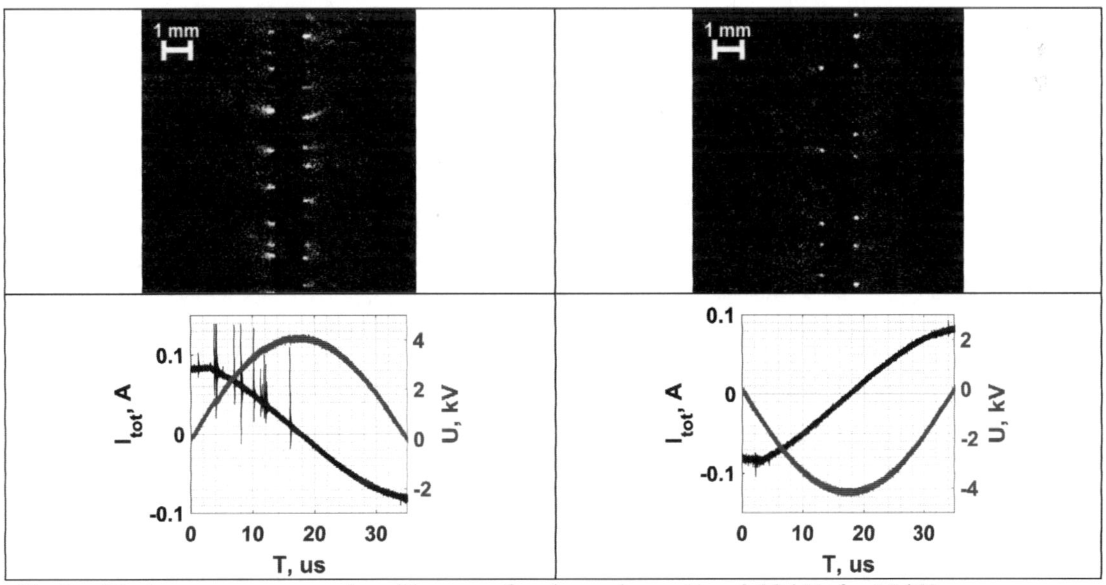

Fig.3 – Discharge structure and oscillograms for sine voltage. U_m= 3.92 kV, f = 15 kHz.

Fig. 3 shows pictures of the discharge structure and oscillograms for sine voltage for the same electrode configuration and near discharge conditions. Contrary to an impulse case the discharge patterns as well as the current oscillograms show discrete microdischarge channels and current impulses for both half-cycles. The comparison with the impulse voltage case indicates the following features of the sine discharge that are different from the impulse voltage case: a) weak and short microdischarge channels during positive half-cycle and small values of corresponding discharge currents; b) discrete character of microdischarge current pulses; c) extremely weak microdischarges during negative half-cycle.

All the above presented discharge characteristics correspond to one value of voltage impulse duration $\tau_{imp} = 30$ μs. The change of τ_{imp} value leads to a marked change of the discharge structure and current values. The integral pictures of the discharge for different τ_{imp} values were done using a photo camera (exposition 1/100 s) and are shown in Fig. 4 together with $I_{dis} = f(\tau_{imp})$ curves. As it is seen the character of τ_{imp} influence depends on the impulse polarity and the moment of the discharge appearance, i.e. the rising or falling slope of the voltage impulse.

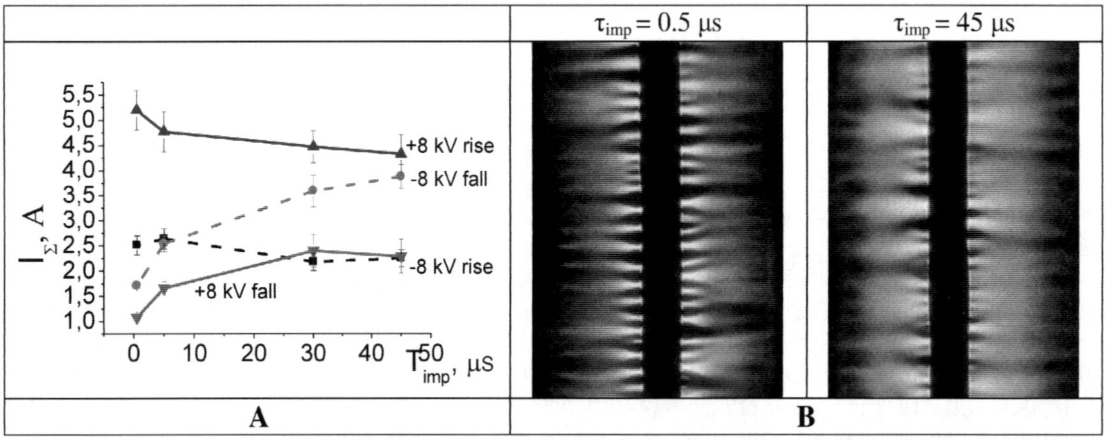

Fig. 4 – Discharge current amplitudes (A) and photos of the discharge (B) for different τ_{imp} values of impulse voltage, U_m= –8 kV. Photos are done in 1 min after discharge ignition. Exposure time 1/100 s.

It has been shown in [6] on base of experimental investigation of traditional barrier discharge driven by periodic impulse voltage of different impulse duration τ_{imp} that a change of τ_{imp} value leads to different discharge formation. An increase of τ_{imp} from 0.2 to 10 μs gives a reduce of the discharge current for rising voltage slope and a change of the current form and an increase of its amplitude at the falling slope of the voltage curve. This result is explained in [6] by a comparability of drift time of charged particles and the impulse duration which leads to a formation of volume charge in the gap. This charge changes the electric field distribution in the gap and the conditions of the subsequent discharge appearance. It is assumed in the present work that the above proposition can be fully applied to SDBDs where the volume charge in the pre-ionization stage is supported by the surface charge on the barrier surface. In this case it is possible to explain the difference in the surface discharge structure and current values with the τ_{imp} increase. Besides, it must be noted that the increase of τ_{imp} for a constant value of voltage frequency leads to a decrease of the pause between subsequent voltage pulses, i.e. the time between the discharge at the impulse falling slope and the subsequent discharge at the rising slope of the following voltage impulse.

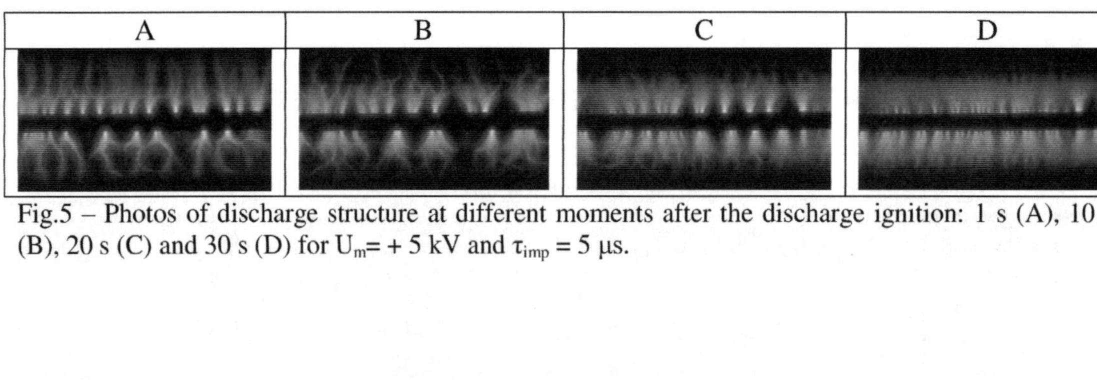

Fig.5 – Photos of discharge structure at different moments after the discharge ignition: 1 s (A), 10 s (B), 20 s (C) and 30 s (D) for U_m= + 5 kV and τ_{imp} = 5 μs.

An additional feature of the surface discharge has appeared when photos of the discharge were done at different moments after the discharge ignition (Fig.5). It is seen that the discharge structure changes with time of discharge existence. As there was no special cooling of the electrode this change may be attributed to some heating of the gas after several seconds. The result means that the discharge characteristics may be quite different when a single impulse or periodic impulse voltage are applied.

A three electrode arrangement was used in the present work to measure the current extracted from the surface discharge plasma layer by a dc potential of the third electrode. It was found out that the voltage impulse duration has an additional influence on the value of the extracted current I_d. The measured extracted current is shown in fig. 6 A for different negative impulse voltage values U_m. Fig. 6 B is for different extracting dc voltage U_d, in this case impulse voltage U_m is +8 kV, frequency f = 15 kHz. These results are a direct consequence of the τ_{imp} influence on the SDBD plasma formation. The I_d strong dependence on U_d for a three-electrode arrangement has been shown in [7]. The character of τ_{imp} influence on I_d varies with different U_d polarity as it is seen in Fig.6 B. With positive U_d polarity the quantity of negative species (negative ions and electrons) that constitute the extracted current just slightly increase for τ_{imp}>20 μs while the extracted current of positive ions becomes 1.8 times higher with an increase of τ_{imp} from 0.5 μs up to 10 μs with negative d.c. voltage. Further increase of τ_{imp} does not change I_d which means that the main quantity of positive ions is produced during first 10 μs of voltage application.

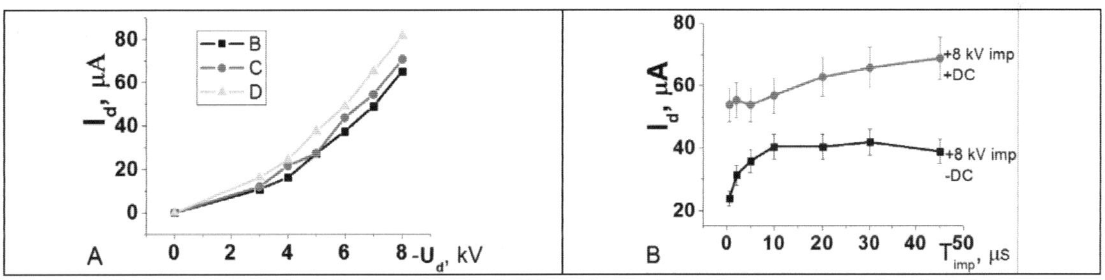

Fig. 6 – Influence of positive d.c. potential of third electrode and voltage impulse duration τ_{imp} on extracted current I_d. In Fig. 6A curve B is for -5.5 kV, C -6 kV, D -7 kV and τ_{imp} = 5 μs

4. Conclusions

The SDBDs driven by periodic impulse high voltage have peculiarities that differ this type of discharge from the sine voltage case:

- the channel structure of the impulse surface discharge is more pronounced especially at the rising slope of the voltage impulse of both polarities;
- the discharge current in the oscillograms is presented by a single pulse for all voltage impulses with slope steepness of ≥ 50 V/ns used in the present work;
- an increase of the voltage impulse duration from 0.5 ns to 50 ns can change the discharge structure as well as the discharge characteristics. This effect can be attributed to a change in the pre-ionization phase of the discharge.

Acknowledgement. The work has been done with financial support of Russian Fund of Basic Researches, grant No 15-08-4384.

References

[1] Fridman A, Chirokov A and Gutsol A 2005 *J. Phys. D: Appl. Phys.* **38** R1–24
[2] Chmielewski A G 2011 *Monitoring, Control and Effects of air Pollution* (Rijeka: InTech)
[3] Bednar N, Matović J, Stojanović G 2013 *Journal of Electrostatics*, **71** 1068-75
[4] Stepanyan S, Starikovskiy A, Popov N and Starikovskaia S 2014 *Plasma Sources Sci. Technol.* **23** 045003

[5] Malashin M, Moshkunov S, Khomich V et al. 2016 *Instruments and Experimental Techniques* **59** 226
[6] Höft H, Becker M, Loffhagen D and Kettlitz M 2016 *Plasma Sources Sci. Technol.* **25** 064002
[7] Malashin M, Rebrov I, Nebogatkin S et al. 2016 *Eur. Phys. J. Appl. Phys.* **75** 24705

Positive Streamer in the Surface Dielectric Barrier Discharge in Air: Numerical Modelling and Analytical Estimations

V Soloviev[1], **V Krivtsov**[2]

[1] Moscow Institute of Physics and Technology, 141700 Dolgoprudniy, Moscow Region, Russia
[2]Dorodnicyn Computing Center RAS, 119333 Moscow, Russia

E-mail: vic__sol@mail.ru

Abstract. According to performed numerical simulation of the surface dielectric barrier discharge driven by positive polarity nanosecond voltage pulse the discharge in this case evolves as a streamer "flying" above the dielectric surface. The distance between the streamer and dielectric surface does not depend on dielectric barrier parameters and applied voltage value. The developed analytical model for surface streamer evolution confirms these results and explains the physics of this phenomenon. The electric field in front of a stationary streamer head is constant and defined only by ionization rate constant of the gas and its density.

1. Introduction

The study of the surface dielectric barrier discharge (SDBD) in atmospheric air is still of the considerable interest because of its promising applications for flow control aerodynamics [1-3] and for plasma assisted ignition and combustion [4,5]. The SDBD driven by nanosecond voltage pulse is a preferable object of investigation for discharge physics understanding, because this discharge develops as a single microdischarge on the leading front of the voltage pulse, in contrast to a set of microdischarges relevant for a case of AC applied voltage. The advantage of studying the one microdischarge generation is in well known initial conditions and, consequently, in the possibility of more reliable interpretation of simulated results and experimental data.

The effectiveness of the nanosecond SDBD application for fuel ignition is strongly affected by the mode of the discharge: whether it is quasi-uniform or filamentary (constricted) mode of discharge development. In quasi-uniform mode the ignition occurs in a spot near the electrode edge [6], whereas in filamentary mode it is spread to longer distance along the discharge filament [7]. The physics of the SDBD transition from one mode to another is not understood yet. In order to get it, more detailed data regarding the discharge spatial structure and temporary evolution is needed.

According to performed numerical simulations [8-11] a single microdischarge in the SDBD evolves as a sliding transient glow discharge for negative exposed electrode polarity and as a streamer for positive one. Whereas for negative polarity the discharge is adjoined to the dielectric surface, for positive polarity it propagates at some distance above the surface. The gap between the streamer and the dielectric surface is characterized by positive ion charge surplus and extra-high electric field value at the level of 300 kV/cm for atmospheric air [9,12]. The existence of high electric field layers in the SDBD was implicitly confirmed in experiment [12].

This work is addressed to enhance our understanding of one aspect of the SDBD physics – the physics of positive streamer propagation. For this purpose the simplified analytical model for positive streamer maintenance condition is developed and the results of this model are compared with the results of numerical simulations. Both in numerical and analytical study we consider the streamer as a 2D sheet of plasma relevant for quasi-uniform SDBD development.

2. Numerical simulation of positive streamer in air

The 2D numerical simulation for SDBD streamer evolution in atmospheric air has been done using the numerical model developed in [8]. The results in the form of the extra positive charge density contours above the dielectric surface in units of 10^{14} cm^{-3} are shown in Fig.1 and demonstrate the streamer "flying" above the dielectric surface. Calculations were done for stepwise voltage amplitude $V = 5$ kV, dielectric thickness $d = 0.1$ mm, and dielectric relative permittivity $\varepsilon = 2.7$.

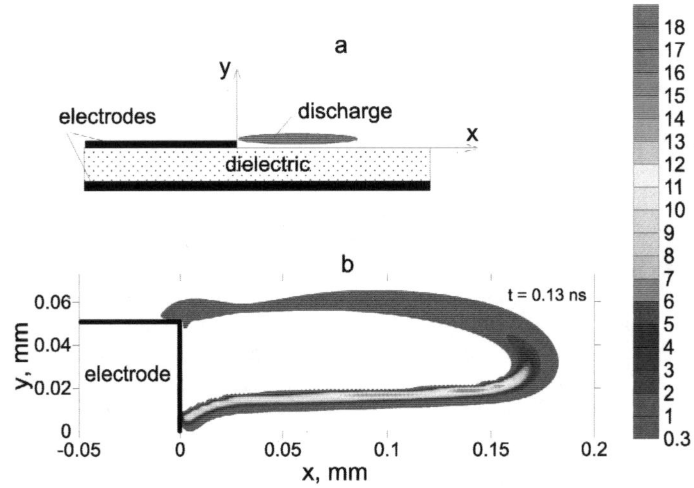

Fig.1 The scheme (a) of electrode layout and (b) the contours of positive charge surplus in 10^{14} cm^{-3}.

3. Analytical model

Numerical simulation has shown that all the quantities inside a streamer body, - the E-field and the charged particle densities, are almost constant in the streamer cross section and strongly vary only in the δ - layer near the streamer boundary. Accordingly, for analytical estimation of the quasi-steady streamer maintenance condition we can use 1D approximation for charged particles transport equations and for Poisson equation for electric field. The spatial coordinate s belongs to E-field line at the center of the streamer as shown in schematic picture of the streamer head in Fig.2a, where h is a streamer distance from the dielectric surface.

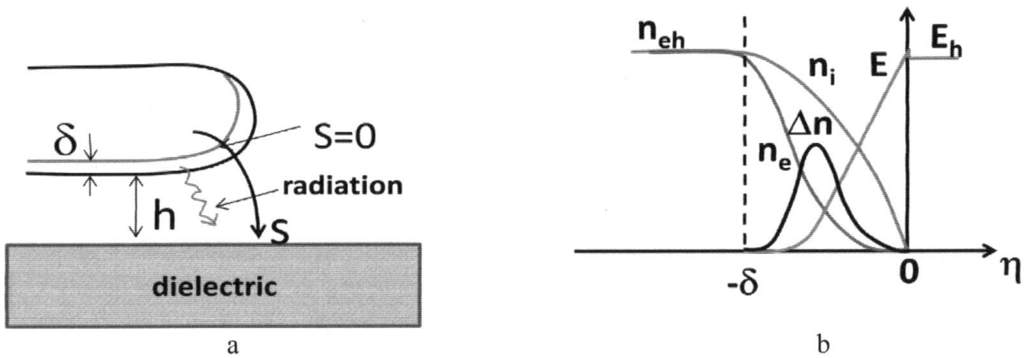

Fig.2 Schematic picture (a) for streamer head and (b) for qualitative solution inside the δ - layer.

The simplified system of equations reads

$$\frac{\partial n_e}{\partial t}-\frac{\partial n_e v_e}{\partial s}=v_i n_e,\quad \frac{\partial n_i}{\partial t}=v_i n_e,\quad \frac{dE}{ds}=4\pi\left(n_i-n_e\right). \tag{1}$$

Here n_e = electron density, n_i = positive ion density, $\Delta n = n_i - n_e$ is a positive ion surplus, $v_e = K_e E$ is electron drift velocity, K_e = electron mobility, v_i = gas ionization frequency. We neglect the negative ion formation and the drift of ions, which are not important in the streamer head formation process [13]. The Eqs.(1) are the same as in simplified model of volumetric streamer propagation. After the transition to new coordinate $\eta = s - v_s t$ moving with streamer head with velocity v_s these equations have the same solution obtained in [14]. According to this solution the electron-ion multiplication mainly occurs inside the δ- layer. The electron-ion density reaching the value n_{eh} behind the streamer head (Fig.2b) and the δ value read [14]

$$n_{eh}=\frac{v_i(E_h)v_s}{4\pi e_e K_e v_{eh}(1+v_s/v_{eh})}, \tag{2}$$

$$\delta \approx \frac{2}{\alpha_T(E_h)}. \tag{3}$$

The electric field inside the δ- layer falls down from E_h value in front of the streamer head to almost zero in the streamer body. The Eqs.(2) and (3) show that electron-ion density n_{eh} and δ thickness are the functions of electric field E_h in front of the streamer head. Here $\alpha_T(E_h)=v_i(E_h)/K_e E_h$ is Townsend coefficient, and $v_{eh} = K_e E_h$ is electron drift velocity in electric field E_h.

The commonly used expression for $\alpha_T(E)$ is

$$\alpha_T(E)=\frac{1}{\lambda_{im}}\exp\left(-E_c/E\right),\ \lambda_{im}=\frac{1}{Ap},\ E_c=Bp, \tag{4}$$

where A and B are the constants defined by gas properties, the book [14] recommends A=15cm^{-1}Torr^{-1}, B = 365V/cm/Torr for air, p is a gas pressure. The λ_{im} means the asymptotic for $E\rightarrow\infty$ value of ionization length and E_c is the characteristic E- field value describing the ionization rate saturation for growing E. The $\lambda_{im}=0.9\cdot10^{-4}$ cm and the $E_c = 277$ kV/cm in atmospheric air.

Following the approach realized in [15] for volumetric streamer between pin and plain electrodes we formulate the electron balance equation near the streamer head. The region between the dielectric surface and the 2D-streamer bottom boundary near the streamer head may be considered as a flat region with all parameters variation only in the normal to dielectric surface direction s. In this approximation the number of electrons generated due to gas ionization inside a gap between dielectric surface and streamer per unit streamer length is

$$N_e(\delta)=\int_{\delta}^{\delta+h} dN_{ph}\left[\exp\left(\int_0^{\delta}\alpha_T(s)ds\right)+1\right], \tag{5}$$

where

$$dN_{ph}=\alpha_{ph}ds\frac{1}{2}\frac{\tau^*}{\tau_R}N_e\alpha_{ex}\exp\left(-\alpha_{ph}(s-\delta)\right) \tag{6}$$

is the number of photoelectrons born on ds length due to UV radiation from the discharge region. The photo-ionization occurs because O_2 molecules are ionized by radiation of $N_2(b^1\Pi_u, b'^1\Sigma_u^+, c'^1_4\Sigma_u^+)$ excited molecules in the band 98.0-102.5 nm [16]. Here α_{ph} = photo-absorption coefficient for this radiation by O_2 molecules, $\alpha_{ex} = 0.07$ is a ratio of excitation rate of aforementioned excited N_2

molecules by electron impact to their ionization rate, τ^* and τ_R are the total and radiative lifetimes of excited N_2 molecules [8,16], N_e is the total number of elections inside a discharge region initialized by N_{ph} photoelectrons and it is defined by equation

$$N_e = N_e(\delta)\exp\left(\int_0^\delta \alpha_T(s)ds\right). \tag{7}$$

The steady discharge development implies that the number of electrons born in multiplication of N_{ph} photoelectrons is greater than N_e number necessary to create these N_{ph} photoelectrons. This condition leads to inequality

$$\chi_{ph}N_e\int_\delta^{\delta+h}\exp\left(-\alpha_{ph}(s-\delta)\right)\left[\exp\left(\int_\delta^s \alpha_T(\xi)d\xi\right)+1\right]ds\cdot\exp\left(\int_0^\delta \alpha_T(\xi)d\xi\right)\ge N_e, \tag{8}$$

where $\chi_{ph} = 0.5\tau^*\alpha_{ex}\alpha_{ph}/\tau_R \approx 0.1\,\text{cm}^{-1}$ and $\alpha_{ph} \approx 100\,\text{cm}^{-1}$ in atmospheric air. After integration the expression (8) transforms to inequality

$$F(E_h,h,\delta) = \frac{\chi_{ph}}{\alpha_T(E_h)}\exp\left(\alpha_T(E_h)h\right)\cdot\exp\left(\int_0^\delta \alpha_T(s)ds\right)\ge 1. \tag{9}$$

For three unknowns E_h, h, δ we have only two conditions: the Eq.(3) and the inequality (9).

The third required condition is the δ - layer formation condition. According to Poisson equation $E_h \approx 4\pi e n_i \delta$. If $F(E_h,h,\delta)\ge 1$, then n_i grows up inside a δ - layer. This process occurs at a constant E_h and results in δ decreasing. It continues until the potential difference ΔV_c on the δ- layer reaches its minimum value. Assuming linear $E(s)$ dependence inside a δ - layer and accounting for Eq.(3) and (4) we get

$$\Delta V_c \approx \frac{E_h\delta}{2} = \frac{E_h}{\alpha_T(E_h)} = E_h\lambda_{im}\exp\left(\frac{E_c}{E_h}\right). \tag{10}$$

From the solution of equation $\partial\Delta V_c/\partial E_h = 0$ for ΔV_c extremum condition we get

$$E_h \approx E_c. \tag{11}$$

This value for electric field in front of the streamer head has been confirmed by numerical simulations both for surface [9,12] and volumetric streamers [17,18].

Combining Eqs.(3), (11) and inequality (9) we get the inequality $F(E_h,h,\delta)\big|_{E_h=E_c,\,\delta=2/\alpha_T(E_c)} \ge 1$ for h value determination, which gives

$$h \ge -e\lambda_{im}\left(\ln\left(e\lambda_{im}\chi_{ph}\right)+1\right). \tag{12}$$

For atmospheric air $h \ge 30\lambda_{im} \approx 0.03$ mm and coincides with numerical simulations [8,9,12].

4. Conclusions

The proposed simplified analytical model for quasi-steady streamer maintenance condition permitted to find out the distance at which the streamer moves above the dielectric surface in the surface dielectric barrier discharge (SDBD) driven by positive polarity voltage pulse. This distance is managed only by the gas ionization rate constant, namely, it is primarily defined by asymptotic for infinite *E*-field ionization length of the gas. For atmospheric air the streamer remoteness from the dielectric surface is around 0.03 mm.

The electric field in front of the quasi-steady streamer head is managed by another parameter of the gas ionization rate constant (Townsend coefficient); it is equal to characteristic electric field value, at which the Townsend coefficient starts to saturate in the growing up electric field. It should be emphasized that this result is valid not only for positive surface streamer in SDBD, but for volumetric streamer between pin and plain electrodes as well. The value of this electric field in atmospheric air is near 300 kV/cm.

Both analytical results for streamer remoteness from the dielectric surface and for electric field value in the streamer head are confirmed by numerical simulation.

Acknowledgements
This work was supported by Russian Ministry of Science and Education (project No 3.9279.2017).

References

[1] Benard N and Moreau E 2014 *Exp Fluids* **55** 1846
[2] Corke T C, Post M L, and Orlov D M 2009 *Exp Fluids* **46** 1
[3] Leonov S B, Adamovich I V, and Soloviev V R 2016 *PSST* **25** 063001
[4] Starikovskiy A and Aleksandrov N 2013 *Prog. Energy Combust. Sci.* **39** 61
[5] Starikovskaia S M 2014 *J. Phys. D: Appl. Phys.* **47** 353001
[6] Anokhin E M, Kuzmenko D N, Kindysheva S V, Soloviev V R, and Aleksandrov N L 2015 *PSST* **24** 045014
[7] Shcherbanev S A, Popov N A, and Starikovskaia S M 2017 *Combustion and Flame* **176** 272
[8] Soloviev V R and Krivtsov V M 2009 *J. Phys. D: Appl. Phys* **42** 125208
[9] Soloviev V and Krivtsov V 2014 *Plasma Physics Rep.* **40** 65
[10] Soloviev V R, Krivtsov V M, Shcherbanev S A, and Starikovskaia S M 2017 *PSST* **26** 014001
[11] Babaeva N Yu, Tereshonok D V, and Naidis G V 2016 *PSST* **25** 044008
[12] Stepanyan S A, Soloviev V R, and Starikovskaia S M 2014 *J. Phys. D: Appl. Phys.* **47** 485201
[13] Soloviev V R, Konchakov A M, Krivtsov V M, and Aleksandrov N L 2008 *Plasma Physics Reports* **34** 594-608
[14] Bazelyan Ed M, Raizer Yu P *Spark Discharge* CRC Press Boca Raton New York, 1997
[15] Pancheshnyi S V, Starikovskaia S M, and Starikovskii A Yu 2001 *J. Phys. D: Appl. Phys.* **34** 1
[16] Zhelezniak M B, Mnatsakanian A Kh, and Sizykh S V 1982 *High Temp.* **20** 423
[17] Babaeva N Yu and Naidis G V 1996 *J. Phys. D: Appl. Phys.* **29** 2423
[18] Pancheshnyi S V and Starikovskii A Yu 2004 *PSST* **13** 1

IOP Conf. Series: Journal of Physics: Conf. Series **927** (2017) 012060 doi:10.1088/1742-6596/927/1/012060

Thermal steam plasma decomposition of organochlorine compounds

A V Surov[1], D I Subbotin[1,2,3], V E Popov[1], S D Popov[1], A I Litvyikova[2], Gh V Nakonechniy[1], E O Serba[1], N V Obraztsov[4]

[1]Institute for Electrophysics and Electric Power of Russian Academy of Sciences (IEE RAS), Dvortsovaya emb. 18, 191186, St.Petersburg, Russia
[2]Saint-Petersburg State Institute of Technology, Moskovsky pr. 26, 190013, St. Petersburg, Russia
[3]Saint-Petersburg State University, Universitetskaya emb. 7-9, 199034, St. Petersburg, Russia
[4]Peter the Great St. Petersburg Polytechnical University, Polytechnicheskaya 29,195251, St.Petersburg, Russia
alex_surov@mail.ru

Abstract. For the almost complete processing of organochlorine compounds are required a high temperature, hydrogen to produce hydrogen chloride and a high degree of mixing. Reforming chlorobenzene by steam and carbon dioxide in the presence of methane using a three-phase AC plasma torch was carried out. Soot composition was analyzed by energy dispersive X-ray analysis. The yield of soot was 0.84% wt. of raw materials, the content of chlorine in the soot was 2.08% by wt.

1. Introduction

The world has accumulated a large amount of chlorine-containing organic waste. Some of them cannot be processed by traditional incineration [1]. One of promising methods for the processing of these materials may be plasma thermal processing, where an electric arc plasma torch can be successfully used as plasma source. Nowadays plasma generating installations using electric arc can be divided into two categories: direct current (DC) and alternating-current (AC) plasma torches. The most widely used are DC devices, overview of the various modern models is presented in [2], AC systems designed for industrial applications are presented in [3]. A variety of gases can be used in plasma torches as the plasma forming medium, of particular interest are devices that operate on the steam [4-7], there are some plants in which the electric arc is in contact with liquid water [8].

There are a number of plasma methods for processing chlororganic substances using nonequilibrium plasma: DBD [9], corona discharge [10], glow discharge [11], etc. It is typically considered the interaction of chlorine-containing substances with the thermal plasma of air [12] or inert gases [13]. In this case the active oxygen concentration is not high, but it is possible to significantly increase the temperature due to the exothermic reaction. Thus chemical bond C-Cl must break and new stable chemical compound forms (for example, HCl). This process is possible only with an excess of hydrogen, which is not enough in aromatic compounds. In this case, thermal steam plasma may be used:

$$C_xH_yO_zCl_n + (x - z)H_2O = xCO + \frac{y+2x-z-n}{2}H_2 + nHCl \quad (1)$$

$$C_xH_yO_zCl_n + \frac{4x-z+n-y}{2}H_2 = (x - z)CH_4 + zCO + nHCl \quad (2)$$

Usage of steam instead of air gives several advantages:

1. No nitrogen involved in the reaction and consequently no risk of formation of toxic nitrogen compounds.

2. The volume of the waste gases is 2-3 times lower than at air plasma processes that reduces the cost of the gas cleaning.

3. Steam is highly effective agent which prevents the formation of soot reaction.

If the original compound contains an insufficient amount of hydrogen for binding chlorine, additional quantity of hydrogen may be produced in the high temperature zone by reaction of steam with methane. The required hydrogen may be produced by thermal steam and carbon dioxide plasma reforming of methane [14]:

$$0.775CH_4 + 0.55H_2O + 0.225CO_2 = 2.1H_2 + CO \quad (3)$$

The specific energy consumption is about 37.5 MJ/kg of converted methane.

The article deals with the processing of organochlorine substances by the example of the decomposition of chlorobenzene using an AC plasma torch operating on a mixture of steam, carbon dioxide and methane.

2. Experimental installation

The investigation was carried out on a laboratory facility shown in Fig. 1. Main part of facility was AC plasma torch tested on a mixture of steam, methane and carbon dioxide [15]. It is a plasma torch with rod electrodes and vortex stabilization of the electric arc and plasma-forming mediums two inputs (the electrode zone and the arc zone). Carbon dioxide (shielding gas) and methane (for power control) were fed in the electrode zone and vapor of organic substance, steam were fed in the arc zone.

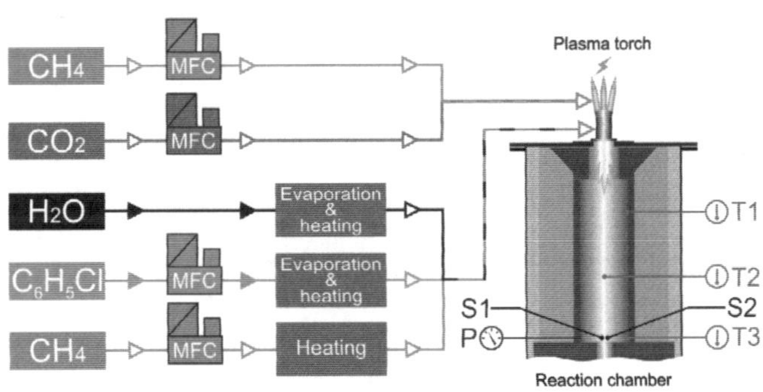

Fig. 1. Scheme of the experimental facility. MFC – mass flow controller, T – temperature sensor, P – pressure sensor, S – sampling point

Chlorobenzene, the simplest aromatic compound, was chosen as an organic matter. When heated, chlorobenzene easily forms soot, so its decomposition is a major challenge.

Sampling of the reaction products was carried out at the points S1 and S2. The composition of the product gas was determined by mass spectrometry Cirrus 300. Sampling of soot was carried out by gravimetric analysis (ISO 9096) using quartz filter. Soot was investigated by X-ray diffractometer, Fourier transform infrared spectrometer and scanning electron microscope.

3. Results and discussion

Flow rates of plasma forming mediums were the following: CH_4 - 0.3 g/s, H_2O - 2.9 g/s, CO_2 - 2.9 g/s, C_6H_5Cl - 4.0 g/s. Power of plasma torch increased with increasing flow rate of chlorobenzene (from 92 to 106 kW – Fig. 2). This is due to the energy consumption at the steam reforming and the increase in heat capacity and thermal conductivity (for account of hydrogen formation).

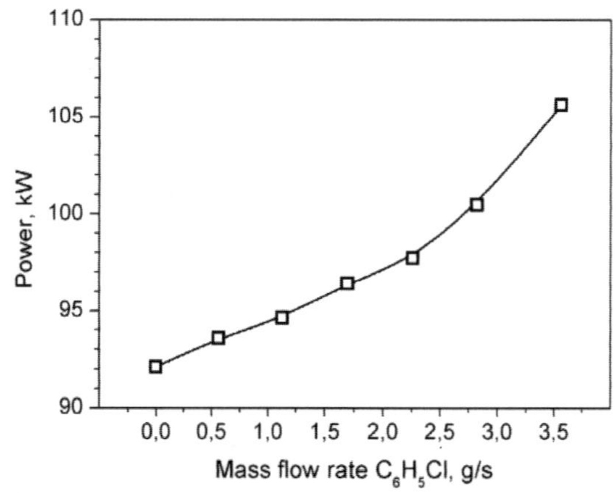

Fig. 2. Power of plasma torch versus mass flow rate of chlorobenzene.

The average temperature in the plasma reactor was 1250°C. Due to the produced gas vortex inside the reactor components are uniformly mixed. The sampling of the product gas and the soot was carried out for 10 minutes. The composition of the dry product gas is shown in Table 1.

Table 1 Composition of the syngas

Substance	CH_4	H_2	CO_2	CO	HCl
Concentration, mol %.	0.04	45.05	3.13	45.51	6.27

The main products of the reaction are hydrogen and carbon monoxide, while other chlorine-containing compounds, except for hydrogen chloride, were not detected. This indicates the complete decomposition of chlorobenzene. The produced product gas can be purified from hydrogen chloride and used to produce thermal and electrical energy, as well as for the synthesis of organic substances.

Energy dispersive X-ray analysis showed (Table 2) that the sample of filter with soot contains carbon, chlorine and products of erosion of electrodes and the plasma torch case (made of stainless steel).

IR-spectroscopy confirmed the presence of chlorine-containing compounds (Fig. 3). The peaks in the spectrogram correspond to C-Cl, O-H, C-O, Cl-O the chemical bonds. This indicates the formation of complex solid chlorine-containing organic substances. Despite the heated filter is used, soot contains water and, possibly, phenolic compounds.

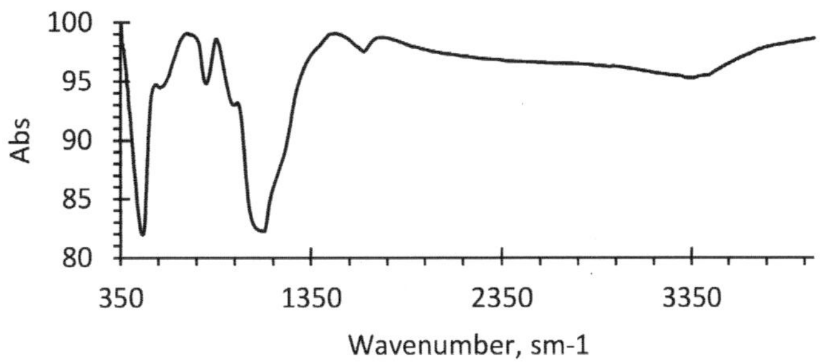

Fig. 3. IR-spectrum of soot with quartz filter.

Table 2 Composition of soot with quartz filter material

Chemical element	Concentration, % wt.
C	43.39
O	28.88
Na	0.30
Al	0.58
Si	16.71
S	0.18
Cl	1.31
K	0.13
Cr	0.59
Mn	0.35
Fe	4.84
Ni	0.30
Cu	2.44

XRD-ray analysis confirmed that the soot is predominantly in an amorphous state. The peaks on the XRD pattern correspond to Al_2O_3 cell (Fig. 4).

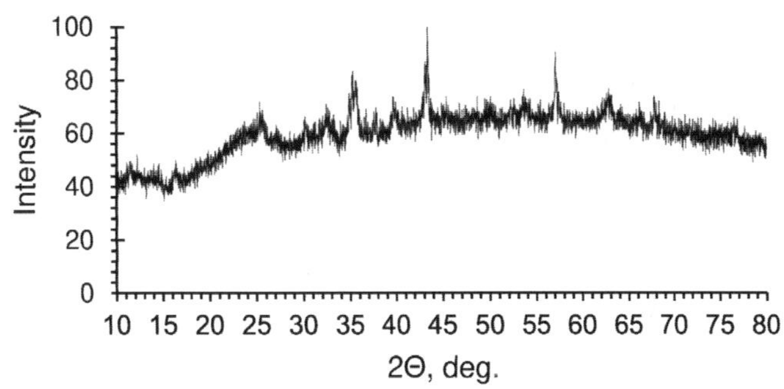

Fig. 4. XRD pattern of soot with quartz filter.

When calculated on the dry product gas soot yield was 1.31% mass. of raw materials, and the chlorine content in the soot - 1.61 wt.%.

The feed residence time in the plasma reactor is 0.1 s, which is not enough for the complete conversion of chlorobenzene. Furthermore, so as stoichiometric ratios were selected, local unevenness of components mixing is very likely possible. Thus soot formation reaction becomes more likely.

Probably, for the decomposition of chlorine-containing soot it is require a longer residence time and a greater excess of steam.

4. Conclusion

The simplest chlorine-containing aromatic compound (chlorobenzene) was selected for the investigation. The result revealed that the high temperature (above 1250 °C) does not provide complete decomposition of chlorobenzene to hydrogen chloride. Sooting is explained small excess oxidant, competing reaction of steam with methane, and small residence times. Further research should focus on determination of process conditions with minimal soot formation.

5. Acknowledgements

The work is partially supported by the Russian Foundation for Basic Research, grant No. 15-08-05909-a.

References

[1] Zhou H, Meng A, Long Y, Li Q and Zhang Y 2015 *Waste Management* **36** 106-118
[2] Mostaghimi J, Boulos M 2015 *Plasma Chem. Plasma Proc.* **35** 421-36
[3] Fulcheri L, Fabry F, Takali S and Rohani V 2015 *Plasma Chem. Plasma Proc.* **35** 565-85
[4] Zukov M F, Zasypkin I M, Timoshevskii A N, Mikhailov B I and Desyatkov G A 2007 *Thermal plasma totches: Design, Characteristics, Applications* (Cambridge International Science Publishing Ltd) p 596
[5] Grigaitienė V, Snapkauskienė V, Valatkevičius P, Tamošiūnas A and Valinčius V 2011 *Catalysis Today* **167** Issue 1 135-140
[6] Kim S-W, Park H-S and Kim H-J 2003 *Vacuum* **70** Issue 1 59-66
[7] Rutberg Ph G, Kuznetsov V A, Serba E O, Popov S D, Surov A V, Nakonechny Gh V and Nikonov A V 2013 *Applied Energy* **108** 505-514
[8] Hlina M, Hrabovsky M, Kavka T and Konrad M 2014 *Waste Management* **34** Issue 1 63-66
[9] Vanraes P, Ghodbane H, Davister D, Wardenier N, Nikiforov A, Verheust Y P, Hulle S W H, Hamdaoui O, Vandamme J, Durme J, Surmont P, Lynen F and Leys C 2017 *Water Research* **116** 1-12
[10] Yan X, Zhu T, Fan X and Sun Y 2014 *Chemical Engineering Journal* **245** 41-46
[11] Lu Q, Yu J and Gao J 2006 *Journal of Hazardous Materials* **136** Issue 3 526-531
[12] Rutberg Ph G, Bratsev A N, Kuznetsov V A, Popov V E and Shtengel' S V 2011 *Biomass and Bioenergy* **35** Issue 1 495-504
[13] Opalińska T, Wnęk B, Witowski A, Juszczuk R and Bartusek S 2016 *Journal of Hazardous Materials* **318** 282-290
[14] Rutberg Ph G, Kuznetsov V A, Popov V E, Popov S D, Surov A V, Subbotin D I and Bratsev A N 2015 *Applied Energy* **148** 159-168
[15] Surov A V, Popov S D, Popov V E, Subbotin D I, Serba E O, Spodobin V A, Nakonechny G V and Pavlov A V 2017 *Fuel* **203** 1007-14

IOP Conf. Series: Journal of Physics: Conf. Series **927** (2017) 012061 doi:10.1088/1742-6596/927/1/012061

High voltage AC plasma torch operating on vapours of organic substances

A V Surov[1], S D Popov[1], E O Serba[1], V A Spodobin[1], Gh V Nakonechniy[1], A V Pavlov[1], A V Nikonov[1], D I Subbotin[1,2,3], V E Popov[1], N V Obraztsov[1,4]

[1]Institute for Electrophysics and Electric Power of Russian Academy of Sciences (IEE RAS), Dvortsovaya emb. 18, 191186, St.Petersburg, Russia
[2]Saint-Petersburg State Institute of Technology, Moskovsky pr. 26, 190013, St. Petersburg, Russia
[3]Saint-Petersburg State University, Universitetskaya emb. 7-9, 199034, St. Petersburg, Russia
[4]Peter the Great St. Petersburg Polytechnical University, Polytechnicheskaya 29,195251, St.Petersburg, Russia
alex_surov@mail.ru

Abstract. The article deals with a three-phase high-voltage AC plasma torch with separate supply of gases and vapours during its operation on a mixture of steam, carbon dioxide, methane and organic vapours. Increase in flow rates of organic vapours and methane leads to increase in the arc voltage drop from 875 to 1150 V and electric power from 80 to 102 kW.

1. Introduction

Plasma torches have found wide application in metal cutting, heat treatment of surfaces, solid waste processing, reforming of natural gas, and also processing of toxic substances. The most complex task of plasma chemistry is a high conversion of organic matter due to high temperature, good mixing of reaction components and the formation of active radicals [1]. In this case, air direct current plasma torches are most often used [2]. In addition, the processing of waste by nonequilibrium plasma is investigated: gliding discharge [3], DBD [4], RF plasma [5]. Particular interest is plasma processing with steam, which leads to the decomposition of complex heteroatomic organic substances (e.g., chlorinated) with the formation of an appropriate hydride (e.g., hydrogen chloride) [6]. Steam plasma torches were used for gasification of municipal solid waste [7], treatment of polychlorinated biphenyls [8], perfluorinated compound [9], reforming natural gas [10], etc.

The atmospheric pressure electric arc plasma torch was used to obtain active plasma radicals (O, H, OH) from steam for reforming glycerin [11]. The power of the plasma torch is from 45 kW to 61 kW (current 200 A, voltage 225 - 305 V), steam flow rate is 2.4-4.5 g/s at 500 K.

Another water-argon DC plasma torch was used for processing toxic substances. This plasma torch is about 110 kW with a W-Th cathode, in which the arc is stabilized by argon and a vortex of liquid water. The flow rate of argon is 22 l/min, the flow rate of evaporating water is about 18 g/min [12]. The anode of the plasma torch is a rotating disk with water cooling, which is located outside the arc chamber.

Also, a CF_4 decomposition system including a steam atmospheric pressure DC plasma torch was investigated. The anode was a copper nozzle, and the cathode was made from a hafnium rod 1.0 mm in diameter [13]. In the plasma torch there are two inputs: liquid water, which evaporates, forms a plasma-forming gas in the anode region; CF_4 and O_2 are fed to the end of the nozzle. The electric current of the arc was from 7 to 9 A, and the voltage was from 110 to 150 V, the mass flow of water and CF_4 was 75 mg/s and 3.75-15 mg/s, respectively.

However, at the moment there is no information on plasma torches operating on vapours of organic substances. Those plasma torches will be most effective for the decomposition of organic substances, because in them all the reaction components will pass through the high-temperature zone.

The article deals with a new three-phase AC plasma torch operating with a mixture of steam, carbon dioxide, methane and organic vapors as a plasma-forming medium.

2. Investigated plasma torch

The investigations were carried out on the facility consisting of a high-voltage three-phase AC plasma torch (Fig. 1), a steam generator with a superheater and a power system with an electrical measuring circuit (Fig. 2). The plasma torch is three cylindrical discharge channels with two injections of plasma-forming gases and three rod electrodes [14]. A shielding gas (carbon dioxide) is supplied to the near-electrode zone, and steam, methane and vapors of a mixture of organic substances are fed into the arc burning zone. The composition of the mixture of organic substances is presented in Table 1.

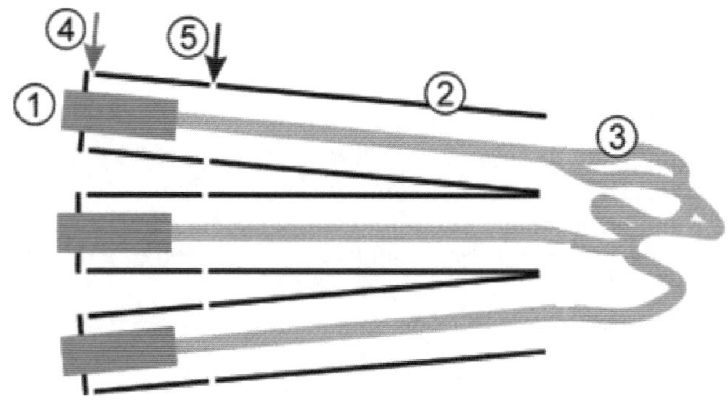

Figure 1 The three-phase AC plasma torch
1 - electrode; 2- arc channel; 3 — electric arc;4 — shielding gas (carbon dioxide); 5 - plasma-forming gas (steam, methane, the mixture of organic substances).

Table 1. Composition and properties of the mixture of organic substances.

Parameters	Units	Value
Toluene	% wt	50
Butanol	% wt	15
Butyl acetate	% wt	10
Ethanol	% wt	10
Acetone	% wt	7
Ethyl cellosolve	% wt	8
Initial boiling point	°C	59
The auto-ignition temperature	°C	403

The selected composition of the mixture of organic substances complicates the production of vapours, so the heating of this mixture was carried out in two stages in different heat exchangers: heating to the boiling point and evaporation of the mixture.

The plasma torch is connected to the power supply and the measuring circuit (Figure 2). The electrical circuit of the power supply system consists of a circuit breaker (SF), a contactor (KM), capacitive reactive power compensators, current-limiting reactors (LL), a step-up transformer (T), measuring transformers of current (TI) and voltage (TU) and secondary converter of sensor (DI and DU) [15]. Losses from ohmic heating in this system do not exceed 2-3% of the total power. The open-circuit voltage of the power source is constant and equal to 10 kV.

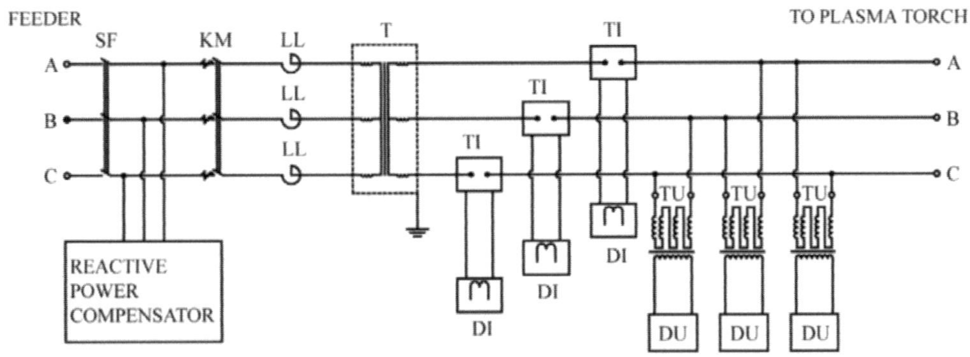

Figure 2 The electric scheme of the power supply system for the three-phase AC plasma torch.

3. Results and discussion

The dependence of the parameters of the plasma torch operation on the flow rate of a mixture of organic substances is studied. CO_2 was supplied to the near-electrode zone, and a mixture of organic substances was fed into the arc burning zone together with steam and CH_4. The steam flow rate is 2.9 g/s, the CO_2 flow rate is 2.9 g/s, the CH_4 flow rate is 0.3 g/s, the flow rate of the mixture of organic substances is varied from 0 to 3 g/s. The plasma torch worked on the environment at atmospheric pressure. Figure 3 shows the dependences of the voltage drop on the arc and the power of the plasma torch on the flow rate of a mixture of organic substances. The effective current value in the arcs was ~ 52 A.

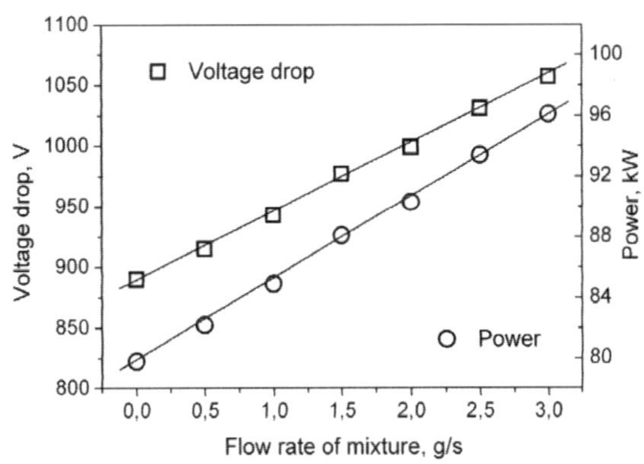

Figure 3 Dependences of the power of the plasma torch and the voltage drop on the electric arc on the flow rate of the mixture of organic substances (at constant flow rates of other components of plasma forming mixture)

Also, the dependences of the operation parameters of the plasma torch on the CH_4 flow rate into the arc burning zone with a constant flow rate of a mixture of organic substances were obtained. The steam flow rate was 2.9 g/s, the CO_2 flow rate was 2.9 g/s, the CH_4 flow rate varied from 0.3 to 1.2 g/s, the organic matter flow rate was 3 g/s. Figure 4 shows the dependences of the voltage drop on the arc and the power of the plasma torch on the flow rate CH_4.

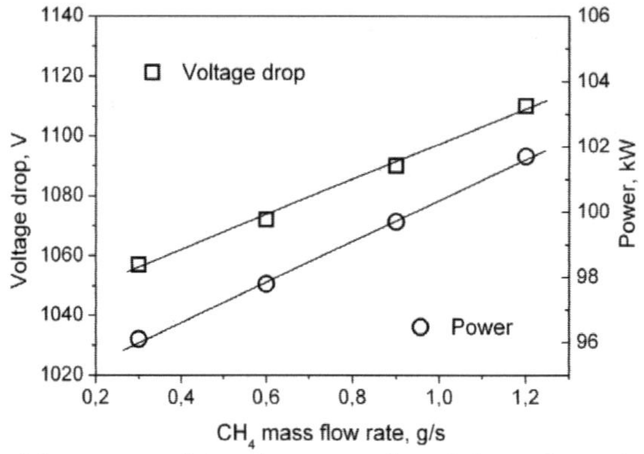

Figure 4 Dependences of the power of the plasma torch and the voltage drop on the electric arc on the methane flow rate (at constant flow rates of other components of plasma forming mixture)

As can be seen from Figures 3 and 4, the voltage drop of the arc of the plasma torch increases with increasing flow rate of mixture vapors of the organic substances and CH_4. This is due to the formation of hydrogen during the reforming of organic substances by steam and carbon dioxide at high temperature:

$$C_6H_5CH_3+6H_2O=6CO+10H_2 \quad (1)$$

$$C_6H_5CH_3+6CO_2=12CO+4H_2 \quad (2)$$

In comparison with carbon dioxide hydrogen has a high heat capacity and thermal conductivity, which leads to an increase in heat losses from the arc. The change in the power of the plasma torch is also important for the processing of organic substances. Depending on the composition of the mixture of organic substances, different energy inputs are required. This value can be changed by feeding vapours of organic substances into the plasma torch or into the plasma chemical reactor located behind the plasma torch. In addition, the steam plasma produced in the plasma torch effectively decomposes the soot formed in the colder parts of the plasma chemical reactor. This fact is confirmed for the most thermally stable hydrocarbon (methane) [16]. Due to the protective gas (CO_2), hydrogen does not interact with the material of the electrodes. In this case, its service life can significantly increase.

4. Conclusion

During the tests it was found that the developed plasma torch operating on a mixture of carbon dioxide, steam, methane and vapours of the mixture of organic substances has a power of 80-102 kW. The power can be changed by changing the flow rates of a mixture of organic substances (80-96 kW) and methane (96-102 kW). This plasma torch is suitable for plasma steam reforming of various organic substances, including their mixtures. Carbon dioxide protects rod electrodes from aggressive steam and participates in reforming. Therefore, the process proceeds without the formation of by-products. The plasma torch operates in a wide range of flow rates of plasma-forming media, which allows achieving of a low concentration of soot in reaction products.

5. Acknowledgments

The work is partially supported by the Russian Foundation for Basic Research, grant No. 15-08-05909-a.

References

[1] Nizio M, Albarazi A, Cavadias S, Amouroux J, Galvez M E and Costa P 2016 International

Journal of Hydrogen Energy 41 Issue 27 11584-11592

[2] Mazzoni L, Janajreh I 2017 International Journal of Hydrogen Energy 42 Issue 30 19446-19457

[3] Opalińskaa T, Wnek B, Witowski A, Juszczuk R, Majdak M, Bartusek S 2016 Journal of Hazardous Materials 318 282-290

[4] Kundu S K, Kennedy E M, Mackie J C, Holdsworth C I, Molloy T S, Gaikwad V V,. Dlugogorski B Z 2016 Chemical Engineering Journal 284 412-421

[5] Hsieh L-T, Lee W-J, Chen C-Y, Wu Y-P G, Chen S-J, Wang Y-F 1998 Journal of Hazardous Materials 63 Issue 1 69-90

[6] Surov A V, Popov S D, Popov V E, Subbotin D I, Obraztsov N V, Kuchina J A, Serba E O, Nakonechny Gh V, Spodobin V A, Pavlov A V and Nikonov A V 2017 Journal of Physics: Conf. Series 825 Issue 1 012016

[7] Zhang Q, Dor L, Biswas A K, Yang W and Blasiak W 2013 Fuel Processing Technology 106 546-554

[8] Kim S-W, Park H-S and Kim H-J 2003 Vacuum 70 Issue 1 59-66

[9] Kim D, Park D W 2008 Surface and Coatings Technology 202 Issues 22–23 5280-5283

[10] Rutberg Ph G, Bratsev A N, Kuznetsov V A, Nakonechny G V, Nikonov A V, Popov V E, Popov S D, Serba E O, Subbotin D I and Surov A V 2014 Technical Physics Letters. 40 Issue 9 725-729

[11] Tamošiūnas A, Valatkevičius P, Gimžauskaitė D, Valinčius V and Jeguirim M 2017 International Journal of Hydrogen Energy 42 Issue 17 12896-12904

[12] Mašláni A, Sember V and Hrabovský M 2017 Spectrochimica Acta Part B: Atomic Spectroscopy 133 14-20

[13] Narengerile, Saito H and Watanabe T 2009 Thin Solid Films 518 Issue 3 929-935

[14] Rutberg Ph G, Kuznetsov V A, Popov V E, Popov S D, Surov A V, Subbotin D I and Bratsev A N 2015 Applied Energy 148 159-168

[15] Rutberg Ph G, Nakonechny Gh V, Pavlov A V, Popov S D, Serba E O and A V Surov 2015 J. Phys. D: Appl. Phys. 48 245204

[16] Surov A V, Popov S D, Popov V E, Subbotin D I, Serba E O, Spodobin V A, Nakonechny G V and Pavlov A V 2017 Fuel 203 1007-14

Laboratory demonstration in the air red and blue mini-jets

V F Tarasenko[1,2], V A Panarin[1], V S Skakun[1] and E A Sosnin[1,2]

[1]Institute of High Current Electronics, 2/3 Akademichesky Ave., 634055 Tomsk, Russia

[2]National Research Tomsk State University, Lenina Ave. 36, 634050 Tomsk, Russia

E-mail: VFT@loi.hcei.tsc.ru

Abstract. The properties of apokamps (miniature plasma jets) with length up to 60 cm at low pressures are studied. It is shown that several properties of apokamps resembles that ones for large-scale transient luminous events (sprites, starters and blue jets) observed in the Earth upper atmosphere. The images and spectra of apokamps in these conditions are presented. For the first time red downward directed mini jets in air at pressures of fraction of Torr were obtained.

1. Introduction

In the past decade, there has been considerable progress in the study of discharges in the upper atmosphere [1-6]. Experimental studies using a high-altitude stations and international space station [6] revealed several important regularities in the formation of blue jets, red sprites, and other types of atmospheric transient luminous events. In particular, some regularities in the formation of blue jets and red sprites were defined. The observations show a multitude of blue, kilometer-scale, discharges at the cloud top layer at ~18 km altitude and a pulsating blue discharge propagating into the stratosphere reaching ~40 km altitude [6]. The blue jets formed from the top of the thunderclouds. Red sprites as distinct from blue jets propagates not only up from the initial plasma, but also downwards in the direction of increased negative charge. The laboratory experimental studies of atmospheric discharges are strongly complicated because of their huge dimensions reaching tens and hundreds of kilometers.

In our works [7-9] we describe a new discharge type with plasma jets (apokamps) formation under atmospheric air pressure. They had a blue color and formed at a bend point (natural or induced) of the discharge channel in the electric field strengthening point. We termed this discharge apokampic (in Greek. από – from and καμπη – bend, turn), and, accordingly, plasma jets germinated in this case – apokamps. Apokamps are directed almost perpendicular to discharge channel. Apokamp is distinguished from other types of plasma jets [10] by its forming in open space without dielectric walls and absence of gas force-feed to discharge area. Details about apokamp properties and formation conditions published in [11-14]. Particularly, it has been shown, that apokamp close to the blue jets in appearance and emission spectra at pressures at hundreds of tens of Torr. The velocities of propagation of apokamps have been measured at various pressures. It has been found that the average values of these velocities are about the velocities of propagation of starters and blue jets in the atmosphere of the Earth. It has been shown that jets (apokamps) with the maximum length are observed in the pressure range corresponding to the altitudes of appearance and propagation of starters and blue jets. But downward directed apokamps formation has not been studied. In addition, there is no studies of laboratory-scale apokamps at a pressure values less than 9 Torr.

In this work, we present the downward directed apokamps at air pressure of shares-units of Torr. Let us note that downward propagation to the Earth is characterized for lower part of red sprites [1-4, 15], and as it has been found in [14] the apokamp color at decreasing pressure changes from blue to red.

2. Experimental setup

The study of pressure and voltage influence in apokamp were carried out with the use of one generator and two discharge chambers, having a height of 25 and 60 cm. They are made of quartz tubes and the electrodes in tubes could be located in different positions relative to the Earth's surface. This makes possible the observation of the diffuse jets (apokamps) formation as in straight up and down.

Downward directed apokamps were observed for the first time. Figure 1 shows a setup that we assembled for this aim.

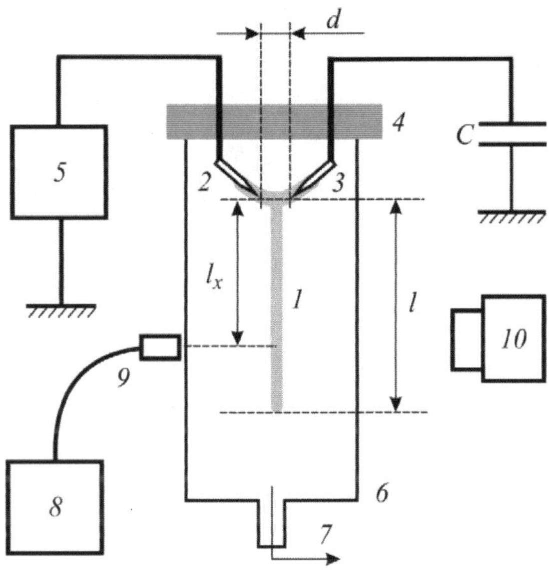

Figure 1. Schematic of the setup: *1* – apokamp; *2* – high-voltage electrode; *3* – floating potential electrode; *4* – dielectric flange; *5* – power supply; *6* – quartz bulb; *7* – gas inlet/outlet; *8* – spectrometer; *9* – collimator with optical fiber; *10* – photo camera. C = 5.6 pF – capacitor for increasing the voltage on the electrodes.

In order to form a plasma jet – apokamp *1*, we used a pulse periodic discharge between cylindrical steel electrodes *2* and *3*, which had a diameter of 1.8 mm and formed a gap of d = 10 mm. The electrodes had needle-shaped ends and the angle between them was 120°. The inclination of the electrodes makes the electric field under the electrodes greater than above ones. The electrode *2* was connected to the secondary coil of the pulse transformer of a generator of high-voltage pulses *5*. The generator formed voltage pulses U_p with a positive or negative polarity with the repetition frequency f up to 50 kHz, the FWHM duration of the pulse $\tau_{1/2}$ = 1.5 μs, amplitude U_p < 13 kV, and the pulse front ~0.8 μs. The electrode *3* had capacitive decoupling with the ground. With an increase in U_p, a discharge channel was first formed and apokamp with the length l appeared. The maximum current of the discharge increased with U_p and was ~ 0.2 mA at U_p = 11 kV and atmospheric pressure of air. When the pressure of air was reduced to tens of Torr, the amplitude of the current was doubled. The time profile of the current and voltage detected by a TDS 3034 four-channel oscilloscope (Tektronics Inc.). The spectra of apokamp was registered in different points on the distance l_x from the discharge channel. To obtain the total picture of luminescence, we used a Canon PowerShot SX60 HS camera in the mode of serial frame-by-frame recording with an exposure time of no less than 0.125 s.

3. Experimental results and discussion

In preliminary experiments, the electrodes in the discharge chamber (as in [7-9, 11-14]) was located at the chamber bottom and apokamps propagates upwards. In this case, apokamps color was blue at elevated pressures (see Fig. 2 (a)). Each apokamp consisted of bright branch forming from discharge channel and streamer launching from branch with average speed of about 200 km/s [8]. The dynamics of development of the apokamp was recorded by an HSFC-PRO four-channel high-speed camera (PCO AG) with the minimum duration of one frame of 3 ns. The camera was switched on with various time decays with respect to the time of appearance of the discharge.

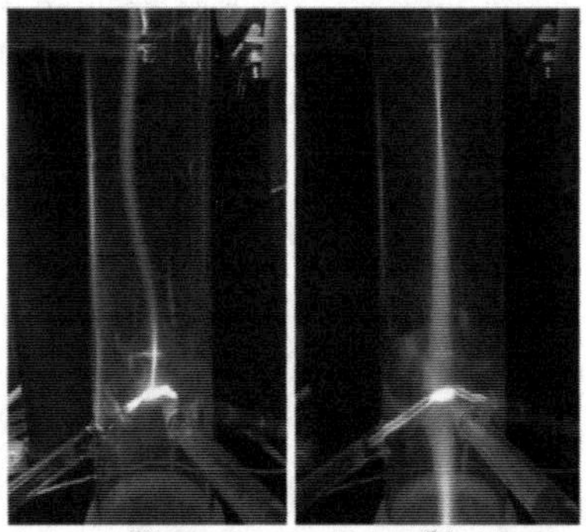

Figure 2. The color and shape of apokamp in air at 120 (a) and 15 (b) Torr. The vertical frame size is 22.2 cm. The electrodes are at the bottom. $U_p \sim 10$ kV.

The color of apokamp at pressure decreasing up to 30 Torr becomes red. In addition, with pressure decreasing the red color downward directed apokamp arise simultaneously with upward directed apokamp (Fig. 2 (b)).

In the current study, we focused on downward directed apokamps forming in discharge chamber shown in Fig. 1. Plasma jets were formed only at relatively low-pressure values and had a red color. At elevated pressures (> 30 Torr) the discharge was transferred to the sparks, that finished on wall of quartz bulb and stable apokamps failed to form. Figure 3 shows the effect of discharge voltage (a, b) and pressure (c, d, e) on apokamps length.

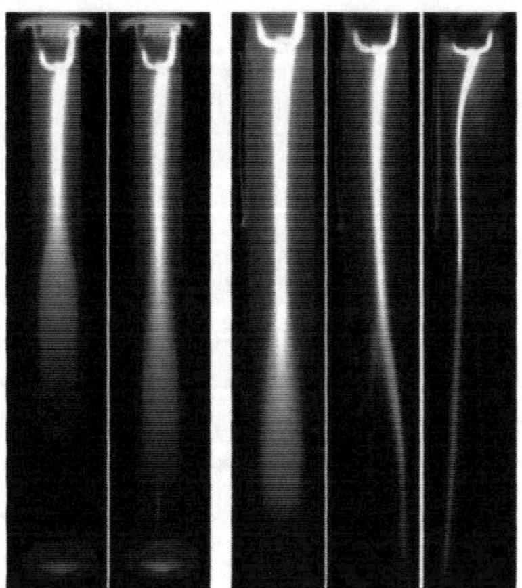

Figure 3. The effect of discharge voltage (a, b) and pressure (c, d, e) on apokamp length and shape in air at $p = 0.1$ Torr, $U_p = 3.9$ kV (a); $p = 0.1$ Torr, $U_p = 6.5$ kVкВ (b); $p = 0.1$ Torr, $U_p = 4.9$ kV (c); $p = 3$ Torr, $U_p = 6.3$ kV (d); $p = 9$ Torr, $U_p = 7.2$ kV (e). The vertical frame size is 72.8 (a, b) and 54.6 (c, d, e) cm. The electrodes placed at the top.

The following main properties of the plasma jets are evident from obtained experimental data. As one would expect, the increasing of voltage pulse amplitude leads to apokamp length increasing (Fig. 3 (a, b)). The pressure relief leads to increasing of apokamp length and diameter (Fig. 3 (c, d, e)). At elevated pressure values are clearly visible two parts of plasma jet. The first part is the bright reddish branch connected to discharge channel, and the second one is blue color jet (Fig. 3 (e)). The diameters of these parts are the same. It should be noted that the natural color of the sprites is also changed from red to blue during their propagation to the Earth [5, 15].

Figure 4 show the apokamp emission spectrum in logarithmic scale. The second positive system spectral band of nitrogen have maximal intensity. Visually we observe a blue color. Although at low-pressure values the second positive system band intensity is much more than the first positive system, visually the color of the apokamp at low pressures is red.

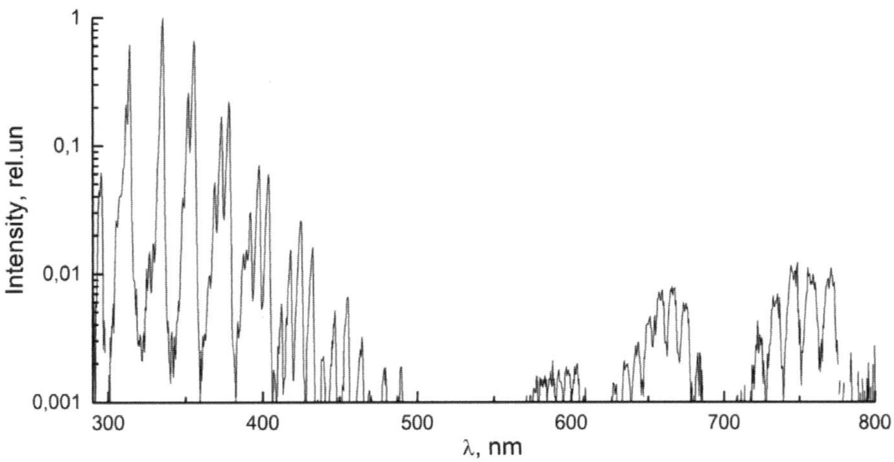

Figure 4. Apokamp spectrum in air at p = 0.1 Torr.

We are going to continue our studies and plan to increase the discharge chamber height (of a meter in size) and create a physical model that will describe the color, shape, and velocity of plasma jets.

4. Conclusion

To summarize, the possibility of red sprites and blue jets analogues forming in laboratory conditions was demonstrated. Red mini sprites and mini blue jets (apokamps) were obtained at voltages of units-tens kV by the use of pulse-periodic discharge with a pulse duration of positive polarity FWHM of 1.5 μs. The downward directed red color apokamps has been formed. The apokamp color depends on air pressure and changed from blue at elevated pressures to red when the pressure units of Torr. Moreover, at the same pressure the apokamp color varies lengthwise from red to blue when the distance from electrodes increase.

5. Acknowledgments

The work is performed in the framework of the State task for HCEI SB RAS, project #13.1.3.

References

[1] Cummer S A, Jaugey N, Li J, Lyons W A, Nelson T E and Gerken E A 2006 *Geophys. Res. Lett.* **33** L04104
[2] McHarg M G, Stenbaek- Nielsen H C and Kammae T 2007 *Geophys. Res. Lett.* **34** L06804
[3] Pasko V P 2007 *Plasma Sources Sci. Technol.* **16** S13–S29
[4] Raizer Y P, Milikh G M and Shneider M N 2010 *Journal of Geophysical Research* **115** A7
[5] Siingh D, Singh R P, Kumar S, Singh A K, Singh A K, Patil M N and Singh Sh 2015 *J. Atmos.*

Solar-Terrestrial Phys. **134** 78
[6] Chanrion O, Neubert T, Mogensen A, Yair Y, Stendel M, Singh R, and Siingh D 2017 *Geophys. Res. Lett.* **44** 496
[7] Skakun V S, Panarin V A, Pechenitsyn D S, Sosnin E A and Tarasenko V F 2016 *Russian Physics Journal* **59** 707
[8] Sosnin E A, Skakun V S, Panarin V A, Pechenitsin D S, Tarasenko V F and Baksht E K 2016 *JETP Letters* **103** 761
[9] Sosnin E A, Baksht E K, Panarin V A, Skakun V S and Tarasenko V F 2017 *JETP Letters* **105** 641
[10] Lu X, Naidis G V, Laroussi M, Reuter S, Graves D B and Ostrikov K 2016 *Phys. Rep.* **630** 1
[11] Panarin A A, Skakun V S, Sosnin E A and Tarasenko V F 2017 *Optics and Spectroscopy* **122** 168
[12] Sosnin E A, Panarin V A, Skakun V S, Baksht E Kh and Tarasenko V F 2017 *Eur. Phys. J. D.* **71** 25
[13] Tarasenko V F, Sosnin E A, Skakun V S, Panarin V A, Trigub M V and Evtushenko G S 2017 *Physics of Plasmas* **24** 043514
[14] Panarin A A, Skakun V S, Sosnin E A and Tarasenko V F 2017 *Atmospheric and Oceanic Optics* **30** 243 (In Russian)
[15] Passas M, Sánchez del Río J, Luque A and Gordillo-Vázquez F J 2014 *IEEE Transactions on Plasma Science* **42** 2664

IOP Conf. Series: Journal of Physics: Conf. Series **927** (2017) 012063 doi:10.1088/1742-6596/927/1/012063

Generation of runaway electrons beams during the breakdown of high-pressure gases

V F Tarasenko[1,2], A G Burachenko[1] and E Kh Baksht[1]

[1]Institute of High Current Electronics, 2/3 Akademichesky Ave., 634055 Tomsk, Russia
[2]National Research Tomsk Polytechnic University, Lenina Ave. 30, 634050 Tomsk, Russia

E-mail: VFT@loi.hcei.tsc.ru

Abstract. Generation of run-away electrons in SF_6, CO_2, argon and nitrogen at high and super high pressures is studied. Super-short avalanches electron beams (SAEB) was obtained and measured with a collector at pressures up to 0.3, 0.7, 1.0 and 1.2 MPa in SF_6, CO_2, argon and nitrogen, respectively. The SAEB duration was shown to be ~60 ps (FWHM) and gas composition has only minor effect on the duration. It was found that in a gap of 4 mm in SF_6, CO_2, argon and nitrogen at pressure up to 0.3, 0.7, 1.0 и 1.2 MPa the voltage pulse duration (FWHM) and amplitude increase with pressure.

1. Introduction

By now it is shown that in high electric field generation of run-away electrons and x-rays due to electron action on an anode and ambient gas is a fundamental physical process, see reviews [1, 2] and monographs [3, 4], and references therein. Due to the run-away electrons and x-rays diffuse discharges can be formed in non-uniform electric fields at high gas pressure [5, 6]. Beams of run-away electrons are used for investigation of cathodoluminescence [7]. However, on many issues concerning the mechanism of runaway electron generation and parameters of runaway electron beams, the results obtained in different scientific groups differ substantially [1-6]. Besides, only two scientific groups have recorded the generation of beams of runaway electrons in various gases at pressures significantly exceeding the atmospheric one (>0.1 MPa). By now, only one scientific group have managed to detect runaway electron beams with collectors downstream of anode foils at a pressure of ≥1 MPa [8-10]. The highest pressure at which a runaway electron beam was detected with a collector was 1.2 MPa for helium [8] and for nitrogen [10]. Three main factors make difficult the detection of a runaway electron beam at a diode pressure higher than 1.0 MPa. First, increasing the gas pressure decreases the beam amplitude. Second, if we increase the gas pressure, we have to complicate the design of a gas diode. Third, it is required to use generators with a voltage amplitude of hundreds kilovolts and rise time shorter than one ns. Note that in work [11] of second scientific group, a runaway electron beam in nitrogen at pressure up to 4 MPa was detected from a scintillator luminescence, but neither the beam current amplitude and nor its duration dependence on pressure and other parameters were measured.

In this work, we present the SAEB parameters measured with collectors downstream of anode behind foils in SF_6, CO_2, argon and nitrogen at a pressure of 0.1–1.2 MPa. The SAEB parameters were measured with temporal resolution of ≅50 ps.

2. Experimental setup

The setup with new modification pulser SLEP-150M [9, 10] was used in our experiments. Inner diameter of the transmission line and gas filed diode were reduced to 30 mm, which allows to work at high pressures. The output of the SLEP-150M pulser (part of transmission line), a gas filed diode, and a collector are shown on the figure 1.

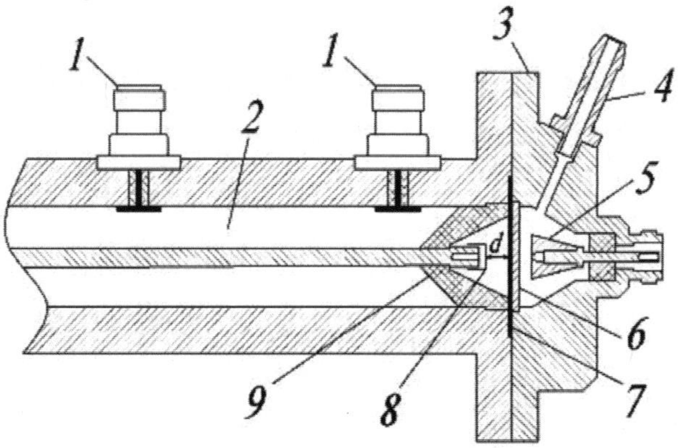

Figure 1. Schematic of the output of the SLEP-150M pulser, gas filed diode, and collector with the 13-mm-diameter receiving area: (*1*) capacitive voltage dividers, (*2*) transmission line of the pulser, (*3*) housing of the collector, (*4*) connection for gas pumping, (*5*) receiver part of the collector, (*6*) mesh, (*7*) foil, (*8*) tubular cathode in gas filed diode, (*9*) insulator.

SLEP-150M formed a voltage pulse with a duration of ~1 ns (FWHM) on a matched load. The voltage rise time was ~ 250 ps at a level of 0.1-0.9. The wave impedance of transmission line *2* was 100 Ohm. The voltage amplitude of incident wave in the transmission line was ~120 kV. The parameters of a SAEB were measured with the cathode that was made as a stainless steel foil tube of diameter 6 mm and thickness 100 μm. The anode of the gas diode was formed by 60-μm AlBe foil *7*, reinforced with a grid *6* from the side of the collector *5*. The collector receiving part *5* was 13 mm in diameter. The measuring equipment also included the capacitive voltage dividers *1*. The signals from the dividers and collector were transmitted to a LeCroy WaveMaster 830Zi-A real-time digital oscilloscope (bandwidth 30 GHz, sampling increment 12.5 ps) via 5D-FB PEEG high-frequency cables (Radiolab) of length 1 m. High-frequency attenuators (Barth Electronics, model 142-NMFP) with a bandwidth of up to 30 GHz were used in measurements. The voltage and the SAEB current were measured simultaneously in each pulse. The time accuracy of SAEB was no worse than 50 ps. The SAEB generation time with respect to the voltage pulse was determined from capacitive current of collector for revealing one the foil was removed. This procedure is described in detail in [12].

3. Experimental results and discussion

We studied the SAEB generation for gap width d=4 mm. The gap was decreased for increasing the parameter U_b/pd, where U_b is the maximum (breakdown) voltage across the gap, p is the gas pressure. In the range of pressures from 0.1 to 1.2 MPa, we recorded waveforms of the gap voltage from the capacitive divider located near the gas diode, incident voltage wave in the transmission line, and SAEB current. The results are presented in figures 2, 3, 4, and 5. Figure 2 shows the breakdown voltage as a function of gas pressure. The breakdown voltage increases with pressure for all gases under study. The parameter values for each gas in figures 2, 3, 4 and 5 are presented for pressure range wherein the collector measured SAEB behind the foil. The voltage pulses duration (FWHM) for different gas pressures are presented in figure 3. The voltage pulse duration increases with the gas pressure. Peak duration in conditions of the SAEB current registration was obtained in N_2 at 1.2 MPa It is it follows from curves in figure 3 that during initial stage of gas breakdown a diffuse discharge is formed in the gap. In spark discharge the voltage pulse duration becomes shorter [9, 10].

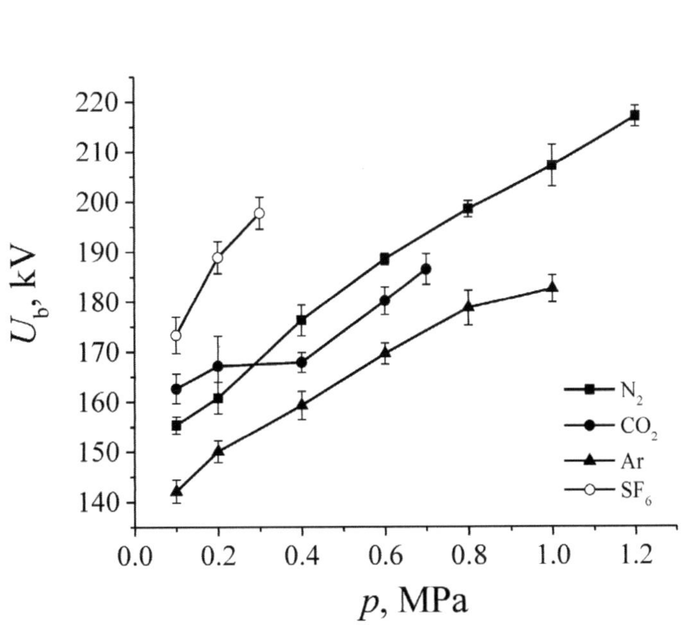

Figure 2. Breakdown voltage versus gas pressure for N_2, CO_2, Ar and SF_6. $d = 4$ mm.

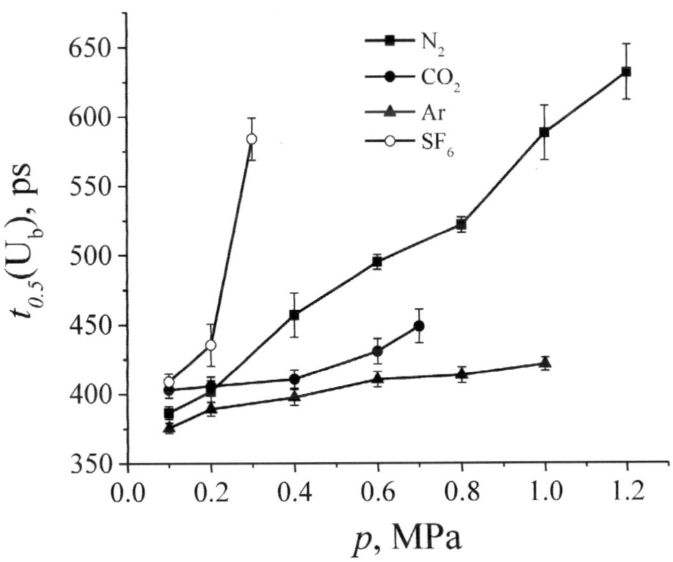

Figure 3. Voltage pulse duration (FWHM) as functions of gas pressure for N_2, CO_2, Ar and SF_6. $d = 4$ mm.

The SAEB current amplitudes and its durations (FWHM) for different gases and its pressures up to 1.2 MPa are shown in figure 4 and figure 5, respectively. The SAEB current amplitudes in N_2 and CO_2 at 0.1 MPa were similar. At higher pressures, maximal SAEB amplitudes were obtained in nitrogen. The SAEB amplitude was minimal in Ar, probably due to low breakdown voltage, see figure 2. However, at 0.2-0.3 MPa minimal SAEB amplitude was measured in heavy gas SF_6 when the breakdown voltage is relatively high. The SAEB current and the voltage were synchronized reasoning that the maximum SAEB amplitude under optimum conditions roughly corresponds to the breakdown (maximum) voltage across the gap [9, 10, 12]. It is seen from figure 5 that the SAEB has duration (FWHM) of about 60 ps and that its decay is shorter than the decay of the gap voltage. This is because most of the runaway electrons are produced on the front of an ionization wave moving from the sharp-ended cathode to the anode [13, 14]. Once the ionization front reaches the anode, the electric field in the gap becomes lower its critical value, and the generation of runaway electrons ceases.

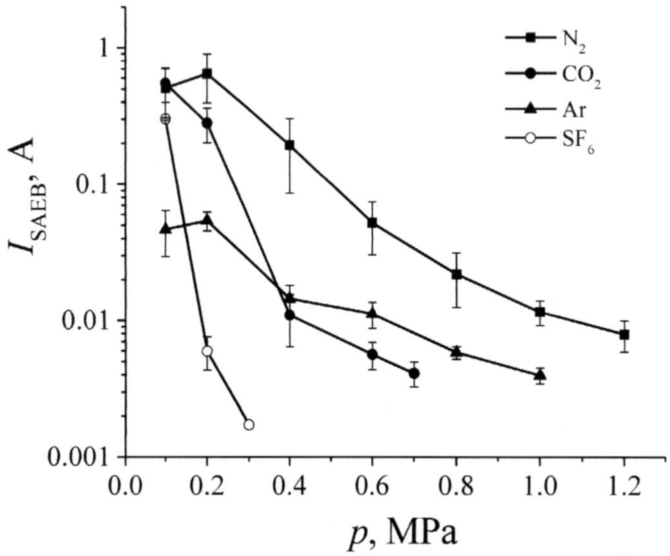

Figure 4. SAEB amplitude vs different gases pressure at $d = 4$ mm.

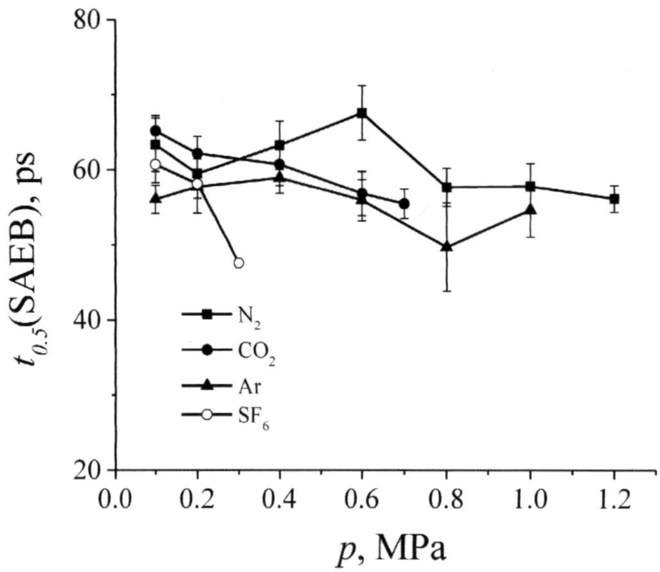

Figure 5. SAEB pulse duration (FWHM) vs different gases pressure at $d = 4$ mm.

The presented dependences are rather typical for high pressures [8–10]. Although the breakdown voltage at $d = 4$ mm increases (figure 2), the SAEB amplitude decreases (figure 4). The highest SAEB amplitudes were obtained in nitrogen at p = 0.1-0.2 MPa, which is the minimum pressure in this experiments. Note that the collector detects only part of the runaway electrons, i.e., those arriving at its receiving area of diameter 13 mm. Moreover, the AlBe foil absorbs the electrons with energies less than 40 keV.

4. Conclusion

Thus, we studied the generation of a supershort avalanche electron beam (SAEB) in different gases (SF_6, CO_2, argon and nitrogen) at high pressures. The beam parameters were measured with a collector for the first time at an argon pressure up to 1.0 MPa, at a CO_2 up to 0.8 MPa and at a SF_6 up to 0.3 MPa. Our experiments demonstrate that at SF_6, CO_2, argon and nitrogen pressure higher than 0.1 MPa, the mechanism of SAEB generation remains the same. Most of the runaway electrons are produced between the ionization front, which bridges the gap, and the anode, and this is in the agreement with experimental [1, 13, 14] and theoretical data [15].

5. Acknowledgments
The work is performed in the framework of the State task for HCEI SB RAS, project #13.1.3.

References

[1] Tarasenko V F, Baksht E Kh, Burachenko A G, Kostyrya I D, Lomaev M I and Rybka D V 2008 *Plasma Dev. Oper*. **16** 4 267
[2] Levko D, Krasik Y E and Tarasenko V F 2012 International Review of Phys. **6** 2 165
[3] Tarasenko V F ed 2016 *Generation of Runaway Electron Beams and X-Rays in High Pressure Gases, Volume 1: Techniques and Measurements* (New York: Nova Science Publishers, Inc)
[4] Tarasenko V F ed 2016 *Generation of Runaway Electron Beams and X-Rays in High Pressure Gases, Volume 2: Processes and Applications* (New York: Nova Science Publishers, Inc)
[5] Babich L P, Loiko T V and Tsukerman V A 1990 *Phys. Usp.* **33** 521
[6] Tarasenko V F ed 2014 *Runaway Electrons Preionized Diffuse Discharges* (New York: Nova Science Publishers, Inc)
[7] Yamamoto N ed 2012 *Cathodoluminescence* (Rijeka, Croatia: In Tech)
[8] Baksht E H, Lomaev M I, Rybka D V, Sorokin D A and Tarasenko V F 2008 *Tech. Phys.* **53** 1560
[9] Tarasenko V F, Burachenko A G and Baksht E Kh 2017 *J. Phys.: Conf. Ser.* **869** 012039
[10] Burachenko A G, Tarasenko V F and Baksht E Kh 2017 *High Voltage* **2** 2 56
[11] Ivanov S N 2013 *J. Phys. D: Appl. Phys.* **46** 285201
[12] Burachenko A G and Tarasenko V F 2010 *Tech. Phys. Lett.* **36** 1158
[13] Tarasenko V F, Beloplotov DV and Lomaev M I 2015 *Plasma Physics Reports* **41** 832
[14] Tarasenko V F, Baksht E Kh, Beloplotov D V, Burachenko A G and Lomaev M I 2016 *Plasma Physics Reports* **42** 369
[15] Kozyrev A V, Kozhevnikov V Yu and Semeniuk N S 2016 *Matter and Radiation at Extremes* **1** 264

The emergence and development of spark channels in the plasma column of a gas discharge between water-solution cathode and a copper anode

G K Tazmeev[1], B A Timerkaev[1] and A K Tazmeev[2]

[1]Kazan National Research Technical University named after A.N. Tupolev, 10 K. Marx str., Kazan, 420111, Russian Federation
[2]Kazan Federal University Naberezhnye Chelny Institute, 68/19 Mira Street, Naberezhnye Chelny, 423810, Russian Federation

E-mail: tazmeev@icloud.com

Abstract. A gas discharge in air between water-solution cathode and a copper anode was experimentally studied. The conditions of the emergence and development patterns of spark channels in the plasma column of discharge were identified.

1. Introduction

In recent years, interest to the gas discharges with liquid electrodes has increased [1]. Much attention is paid to the variant of the liquid-phase cathode. In this variant, like a glow discharge, a volumetric plasma column is formed above the cathode. Burning modes with spark channels were detected in [2] and some integral characteristics of the discharge were investigated. The purpose of this work was a more detailed study of these regimes.

2. Experiment

The scheme of the experimental setup is shown in figure 1. The water solution is expiring from the tube *1* that installed vertically. The tube was made of dielectric material. Its internal diameter was 70 mm. A graphite plate for supplying a negative potential from the power source was mounted. Above of the tube was a round copper anode with a diameter of 100 mm. The anode was installed at different altitudes. Thus, the length of the discharge gap l varied within 4-7 cm. The water solution was played the role of a liquid cathode. It was a solution of sodium chloride in distilled water. In the experiments, solutions with a concentration by mass of 0.5 to 0.05% were used. Their specific electric conductivity σ was within the range of 2-9 mSm/cm. The mass flow rate of the water solution varied in the range 5-30 g/s. The water solution was cooled by forced circulation through a heat exchanger.

The discharge was supplied from a three-phase two-half-period rectifier with an output volt-age of 3000 V. The voltage pulsations were smoothened by means of capacitive-inductive filter *C-L-C*. The current was regulated with ballast resistor *R3* by varying its resistance by a step from 50 to 200 Ω. High-speed video shooting was made by the Photron FASTCAM SA4 camera in the mode of 10000 frames per second.

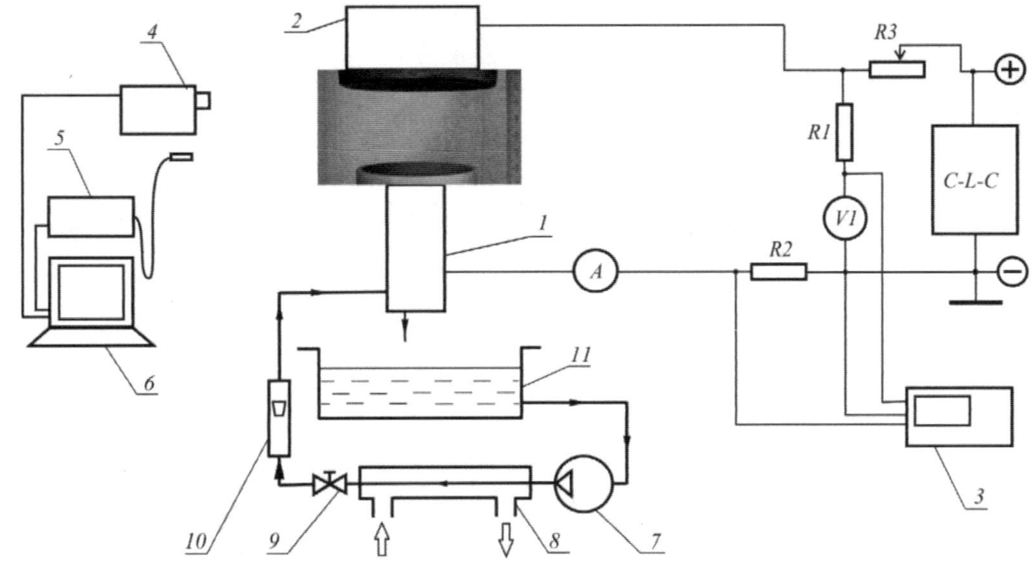

Figure 1. Experimental setup. *1* – cathode assembly; *2* – anode; *3* – oscilloscope; *4* – high-speed video camera; *5* – spectrometer; *6* – computer; *7* – hydraulic pump; *8* – heat exchanger; *9* – valve; *10* – rotameter; *11* – collection container.

3. The experimental results and their analysis

The spark channels are not regular. Their location, geometry and duration was varied randomly. As an example, a series of frames that cover the life span of a single spark channel are shown in figure 2. The first frame is a photograph of the discharge gap. As can be seen, the spark channel is generated near the metal anode and breaks into the direction of the liquid cathode (frames 1-5). After the discharge gap is closed, the spark channel expands (frames 6-9). At the same time, a branching takes place near the cathode. In the period of developed combustion, the binding to the liquid cathode is continuously changing (frames 10-17). In the middle part, bends are formed which are characteristic of a free electric arc. The binding to the anode is targeted and it is stationary. The disappearance of spark channel occurs almost instantaneously (frames 18-20).

It has been recorded that only a few are being developed from the bud. Some of them were lengthened by 2-3 cm and then disappeared (figure 3). The spark channels are spread from the copper anode at speeds of 100-150 m/s.

On video frames, the spark channels were turned out to be scarlet. This colour is characteristic for the Balmer spectral line H_α. And in fact, at the time of the appearance of spark channels, the emission of hydrogen atoms was the predominant emission. At these times, in the panoramic spectra in the visible wavelength region, the spectral line H_α was the most intense.

A considerable amount of water vapor and also sodium is entering into the discharge gap from the liquid cathode. In all likelihood, thermal dissociation of water molecules into atoms takes place inside the spark channel. At the same time, hydrogen atoms can predominate among the radiating atoms (H and Na) in spark channel because the water solution used as the cathode, is highly diluted.

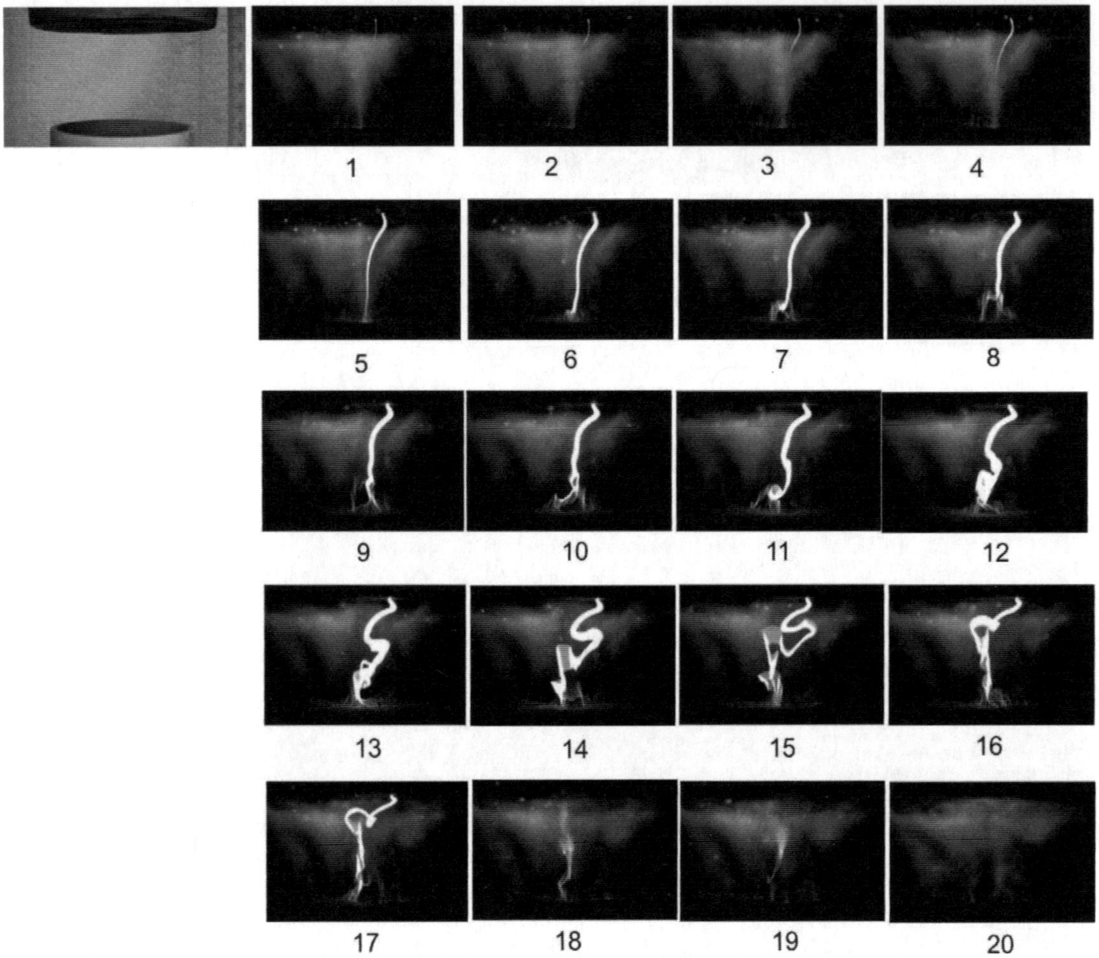

Figure 2. A frame by frame of video shooting of spark channel. The interval between frames was 0.1 ms.

The spark channel of scarlet colour is formed against the background of yellow radiation of sodium atoms. Such a picture indicates that, despite the appearance of spark channel, the space above the liquid cathode remains current-conducting in the larger part. In a short time of the existence of spark channel (on the order of several milliseconds), massive ions do not have time to reach the electrodes and continues their relatively slow motion, creating an ion current. Inside of spark channel, current flows is predominantly provided by electrons, which are much lighter than ions.

The origin of spark channels, apparently, is also due to the uneven distribution of electrons. Spark channels are appears where the concentration of electrons is high. The appearance of spark channels near the metal anode indicates that the current is closed to the positive electrode mainly by electrons.

The appearance of spark channels was recorded on oscillograms as abrupt current changes. In the intervals of stable combustion (without spark channels), the discharge current was 10-15 A, and the voltage is 2400-2700 V. During the existence of spark channels, the current was increased to 30-35 A. At the same time, the voltage was varied not so much. It decreased by 200-500 V at the time of the appearance of spark channels.

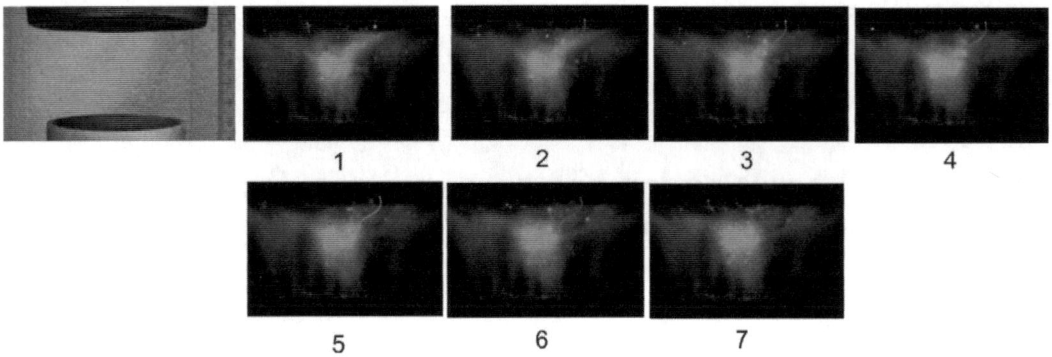

Figure 3. Incomplete development of the channel. The interval between frames was 0.1 ms.

4. Conclusions
The occurrence of spark channels occurs under certain conditions. The specific electrical conductivity of the water solution serving as the cathode should be small. The discharge current must be considerable. In the case of a water solution of sodium chloride, the spark channels are necessarily appears at σ = 2-5 mSm/cm and I > 5 A.

References
[1] Bruggeman P J et al. 2015 *Plasma Sources Science and Technology* **25** 053002
[2] Tazmeev Kh K and Tazmeev A Kh 2014 *J. Phys.: Conf. Ser.* **567** 012035

IOP Conf. Series: Journal of Physics: Conf. Series **927** (2017) 012065 doi:10.1088/1742-6596/927/1/012065

Transfer of charges and substances in a gas-discharge plasma from the liquid electrolyte cathode containing salts of alkali metal

Kh K Tazmeev[1], B A Timerkaev[2] and G K Tazmeev[2]

[1]Kazan Federal University Naberezhnye Chelny Institute, 68/19 Mira Street, Naberezhnye Chelny, 423810, Russian Federation
[2]Kazan National Research Technical University named after A.N. Tupolev, 10 K. Marx str., Kazan, 420111, Russian Federation

E-mail: tazmeevh@mail.ru

Abstract. A gas discharge in the air between the flowing liquid cathode and a solid anode was studied experimentally. Aqueous solution of sodium chloride was used as the liquid cathode electrolyte. Mass loss due to evaporating and sputtering was compensated by adding distilled water in a continuous mode. It was found that the specific electric conductivity of the aqueous solution is almost unchanged. The regularities of changes in the composition of an anionic solution were discovered.

1. Introduction

The work is devoted to the study of a gas discharge, which is excited between a liquid cathode and a metallic anode at atmospheric pressure. The advantage of such discharges is that they make it possible to obtain a nonequilibrium plasma at atmospheric pressure by relatively simple methods. Therefore, interest in them is growing steadily every year. In this case, the plasma produced in a gas discharge has broad practical applications [1-6].

2. Experiment

The experiments were carried out on an installation described in [7]. Electric power was supplied from sources based on an inverter converter. Power supplies of this type have the advantage that they allow maintaining current with high accuracy. According to the technical characteristics of the sources, the errors of current stabilization do not exceed 1%.

A solution of sodium chloride in distilled water as a liquid cathode, was used. The volume of the aqueous solution of the Vs in the hydraulic system of the experimental setup was kept constant. In experiments, the constancy of the Vs volume was ensured by compensating of the loss of the solution by adding distilled water. It was happening in a continuous mode. The physicochemical parameters of the aqueous solution (pH, specific electric conductivity and content of NaCl) were measured by the laboratory instrument ANION 4150.

3. The experimental results and their analysis

In the experiments it was found that after the action of the gas discharge the pH of the aqueous solution is increases. The alkalization of the aqueous solution was occurred. This process can be explained as follows. During the combustion of the discharge, some of the Cl^- anions are re-moved from the solution. They are replaced by hydroxyl ions OH^-. Since the solution contains Na^+, cations, it can be considered that the action of a gas discharge leads to the formation of sodium hydroxide in a dissociated state.

Adding hydrochloric acid to the solution, it can be neutralized. In this case, a reaction will occur with the formation of sodium chloride

$$NaOH + HCl = NaCl + H_2O.$$

This process was performed by titration. After exposure of gas discharge, a sample was taken and an acid (titrant) was added in, such an amount that completely restored the original pH value. It was found that the content of NaCl was also restored. The specific electrical conductivity of the aqueous solution remained practically unchanged. The deviation from the initial value was insignificant.

The processes of charges transfer and substance can be described as follows. Under the influence of the discharge, the solution is sprayed and the droplets are transformed into vapour. It must be assumed that under conditions of the discharge, the liquid – gas phase transition, proceeds rapidly and has an explosive character. In this case, the anions and cations that are carried out as a part of the drops from the aqueous solution is spread away. In the electric field of a gas discharge Cl^- ions are drifting toward the anode, and thus they are removed. The amount of chlorine in aqueous solution is reduced. Na^+ ions are drifting to the cathode and returned to the aqueous solution. On the surface of the graphite plate which is mounted inside of the aqueous solution for supplying thereto negative potential, Na^+ ions are restored. The metallic sodium is reacts instantly with water, forming a hydroxide of NaOH, which in turn dissociates into Na^+ и OH^- ions. Thus, Cl^- anions that retired from the aqueous solution are replaced by the same amount of hydroxyl OH^- ions. When the pH is restored, an excessive amount of hydroxyl OH^- ions is neutralized and Cl^- ions are returned into aqueous solution. In this case, the NaCl content in the solution is restored, i.e. the initial variant of the aqueous solution is obtained. Hence it can be concluded that the amount of sodium in the aqueous solution does not change. Na^+, ions, released from the aqueous solution as a part of drops, practically all return.

Negative chlorine ions will not be return back, unlike sodium ions. Let us consider how changing their content in the aqueous solution. In the volume of the sample υ there is Cl^- in the amount of $c\cdot\upsilon$. Here c – is the unknown molar concentration. In a freshly prepared solution with a volume equal to the volume of the sample υ, the amount of Cl^- ions is $c_0\cdot\upsilon$. During the combustion of the gas discharge, part of NaCl was converted into NaOH in a ratio of 1:1. In the same ratio, chloride ions Cl^- were replaced on hydroxyl ions OH^-. The loss of chloride ions Cl^- in the sample is $c_0\cdot\upsilon$ - $c\cdot\upsilon$.

When titrating the loss is restored by ions contained in the titrant: $c_t\cdot\Delta\upsilon$. Here $\Delta\upsilon$ is the volume of the titrant. In this way: $c_0\cdot\upsilon - c\cdot\upsilon = c_t\cdot\Delta\upsilon$. From here

$$\frac{c}{c_0} = 1 - \frac{c_t}{c_0}\cdot\frac{\Delta\upsilon}{\upsilon}. \qquad (1)$$

Provided that small portions (droplets) are removed from the aqueous solution without changing the composition, it is possible to obtain the formula

$$\frac{C}{C_0} = \exp(-\Delta V / V_S). \qquad (2)$$

Here ΔV – volume of distilled water, which is needed to compensate for loss of the aqueous solution. This condition is realized in the absence of a gas discharge ($I = 0$), when the aqueous solution flows out of the hydraulic system and at the same time distilled water is added to it. As can be seen from figure 1, the results of such "idle" experiments (marked by squares) are in complete agreement with the calculation according to formula (2). At the same figure shows the experimental data (marked with circles) which processed using formula (1). They are significantly different from "idle". Thus, when the combustion of gas discharge, content of chlorine in the aqueous solution is decreases much more slowly than in the "idle" experiments. This is a very important experimental fact that deserves attention.

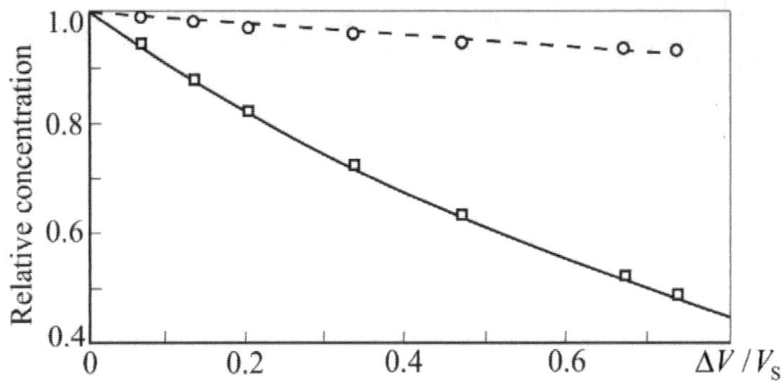

Figure 1. Change in the concentration of ions in an aqueous solution. Solid line – calculation by formula (2). V_s = 15.0 l. I = 11.0 A.

Suppose that in the near-boundary region "aqueous solution-plasma" the current is completely ionic. Positive Na^+ ions are going to the cathode (into the aqueous solution), and the negative Cl^- and OH^- ions leave from it. Let the concentration of Na^+ ions in the drops is the same as in the aqueous solution, i.e. $C^+ = C$. Under this condition, the positive ion current can be calculated by using the rate of decrease m of the aqueous solution.

$$I^+ = \frac{q^+}{t} = \frac{N^+ \cdot e}{t} = \frac{\nu^+ \cdot N_A \cdot e}{t} = \frac{C^+ \cdot \Delta V \cdot N_A \cdot e}{t} = \frac{C^+ \cdot N_A \cdot G}{\rho}. \tag{3}$$

Here q^+, N^+ и ν^+ – the total charge, the total number and the number of moles of positive ions. G and ρ – mass flow and density of the aqueous solution. For an approximate estimation of ρ can be taken as that of water, because the solution is weakly concentrated.

The concentration of sodium in the aqueous solution is practically unchanged. Therefore, in the first approximation, we can assume that $C^+ = C_0$.

Let us consider an example of calculation for the case I = 11.00 A and C_0 = 0.1 mol/l. In this case, the calculation gives I^+ = 6.84 A, which is slightly more than half of the total current I = 11.00 A. Let us analyze the result obtained. To begin with, in general, the aqueous solution is electrically neutral. The molar concentrations of positive and negative ions in it are the same $C^+ = C^-$. The aqueous solution maintains electroneutrality in the process of gas discharge exposure. Consequently, the drops do not carry out ions of any sign in an excessive amount. They are also electrically neutral. The condition $C^+ = C^-$ also is being implemented in the drops. In this case, as follows from formula (3), the ion current I^+ is proportional to C^+. Similarly, I^- is proportional to C^-. Therefore, the condition $I^+ = I^- = I/2$ must be fulfilled. Thus, the calculations of I^+ give an overestimated result.

Let us accept I^+ = 5.5 A. Using formula (3), it is possible to find the corresponding value of the molar concentration C^+ = 0.08 mol/l. Thus, $C^+ < C_0$. It turns out that the part of the aqueous solution which is carried away by the drops as if diluting by water. The essence of this seeming effect is quite understandable. In fact, along with the drops ablation is decreased the loss of aqueous solution through an evaporation process. The volume of additional pure water enters in ΔV as the volume of the droplets phase. Therefore, calculation by formula (3) gives an underestimate value of C^+.

The picture of transport phenomena in a simplified form can be represented, as shown in figure 2. Distilled water with volume ΔV is enters the hydraulic system. Under the influence of a gas discharge, the aqueous solution decreases in the same volume. The loss occurs in the form of drops and water vapor. The droplets phase is 4/5 parts, and the water vapor is 1/5 part.

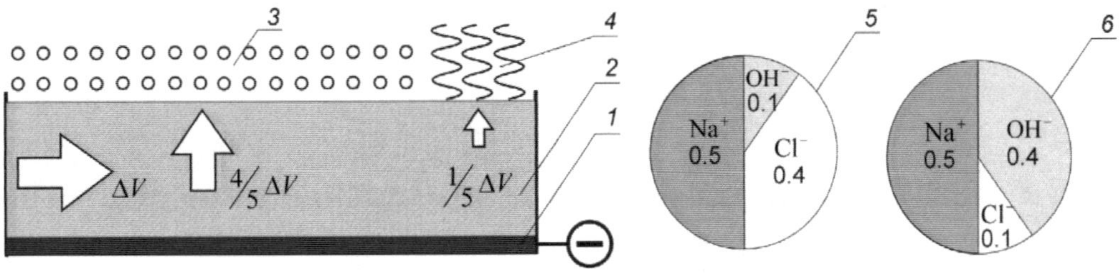

Figure 2. A simplified picture of transport phenomena. *1* – graphite plate; *2* – aqueous solution of sodium chloride; *3* – droplets phase carried away from the cathode of solution; *4* – water vapour. I = 11.0 A. V_s = 15.0 l. $\Delta V/V_s = 2/3$.

The circular diagrams in this figure reflect the molar content of ions in the aqueous solution (5) and in the drops (6). The figures indicate the proportions of cations and anions in the total numbers of ions.

4. Conclusions

The substance comes from the water-solution cathode into the gas-discharge plasma in the form of drops and vapour, the droplets phase predominates substantially. Cations and anions contained in the aqueous solution are transferred in the drops. The cations of the alkali metal, in this case sodium, are almost completely returned to the water-solution cathode. The proportions of different anions in drops depend on their mobility. In a larger number in drops fall anions with higher mobility.

References

[1] Samukawa S, Hori M, Rauf S, Tachibana K, Bruggeman P, Kroesen G, J Whitehead C, Murphy A, Gutso A and Starikovskaia S 2012 *J. Phys. D: Appl. Phys.* **45** 253001
[2] Maksimov A I and Khlyustova A V 2009 *High Energy Chemistry* **43** 149
[3] Webb M R and Hieftje G M 2009 *Anal. Chem.* **81** 862
[4] Tazmeeva R N, Ziganshin R R and Tazmeev Kh K 2008 *Vestnik mashinostroeniya* No. 10 87
[5] Valiev R I, Gaisin A F, Gaisin F M, Gumerov A Z, Nasibullin R T, Sadriev R S, Sarimov L R and Khafizov A A 2014 *Izvestiya VUZov. Fizika* **57** No. 3-3 66
[6] Rybkin V V, Shikova T G and Titov V A 2008 *High Energy Chemistry* **42** 485
[7] Tazmeev Kh K, Arslanov I M and Tazmeev G Kh 2014 *J. Phys.: Conf. Ser.* **567** 012001

Surface modification of additive manufactured metal products by an intense electron beam

A D Teresov, N N Koval, Yu F Ivanov, E A Petrikova, O V Krysina

Institute of High Current Electronics SB RAS, 634055, Tomsk, Russia

E-mail: tad514@sibmail.com

Abstract. On the example of VT6 titanium alloy it is shown that successive surface modification of additive manufactured metal specimens in vacuum at an argon pressure of $3.5 \cdot 10^{-2}$ by ten pulses with 200 μs, 45 J/cm^2 and then by three pulses with 50 μm, 20 J/cm^2 provides a considerable decrease in their porosity and surface roughness (20 times for R_a) while their surface microhardness, friction coefficient, and wear level remain almost unchanged. After electron beam irradiation, the ultimate tensile strength of the material increases 1.33 times, and its tensile strain 1.18 times. For specimens obtained by conventional metallurgy and irradiated in the same modes, no such effects are observed.

1. Introduction

Today, 3D printing is much used in different technologies such as those of aircrafts, space crafts, submarines, tools, prostheses, implants, jewelry, etc. [1–4]. Additive manufacturing uses computer models to create 3D objects by joining materials layer by layer rather than by removing some, for which it differs from conventional mechanical technologies. Additive processes are classified by material (liquid, loose, polymer, metal powder, metal wire), process assistance (laser, electron beam), binding method (heat, UV or visible radiation, binding agent), and layer formation (selective laser or electron beam sintering, jet printing). For selective sintering, powder material is uniformly distributed on a platform and selectively sintered with a laser or electron beam according to a current model cross-section. For jet printing, material is deposited directly to the point of energy delivery [5].

Now, metal objects are most often grown using 3D printers based on selective laser or electron beam sintering. The formation of an object occurs in a thin (50–100 μm) layer of metal powder where individual particles are sintered under the action of heat. The problem common to all additive technologies concerns porosity in a synthesized material. Some studies show that porosity varies with material and fusion mode [6]. For example, for aluminum and titanium alloys, it can reach 4–5 % and 2 %, respectively, and for steels, it is less than 0.2 %. Porosity can be minimized by repeated fusion of layers [7, 8], i.e., each layer is sintered twice with a laser, which decreases this parameter an order of magnitude but almost doubles the growth time of an object. Another problem in additive technologies concerns anisotropy (structural, mechanical) and high internal stress, which is inevitable in the layer-by-layer synthesis of an object. At the interface of an object and powder, a highly rough porous layer is often formed due to insufficiently fused powder particles.

Here we report on the use of a submillisecond pulsed electron beam for decreasing the surface roughness and porosity of VT6 titanium alloy specimens manufactured by selective electron beam sintering of powder particles of size 40–100 μm.

2. Material and research techniques

The test specimens were VT6 titanium plates of dimensions 15×15×2 mm obtained from metal powder particles of size 40–100 μm by selective electron beam sintering in vacuum (Arcam A2X, Sweden). For comparison, we used VT6 specimens cut from a bar of diameter 15 mm parallel to its longitudinal axis; their preliminary thermomechanical treatment conformed to GOST 22178-76. Part of the specimens was irradiated with an intense pulsed electron beam on the SOLO setup [9] in two successive modes: ten pulses with an electron energy of 15 keV, energy density of 45 J/cm^2, duration of 200 μs, and repetition frequency of 0.3 s^{-1} (mode I) and three pulses at 15 keV, 20 J/cm^2, 50 μs, and 0.3 s^{-1} (mode II). The operating pressure was $3.5 \cdot 10^{-2}$ (Ar). Mode I greatly decreases the surface roughness and porosity of e-beam sintered VT6 specimens [10]. This mode was used for pre-polishing, and mode II for further finishing.

The surface roughness of the specimens was measured on an MNP-1 optical profilometer (sampling length 0.8 mm, no less than ten measurements per specimen), and their surface microhardness was measured on an PMT-3M device (no less than ten points in different regions at a load of 0.5 N).

The specimens were tested for tension on an Instron 3369 machine at a rate of 0.2 mm/min and temperature of 20 °C. Their gage section parallel-sided lengthwise and right-angled at the ends had an initial thickness, width, and length of 2.0 mm, 2.0 mm, and 10 mm, respectively.

The specimens were also tested for dry friction on a TRIBOtechnic ball-on-disk tribometer with a hard WC–Co ball of diameter 6 mm. The conditions were the following: room temperature, relative humidity 50 %, normal indenter load 3 N, sliding velocity 2.5 cm/s, track diameter 4 mm, sliding distance 15 m. The wear volume was determined after profilometry of a formed track.

The surface structure was examined on a Philips SEM-515 scanning electron microscope (U_{max}= 30 kV).

3. Research results and discussion

Figure 1 shows the surface structure of additive manufactured VT6 specimens before and after pulsed electron beam treatment. Their initial surface is rough and porous and reveals stuck powder particles. After pulsed electron beam irradiation, both in mode I and II, their surface layer assumes a homogenous polycrystalline structure, its porosity decreases, and powder particles are indistinguishable.

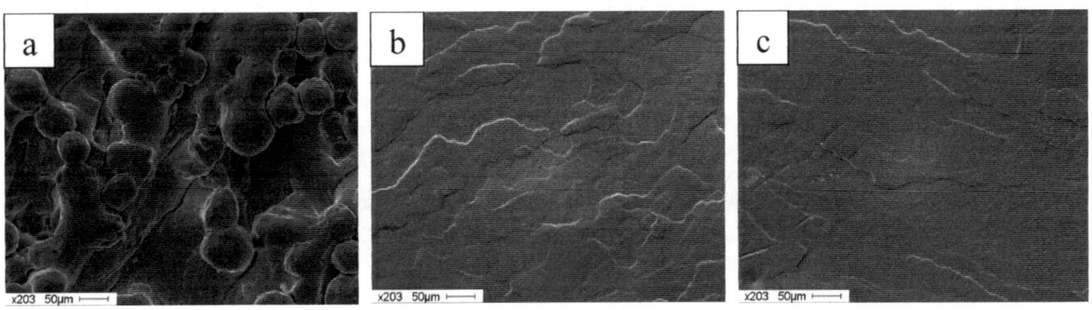

Figure 1. Surface structure of additive manufactured VT6 specimens before irradiation (a), after irradiation in mode I (b), and after successive irradiation in mode I and II (c).

The surface roughness of the additive manufactured VT6 specimens before and after pulsed electron beam irradiation is presented in Table 1. As can be seen, R_a after irradiation decreases ~20 times and R_z decreases ~19 times, and an additional decrease in R_a and R_z is provided by irradiation in mode II.

The surface microhardness of the specimens is presented in Table 2. The measurements show that this parameter after irradiation remains unchanged (measurement error ~10%). The friction coefficient and wear parameters before and after irradiation are close in value (Table 2). After irradiation in mode II, the tribological properties reach the initial level.

Table 1. Surface roughness of additive manufactured VT6 titanium specimens

State	Roughness R_a, μm	Roughness R_z, μm
Initial	20.2±1.5	93.8±10
Mode I	1.3±0.5	6.2±1.5
Mode I+II	1±0.5	5±1.5

Table 2. Surface microhardness and wear of additive manufactured VT6 titanium specimens

State	Hardness, HV_{50}	Fraction coefficient	Wear groove depth, μm	Specific wear rate, 10^{-4} mm³/N·m
Initial	306	0.36	8.5	4.6
Mode I	324	0.41	9.2	5.9
Mode I+II	303	0.32	8.4	5.2

Analyzing the data of tensile tests in Table 3 and Fig. 2, one can see that the highest mechanical properties are found for the additive manufactured specimens after pulsed electron beam irradiation: the tensile strength increases 1.33 times, and the tensile strain 1.18 times. The ductility of the specimens obtained by conventional metallurgy decreases after irradiation in the same modes. Our study shows that the difference in the mechanical properties of the materials is due to differences in their defect structure. After intense pulsed electron beam irradiation, the additive manufactured material reveals a polycrystalline structure with a grain size of 15–60 μm and a cellular structure with a cell size of 0.5–1.2 μm in the grain volume. The material obtained by conventional metallurgy assumes a polycrystalline structure with a grain size of 50–800 μm and a rough lamellar structure in the grain volume.

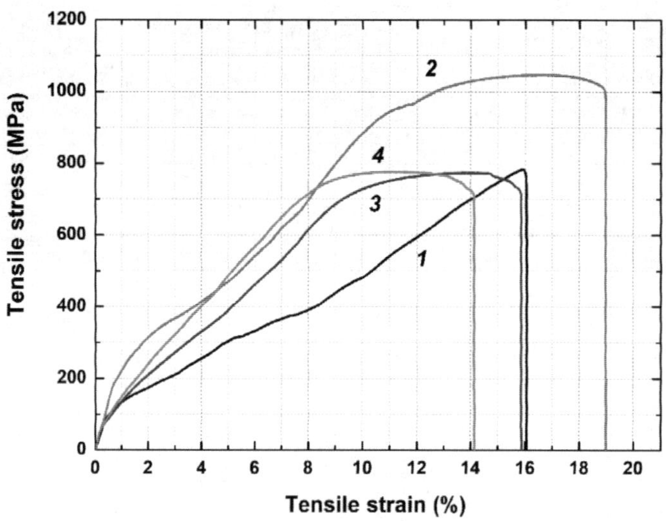

Figure 2. Strain hardening curves for VT6 titanium alloy under tension: *1* – additive manufacturing; *2* – additive manufacturing and irradiation; *3* – conventional metallurgy; *4* – conventional metallurgy and irradiation.

Table 3. Tensile data for additive manufactured (AM) and casted (C) VT6 titanium specimens

State	Ultimate tensile strength, MPa	Tensile strain, %	Displacement at ultimate tensile strength, mm
Initial (AM)	783.6	16.1	1.59
Mode I+II (AM)	1047.2	19	1.67
Initial (C)	774.5	15.9	1.39
Mode I+II (C)	776.8	14.2	1.12

4. Conclusion

Thus, on the example of VT6 titanium alloy, it is shown that pulsed electron beam irradiation provides the formation of a homogeneous polycrystalline structure and greatly decreases the porosity and surface roughness in electron beam sintered materials: ~20 times for R_a. The strength and tribological properties remain the same as in the initial material. The highest tensile strength and ductility is found in the additive manufactured specimens exposed to pulsed electron beam irradiation: these parameters increase 1.33 and 1.18 times, respectively. The ductility of the material obtained by conventional metallurgy decreases after irradiation in the same modes. The proposed method of successive pulsed irradiation at an energy density of >30 J/cm^2 and ≤ 20 J/cm^2 can modify the mechanical properties and greatly decrease the surface roughness of additive manufactured materials with $R_a > 3$ μm.

Acknowledgments
The work was supported by the Russian Science Foundation (project No. 14-29-00091).

References

[1] Gibson I, Rosen D, Stucker B 2016 *Additive manufacturing technologies: 3D printing, rapid prototyping, and direct digital manufacturing* (Moscow: Technosfera) 656 [in Russian]
[2] Shishkovsky I V 2015 *Basics of high-resolution additive technologies* (St. Petersburg: Piter Publ.) 348 [in Russian]
[3] Evans B 2012 *Practical 3D printers: The science art of 3D printing* (New York.: Apress) 306
[4] Valetov V A 2015 *Additive technologies: state and prospects* (St. Petersburg: ITMO Univ.) 63 [in Russian]
[5] Sachs E M, Cima M J, Williams P, Brancazio D and Cornie J 1992 *J. Eng. Ind.* **114** 4 481–488
[6] Louvis E et. al. 2011 *Journal of Materials Processing Technology* **211** 2 275–284
[7] Yasa E et. al. 2011 *Rapid Prototyping Journal* **17** 5 312–327
[8] Yasa E, Kruth J 2011 *Advances in Production Engineering & Management* **6** 4 259–270
[9] Koval N N, Ivanov Yu F 2008 *Izv. Vyssh. Uchebn. Zaved. Fiz.* **5** 60–70 [in Russian]
[10] Teresov A D, Koval N N, Ivanov Yu F, Petrikova E A 2016 *Izv. Vyssh. Uchebn. Zaved. Fiz.* **59** 9/2 264–266 [in Russian]

The low-cost microwave plasma sources for science and industry applications

V N Tikhonov[1,2], S N Aleshin[1], I A Ivanov[1,2] and A V Tikhonov[1]

[1] Russian Institute of Radiology and Agroecology, Kievskoe shosse, 109th km, Obninsk, Kaluga region, 249032, Russian Federation
[2] Scientific and production enterprise "Agroecoteh" LLC, Kievskoe shosse, 109th km, Obninsk, Kaluga region, 249032, Russian Federation

E-mail: v.n.tihonov@yandex.ru

Abstract. Microwave plasma torches proposed in the world market are built according to a scheme that can be called classical: power supply – magnetron head – microwave isolator with water load – reflected power meter – matching device – actual plasma torch – sliding short circuit. The total cost of devices from this list with a microwave generator of 3 kW in the performance, for example, of SAIREM (France), is about 17,000 €. We have changed the classical scheme of the microwave plasmathrone and optimised design of the waveguide channel. As a result, we can supply simple and reliable sources of microwave plasma (complete with our low-budget microwave generator up to 3 kW and a simple plasmathrone of atmospheric pressure) at a price from 3,000 €.

1. Introduction

Presently there are needs for cost-efficient, simple and reliable sources of Low-temperature plasma. However wide application of thermal & non-thermal plasma and also plasma-chemical technologies, with "pure" microwave plasma torches is prevented by their high cost.

To date, the most usage was made with three types of plasma sources [1]

1) electric arc plasma sources – with AC or DC power supply,
2) high-frequency (HF)
3) ultra high frequency (UHF), microwave (MW) plasma torches.

Historically, in the first industrial plasma sources have been used the arc discharges of DC and industrial frequency AC, currently they are the most extensively studied and widely used in industry.

The use of inductively coupled plasma began with invention of high-power HF vacuum tubes generators (in the range from 60 kHz to 60 MHz) It does gave the possibility to isolate the discharge from the electrodes and the walls of the discharge chamber to obtain a pure electrodeless plasma that does not have thermal and electrical contact with the elements of the plasma torch.

Next step in plasma technology was using of the frequency bands 915 and 2450 MHz allowed for industrial applications. Powerful generators on magnetrons are now widely used as power sources for microwave plasma torches.

2. Design low cost microwave generators for plasma torches

The cost of microwave generators for industrial heating can be estimated as $\sim 1.5 \div 2$ € per watt power [2]. An impedance coupling structures could even double this figure. At the same time, the price of a microwave module of household microwave oven may be less than 0.1 € per watt. That price includes: magnetron, complex housing, high voltage transformer, ventilation, locking and control system, ets.

Therefore, it is quite reasonable to use the main components and accessories of household microwave oven as a cheap source of microwave energy for plasma torches. Mass production and competition of leading manufacturers in the global market provide high reliability of their products at very low cost.

For implementation these advantages it was necessary to solve several problems:

2.1. Really continuous mode of magnetron generation

The power supply circuit of the magnetron of a conventional microwave oven consist of a high voltage transformer loaded with a half-wave rectifier with a voltage doubling. Therefore, the envelope of the microwave signal is a microwave energy pulse of length about 1 ms with a repetition period of 2 ms that is half of the supply voltage period is expended on the energy storage process in the capacitor. This mode is not suitable for powering the plasma devices operating at atmospheric pressure. During the pause (~1 ms), ions in the discharge region can to recombine and the discharge region is blown off the discharge from chamber by working gas stream.

We have developed and designed a special scheme for high-voltage magnetron supply from a three-phase AC network using three step-up transformers, operating in the "three quarter-bridge with doubling, in parallel" configuration [3]. This scheme provides a ripple level of about 15% without the use of any additional circuitry techniques and tools, and it is more than enough to maintain a continuous plasma discharge at atmospheric pressure.

2.2. Ability to control the level of continuous power

This problem we solved by a discrete change in the ratio of the parameters of capacitive storage devices in voltage doublers for each phase. In this way can be provided of several discrete levels (from 0.4 to 3 kW) of the continuous output power of a magnetron microwave generator can be provided.

This power supply has a switchable or adjustable current of the cathode heating prolong the service life of the magnetron. It is also possible to equip the power supply with devices for measuring of the filament and anode current of the magnetron. A source with four levels of working power, with smooth control of the filament current and direct measurement of the anode current is shown in (figure 1).

Figure 1. High-voltage magnetron power supply with four levels of operating power.

Figure 2. Magnetron OM 75P (Samsung) converted to water cooling.

IOP Conf. Series: Journal of Physics: Conf. Series **927** (2017) 012067 doi:10.1088/1742-6596/927/1/012067

2.3. Increase the effectiveness of the magnetrons cooling system

To operate at higher power levels, we developed a technology for modifying the serial microwave magnetrons used in domestic and industrial microwave ovens from air to water cooling. Was designed and made the water radiator in the form of a split monoblock made of aluminum alloy, which has a fairly good thermal contact with both magnets and a tight fit to the magnetron body due to the power clamping of the split structure (figure 2).

For example, magnetron OM 75P (Samsung) with a rated output power of 1 kW, equipped with a monoblock water cooler, can generate a power of more than 2.5 kW during extended continuous operation. Moreover, the heat load at each of the three transformers of the power supply has decreased to the value of about 70% from the regular mode.

3. Decrease of high cost of components microwave plasma torches

It is easy to see that the microwave plasmathrones offered in the world market is built according to the following scheme: power supply – magnetron head – microwave isolator with water load – reflected power meter – matching device – actual plasma torch – sliding short circuit (figure 3).

With all the obvious advantages of such a classic approach, one can not ignore the high cost of each component of the scheme individually and entire installation as a whole. Thus, the total cost of only three items from the above list, namely: the matching device, the directional coupler and the circulator with the load, produced of UIY Tecnology Co., Ltd, China, is about 8,000 €. The total cost of the SAIREM (France) microwave plasma system with a microwave generator of 3 kW, presented on (figure 3), is about 17,000 €.

3.1. Design of low cost of the waveguide system

Continuing our traditional principle of minimal budgetary development, we have simplified the design of the waveguide path as much as possible – even at the cost of a possible reduction in some of the technical characteristics of the microwave plasma torch. We left only the most necessary components from the classical scheme: power supply – magnetron head – waveguide system – plasma source – water load and simplified the design of the waveguide path as much as possible (figure 4).

As we see, the waveguide system in our case is an all-welded stainless steel construction based on a standard rectangular profile of 100*50*2 mm. At the end of the waveguide, a bend in the H-plane is made and a resonant water load is located here. A matching inductive diaphragm is located along the line of the break of the waveguide.

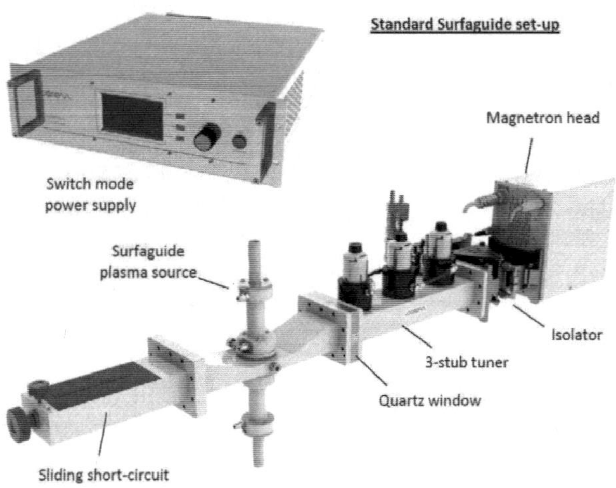

Figure 3. Standard Surfaguide installation – SAIREM (France).

Figure 4. Magnetron head and waveguide system.

Although the terminal dummy load, located after the plasma torch instead of a short circuit, absorbs some of the useful power and reduces twice the electric field in the waveguide (worsening conditions for ignition of the discharge), it is needed to protect the magnetron from the reflected wave in the absence of a plasma discharge. In the classical configuration, these functions are performed by the circulator with its load and the matching device with the corresponding system for measuring the incident and reflected waves in the supply waveguide.

It was shown by subsequent experience that the lack of the above structure should be regarded as an almost inability to adjust the waveguide channel parameters. In particular, the choice of the optimal location of the plasma torch relative to a matching diaphragm turned out to be quite a challenge.

3.2. Efficiency of the microwave power injection. Adjustment device

We have corrected the design of the water load using a movable half-wave Teflon transformer to be able to adjust value of the reflection. If such transformer located in a node of the electric field of a standing wave (which is formed between the diaphragm and a short-circuit of the waveguide), it minimally affects the electric field in the load and its input resistance. When the transformer is displaced towards one or another antinode of the standing wave in the load, the complex input impedance of the load is rearranged, and the maximum of the standing component of the electromagnetic wave in the main waveguide is displaced too. The design of this trimmer is simple; it does not contain any movable electrical contacts or even no metal parts at all (figure 5).

As shown by the measurements, with such adjustment it is possible to ensure the efficiency of the microwave power injection directly into the plasma up to 90% and more.

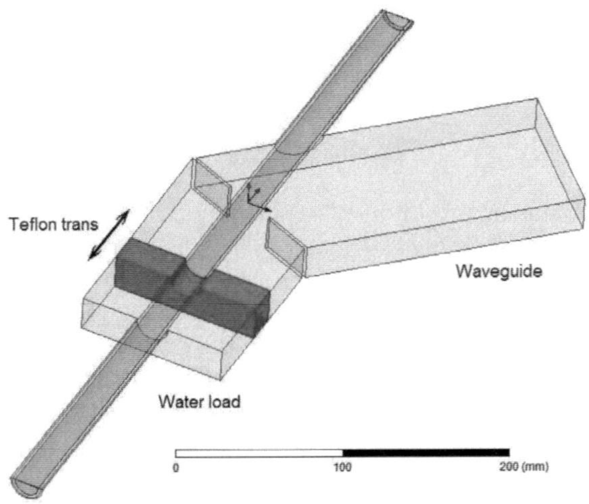

Figure 5. Resonant water load with a movable half-wave Teflon transformer (left). Magnetron head and waveguide system with upstream plasma torch and adjustable water load (right).

3.3. Pipe fittings and cooling systems

Continuing our traditional principle of maximum use of the ready-made components, we currently supply our microwave plasma systems with plasma torches, built on the basis of standard threaded stainless steel and brass fittings for metal pipes. Availability of fittings, at an affordable price and the huge assortment allows quickly select the configuration of the burner for a specific task. (figure 6).

As the plasma temperature can reach several thousand degrees, water cooling of the support branch pipe or the entire adjacent section of the waveguide make possibilities for long duration technological processes. (figure 7)

Figure 6. Stainless steel and brass pipe fittings in the designs of plasma torches.

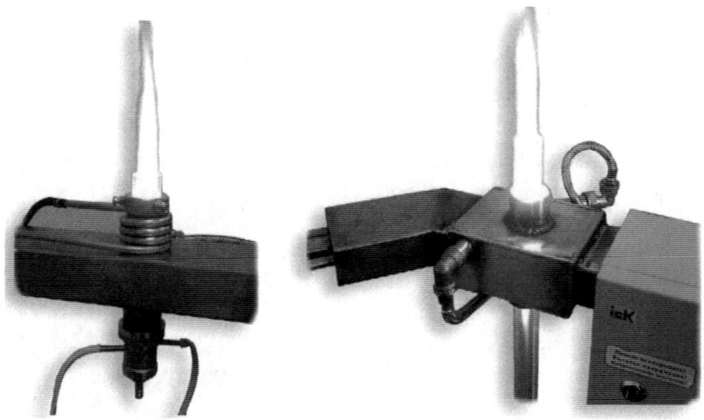

Figure 7. Options for additional cooling of the elements of the plasma torch construction.

3.4. The primary ignition of a microwave discharge.

We have found a method for initiating a discharge by introducing a pointed graphite rod with a copper coating into the discharge chamber along the axis of the plasmathrone. To do this, it is necessary to observe several design conditions associated with achieving a resonance along the length of the graphite rod. Automatic ignition mode can be provided with this method, if needed.

The limited volume of the article does not allow us to talk about our developments in the introduction of powders into the plasma, the sources of nonthermal plasma, the plasma torch with a remote dielectric wall, and others exciting possibilities given by our design low cost microwave generators and plasma torches.

4. Conclusions

All of the above has allowed us to develop and produce simple, reliable, inexpensive and highly effective microwave plasma torches for different purposes with the ability to implement continuous technological processes around the clock.

References

[1] *Encyclopaedia of Low-Temperature Plasma.* Entrance Volume, Book 2. Ed. by V. E. Fortov. (Moscow: MAIK Nauka/Interperiodika, 2000) [in Russian]
[2] Tikhonov V, Ivanov I, Kryukov A and Tikhonov 2015 *J. Prikl. Fiz.* **5** 102-6
[3] Tikhonov V, Pugashkin D and Chetokin Ya 2011, RF Patent No. 2480890

IOP Conf. Series: Journal of Physics: Conf. Series **927** (2017) 012068 doi:10.1088/1742-6596/927/1/012068

Discharge creeping along the surface in the process for producing nanomaterials

B A Timerkaev[1], A A Andreeva[1,2] and A O Sofronitskiy[1]

[1] Kazan National Research Technical University named by A.N.Tupolev, 420111, K.Marx str., 10, Kazan, Russian Federation
[2] Kazan Federal University, Institute of Engineering, 420008, Kremlevskaya St. 18, Kazan, Russian Federation Kazan,

E-mail: artempic8@mail.ru, btimerkaev@gmail.com

Abstract. In this paper, we propose a new principle of assembling carbon nanoparticles in the plasma of a glow discharge creeping along the surface. In this paper, it is shown that carbon nanoparticles (fullerenes and nanotubes), as well as light fractions of oil, can be produced by means of a glow discharge on the surface of the fuel oil. Single-walled carbon nanotubes of about 10 μm in length were obtained.

1. Introduction

The organization of the interaction of a glow discharge with dielectric liquids causes certain difficulties due to the fact that the plasma of a glow discharge always deviates upward, and the liquid is always at the bottom. Therefore, there must be a force of a different nature, which exceeds the strength of Archimedes and presses the plasma to the surface of the liquid. As such a force can act as an Ampere force, which acts on the current of a glow discharge in a magnetic field. In the presence of a magnetic field of the corresponding direction, the plasma of a glow discharge can be pressed against the surface of the liquid. Since the directions of motion of electrons and ions in a glow discharge plasma are opposite, the directions of action of the Lorentz force on them will have the same direction. The discharge will run near the surface. In this paper, we studied the characteristics of such a discharge, which we called a "creeping discharge".

A new principle of assembling carbon nanoparticles in the plasma of a glowing discharge is proposed. Electrons accelerated in a strong electric field and directed by the action of the Lorentz force to the surface of a hydrocarbon feedstock are capable of breaking hydrocarbon molecules into constituent elements. Thus, after the action of high-energy electrons on molecules of liquid hydrocarbons, both atomic carbon and hydrogen appear, as well as a great variety of the results of their synthesis. Including volatile and light hydrocarbons. Going through the plasma of a glow discharge they continue to be bombarded with electrons, enriching the elemental composition of the products of decomposition of hydrocarbons. At the same time, carbon atoms in a glow discharge acquire a charge and are directed to the electrodes, where, basically, the synthesis of carbon nanomaterials occurs. The processes of the action of electric discharges on hydrocarbons were previously studied in [1-8].

2. Experiment

The schematic diagram of the experimental setup is shown in Figure 1. The installation includes: an adjustable high-voltage power supply, a block of ballast resistances, a measuring equipment, a vacuum system, a reaction chamber, an electrode system, a working fluid reservoir, magnets.

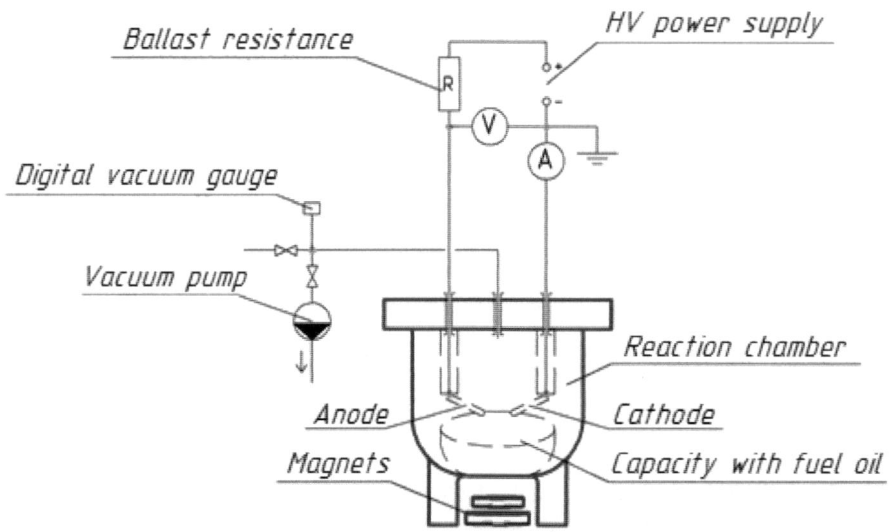

Figure 1. Schematic diagram of the installation.

Before the beginning of the experiment, a vacuum is created in the flask by means of a pump to a value of 2 Torr. Then a voltage is applied to the electrodes, which is increased until the discharge appears. The magnetic field presses the discharge to the surface of the fuel oil, and the discharge, as it were, spreads on the surface of the fuel oil. In the course of the experiment, the black oil in the tank begins to intensively transform into a gas vapor phase of white color. That is, electrons accelerated in a strong electric field acquire sufficient energy to break up the hydrocarbon atoms into their constituent elements. Thus, after the action of high-energy electrons, molecules on liquid hydrocarbons appear as atomic carbon, hydrogen, and also a great variety of the results of their synthesis. Including volatile and light hydrocarbons. Going through the plasma of a glow discharge they continue to be bombarded with electrons, enriching the elemental composition of the products of decomposition of hydrocarbons. At the same time, carbon atoms in a glow discharge acquire a charge and are directed to the electrodes, where, and mainly, the synthesis of carbon nanomaterials occurs. Due to the conversion of fuel oil during the experiment into the gas-vapor phase, the pressure in the flask gradually increases. On the rate of increase in pressure, one can judge the intensity of the process of conversion of fuel oil into light and volatile fractions. For this, it is necessary to know the composition of the gaseous phase, the increment in pressure, and the mass of carbon nanomaterials formed on the electrodes. It is also necessary to weigh the fuel oil before and after the experiment. With a discharge power of the order of 150 W for 10 minutes, the pressure in the chamber rises to 100 Torr.

The glow discharge arising between the electrodes, judging from the current-voltage characteristic, is a hindered glow discharge. The current-voltage characteristic of the discharge is shown in Figure 2. It is seen that even at sufficiently high currents the discharge remains volumetric with a voltage of the order of 800 V. This is most likely due to the action of the magnetic field on the discharge. The Lorentz force, which acts on the discharge electrons, additionally expands the current region and does not allow the discharge to be contrapulated.

For a visual demonstration of the "creeping discharge on the surface," Figure 3 shows a photograph. In this case, a ceramic substrate was used as a dielectric, by which the discharge "floats".

In the case of the interaction of a "creeping discharge" with hydrocarbon raw material, the color of the discharge became yellow-orange.

3. Study of samples of soot formed

During the experiment with hydrocarbon raw materials, carbon black formed on the electrodes, in which, as further studies have shown, large numbers of fullerenes are contained. To extract fullerenes from soot, the Soxhlet apparatus was used. The obtained extract was studied by infrared spectroscopy.

It is known that, the lower the symmetry of fullerene, the more absorption frequencies we can observe. Because of the high symmetry of the C60 molecule, only 4 absorption peaks appear in the IR spectrum at 1429, 1183, 577 and 528 cm-1 and a width varying in the range 3-10 cm-1. The lower the symmetry of fullerene, the greater the absorption frequencies, respectively, in the C70 molecule, 11: 1460, 1430, 1414, 1134, 795, 674, 642, 578, 565, 533 and 458 cm-1 are observed.

The IR absorption spectra of higher fullerenes are more complex than those of C60 and C70 molecules. The absorption bands of the C76 molecule correspond to wavelengths of 230, 286, 328, 350, 378, 405, 455, 528, 564, 574, 642, 709, 768 nm. In the absorption spectrum of the C84 molecule, the bands 280, 320, 380, 393, are singled out. 476, 566, 616, 668, 760 and 912 nm.

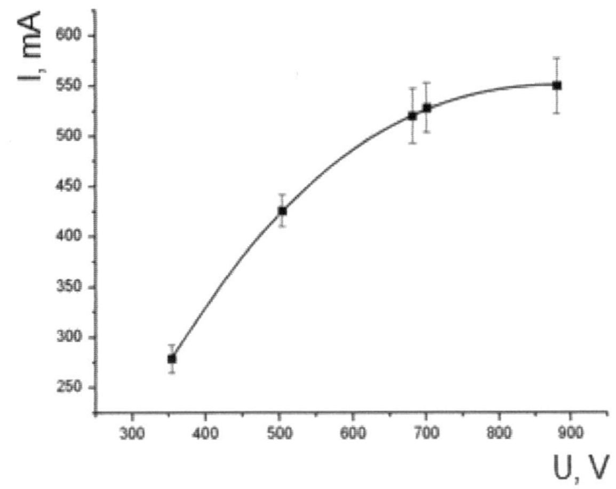

Figure 2. Volt-ampere characteristic of the discharge.

With this process, the discharge power is about 300 watts.

Figure 3. A photograph of a discharge on the surface of a dielectric.

The carried out investigations of the obtained carbonaceous mass formations by the spectroscopic method and the IRS method showed the presence of fullerenes C60, C70 and higher fullerenes in the soot formed.

4. Results

A new type of discharge with an increasing volt-ampere characteristic, which is called a "creep-discharge discharge," is studied. Surface discharging on the surface is ideal for organizing the interaction of a discharge with a dielectric surface. As an example, the interactions of such a discharge with a ceramic surface and a fuel oil surface have been studied. It follows from the conducted researches that such a discharge is suitable for creating technologies for the synthesis of carbon nanoparticles, gasification of coal, as well as for deepening the processing of fuel oil and oil sludge. The target products of processing heavy hydrocarbons are volatile and light oil fractions, as well as carbon nanomaterials.

References

[1] Timerkaev B A, Sofronitskiy A O and Andreeva A A. Carbon nanotubes formation in the decomposition of heavy hydrocarbons creeping along the surface of the glow discharge. 2016 *Journal of Physics: Conference Series,* Volume **669**, conference 1, 012062. doi:10.1088/1742-6596/669/1/012062

[2] Timerkaev B A, Andreeva A A and Sofronitskiy A O. Discharge creeping along the surface in the process of cleaning and strengthening of the materials surface. 2017 *Journal of Physics: Conference Series,* Volume **789**, conference 1, 012063. doi:10.1088/1742-6596/789/1/012063

[3] Sadikov K G, Sofronitskiy A O and Dautov I G. Functional plasma sprayed coatings on magnesium ceramic substrates. 2017 *Journal of Physics: Conference Series,* Volume **789**, conference 1, 012043. doi:10.1088/1742-6596/789/1/012043

[4] Timerkaev B A, Ahmetov M M, Zalyaliev B R, Petrova O A and Israfilov D I. Longitudinal distribution of electrical parameters in normal glow discharge. 2014 *Journal of Physics: Conference Series,* Volume **567**, conference 1, 012036. doi:10.1088/1742-6596/669/1/012036

[5] Saifutdinov A, Timerkaev B, Zalyaliev B. Control of the glow discharge parameters at low pressures by means of a transverse supersonic gas flow. 2016 *High Temperature* **54** 669

[6] Galeev I G and Asadullin T Ya. Obtaining fullerene-containing soot during combustion of gaseous hydrocarbons in an external electric field. 2016 *Journal of Physics: Conference Series,* Volume **669**, conference 1, 012016. doi:10.1088/1742-6596/669/1/012016

[7] Galeev I G and Asadullin T Ya. The magnetic field application for the gas discharge plasma control in processes of surface coating and modification. 2017 *Journal of Physics: Conference Series,* Volume **789**, conference 1, 012003. doi:10.1088/1742-6596/669/1/012003

[8] Galeev I G and Asadullin T Ya. Improving of stability of the volumetric glow discharge in the gas flow. 2017 *Journal of Physics: Conference Series,* Volume **789**, conference 1, 012012. doi:10.1088/1742-6596/669/1/012012

Creation of relief coatings on the surface of silicate materials in the plasma of radio-frequency induction discharge at low pressure

A.V. Trofimov[1], **E.F. Voznesensky**[1], **I.S. Miftakhov**[1], **A.I. Nagmutdinova**[1], **I.V. Krasina**[1], **F.S. Sharifullin**[1] **and E.A. Skidchenko**[1,2]

[1] Department of Plasma Technology and Nanotechnology of High Molecular Materials, Kazan National Research Technological University, Kazan, 420015, Russia

[2] Skolkovo Institute of Science and Technology, Skolkovo Innovation Center, Nobelya str., Building 3, Moscow, 143026, Russia

m_storm.zip@mail.ru

Abstract. The paper shows the possibility to create relief coatings based on powdered nonmetallic materials (SiO_2 and Al_2O_3) on a silicate substrate in the radio-frequency induction plasma at the low pressure. The features of the structure and properties of the obtained coatings have been studied.

1. Introduction

Nanorelief coatings are currently used to produce materials with great contact angle of wetting values on the "lotus effect" principle. To obtain this result, the combined effect of the surface nanoroughness and its chemical structure should be used. The superhydrophobic properties can only be achieved by the dimensions selecting of the surface relief. Creation of the superhydrophobic coatings on the heterogeneous materials was described in details in the works of L.B. Boinovich [1-3]. It is also known, that silicate and ceramic materials with developed surface are in demand for catalysts carriers' creation. [4].

The study shows investigation in the possibility to obtain the developed surface on the silicate glass by application and thermal fixing of dispersed material under gas radio-frequency induction (RFI) low prēssure discharge treatment. Al_2O_3 and SiO_2 powders with different dispersion were used as coating materials.

2. Materials and methods

The household and quartz glass plates, glass fibers and glass microspheres were used as substrates. The powders with different dispersion were used as a coating material,: microdispersed SiO_2 with particle dimension 0.5 - 5 μm and nanodispersed SiO_2 with a particle dimension 50 - 300 nm were obtained by the gas-phase method [5], as well as α-Al_2O_3powder after mechanical dispersion with particles dimension 0.5 - 100 μm. Granular stearic acid was used as an additional non-toxic chemical water-repellent agent.

Experimental RFI plasma set was employed for the powders deposition on the substrate (Fig. 1). Technical parameters of the treatment were the following: ions energy of 30 eV; density of electron current of 25 A/m^2; pressure of 60 Pa; plasma-forming gas – Ar and a mixture of Ar-propane (90:10).

The powders were injected into the RFI discharge through the plasmatron together with the working gas.

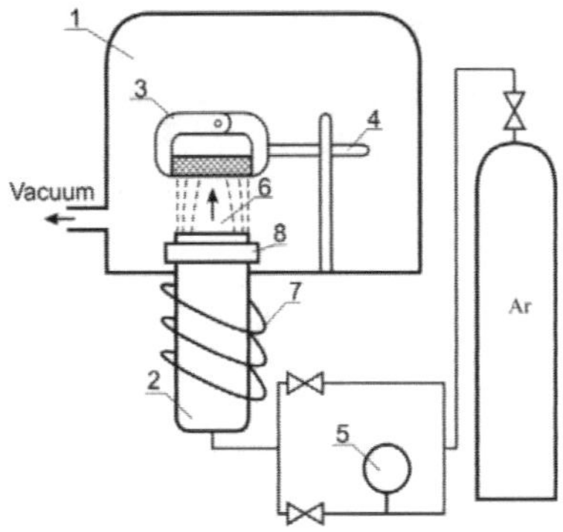

Figure 1. The scheme of radio-frequency induction installation

The process of relief coating applying on the substrate included the following steps (Fig. 1):

1. The billet was installed into the operating vacuum chamber 1 of the radio-frequency induction installation above the cut of the quartz plasmatron 2 in the holder 3 at a predetermined height (adjustable by the bracket 4 in the range of 30 - 60 mm).
2. The system was evacuated to a pressure of 10 - 100 Pa.
3. The discharge was ignited and its specific technological parameters were set; the billet was heated above the plasma jet 6 to a temperature of 250 - 300 °C for cleaning, surface activation and internal strains removal.
4. The coating material of the selected fraction was supplied through a powder dispenser 5 with a flow of plasma-forming gas into the working zone through a plasmatron, placed in an inductor 7 with an additional electrode 8.
5. The samples of the catalysts heated in the discharge core to temperatures close to the melting point were thermally deposited on the heated substrate.
6. For additional cleaning and coating fixing the billet was treated in the discharge for 15 - 60 min after the powder supply was stopped.
7. After the discharge was switched off the billet was blown by an inert gas stream.

3. Results and discussion

The coating applying was resulting in the formation of a regular micro-grained relief with the sizes of elements of 0.5 - 1.5 µm. The microscopy results of the flat glasses with coatings are shown in Fig. 2, 3; on glass fabric fibers – in Fig. 4.

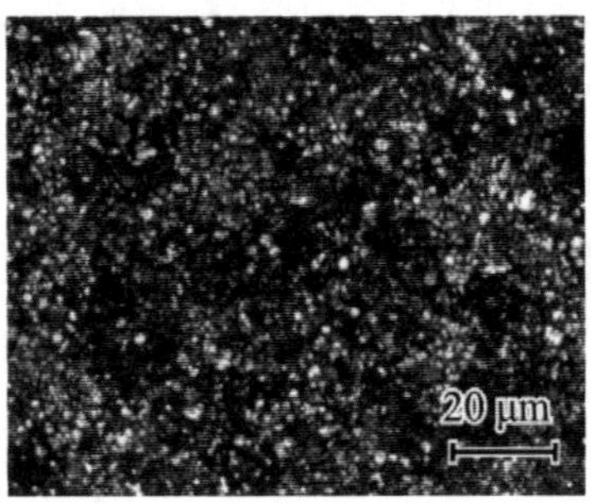

Figure 2. Optical image of Al_2O_3 relief coating, ×2000

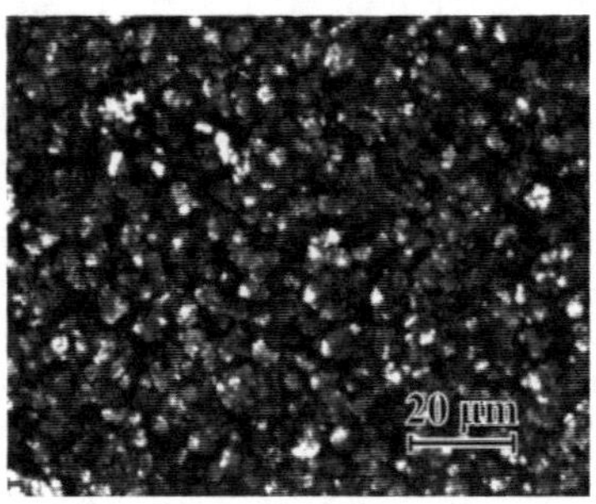

Figure 3. Optical image of SiO_2 relief coating, ×2000

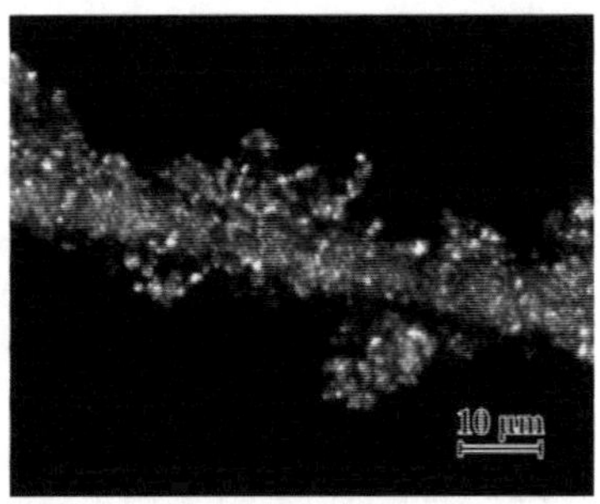

Figure 4. Optical image of glass fiber with SiO_2 coating, ×4000

The texture, obtained on the glass surface, is characterized by high regularity and granularity. A uniform multimodal relief was formed on glass fiber by agglomerates of particles and individual particles with the details dimensions of 0.5 - 5 µm (Fig. 4).

The possibility to develop the glass microspheres surface is also considered. It was proposed to create a microtexture on the glass spheres (d = 50-500 μm) surface by depositing and fixing silicon dioxide nanoparticles under low pressure RFI plasma treatment. Microspheres and nanoparticles were mixed, according to 1:10 mass ratio, and placed in a hopper for powder feed (item 5, Fig. 1) into the discharge, and caught by a trap filter above the plasmatron cut. The microstructure of the samples was studied by SEM method; the results of the study are shown in Fig. 5. The SiO_2 relief coating with size details of 1-5 μm was formed on the glass microsphere.

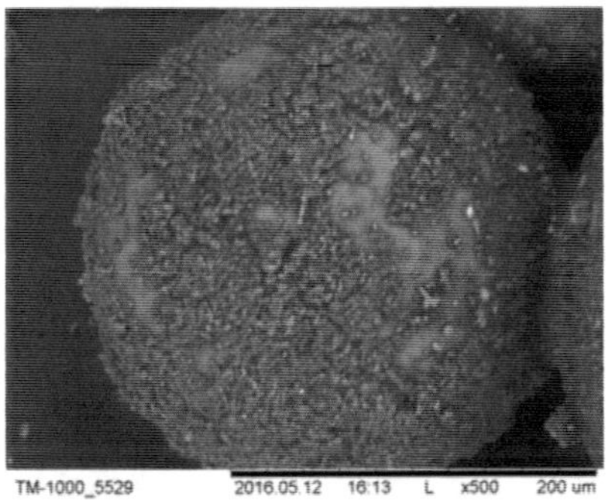

Figure 5.SEM-image of the surface of a glass microsphere with SiO_2 coating, ×500

The results of the created relief on the hydrophobic properties, was investigated by using wetting angle determination method on the KRUSS DSA 30 installation. The results (Fig. 6) demonstrate increased contact angle of wetting on the SiO_2 coating with the stearic acid additionally applied.

a) 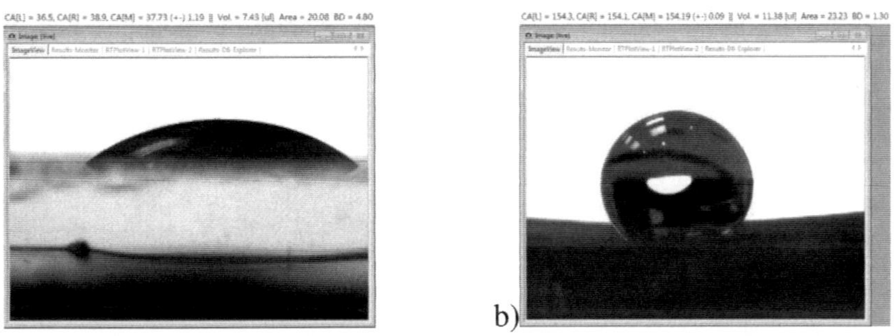 b)

Figure 6. The image of water drop on a glass sample: a – without coating; b – with SiO_2 coating and using of stearic acid

The structure on the surface of silicate glass, created by RFI plasma treatment, enables to strengthen the hydrophobic properties of the surface, in particular, the value of the wetting angle (θ) increased from 37.73° to 154.19°. Therefore, it is advantageous to obtain the superhydrophobic coatings on the silicate materials surface by given technique.

The phase composition of the Al_2O_3 coating was studied by X-ray diffraction method. It was established that with Ar as plasma-forming gas the predominant phase is α-Al_2O_3, whereas in the mixture of Ar-propane gases, the γ-Al_2O_3 phase appears in the coating material (Fig. 7). The X-ray diffraction results demonstrate the following phase composition of the coating: 15% – γ-Al_2O_3, 15% – α-Al_2O_3, 70% – crystalline γ-Al_2O_3. The chemical effect of the plasma was manifested in a significant crystallization of the sample.

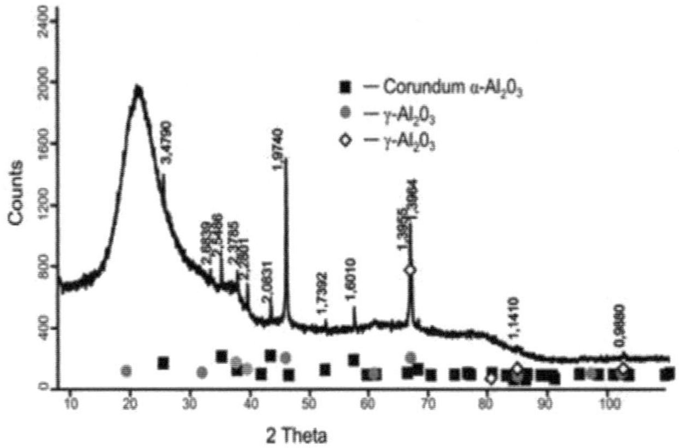

Figure 7. Diffractogram of the Al_2O_3 coating

4. Conclusion

Overall, the possibility to obtain the relief non-metallic coatings on silicate substrates with different geometry in radio-frequency induction plasma discharge at low pressure has been evidenced with a number of experiments. These coatings can be applied to create catalysts carriers, as well as to impart superhydrophobicity based on the "lotus effect" to silicate materials.

References

[1] Boinovich L.B. Production and application of superhydrophobic films to prevent leakage currents / L.B. Boinovich, A.M. Emelyanenko, A.M. Muzafarov, A.M. Myshkovsky, A.S. Pashinin, A.Yu. Tsivadze, D.I. Yarov // Strides in chemistry and chemical technology. Volume XXII. 2008. № 9 (89).P17-23.

[2] GnedenkovS.V., EgorkinV.S., SinebryukhovS.L., VyalyI.E., PashininA.S., EmelyanenkoA.M., BoinovichL.B./Superhydrophobiccompositecoatingsonthesurfaceofmagnesiumalloy // BulletinoftheFarEasternBranchoftheRussianAcademyofSciences. 2013. №5 (171) P.3-11.

[3] Gnedenkov S.V., Egorkin V.S., Sinebryukhov S.L., Vyaly I.E., Emelianenko A.M., Boinovich L.B./Superhydrophobic protective coatings on aluminum alloy // Bulletin of the Far Eastern Branch of the Russian Academy of Sciences. 2014. №2 (174) P.52-61.

[4] Nguyen, T.T. Co_3O_4 Catalysts on CeO_2-ZrO_2 Supports and Co_3O_4-CeO_2 Catalysts on Al_2O_3/SiO_2 Supports for the Oxidation of Propylene / T.T.Nguyen , P.T.M.Pham, T.M.Le, I.V.Driessche // Journal of Chemistry. – 2015. – V. 2015. – P. 1–11.

[5] Katnov, V.E. Investigation of the influence of the technological parameters of the preliminary grinding on the properties of the SiO_2 powder used in the plasmochemical synthesis of nanoparticles / V.E. Katnov, P.V. Grishin, R.R. Katnova, S.N. Stepin // Bulletin of the Kazan Technological University. - 2013. - V.16, No.5. - P.69-71.

On transition from diffuse mode to the constricted one with high-current cathode spot in overvoltage open discharge in D2

Yu S Akishev[1,2], V B Karalnik[1], M A Medvedev[1], A V Petryakov[1], N I Trushkin[1], A G Shafikov[1]

[1] SRC RF TRINITI, 108840, Moscow, Troitsk, Pushkovykh street, 12, Russia
[2] NRNU MEPhI, 115409, Moscow, Kashirskoe shosse, 31, Russia

Abstract. So called "open discharges" in a narrow gap between the solid cathode and grid anode are widely used for generation of the pulsed high-current electron beams with energy up to 100 keV. The need to get high-energy e-beams leads to the necessity in using of strong overvoltage of the short gas gap with the reduced electric field of the order of 10^5 Td or higher. The discharge under strong overvoltage is unstable and tends to transit into high-current regime with low voltage. In the case of the open discharge in D2 at low pressure (about 0.5-2 Torr) and powered by stepwise voltage with amplitude up to 25 kV we revealed that this discharge exhibits two diffuse regimes which follow one by one and finally transits into the constricted mode with formation of high-current spots on the cathode. The physical properties of these gas discharge regimes have been explored in detail with the usage of the fast multi-frame camera synchronized with the current and voltage of discharge. Our findings promote more insight into physics of the overvoltage open discharge generating the e-beams with energy up to 25 keV.

1. Introduction

Strong overvoltage pulsed discharges operating at low pressures (P < 10 Torr) in a narrow gap (d < 1 cm) between the solid cathode and grid anode are often called as "open discharges". The reduced electric field E/N in these discharges reaches of huge magnitude of 10^5 Td and even more. Such electric fields correspond to an appearance of run-away electrons which are able to form the high-current electron beams with a high efficiency. This is a reason why the strong over-voltage pulsed discharges are widely used for generation of e-beams with energy up to 70 keV. A lot of useful information about properties of the open discharges powered with the voltage up to 100 kV can be obtained from [1]. Unfortunately, the discharge under strong overvoltage is unstable and tends to transit quickly into high-current regime with low voltage which does not generate the e-beam. Note that up to now, the physical mechanisms providing the existence of the overvoltage open discharges and their transition into low-voltage regimes are the subjects for discussions. We have studied these mechanisms by example of the strongly overvoltage open discharge in short gap (3 mm) in D2 at low pressure (2 Torr) powered with the voltage of amplitude up to 25 kV. Why we used the voltage of such amplitude? A reason is that the high-energy e-beams (about 70 keV) generate the hard bremsstrahlung X-radiation which has a high penetrability and brings a serious bio-danger. Due to that the installations using the e-beams of such high energy represent bio-danger at their service and demand the application of special measures of bio-protection. In contrast, the e-beam of 25 keV in energy does not generate the hard bremsstrahlung X-radiation. Because of that the open discharges powered with the voltage up to 25 kV and generating the high-current e-beams with energy up to 25 keV are environmental friendly and of great interest to practical applications. Despite the importance of these discharges for practice, their physical properties have been explored insufficiently. Our findings promote more insight into mechanisms of the instability of the overvoltage open discharge powered with the voltage up to 25 keV.

2. Experimental setup

We have designed the three-electrode system which provides a higher reproducibility of the parameters of the open discharge compared to traditional two-electrode system. The electrical scheme of the three-electrode system used for generation of the open overvoltage discharge is depicted schematically in Figure 1a. The pulsed over-voltage discharge has been formed in a narrow gap between the solid cathode (1) and the grid anode (2). The distance between these electrodes is equal to 3 mm. This discharge is ignited with a high voltage transferred by high-current thyratron T from the capacitor C1 through the ballast resistor R1 with a resistance of 44 Ohm. There was a complimentary discharge between the cathode (1) and additional grid anode (3). The distance between cathode and the auxiliary grid anode is equal to 16 mm. The geometrical transparency of each grid was about 70%. The auxiliary low-current and low-voltage discharge was fed by the capacitor C2 through the ballast resistor R2 with a resistance of 1 MOhm and used to form the preionization of gas in the gap of the main discharge. High-voltage diode D prevents the transfer of a negative high voltage from capacitor C1 to the low-voltage capacitor C2.The applied high voltage was measured by HV divider PINTEK HVP-39 (1000:1, 40 kV, 200 MHz). The discharge current was measured by a low-inductive shunt with resistance of 0.024 Ohm. All electrical signals were recorded by the digital oscilloscopes such as Tektronix TDS 520, Tektronix TDS 2012 and Tektronix DPO2024.

Figure 1b shows total lay-out of the components of the experimental setup used for investigation of the transitions of unstable regimes of the overvoltage open discharge by the fast multi-frame camera equipped with the intensifier. The optical system was focused in such a way to take the image of the discharge zone including its cathode area. The three-electrode system was mounted inside the quartz tube. Before each experiment, the tube was pumped out up to the pressure of $P=10^{-3}$ Torr and then filled with deuterium of high purity (99.99%) up to the required pressure. To avoid the accumulation in the tube of the impurities happening due to plasma chemistry and inleakage from ambient air, the deuterium was constantly pumped over through a tube with a low gas flow rate.

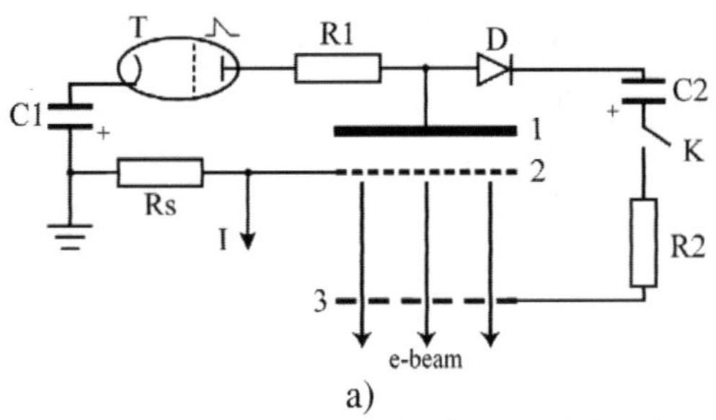

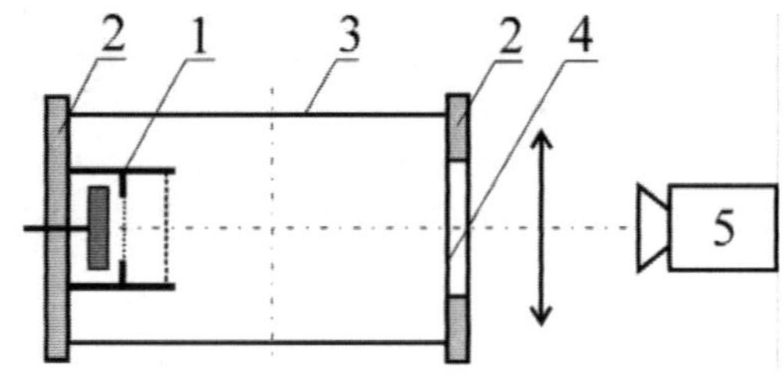

Figure 1. a) The electrical scheme of the three-electrode system generating a high-current e-beam with energy up to 25 keV. 1 - the solid cathode; 2 - the grid anode of the main discharge; 3- the grid anode of the auxiliary discharge; R1=44 Ohm, R2 =1 MOhm, Rs = 0.024 Ohm; C1 = 12.5 nF, C2 = 100 μF. b)1- the electrode system; 2 – flanges; 3 - quartz tube; 4- quartz window; 5- fast multi-frame camera.

3. Experimental results and discussion

Figure 2 shows the Current-Voltage characteristic of the open discharge in the diffuse overvoltage regime which generates high-energy e-beam. This characteristic can be approximated by cubic dependence $I \sim U^3$. However, the higher applied voltage and current, the shorter duration of the overvoltage regime which is the most interesting for practice. Therefore there is an insistent need in

IOP Conf. Series: Journal of Physics: Conf. Series **927** (2017) 012070 doi:10.1088/1742-6596/927/1/012070

studying of the mechanism of instability of this regime. The experimental data showing in detail the spatial-temporal evolution of the open discharge up to its transition into the constricted low-voltage regime with the high-current cathode spots are presented in figures 3-5. This set of data obtained for different gas pressures can be very useful for development of the physical mechanisms responsible for transition of the open discharge from one regime into another.

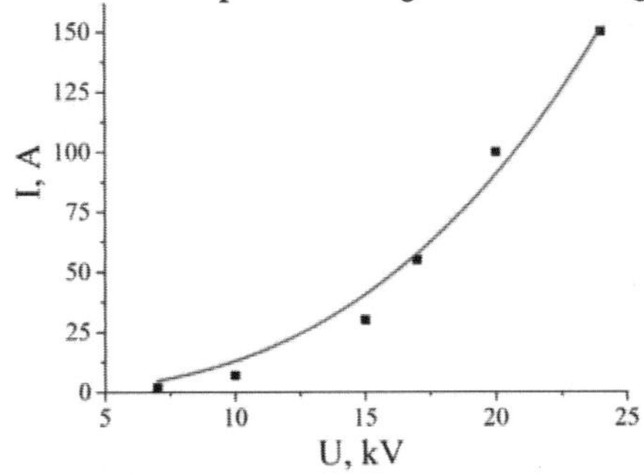

Figure 2. The Current-Voltage characteristic of the open discharge in the diffuse overvoltage regime. The solid curve is the approximation I ~ U^3. Deuterium, P = 2 Torr, the auxiliary low-current discharge is powered with voltage U* = 1 kV.

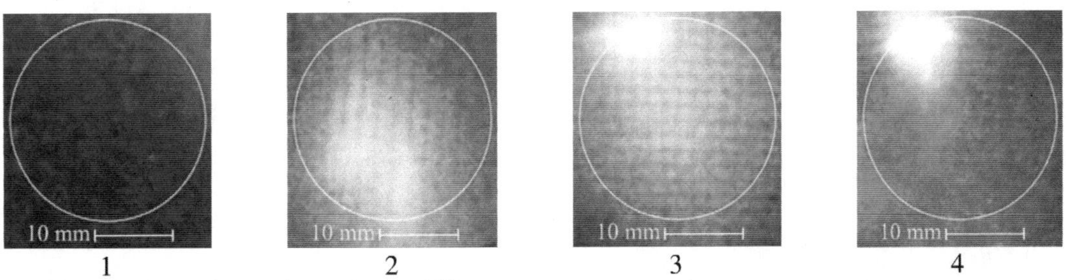

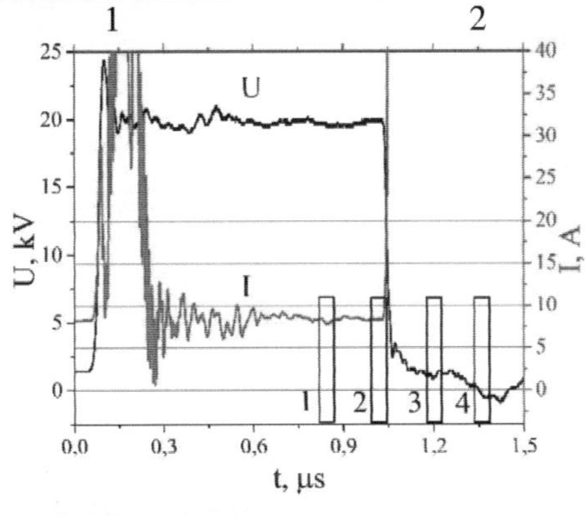

Figure 3. The set of the discharge images taken by a fast multi-frame camera and correlated with the current-voltage waveform. The white circle in each shot depicts the boundary of the discharge area on the cathode. The exposure time of each shot is 50 ns. The enumerated time moments marked in the current-voltage waveform correspond to those at which the enumerated images were taken. Deuterium, P = 0.5 Torr, U = 20 kV, U* = 2.5 kV.

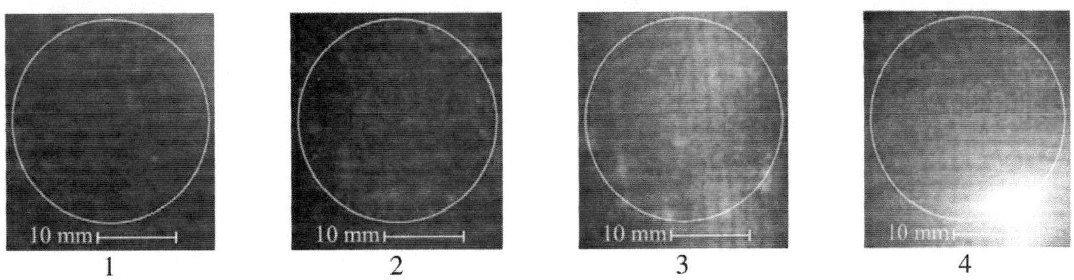

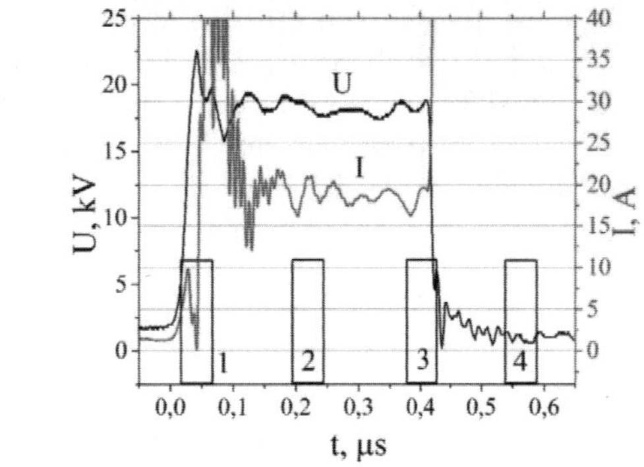

Figure 4. The set of the discharge images taken by a fast multi-frame camera and correlated with the current-voltage waveform. The white circle in each shot depicts the boundary of the discharge area on the cathode. The exposure time of each shot is 50 ns. The enumerated time moments marked in the current-voltage waveform correspond to those at which the enumerated images were taken.
Deuterium, P = 1 Torr, U = 20 kV, U* = 2 kV.

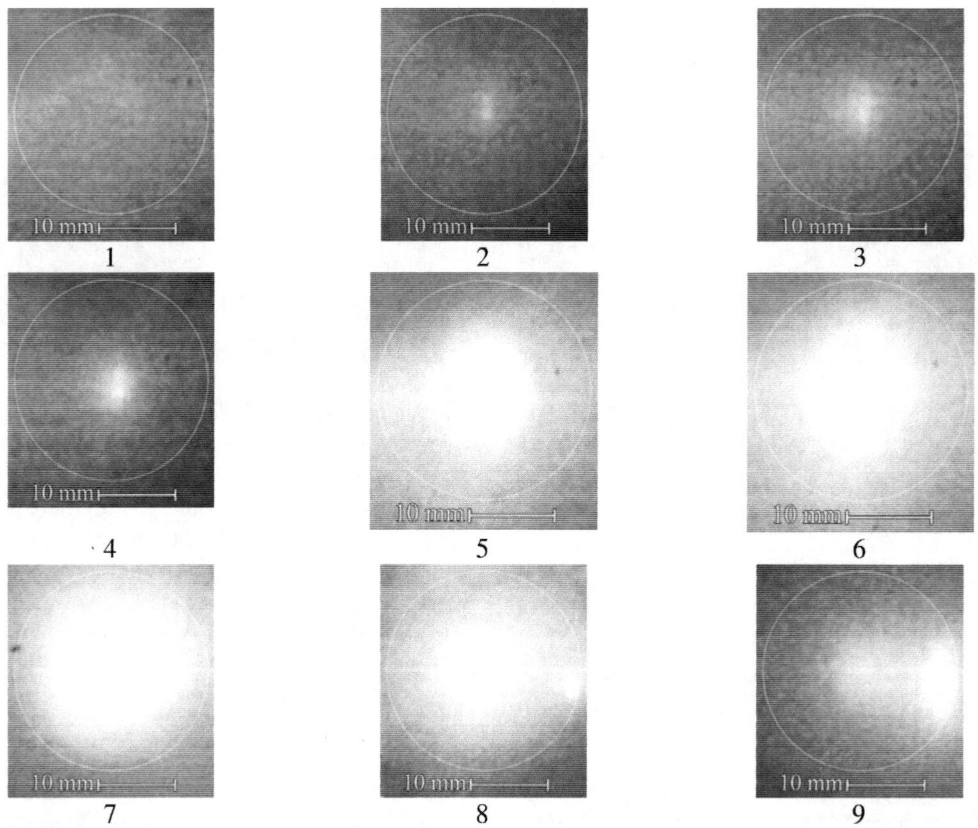

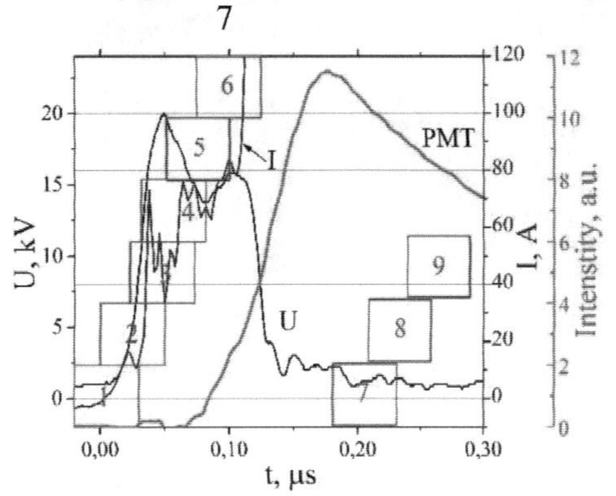

Figure 5. The set of the discharge images taken by a fast multi-frame camera and correlated with the current-voltage waveform. The white circle in each shot depicts the boundary of the discharge area on the cathode. The exposure time of each shot is 50 ns. The enumerated time moments marked in the current-voltage waveform correspond to those at which the enumerated images were taken.
Deuterium, P = 2 Torr, U = 20 kV, U* =1 kV

Close examination of all figures 3-5 leads to the conclusion that the cathode current spots have a non-stationary in time and chaotically in space behavior – small spots can appear and disappear many times in different places of the cathode area before the formation of a single high-current spot at the constricted low-voltage regime. Besides, the existence of non-stationary cathode spots in itself does not influence the diffuse overvoltage regime and its duration. A reason is that the cathode spots in itself do not shunt the gap - they can only initiate the streamers (high-conductive current filaments) which can propagate from cathode towards the anode. In general, after the arrival of streamers to the anode the gap shunting happens and high-voltage regime transits into the low-voltage mode [2]. Note, the propagation of streamers towards the anode requires the existence in the gap of the intensive direct ionization of neutrals by electron impact. However, the extremely high electric field at the cathode in the overvoltage regime leads to that the electrons have become the run-away electrons which have the extremely low efficiency of the impact ionization. This is a reason why the propagation of streamers towards the anode is impossible in the overvoltage regime.

So, the non-stationary and chaotic cathode current spots can change a bit the transverse structure of the open discharge in the overvoltage regime but not eliminate this regime at all. Of course, the cathode current spots can influence the homogeneity of transverse structure of the e-beam generated by the open discharge as well. Based on the mentioned above, we can state that, in our opinion, the mechanism responsible for transition of the overvoltage regime into abnormal glow mode is not connected with the cathode spots but is related with the propagation of the plane ionization wave from the anode towards the cathode. After the wave approaches the cathode, the voltage drop across the gap will sharply decrease and the discharge will pass into the transverse uniform abnormal glow mode. However, just the cathode spots are responsible for shunting the discharge gap in the abnormal glow mode and its transition into the very low-voltage constricted mode with, as a rule, single high-current cathode spot.

4. Conclusion

The open discharge in D2 at low pressure (P = 0.5-2 Torr) and powered by stepwise voltage with amplitude up to 25 kV exhibits two diffuse regimes which follow one by one and finally transits into the constricted mode with the formation of high-current spots on the cathode. The physical properties of these gas discharge regimes have been explored in detail with the usage of the fast multi-frame camera synchronized with the current and voltage of discharge. Based on the results presented above, we can conclude that the mechanism responsible for transition of the overvoltage regime into the transverse uniform abnormal glow mode is not connected with the cathode current spots observed in the overvoltage regime but is related with the propagation of the ionization wave from the anode towards the cathode. Nevertheless, the cathode spots play essential role in the destabilization of abnormal glow mode - namely they are responsible for shunting the discharge gap in this mode and its transition into the very low-voltage constricted mode with, as a rule, single high-current cathode spot. In total, our findings promote more insight into mechanisms of the instability of the overvoltage open discharge powered with the voltage up to 25 keV.

Acknowledgments

This work has been carried out thanks to full financial support by the Russian Science Foundation (Grant № 16-12-10458).

References

[1] Sorokin A and Bokhan P 1994 *Technical Physics Letters* **20** 86

[2] Akishev Yu, Grushin M, Kochetov I, Karal'nik V, Napartovich A and N Trushkin 2005 *Plasma Sources Sci. Technol.* **14** 518

IOP Conf. Series: Journal of Physics: Conf. Series **927** (2017) 012071 doi:10.1088/1742-6596/927/1/012071

Effect Crossed Electric and Magnetic Fields of a Glow Discharge

R. Vafin[1], I. Yagafarov[1], Y. Khusainov[1], A. Pesin[2], D. Pustovoytov[2]
[1] Ufa State Aviation Technical University, Ufa - 450000, Russia.
[2] Nosov Magnitogorsk State Technical University, Magnitogorsk - 455000, Russia
Email: vafinrk@mail.ru

Abstract. A new method of processing in a glow discharge plasma in crossed electric and magnetic fields is investigated. This method makes it possible to intensify the process of diffusion saturation of the surface with nitrogen. Volt-ampere characteristics of a glow discharge in crossed electric and magnetic fields are obtained. The effect of ion nitriding in crossed electric and magnetic fields on the hardness and structural-phase composition of tool steels R6M5 and Kh12 is studied.

1. Introduction

In order to raise the endurance of tools their surface is modified by various methods including ion-plasma nitriding characterized by such features as controllability and ecological safety [1-3].
In ion nitriding the parts are heated to the nitriding temperature due to the energy of gas ions bombarding the surface at simultaneous diffusion of nitrogen ions into the treated surface. The processes in the plasma of glow discharge (dissociation, excitation, ionization, etc.) occur in what is known as the region of cathode drop [4]. One of the factors capable to activate these processes may be superposition of magnetic field on the glow discharge, which is used successfully in various gas discharge systems [5, 6].

The aim of the present work consisted in studying the effect of magnetic field on the process of ion nitriding, on the structure, on the phase composition, and on the microhardness of the surface layer of tool steels R6M5 and Kh12.

2. Materials and Research Methods

In order to study the process of ion nitriding and the characteristics of glow discharge in magnetic field we used an experimental installation created on the base of a commercial ELU-5 facility. The facility was equipped with a pulse energy source of type ApEl-M-5PDC, which allowed us to lower the heat load on the electrodes and provided effective arc extinction in a mode of anomalous glow discharge. We placed a standard magnetron (450 x 50 x 100 mm) with magnetic field induction $B = 0.03$ T in the vacuum chamber of the installation (figure 1). The parameters of the discharge were obtained from readings of the display of the switching power supply.

We subjected test pieces of tool steels R6M5 and Kh12 after hardening and tempering at 600°C to nitriding.

The working gas was a mixture of nitrogen, argon, and acetylene (75% N_2 + 20% Ar + 5% C_2H_2). The acetylene was used for deactivating the retained oxygen and preventing formation of an oxide film on the surface of the steel, which hinders formation of nitrided layer.

The pressure of the working gas was varied in the range of 5 - 200 Pa and controlled with the help of a VIT-3 vacuum gage. The working temperature (490 - 510°C) was attained in 2 - 3 min. The temperature of the test pieces was measured using a Termiks optical pyrometer.

A quantitative and qualitative analysis of the structure of the alloys was carried out on a Zeiss Axiovert metallographic microscope.

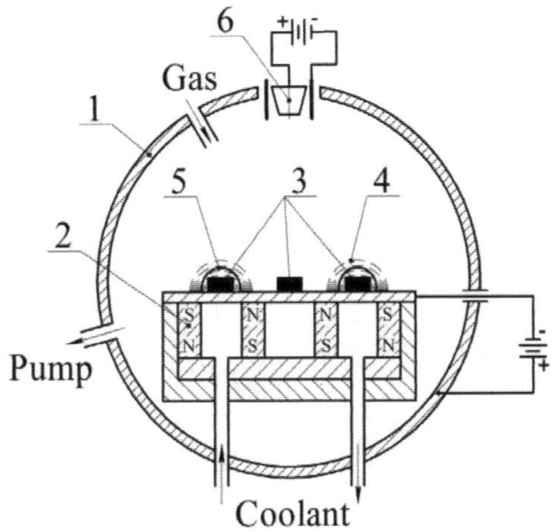

Figure 1. Scheme of the experiment
1 – vacuum chamber, 2 – magnet system, 3 – samples, 4 - lines of magnetic induction, 5 – high-density nitrogen plasma, 6 – arc plasma source.

3. Results and Discussion

Near of the cathode and on the axis of the system, the superposition of magnetic field leads to:

- Decrease in the plasma potential due to the magnetic confinement of electrons;
- Formation of a radial potential well for ions, which prevents their escape in the radial direction.

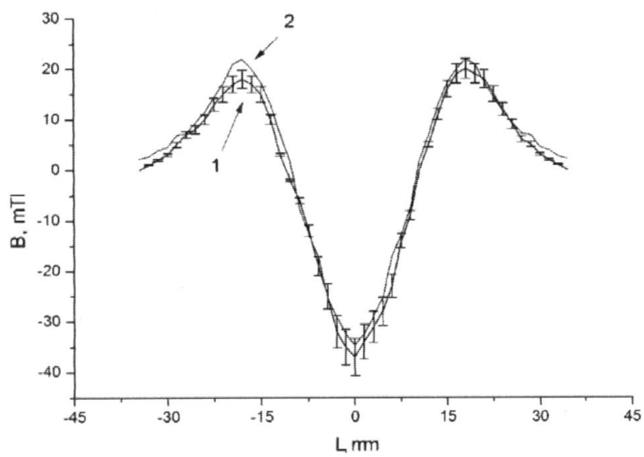

Figure 2. Radial distribution of magnetic field induction at a distance of 6 mm from the cathode:
1 - measured with a teslameter EM4305 (experimental);
2 - modeled by ELCUT using the finite element method (calculated).

Figure 2 shows the radial distribution of the magnetic field induction, which explains the characteristic behavior of a glow discharge in the magnetic field of Figure 3. The region of propagation of electrons is, on the one hand, limited by a relatively high negative potential on the substrate (~400-500 V), and on the other hand by a relatively strong magnetic field. Thus, the electrons are forced to oscillate along the lines of force of the magnetic field, which ensures more efficient utilization of their energy

Figure 4 shows the current-voltage characteristic (VAC) of the discharge obtained at different argon pressures in the chamber. With decreasing pressure, the VAC is shifted to the region of high operating stresses, due to a decrease in the concentration of charged particles. The increase in the residence time of electrons in a glow discharge plasma when the cathode is placed in a magnetic field promotes the displacement of the VAC upward (Figure 4). The VAC of a glow discharge with a magnetic field, in contrast to the conventional circuit, has a steeper rise, which indicates the efficiency of ionization processes [7].

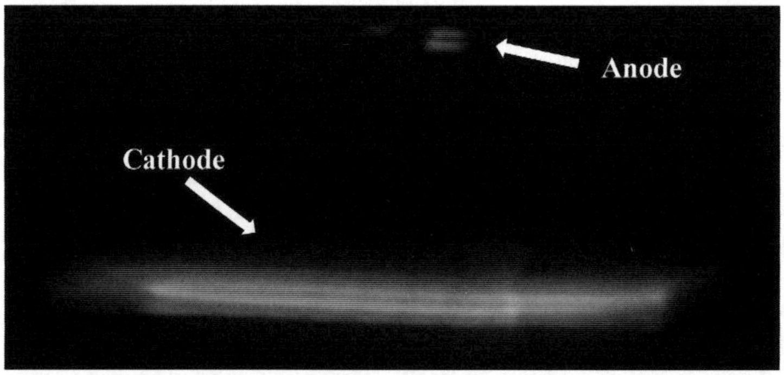

Figure 3. The glow of the near-cathode region of a glow discharge in a magnetic field at a pressure of 80 Pa (N_2 75% + Ar 20% + C_2H_2 5%) U = 420 V, I = 1.8 A.

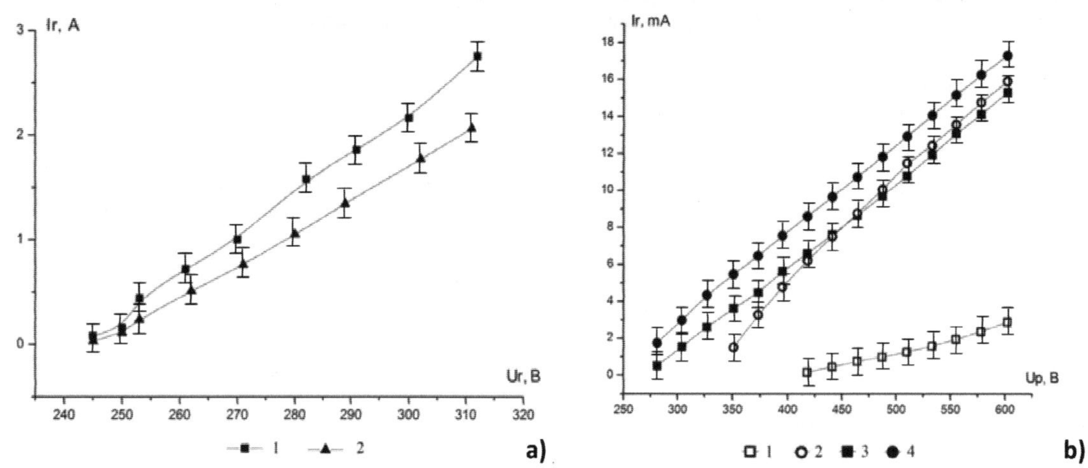

Figure 4. VAC for a glow discharge at various pressures,
pulse mode 50 kHz, 80%, mixture N_2 75% + Ar 20% + C_2H_2 5%:
a) - 100 Pa (1 - in a magnetic field, 2 - without a magnetic field);
b) - 5 Pa (1 - without a magnetic field, 2 - with a magnetic field) and
15 Pa (3 - without a magnetic field, 4 - with a magnetic field)

The influence of the magnetic field on the current-voltage characteristic of a glow discharge was investigated in the pressure range 5-15 Pa, the difference in the discharge current at the same pressure in the vacuum chamber is most noticeable after applying a potential to the electrodes (400 V). A similar discharge behavior is associated with its transition from a normal glow discharge to an anomalous one.

Figure 5 - 8 shows the structure of steels R6M5 and Kh12, subjected to ion nitriding in a glow discharge and ion nitriding in a glow discharge with a magnetic field. Analysis of the structure of steels shows the presence of a dark diffusion zone, its size varies depending on the technological parameters of the glow discharge. The diffusion zone is an α-phase - nitrogen ferrite with a bcc lattice. The transition from the nitrided layer to the underlying layers is smooth, which is one of the basic requirements for the microstructure of nitrided steel.

The magnetic field during ion nitriding of R6M5 steel in a glow discharge increases the thickness of the hardened layer from 13 μm to 84 μm (6 times) at a pressure in a vacuum chamber of 44 Pa (Figure 5-6). The magnetic field in the ionizing nitriding of steel Kh12 in a glow discharge increases the thickness of the strengthened layer from 16 μm to 81 μm (5 times) at a pressure in a vacuum chamber of 44 Pa (Figure 7-8). When a magnetic field is used, favorable conditions for the sorption processes are formed, as a result, a high concentration of the diffusing element on the cathode surface is provided and a high concentration gradient is formed. Thus, the intensification of ion nitriding by a magnetic field is due to the acceleration of the diffusion of active nitrogen in the metal.

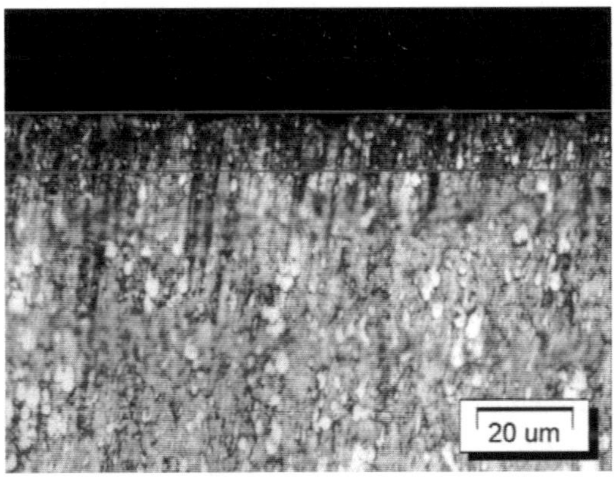

Figure 5. The microstructure of R6M5 steel after nitriding in a glow discharge T = 380 ° C, t = 4 hours, P = 44 Pa, U = 600 V.

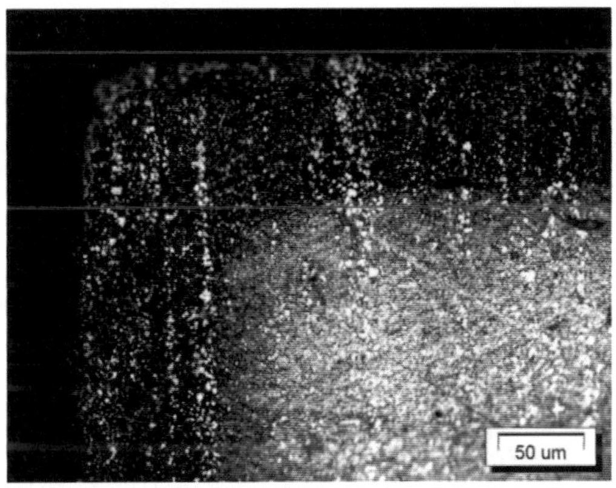

Figure 6. The microstructure of R6M5 steel after nitriding in a glow discharge with a magnetic field T = 500 ° C, t = 4 hours, P = 44 Pa, U = 600 V.

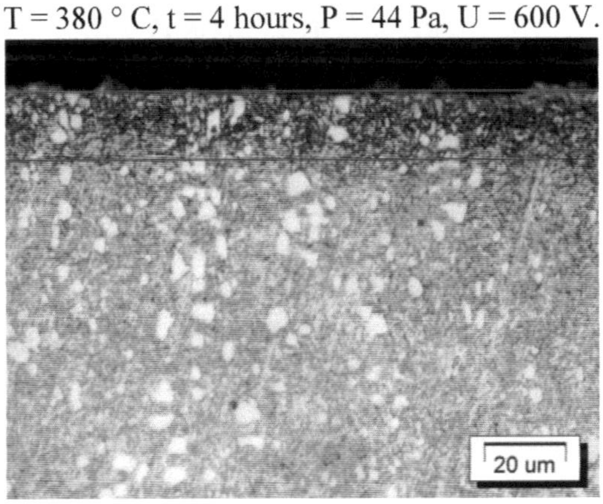

Figure 7. The microstructure of steel Kh12 after nitriding in a glow discharge T = 380 ° C, t = 4 hours, P = 44 Pa, U = 600 V.

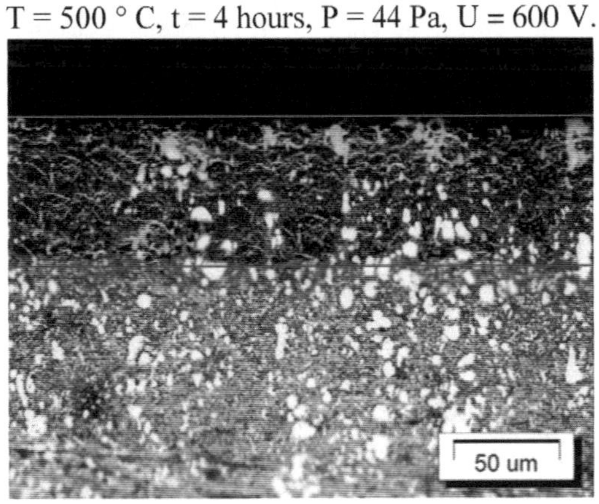

Figure 8. The microstructure of steel Kh12 after nitriding in a glow discharge with a magnetic field T = 500 ° C, t = 4 hours, P = 44 Pa, U = 600 V.

IOP Conf. Series: Journal of Physics: Conf. Series **927** (2017) 012071 doi:10.1088/1742-6596/927/1/012071

4. Conclusion

- VAC of a glow discharge with a magnetic field, has a steeper rise, which indicates the effectiveness of ionization processes.
- The use of a magnetic field in the nitriding of R6M5 and Kh12 steels in a glow discharge increases the thickness of the strengthened layer 5-6 times at 44 Pa, and favorable conditions for sorption processes develop on the cathode surface.

Acknowledgments
The study was supported by the grant from the Russian Science Foundation (#15-19-10030)

References

[1] Goncharenko I.M. 2000 Evolution of the structure and phase composition of hardened 4140 steel in the process of plasma nitriding *5–th International Conference on Modification of Materials with Particle Beams and Plasma Flows* (Tomsk: Institute of Physics and Technology TPU) pp 330–333
[2] Edenhofer B. 1976 The ion nitrating process – thermo chemical treatment of steel and cast materials *Metal and Material Technological* vol 8 №8 pp 421–426
[3] Koval N.N. 2000 Elion nitriding of steels *5–th International Conference on Modification of Materials with Particle Beams and Plasma Flows* (Tomsk: Institute of Physics and Technology TPU) pp 327–329
[4] Brading H.J., Morton P.H, Earweaker G. 1992 Plasma–nitriding with nitrogen, hydrogen and argon gasmixtures: Structure and composition of coating *Surface engineering* vol 8 №3 (Birmingham: Elsevier) pp 206–211
[5] Bradley J.W., Arnell R.D., Armour D.G. 1997 Measurement and modelling of the bulk plasma in magnetron sputtering sources *Surf. and Coat. Technol.* vol.97, (Salford: Elsevier) pp 538–543
[6] Arnell R.D., Kelly P.J. 1999 Recent advances in magnetron sputtering *Surface and Coatings Technology* vol 112 (Salford: Elsevier) pp 170–176
[7] Vafin R.K., Ramazanov K.N. 2010 Surface Modification of Tool Steel during Ion Nitriding in Magnetic Field *10th International Conference on Modification of Materials with Particle Beams and Plasma Flows* (Tomsk: Publishing House of the IOA SB RAS) pp 458–461

IOP Conf. Series: Journal of Physics: Conf. Series **927** (2017) 012072 doi:10.1088/1742-6596/927/1/012072

Modification of polypropylene in the afterglow of the atmospheric pressure discharges in air and argon

D P Vasilkin[1], T G Shikova[1], V A Titov[2], S A Smirnov[1] and N S Khomyakova[1]

[1]Ivanovo State University of Chemistry and Technology, Ivanovo, 153000, Russia
[2]G A Krestov Institute of Solution Chemistry of the Russian Academy of Sciences, Ivanovo, 153045, Russia

Abstract. Polypropylene films were modified in the flowing afterglow of the atmospheric pressure DC discharge in argon and in air. The modification was carried out at a discharge current of 15 mA, a gas flow rate of 24 and 105 m/s and a treatment time of 3 - 30 s. Polymer samples were placed on the distance of 5 - 15 mm downstream the plasma Contact angles for water and ATR-FTIR spectra were used for the film surface characterization. Concentrations of oxygen containing groups in modified polymer layer were estimated on the base of ATR-FTIR data. Modification in the both plasma forming gases results in the decrease of the contact angles and in the formation of oxygen containing groups in the polymer surface layer. Dependencies of contact angles on treatment time, gas flow rate and plasma - polymer distance were obtained. Increasing the treatment time and the gas flow rate results in a higher oxidation degree of the PP. Treatment in the afterglow of the argon plasma has been shown to give the less water contact angles and more densities of oxygen containing groups in polypropilene at the gas flow rate of 105 m/s and the treatment time of 30 s.

1. Introduction

Plasma modification of polymer materials is a promising tool for improving their properties: surface energy, wettability, adhesion to coatings etc. To improve these properties, atmospheric pressure discharges in different gases are used: in air [1-4], argon [5-7], helium [8], mixtures of oxygen and argon [9], oxygen and helium [8], nitrogen and argon [10]. Regardless of polymer and the plasma forming gas, the similar qualitative changes in the surface properties were observed: a decrease of the contact angles, an increase in the concentration of oxygen containing groups and in the surface roughness. However, it is difficult to compare the effectiveness of the modification depending on the gas nature and on the polymer type, since the treatment conditions are strongly different in various studies. These conditions include not only the plasma gas but discharge type, dissipated power, treatment time, gas flow rate and polymer location.

The aim of this study was to compare the results of modifying polypropylene films in the flowing afterglow of the atmospheric pressure direct current discharges in air and in argon.

2. Experimental

Simplified scheme of the experimental setup is shown in figure 1. Direct current (DC) discharge was excited between the nozzle anode with the inner diameter of 470 μm, through which the plasma forming gas was supplied, and a stainless steel plate with the thickness of 1 mm, which was used as a cathode. Interelectrode gap was 1 mm, discharge current was 15 mA. The gas flow rate was 24 and 105 m/s in different experiments.

Polypropylene (PP) films with the 20 μm thickness (TU RB 00204079.164-97, Mogilevkhimvolokno Ltd) were used in our experiments The polymer samples with the sizes of 1,5×5 cm^2 were placed on the distance of 5, 7, 10, 12 and 15 mm downstream the plasma. To characterize modification results, water contact angles measurements and ATR-FTIR spectroscopy (Avatar 360 FT-IR ESP) were used. Elemental analysis of the sample was performed on a scanning electron microscope (Tescan VEGA 3 SBH) with an X-ray energy dispersive microanalysis system (Oxford Instruments X-act).

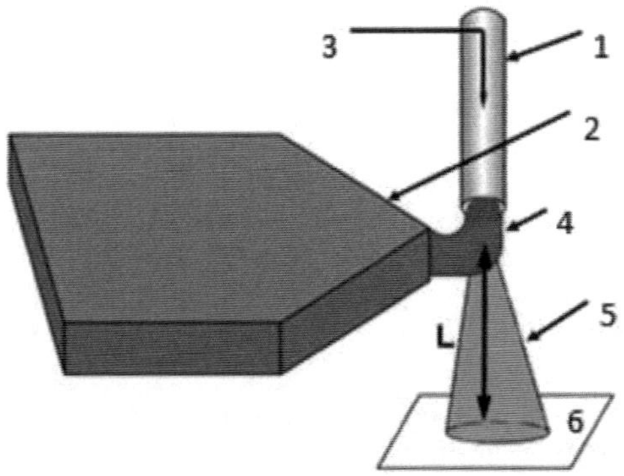

Figure 1. Simplified scheme of the experimental setup: anode (1), cathode (2), gas flow (3), discharge (4), afterglow (5), polymer (6).

3. Results and discussion

Dependences of the PP film contact angle on the treatment time are shown in figure 2. Treatment in the flowing afterglow of the discharges in argon and in air leads to an improvement in the surface wetting: the water contact angles decrease from 89° for the original sample to 58° and 44° for the samples treated with the discharges in air and argon, respectively. The most pronounced changes occur in the first 10-20 seconds of treatment, a further increase in treatment time does not lead to a noticeable change in the contact angle.

Water contact angles obtained at the treatment of PP films with the discharges in air and argon are listed in table 1. These data show that for a fixed exposure time the surface wetting increases with increasing gas flow rate and with decreasing distance between sample and plasma where the active species are generated. The results obtained at the treatment in both gases are similar at the flow rate of V=24 m/s but are different at V=105 m/s: treatment in argon plasma leads to lower contact angles.

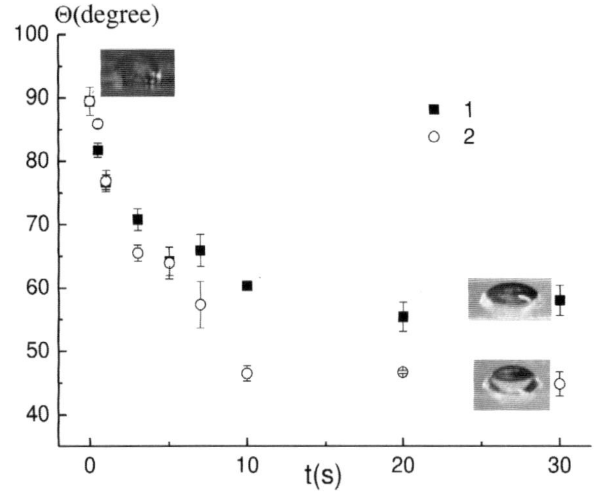

Figure 2. Dependences of the water contact angle on the time of PP film treatment with the discharges in air (1) and in argon (2) at the gas flow rate of 105 m/s and the distance from plasma of 5 mm.

Table 1. Water contact angles for the PP films treated in the flowing afterglow of the air and argon discharges at the treatment time of 30 s.

Gas flow rate (m/s)	Plasma - polymer distance (mm)	Water contact angle (degree)	
		Air	Argon
24	5	51	57
	7	78	78
	10	76	83
	12	85	84
	15	85	86
105	5	58	44
	7	64	49
	10	79	58
	12	75	57
	15	84	66

ATR-FTIR spectra showed the formation of oxygen containing groups in the polymer surface layer (figure 3). Treatment in the both gases results in the increase of absorption in the region of stretching vibrations of the C=O bond (1590-1820 cm^{-1}). When argon plasma is used, there is also an increase in the absorption of stretching O-H bonds (3000-3600 cm^{-1}). Increasing the treatment time and the gas flow rate results in a higher oxidation degree of the PP. It is more pronounced for a discharge in argon (table 2). The relative concentration of oxygen containing groups decrease with the increasing the plasma - polymer distance.

Concentrations of carbonyl and hydroxyl groups in modified layer of PP can be estimated as $c = \Delta A/(\varepsilon_\lambda/l)$, where ΔA is the difference between the absorbance in the treated and initial samples, ε_λ - the extinction coefficient and l is the penetration depth of the IR radiation into the sample. The penetration depth was estimated using the formula: $l = \lambda_1/[2\pi(\sin^2\Theta - n^2_{21})^{1/2}]$, where λ_1 is the radiation wavelength in an optically denser medium, Θ is the angle of incidence, n_{21} is the relative refractive index. The extinction coefficients from references [11-13] were used for the calculation.

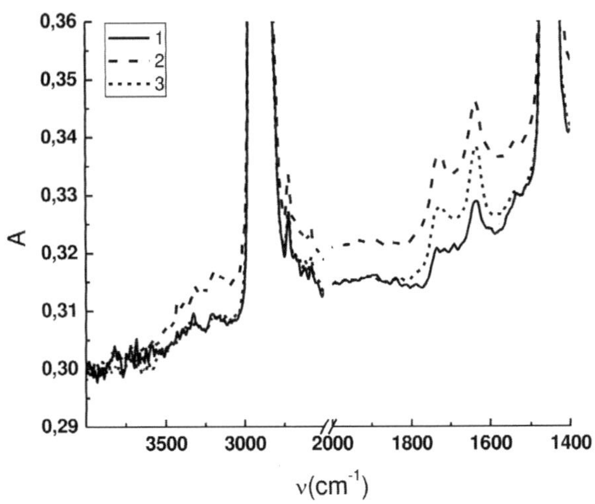

Figure 3. ATR-FTIR spectra of PP films: 1 - initial untreated sample, 2 - sample after the treatment with the argon plasma, 3 - sample after the treatment with the air plasma. Treatment conditions: gas flow rate 105 m/s, treatment time 30 s, plasma - polymer distance 7 mm.

Table 2. Relative absorbance for CO and OH groups normalized to the values for untreated sample.

Treatment conditions		A_{3200}/A_{2838}	A_{1722}/A_{2838}	A_{1633}/A_{2838}
Gas and flow rate	Treatment time (s)			
Untreated sample		1,00	1,00	1,00
Argon, 24 m/s	3	1,03	1,20	0,99
	20	1,21	1,41	1,24
	30	1,59	1,77	1,43
Argon, 105 m/s	3	0,90	1,81	1,39
	20	1,25	2,28	1,44
	30	1,71	2,66	1,58
Air, 24 m/s	3	1,04	2,09	1,50
	20	1,06	2,48	2,04
	30	0,89	2,32	1,85
Air, 105 m/s	3	0,95	1,92	1,59
	20	1,07	2,76	2,00
	30	0,97	2,44	1,72

Note: A_{2838} - the absorbance at the maximum of the band for the C-H in CH_2; A_{1722} - the absorbance at the maximum of the band for the C=O in the carboxyl group, ketones and aldehydes; A_{1633} - the absorbance at the maximum of the band for the C=O in the in the enolic diketones and unsaturated oxyketones.

The results of estimating the concentrations of oxygen containing groups are listed in table 3. These data agree with the results of the X-ray energy-dispersion analysis, which also showed a higher oxygen content in the PP treated with argon discharge at a gas flow rate of 105 m/s (table 4). Data of table 3 shows the values of concentrations for oxygen containing groups to be higher than those obtained at the thermo- and photooxidation of polypropylene [13, 14]. The total concentration of carbonyl groups after thermal oxidation of PP at a temperature of 138 ° C for two hours was 0.128 mole/l [13], and for photo-oxidation for 335 hours - 84 mole/l [14].

Table 3. Concentration of oxygen containing groups after the treatment of PP films in the flowing afterglow of the DC discharges in air and argon.

Gas and flow rate	Treatment time (s)	Concentration of groups (mole/L)			Total concentration (mole/L)
		OH	C=O (1722 cm^{-1})	C=O (1633 cm^{-1})	
Argon, 24 m/s	20	0,165	0,031	0,044	0,240
	30	0,495	0,062	0,088	0,645
Air, 24 m/s	20	-	0,109	0,206	0,315
	30	-	0,109	0,162	0,271
Argon,105 m/s	20	0,165	0,093	0,088	0,346
	30	0,660	0,124	0,118	0,902
Air, 105 m/s	20	-	0,140	0,191	0,331
	30	-	0,109	0,132	0,241

Table 4. Concentration of oxygen atoms in the modified PP films.

Plasma treatment	[C] (mass %)	[O] (mass %)
Untreated sample	97,20	2,80
Air plasma	96,76	3,24
Argon plasma	96,37	3,63

4. Conclusions

The comparative study of the polypropylene film surface modification in the flowing afterglow of the atmospheric pressure DC discharge in argon and air has shown the following. Modification in the both plasma forming gases results in the decrease of the contact angles and in the formation of oxygen containing groups in the polymer surface layer. The contact angles depends on the gas flow rate and on a plasma - polymer distance at the fixed treatment time. Increasing the treatment time and the gas flow rate results in a better wettability and a higher oxidation degree of the PP surface. Treatment in the afterglow of the argon plasma has been shown to give the less water contact angles and more densities of oxygen containing groups in polypropilene at the gas flow rate of 105 m/s and the treatment time of 30 s.

Acknowledgments

The study was supported by the Russian Foundation for Basic Research and the Government of Ivanovo Region through the grant No. 15-42-03124.

References

[1] Hajkova P, Tichý D, Čmelík J and Antoš P 2015 *Acta Polytechnica* **55** 96
[2] Akishev Yu et al. 2008 *J Phys D: Appl Phys* **41** 235203
[3] Noeske M, Degenhardt J, Strudthoff S and Lommatzsch U 2004 *Int J Adhesion & Adhesives* **24** 171
[4] Lommatzsch U, Pasedag D, Baalmann A, Elinghorst G and Wagner H E 2007 *Plasma Process. Polym.* **4** 1041–45.
[5] Cho S-J et al. *2011 Jap. J. Appl. Phys.* **50** 1S1 01AK02
[6] Cheng C, Liye Z and Zhan R-J 2006 *Surf. Coat. Technol.* **200** 6659-65
[7] Fang Z, Yang J, Liu Y, Shao T and Zhang C 2013 *IEEE Trans. Plasma Sci.* **41** 1627–34
[8] Dowling D P, Tynan J, Ward P, Hynes A M, Cullen J and Byrne G 2012 *Int. J. Adhesion & Adhesives* **35** 1–8
[9] Shrestha R, Subedi D P, Shrestha B K and Shrestha A 2016 *Int. J. of Rec. Res. Rev.* **9** 15–20
[10] Knoll A J, Luan R, Bartis E A J, Hart C, Raitses Y and Oehrlein G S 2014 *Appl. Phys. Lett.* **105** 171601
[11] Brand J C D and Eglinton G *Application of Spectroscopy to Organic Chemistry* 1965 (London, Oldbourne Press)
[12] Bellamy L J 1954 *The Infra-red Spectra of Complex Molecules* (Michigan: Springer)
[13] Adams J H 1970 *J. Polym. Sci. Pol. Chem.* **8** 1077
[14] Rabek JF, Ranby B: *Photodegradation, photooxidation and Photostabilization of Polymers.* New York: John Wiley; 1975

On efficiency during the generation of runaway electrons beams in stationary open discharge

V S Voiteshonok[1], A I Golovin[1], A V Turkin[1], A I Shloydo[1]

[1]SSC FSUE Keldysh Research Centre, Russia, 125438, Moscow, Onezhskaya 8

E-mail: kerc@elnet.msk.ru

Abstract. Measurements of energetic efficiency were accomplished in generation of runaway electrons beams for helium, air and water vapor with various cathode materials. The efficiency near 80% was demonstrated for runaway electrons beam generated in helium and water vapour.

Generators of electron beams (GEB), working directly in gas environment, are the advanced means for solving many technological problems [1]. Several variants of the design of such GEB were proposed in [2], also in work [3] and several others, studies of the operation of the GEB were carried out.

Evaluation of efficiency of the formation of an electron beam is proposed in [4]. Energetic efficiency or efficiency is defined as the ratio of the power of the generated electron beam W_n to the total input power W:

$$\eta = \frac{W_n}{W} \tag{1}$$

As a rule, the calorimetric method, based on heating of specially installed target by a beam, is used to measure the efficiency of electron-beam guns [5]. As applied to the open discharge, this method can give significant errors since the target heated by the beam must be installed at a distance much smaller than the beam residual range – otherwise there will be significant energy losses in the gas. However, such installation of the target can affect the operation of the generator, what was experimentally confirmed in [6] (the beam residual range was estimated by the method of [7]).

It should be noted that measurements of the efficiency of electron beams generation in an open discharge in works [8] and [9] by measuring of currents flowing to the anode and to the special collector of electrons are based on an incorrect determination of the coefficient of efficiency, not taking into account the contribution of the secondary electrons. These measurements were criticized in many works, in particular, in [10-13].

In this connection, it was decided to estimate the energetic efficiency by measuring the heating of the electron beam generator itself, assuming that all supplied electric energy, with the exception of the energy carried away by the electron beam, goes to the heating of the generator and the vacuum chamber, due to the heat rejection. In this case, the heat capacity of the generator C_{gen} can be easily determined from the book reference [14] on the heat capacity of constructional material and from results of weighing the details of the generator.

The temperature dependence of the generator body on time can be represented as,

$$\frac{dT_{body}}{dt} = \frac{W - W_n - W_O}{C_{gen}}, \tag{2}$$

where W_o is the rate of heat rejection from the generator.

The power W delivered to the discharge during the operation can be measured by measuring the voltage and current of the discharge supply taking into account the losses on the ballast resistance R. Assuming the uniform heating body, the dependence of the heat rejection rate on temperature can be determined from (2) according to the derivative of temperature of the body during the cooling at once after the power-off.

Thus, the measurement of efficiency includes measuring the power supply of the discharge and the temperature of the generator body for some time during the heating process and measuring the temperature when the generator is cooled after the discharge is turned off. In the second measurement, the heat rejection power W_o is determined as a function of temperature, which makes it possible to determine the beam power W_n from the results of the first measurement. Further, the value of efficiency can be found by the formula (2) for any moment of time of the heating stage.

In the case of the generator transfer to the stationary thermal mode, the power for its heating is equal to the heat rejection power and the efficiency of the generator can be determined by measuring the supplied electric power and the rate of change of the body temperature immediately at once after the power supply is turned off –и T'_{off}. Then from (1) and (2)

$$\eta = 1 - \frac{C_{gen}}{W}\left|T'_{off}\right|.$$

This method of determining of efficiency is not always possible, since the output to the stationary mode can take a long time (hours), and the temperature of the generator body in this case can reach hundreds of degrees. In addition, this method has a greater error than the processing of data collection on the rates of heating and cooling.

The method of processing of experimental data, based on the assumption that the amount of heat given off by the unit of the generator surface per unit of time, according to the Newton-Rikhmann law, is proportional to the difference between the generator and vacuum chamber temperatures. This method, which also takes into account the inconsistency of the input power during the heating stage, is described in [15]. All values of efficiency given below are obtained by using this method of processing.

In the first series of experiments, the efficiency measurements were carried out in helium for various hole diameters in the discharge-limiting insulator, power-supply voltages and cathode materials. In the second series - with a design without an insulator in helium, air and water vapor. In this series, a steel cathode and an anode with a 1 mm diameter hole were used. The gap between the cathode and the anode was 0.23 mm.

All measurements are performed on installation, the description of which is given in [3]. The diagram of the electron-beam plasma generator during the efficiency measurement using an insulator design is shown in Figure 1. The generator consists of: cathode 1, insulator 2, union nut 3, body 4, anode 5.

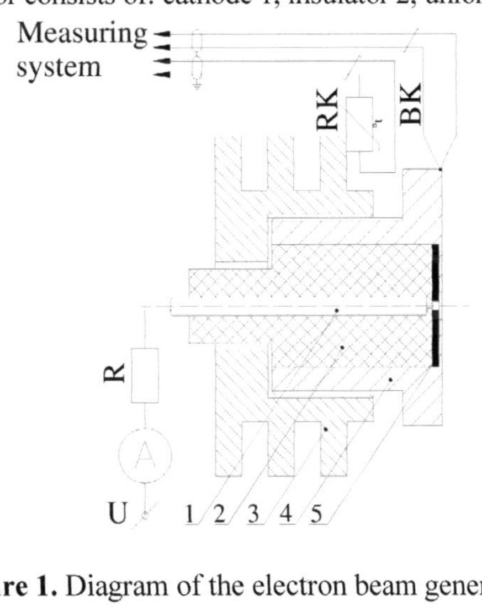

Figure 1. Diagram of the electron beam generator.

The main measurements of the generator temperature were carried out with a thermocouple of chromel-copel type located on the body 4 at the point indicated in Fig. 1. Shielded twisted pair of F/FTP type was used as signal wires. The error in measuring the temperature did not exceed ± 0.5 K.

Additional control measurements by KTY-200 thermistor were made to estimate the non-uniformity of the surface temperature. The thermistor was installed on the union nut 3, separated from the zone of maximum heat-removal by threaded joint. The error in measuring the temperature is not more than ± 0.1 K.

The temperature and pressure of helium during the tests were constant. The content of the admixture of air in helium was in the range of 0.1-3%.

The conditions for the first series of experiments and the obtained results are given in Table 1. The following notations are used: d – the diameter of hole in the insulator, R – ballast resistance, U – power supply voltage, P_{He} – helium pressure in the vacuum chamber. The first column of the table indicates the number of experiment; the same numbering is used later in the text.

Table 1 – Conditions for conducting the first series of measurements and the obtained results

no	*d*, mm	Cathode material	*U*, kV	P_{He}, Pa	Efficiency, %
1	0,5	LaB_6	6,5	1504	51
2	2,3	steel	4,4	494	54
3	1,0	LaB_6	4,4	1513	71
4	1,0	Mo	4,4	1507	48
5	0,3	Cu	8,0	2060	83
6	1,0	Cu	4,4	2025	54
7	1,0	Cu	4,4	1501	53
8	0,5	Cu	4,4	1507	63
9	0,5	Cu	7,0	1504	78

In experiments 5-9, an aluminum disk with a diameter of 200 mm and a thickness of 1.5 mm was placed coaxially on the front end of the generator body (not shown in Fig. 1), which led to an increase in both the heat capacity and the heat dissipation.

As an example, Figure 2 shows the temperature variation of the generator body and the discharge power during the full measurement cycle, including heating, reaching the steady state of thermal conditions, and cooling down after the power-off.

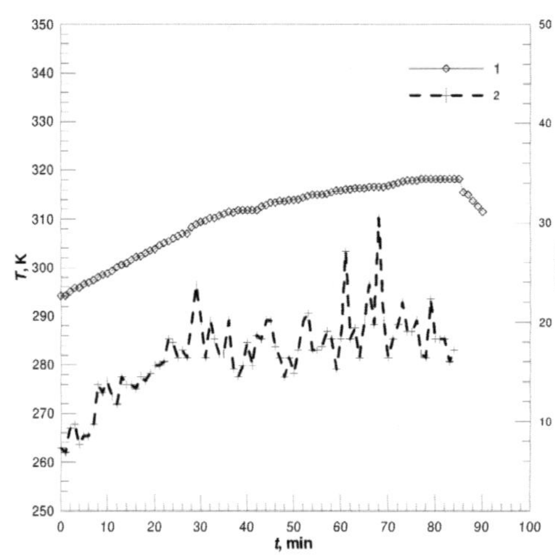

Figure 2. Dependence of temperature (curve 1) and electric power of the generator (curve 2) on time in the measurement 8.

The fragment of the heating process in experiments 1-7 is shown in Fig. 3. It can be seen that in the experiment 5 the operation of the generator was unstable and the power consumption varied from 50 to 150 W. Probably such work of the generator is explained by the small hole diameter in the insulator - only 0.3 mm. In other experiments, the power may spontaneously changed during the operation, probably due to a change in the state of the cathode due to erosion.

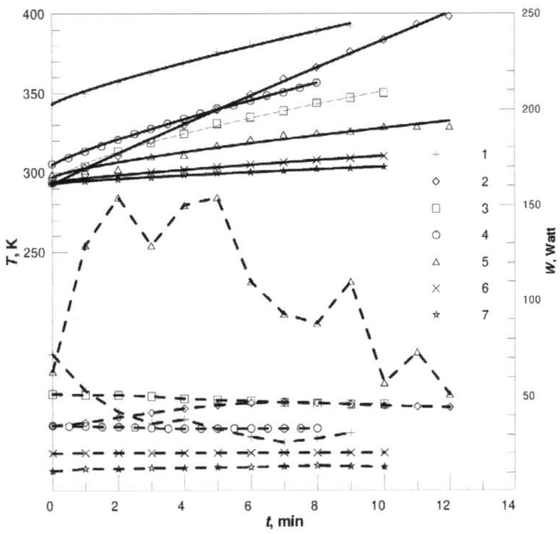

Figure 3. Dependence of temperature (solid lines) and electric power of the generator (dashed lines) on time in the process of heating in measurements 1-7.

Experiments 8 and 9 achieved a steady-state thermal regime; temperature and power in the steady state regime are shown in Fig. 4.

The temperature difference between the thermocouple (BK) and the thermistor (RK) did not exceed 3-5% and decreases as the generator cools down. The relative error in determining of efficiency is estimated at 10%.

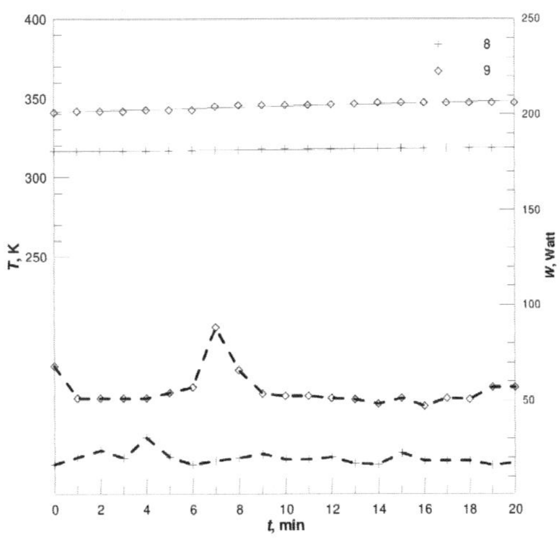

Figure 4. Dependence of temperature (solid lines) and electric power of the generator (dashed lines) on time in steady state regime in measurements 8 and 9.

Based on the results of both the first and second series of experiments, it is seen that with increasing voltage, the coefficient of efficiency also increases. The only exception is measurement 1. Probably, due to some assembly defect in the generator, the parasitic discharge was ignited in this experiment, which led to the decrease in the coefficient of efficiency.

The results of second series of experiments are given in Table 2.

Table 2. Conditions for the second series of measurements and the obtained results.

no	Working gas	*P*, Pa	*U*, kV	Efficiency, %
10	helium	1545	4,4	66
11	helium	951	4,4	59
12	helium	964	6,5	76
13	air	322	4	57
14	air	307	4	59
15	water	300	3	53
16	water	310	6	79
17	water	508	3	46

Values of the efficiency measured at different pressures, but at the same voltage, differ somewhat from each other. However, no dependence is observed, and the difference in the value of efficiency is slightly higher than the error in the evaluation of this parameter.

Unfortunately, measurements in air and water vapor are performed at several other voltage values than measurements in helium. According to Table 2, it can be assumed that the efficiency in experiment in these gases is somewhat higher, especially in water vapor. However, although differences are observed in all measurements, they are close to the error of the method.

References

[1] Golovin A I and Shloydo A I 2016 Modern electron beam generators for technological applications (a review) *Advances in Applied Physics* **4(5)** 439-48

[2] Bobrov V A, Voiteshonok V S, Golovin A I, Golubev M M, Turkin A V and Shloido A I 2014 Electron beam generation RF Patent 2535622

[3] Bobrov V A, Voiteshonok V S, Golovin A I, Golubev M M, Lomakin B N, Turkin A V and Shloydo A I 2013 Analysis of continuous generation of electron beams in medium-pressure gases *Technical Physics* **58** 1205-10

[4] Golovin A 2016 Efficiency of electron beam generation by a stationary open discharge *Prikl. Fiz.* **1** 77-81.

[5] Zav'yalov M A, Kreindel Y E, Novikov A A and Shanturin L P 1989 *Plasma processes in technological electron guns* (Moscow: Energoatomizdat)

[6] Sorokin A 2003 Evaluation of the photoelectron contribution to the efficiency of open discharge *Technical Physics Letters* **29** 836-7

[7] Baranov V F 1974 *Dozimetriya elektronnogo izlucheniya (Dosimetry of Electron Radiation)* (Moscow: Atomizdat)

[8] Bokhan P and Zakrevsky D 2002 High-efficiency electron beam generation in an open discharge without anode grid *Technical Physics Letters* **28** 73-5

[9] Bokhan P and Zakrevsky D 2002 On the mechanism of electron beam generation in an open discharge with anomalously high efficiency *Technical Physics Letters* **28** 454-6

[10] Sorokin A R 2003 Evaluation of the photoelectron contribution to the efficiency of open discharge *Technical Physics Letters* **29** 836-7

[11] Sorokin A R 2006 Electron beam formation in an anomalous glow discharge *Technical Physics* **51** 580-8

[12] Sorokin A R 2003 Atom-bombardment-induced electron emission from a cathode in a glow discharge: Example of an open discharge *Technical Physics Letters* **29** 701-4

[13] Sorokin A R 2003 A comment on the measurement of efficiency of the electron beam formation in an open discharge *Technical Physics Letters* **29** 171-4

[14] Grigoriev I S and Meilikhov Y Z 1991 *Physical values* (Moscow: Energoatomizdat)

[15] Voiteshonok V S, Golovin A I, Egorova E K, Lomakin B N, Turkin A V and Shloydo A I 2017 Experimental Investigation of High-Voltage Glow Discharge Efficiency as a Source of Beams of Run-Away Electrons *High Temp.* **55** 1–8

Features of triple recombination in ultracold plasma

V S Vorob'ev [1,2]
[1] Joint Institute for High Temperatures of Russian Academy of Sciences, Izhorskaya Street 13 Bldg. 2, Moscow 125412, Russia,
[2] National University of Science and Technology "MISIS", Leninskiy prospekt 4, Moscow 119049, Russia

Abstract. We show that in the ultracold plasma, when the coupling parameter characterizes the interaction of electrons and ions exceeds unity, the collisional three-body (e – e – ion) recombination coefficient (TBR) becomes inversely proportional to the plasma frequency, whereas for the small values of this parameter it ~ $T^{-9/2}$ (T is the temperature). We compare the results obtained with different theoretical models and numerical simulation data of the recombination process.

Recently a new unique object has been introduced into the study of TBR in the strong coupling regime, including electron-electron interaction, namely ultracold plasma (UCP) experimentally investigated, for example, in Refs. 1 - 3. These and other experimental data stimulated the appearance of studies in which the TBP process in the strongly coupled plasma was modeled by numerical methods[5 - 9]. These studies showed that with the growth of the coupling parameter, a significant decrease in the recombination coefficient (α_T) occurs in comparison with the well-known expression obtained by Tomson[10] from simple scaling arguments.

Let us to consider the triple recombination process in the spirit of the Thomson theory[10]. Following this theory, the electron gives its average kinetic energy K in time τ to another electron, and if it is in the neighborhood of the ion, it forms a bound state with the latter. Therefore, the rate of change in the electron density can be written as

$$\frac{dn_e}{dt} = -\frac{n_e}{\tau} w \quad ; \quad w = 1 - \exp\left[-\frac{4\pi n_i}{3}\left(\frac{Ze^2}{K}\right)^3 \right] \tag{1}$$

Here w is the probability that the nearest electron is at a distance Ze^2/K from the selected ion with a charge Z participating in the recombination process. $1/\tau = n_e\langle \sigma v_e \rangle \sim n_e e^4 K^{-2}\sqrt{K/m}$, where $\sigma \sim e^4/K^2$ is the cross section of elastic collision of an electron on an electron, $v \sim \sqrt{K/m}$ is the average electron velocity. As a result we find

$$\frac{dn_e}{dt} = -Cn_e^2 \frac{e^4}{\sqrt{mK^3}}\left(1 - \exp\left[-\frac{4\pi n_i}{3}\left(\frac{Ze^2}{K}\right)^3 \right]\right) \tag{2}$$

where C is a numerical constant. For an ideal plasma, the exponent in Eq. (2) is small and this formula will be transformed into the well-known Gurevich-Pitaevskii expression[11, 12], if we let the constant C be equal to $C = \sqrt{2\pi}\Lambda/3$. Here Λ is the Coulomb logarithm for bound states (see Ref.[13] for details). With this constant, it is convenient to rewrite Eq. (3) by introducing a dimensionless recombination coefficient

$$\alpha^* = \frac{dn_e}{dt}\frac{Z^{3/2}}{n_e\omega_p} = -C_1\left(\frac{U}{K}\right)^{3/2}\left(1-\exp\left[-\left(\frac{U}{K}\right)^3\right]\right) \quad (3)$$

where $\omega_p = \sqrt{4\pi e^2 n_e/m}$ is the plasma frequency.

To study TBR process, we consider a pair of ions with a charge Z and an electron nearest to the ions, which has a positive energy and located at a distance *r* of this ion. The total energy of the electron *E*, its momentum *p*, it interacts with the nearest ion and all the other particles located outside the sphere of radius *r*. The distribution function for this case was found in Ref. 14. It has the form of a Gaussian distribution and represents the probability of finding an electron with energy *E* and momentum *p* at a distance *r* from the ion

$$f(E,p,r) = \frac{1}{\sqrt{2\pi D}}\exp\left(-\frac{\left(E - p^2/2m + Ze^2/r\right)^2}{2D}\right)exp\left(-\frac{4\pi n_i r^3}{3}\right)\frac{1}{A} \quad (4)$$

where *A* is the normalizing constant. The function $exp\left(-4\pi n_i r^3/3\right)$ coincides with the Poisson probability of the absence of particles in the cavity *r* (approximation of nearest neighbor-ANN). The dispersion *D* is equal to the mean square of the potential energy due to the interaction of the extracted electron with the rest of the particles located outside the sphere of radius r. According to Ref.14 in ring approximation it has the form $D = e^2T/(r_D + r)$, where $r_D = \sqrt{T/4\pi e^2 n_e}$ is the Debye radius.

The kinetic energies averaged over positive energies are found as

$$K = \int_0^\infty dE e^{-E/T}\int\frac{d\mathbf{p}d\mathbf{r}}{(2\pi\hbar)^3}\frac{p^2}{2m}f(E,p,r)/A \;\; ; A = \int_0^\infty dE e^{-E/T}\int\frac{d\mathbf{p}d\mathbf{r}}{(2\pi\hbar)^3}f(E,p,r) \quad (5)$$

Figure 1a shows the dependences K/T and U/T as a function of the coupling parameter γ. For small values of γ, the values K/T and U/T tend to 3/2 and to 0 respectively, which corresponds to free electrons. With increasing γ, the kinetic and potential energies increase too (in absolute value). The calculations with a Gaussian distribution and ANN give fairly close results. It is important to note that calculations according to Eqs. (4 , 5) that take the dispersion into account are approximated by simple expressions. Thus, curves 3 and 3' in Figure 1 and curve 3 in Figure 2 correspond to approximations

$$K/T \approx 3(1+\gamma)/2 : U/T \approx -3\gamma/2 ; \Gamma^* = |U|/K \approx \gamma/(1+\gamma) \quad (6)$$

The value $\Gamma^* = |U|/K \approx \gamma/(1+\gamma)$ characterizing the ratio of the potential energy to the real kinetic energy in this case tends to unity with increase of γ.

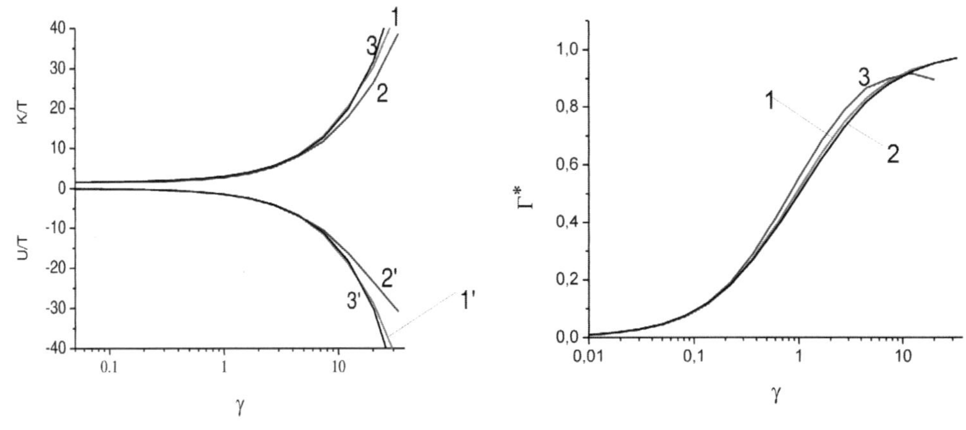

(a) (b)

Figure 1. Dependences of the kinetic, potential energies (a) and their ratio (b) on the coupling parameter γ. Curves 1 and 1' are calculations according to Eqs. (4 - 8) with the dispersion, 2 and 2' are approximation of nearest neighbor, 3 and 3'are approximations (6).

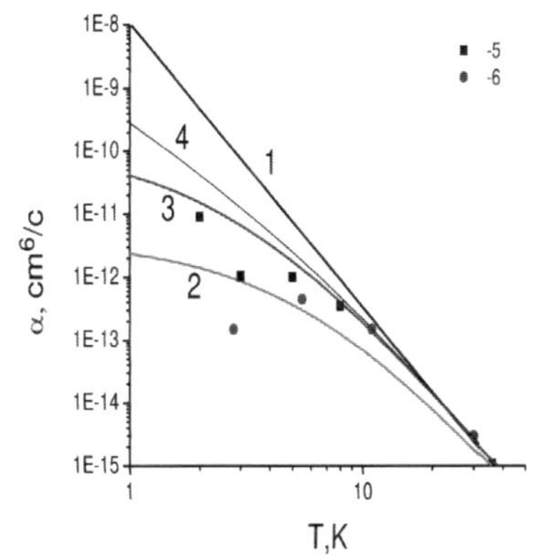

Figure 2. The dependence of the TBR coefficient on the temperature. Curves: 1 - according to diffusion theory [11, 12], 2, 3 are results of this work modelling: 5, 6, 7, 8– Refs. 16, 9, at $n_e = 10^{10}$ and 10^9 cm^{-3} respectively, 4 is result of Ref. (15) at $n_e = 10^{10}$ cm^{-3}. Results of the numerical modeling 5 and 6 are according to Ref. 9, 4.

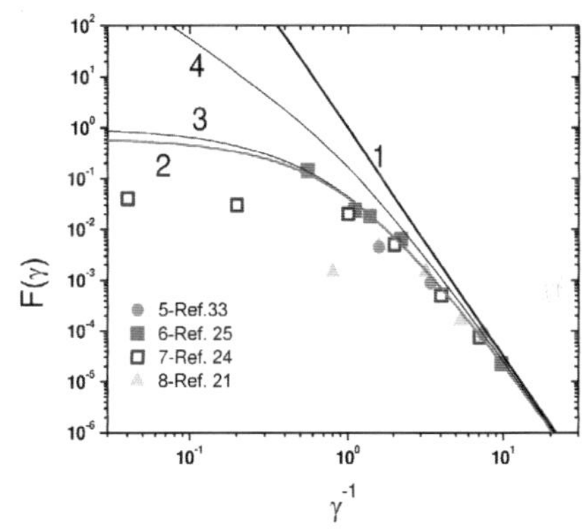

Figure 3. The dimensionless recombination rate as a function of the parameter γ. Curves: 1 – $\gamma^{9/2}$, 2 is Eq. (7), 3 is approximation $(\gamma/(1+\gamma))^{9/2}$ from Ref. 9, 4 – Ref. (15). Numerical modelling: 5, 6, 7, 8– Refs. 16, 9, at $n_e = 10^{10}$ and 10^9 cm^{-3} respectively, 4 is result of Ref. (15) at $n_e = 10^{10}$ cm^{-3}. Results of the numerical modeling 5 and 6 are according to Ref. 9, 4. of the parameter γ. Curves: 1 – $\gamma^{9/2}$, 2 is Eq. (7), 3 is approximation $(\gamma/(1+\gamma))^{9/2}$ from Ref. 9, 4 – Ref. (15). Numerical modelling: 5, 6, 7, 8– Refs. 16, 9, 7, 5 respectively.

Substituting the approximations (8) into Eq. (4), we obtain

$$\alpha * = \frac{dn_e}{dt}\frac{Z^{3/2}}{n_e\omega_p} = -\frac{\Lambda}{\sqrt{24\pi}}\left(\frac{\gamma}{1+\gamma}\right)^{3/2}\left(1-\exp\left[-\left(\frac{\gamma}{1+\gamma}\right)^3\right]\right) \quad (7)$$

The value of $\alpha *$ depends only on the coupling parameter γ. In the ideal case, when $\gamma << 1$, $F(\gamma) = \gamma^{9/2}$. For large values of $\gamma >> 1$, the dimensionless recombination coefficient tends to a constant value. It follows that in an ultracold plasma the recombination time is inversely proportional to the plasma frequency. Figure 2 shows the temperature dependence of TBR, obtained by analytical methods and by numerical simulations.

Four calculated curves and numerical simulation data[5, 8] are shown in Figure 2. The methods of modeling the recombination process developed in these works are described in detail in Ref. (8). The main array of calculated points is enclosed between our curves 2 and 3 with $n_e = 10^{10}$ and 10^9 cm^{-3}. The functions $F(\gamma) = \alpha^*/C$, corresponding to different approximations, and also data of numerical simulations, are plotted in Figure 3.

Acknowledgments
The work was supported by the grant of Russian Science Foundation No. 14-50-00124.

References

[1] Killian T C, Kulin S, Bergeson S D, Killian T C, Kulin S, Bergeson S D, Orozco L A, Orzel C and Rolston S L 1999 *Phys. Rev. Lett.* **83** 4776
[2] Kulin S, Killian T C, Bergeson S D and Rolston S L 2000 *Phys. Rev. Lett.* **85** 318
[3] Killian T C, Lim M J, Kulin S, Dumke R, Bergeson S D and Rolston S L 2001 *Phys. Rev. Lett.* **86** 3759
[4] Lankin A V and Norman G E 2009 *J. Phys. A: Math.Gen.* **42** 214032
[5] Lankin A V and Norman G E 2009 *Contrib. Plasma Phys.* **49** 723
[6] Bannasch G and Pohl T 2011 *arXiv:1109.6456v1* Cited September 29, 2011.
[7] Bannasch G and Pohl T 2011 *Phys. Rev. A* **84** 052710
[8] Bobrov A A, Bronin S Ya, Zelener B B, Zelener B V, Manykin E A and Khikhlukha D R 2011 *JETP* **112** 527
[9] Bobrov A A, Zelener B B, Zelener B V and Khikhlukha D R 2013 *High Temperature* **51** 615
[10] Thompson J J 1924 *Phil. Mag.* **47** 377
[11] Gurevich A V and Pitaevskii L P 1964 *Sov. Phys. JETP* **19** 870
[12] Gurevich A V 1964 *Geomagn. Aeron.* **4** 3
[13] Biberman L M, Vorob'ev V S and Yakubov I T 1987 *Kinetics of Nonequilibrium Low Temperature Plasmas* (Berlin: Springer, Verlag) p 498
[14] Vorob'ev V S and Khomkin A L 1976 *Theoretical and Mathematical Physics* **26** (3) 246–254
[15] Biberman L M, Vorob'ev V S and Yakubov I T 1987 *Dokl. Akad. Nauk SSSR* **296** 577
[16] Maiorov A, Tkachev A N and Yakovlenko S N 1991 *Russ. Phys. J.* **34** 951

Influence of plasma modification on hygienic properties of textile fabrics with nonporous membrane coating

E F Voznesensky[1], R G Ibragimov[1], O V Vishnevskaya[1], V A Sisoev[1], G G Lutfullina[1], N V Tihonova[2]

[1] Department of Plasma Technology and Nanotechnology of High Molecular Weight Materials, Kazan National Research Technological University, Kazan, 420015, Russia
[2] Department of Designing clothes and shoes, Kazan National Research Technological University, Kazan, 420015, Russia

olesya-zef@yandex.ru

Abstract. The work investigated the possibility of using plasma modification to improve the hygienic properties of textile materials with nonporous membrane coating to improve vapor-, air-permeability and water-resistant. Determined that, after plasma modification changes degree of supramolecular orderliness of the polymers nonporous membrane coating and the base fabric.

1. The Introduction

Over the past decades, textile materials with membrane coating have become very popular among consumers due to the expansion of the range of membrane coatings and the operational capabilities of fabrics. Textile materials with membrane coating are used for making workwear used in adverse weather conditions, such as strong wind, rain, snow, frost. Working clothes regulate the thermal state of the body, protecting it from overheating and overcooling, helps to prevent colds, so it must be air- and vapor-permeable, and also meet certain operational requirements: be comfortable, strong enough, waterproof and frost-resistant. Correctly selected for certain conditions, work clothes contribute to the prevention of injuries and occupational diseases [1]. The unique capabilities of products from these materials to protect from the atmospheric precipitation outside, and at the same time to transport the moisture emitted with a body to the external environment, essentially expand borders of comfort in comparison with the traditional waterproof clothing made of the ordinary fabrics with water repellent impregnation. Despite the modern variety of polymeric membranes, the option of combining membrane materials with fabrics and a variety of brands created both by producers of fabrics, and producers of finished products. The key indicators of quality of membrane material are high parameters of water-resistance, vapor-permeability and air permeability. In the domestic market, textile materials with nonporous membrane coating are used for sewing the top overalls. This is due to the low-cost production of such materials, their high strength and resistance to aggressive chemical compounds, compared to textile materials with nonporous membrane coatings, whose production is associated with costly multi-stage processes. However, textile materials with nonporous membrane coating have a rather serious drawback is a low vapor-permeability. The actual way to eliminate this disadvantage is the modification of membrane materials. Analysis of modern technologies used in the textile industry has shown that modification of tissues by nonequilibrium low-temperature plasma (NLTP) is widely used not only in scientific research, but also for solving specific production and technological problems [2-4]. Plasma modification is a universal way to regulate the properties of materials, while it does not degrade their protective characteristics. The purpose of this study was to modify textile materials with nonporous membrane coating in NLTP to improve their hygienic properties.

2. Materials, methods and equipment

The fabrics of the company «Tchaikovsky Textile»: a two-layer polyester fabric with a polyurethane (PU) nonporous membrane coating that is printed on the back side, Climate 3 and a three-layer polyester fabric with PU coating Climate 3+, which additionally has a lining fabric from the membrane side, were chosen as the objects of research [5]. On the fabric side of the fibers, the impregnating composition of the organofluorine preparation of water- and oil-water repellent finish is applied to the tissue.

The experiments were performed on an industrial radio-frequency capacitive plasma (RFC) plant [6]. The industrial plant consists of the following main parts: a vacuum chamber with internal equipment, placed on a single frame base, a sliding door with a cart on which the machine for rewinding fabrics is based, a vacuum evacuation system, a cooling system, a radio-frequency generator (RF generator), a control panel. The rectangular chamber is made of carbon steel and is welded. The walls of the chamber have stiffeners. The camera body is hidden by decorative panels. A more detailed description of this setup is presented in the article [6].

Since we are dealing with materials with diametrically opposite surface properties on both sides, the fabrics were modified in two stages: from the membrane coating in the air medium, and on the side of the base fabric, the material was processed in argon-acetylene medium. Based on previous studies [4-6], it is known that the modification of RF plasma in a nitrogen or air medium increases the hydrophilic properties of the surface and in the hydrocarbon gas medium, hydrophobic properties. Parameters of plasma treatment discharge: power (W_p) 1000-1500 W, working gas pressure (P) 21,6-20,5 Pa, processing time (t) 10-40 min, gas flow rate (G) 0,02-0,1 G/s, air and a mixture of argon-acetylene were used as the plasma-forming gas.

To assess of the changes in the hygienic characteristics of textile materials with nonporous membrane coating after modification in NLTP, the following studies have been carried out: determination of vapor-permeability, air-permeability, water-resistance [7-9]. To study the changes in the supramolecular structure of the membrane layer and the base fabric after the NLTP treatment was used X-ray diffraction analysis (XRD) [10].

3. Results

At the first stage of work two-factor optimization of parameters of two-stage plasma processing of textile materials with nonporous membrane coating has been carried out [11].

The optimum mode of plasma treatment on the side of the membrane coating for the selected materials is W_p = 1200-1500 W, P = 22 Pa, t = 15 min, G_{air} = 34.5 g/s, and on the base fabric side is W_p = 1300-1500 W, P = 22 Pa, t = 7 min, $G_{Ar/C2H2}$ = 29.6 g/s.

The change in values of indicators of hygienic properties of materials after two-stage plasma treatment in optimal modes is presented in table 1.

Table 1. Values of hygienic properties of textile materials with nonporous membrane coating before and after two-stage treatment in NLTP in optimal conditions

Material	Sample	Vapor-permeability, g/m^2 for 24 hours	Air-permeability, mm/s	Water-resistance, Pa
Climate 3	Before modification	1740	2,6	98100
	After modification	1955	3,4	more 100000
Climate 3+	Before modification	2500	5,2	85300
	After modification	2720	8,0	more 100000

The obtained data show that after two-step plasma modification in optimal modes there is an increase in hygienic properties, namely water vapor-permeability material Climate 3 - 12,3% and material Climate 3+ by 8,8%; air-permeability of the material Climate 3 by 31% and material Climate 3+ by 54%; the water-resistance of the material Climate 3 by 1,9% and the material Climate 3+ by 17,2%. Consequently, the plasma treatment of textile materials with nonporous membrane coating for the production of special clothing improves the transport of moisture vapor from the susceptible space to the outer surface of the materials, which ensures the maintenance of normal functions of the human body thermoregulation and the waterproof properties of materials due to ion bombardment of the surface, which presumably leads to the rearrangement of intermolecular van der waals and hydrogen bonds.

Substantial changes in the microstructure of the materials after treatment with NLTP according to the confocal microscopy (CM) data are not observed (Fig.1), the changes appear in some smoothing of the microrelief of the membrane coating.

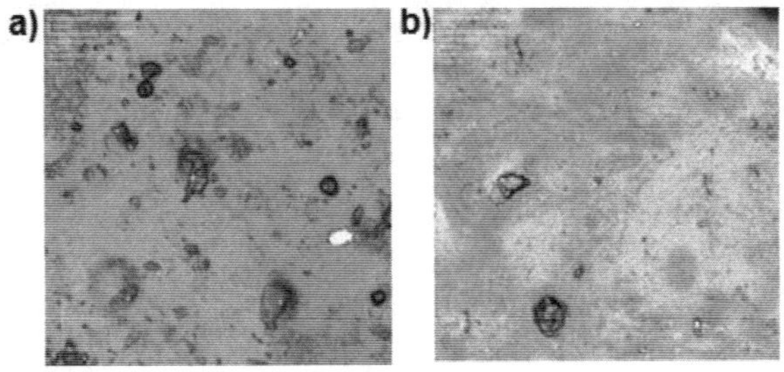

Figure 1. CM images (×200) of surface morphology: PU membrane coating of the material Climate 3 before (a) and after (b) NLTP treatment

The results of the X-ray diffraction analysis (Fig. 2) demonstrate decrease in the intermolecular ordering of the PU coating in the air environment, due to the breakdown of intermolecular hydrogen and van der waals bonds, which is confirmed by an increase in the vapor- and air-permeability of the materials.

The modification of fabric-based in argon-acetylene medium results in degassing the impregnating composition of the organofluorine preparation with water and oil-water repellent finish, fixing the functional groups of the hydrophobizer on the surface of the fabric fibers, ordering the amorphous phase, and increasing the crystalline phase, which explains the increase in the water resistance of the materials.

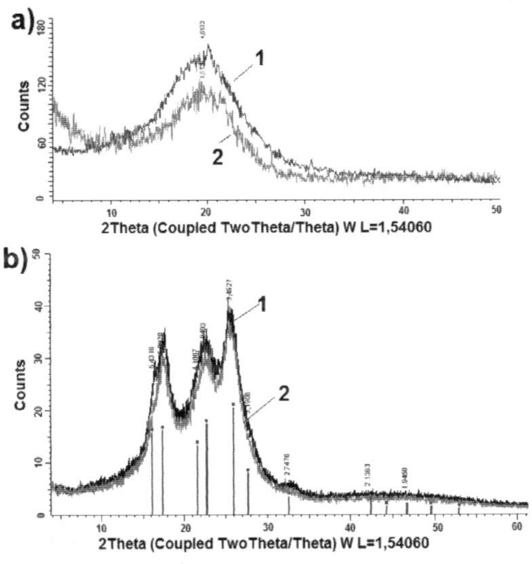

Figure 2. Diffractograms of PU membrane coating (a) and polyester fabric-based material (b) Climate 3 before (1) and after (2) NLTP treatment

4. Conclusions

Textile membrane materials with an increased complex of hygienic properties have been experimentally obtained. As a result of the studies, the following conclusions can be drawn:

1. Optimum regimes of two-stage plasma modification are established: on the side of the membrane coating is W_p = 1200-1500 W, P = 22 Pa, t = 15 min, G_{air} = 34.5 g/s, and on the fabric-based side is W_p = 1300-1500 W, P = 22 Pa, t = 7 min, $G_{Ar/C2H2}$ = 29.6 g/s.

2. After modification in NLTP, the hygienic properties increase, namely: the vapor-permeability of the material Climate 3 by 12,3% and the material Climate 3+ by 8,8%; air-permeability of the material Climate 3 by 31% and material Climate 3+ by 54%; water-resistance of the material Climate 3 by 1,9% and the material Climate 3+ by 17,2%.

3. The effect of NLTP on the membrane coating reduces the degree of ordering of the supramolecular structure of the PU membranes due to conformational transformations and the separation of the supramolecular structure, which is consistent with the results of the permeability of materials due to

intermolecular diffusion, and on the side of the fabric-based leads to the ordering of the amorphous phase and an increase in the crystalline phase, consequence of this is an increase of water resistance.

Through this study, it can be concluded that the plasma treatment is quite gentle modification method for textile fabrics coated nonporous membrane obtained multifunctional materials may be recommended for sewing comfortable waterproof clothing for human.

Acknowledgments
Authors thank professor I. Sh. Abdullin for help in discussing the results of work.

References

[1] Romanov V E 1981 *Sistemnyy podkhod k proektirovaniyu specialnoy odezhdy* [System approach to the design of special clothing]. Moscow (Izdatel'stvo Legkaya i pischevaya promyshlennost) 247 p (In Russian)

[2] Abdullin I Sh, Ibragimov R G, Zaitseva O V, Vishnevsky V V, Osipov N V 2014 *Sovremennye tkani s membrannym pokrytiem* [Modern fabrics with membrane coating]. Kazan (*Vestnik tekhnologicheskogo universiteta* vol 12) pp. 37-41 (In Russian).

[3] Abdullin I Sh, Ibragimov R G, Zaitseva O V, Vishnevsky V V, Osipov N V 2014 *Gidrofobizaciya poverkhnosti materialov s membrannym pokrytiem s pomoschyu plazmennoy obrabotki* [Hydrophobization of the surface of materials with membrane coating by plasma treatment]. Kazan (*Vestnik tekhnologicheskogo universiteta* vol 19) pp. 47-49 (In Russian).

[4] Abdullin I Sh, Ibragimov R G, Zaitseva O V, Vishnevsky V V, Osipov N V 2014 *Tkani s membrannym pokrytiem* [Fabrics with membrane coating]. Saint-Petersburg (*Dizayn. Materialy. Tekhnologiya.* vol 5) pp. 25-29 (In Russian).

[5] Assortment of fabrics with coatings of the company "Tchaikovsky textiles" of fabrics [Electronic resource]. - Access mode: http://www.textile.ru/production/cover.

[6] Ibragimov R G, Voznesensky E F, Grevtsev V A, Vishnevskaya O V, Vishnevsky V V, Osipov N V 2017 *Vliyanie plazmennoy obrabotki na morozostoykost membrannykh materialov* [Effect of plasma treatment on frost resistance of membrane materials]. Kazan (*Vestnik tekhnologicheskogo universiteta* vol 3) pp. 88-90 (In Russian).

[7] Abdullin I Sh, Ibragimov R G, Vishnevskaya O V, Osipov N V 2014 *Modifikaciya poverkhnosti tkani s membrannym pokrytiem neravnovesnoy nizkotemperaturnoy plazmoy* [Modification of the surface of a tissue with a membrane coating by a nonequilibrium low-temperature plasma]. Kazan (Izdatel'stvo KNITU) pp. 54-58 (In Russian).

[8] Abdullin I Sh, Ibragimov R G, Vishnevskaya O V, Vishnevsky V V, Osipov N V 2016 *Regulirovanie vozdukhopronicz aemosti dvukhsloynykh membrannykh materialov s pomoschyu plazmy* [Regulation of air permeability of two-layer membrane materials by means of plasma]. Kazan (Izdatel'stvo KNITU) pp. 123-126 (In Russian).

[9] Abdullin I Sh, Ibragimov R G, Vishnevskaya O V, Vishnevsky V V, Osipov N V 2016 *Povyshenie vodoupornosti materialov s membrannym pokrytiem* [Increase in water resistance of materials with membrane coating]. Ekaterinburg (Izdatel'stvo UrFU) pp. 326-327 (In Russian).

[10] Abdullin I Sh, Ibragimov R G, Vishnevskaya O V, Vishnevsky V V, Osipov N V, Gorelisheva V E 2016 *Issledovanie izmeneniya svoystv poverkhnosti dvukhsloynykh membrannykh materialov posle plazmennoy obrabotki* [Investigation of changes in surface properties of two-layer membrane materials after plasma treatment]. Kazan (*Vestnik tekhnologicheskogo universiteta* vol 8) pp. 75-78 (In Russian).

[11] Abdullin I Sh, Ibragimov R G, Vishnevskaya O V, Vishnevsky V V, Osipov N V 2016 *Optimizaciya i modelirovanie parametrov plazmennoy obrabotki membrannykh materialov* [Optimization and modeling of parameters of plasma treatment of membrane materials]. Kazan (*Vestnik tekhnologicheskogo universiteta* vol 17) pp. 82-84 (In Russian).

Experimental study of heat transfer in a glow discharge

R F Yunusov, M M Garipov

Kazan National Research Technical University named by A.N. Tupolev,420111, K. Marx str., 10, Kazan, Russian Federation

E-mail: optanir@mail.ru

Abstract. The article is devoted to the experimental study of heat transfer in a glow discharge. The longitudinal discharge was ignited in a stream of air moving in a cylindrical discharge chamber (DC). Copper electrodes were located at a distance of 9 cm from each other perpendicular to the axis of the DC. The gas pressure P = (2 - 9) kPa and its flow rate G = (0 - 0.035) g / s were monitored. The discharge current and voltage varied accordingly in the ranges: I = (30-80) mA, U = (1-2) kV. Falling current-voltage characteristics were observed. The temperature of the neutral particles was measured using chromel-alumel thermocouples located along the DC in three sections. The location of the thermocouples made it possible to measure the temperature distribution both along the DC and across it. The discharge shape was monitored photographically and visually. It was found that with an air flow rate G = 0.017 g / s, the discharge loses its axial symmetry and contracts to the wall of the DC. The temperature field of neutral particles also changes significantly.

1. Introduction

A plasma glow discharge is widely used in gas-discharge light sources, as the active medium of gas lasers [1-13], modern technologies for production of various coatings, nanotechnology methods for obtaining unconventional materials [14-19]. Therefore, the study and research of glow discharge plasma parameters is of great scientific and practical interest. An important feature of glow discharge plasma is its thermal non-equilibrium: the electron temperature can be several times higher than the temperature of neutral molecules. For many applications of such an object, select the operating mode, it is important to know the heat transfer processes occurring in the gas discharge. Heat transfer in the gas discharge can be regarded as heat in the gas stream with internal heat sources. For theoretical analysis of these tasks, you must know or obtain the spatial distribution of the power heat sources. In the above work theoretically and experimentally the potential distribution were obtained, the electron density and electric field strength in the case of longitudinal and transverse glow discharges. The purpose of this article is a further experimental study of heat transfer processes in a longitudinal glow discharge.

2. Experiment

The neutral gas temperature measurement in a glow discharge plasma has certain difficulties related to the lack of balance between the neutral and charged particles, the possibility of chemical processes, etc. Definition of methodological error of temperature measurement is part of the general problem of the heat exchange system of bodies of research that are in contact with the environment or bodies. When you enter the radial thermocouple, temperature measurement errors are caused by the heat exchange by radiation between the thermocouple and the wall of the DC, the heat transfer by conduction from the sensor through the wires and thermocouple protection tubes, it's heating due to heterogeneous processes. Taking into account all the factors of the neutral gas temperature measurement, error does not exceed 8%, which is consistent with the available data.

The circuit design and the DC for studying the processes of heat exchange is shown in Figure 1. Copper electrodes located at a distance of l = (0, 08 - 0, 1) m, located in the side branches of the glass tube diameter of 0,01m. The cathode is located on the left, the anode is to the right, and the gas velocity is directed from the cathode to the anode. On the outer surface of the DC, there are nine capillaries directing the thermocouple and pressure measure. The DC joins vacuum hoses by pumping system. Current and discharge voltage are controlled by switch devices, and for detecting the current-voltage characteristics plotter type H-306 was used. The temperature of the outer surface of the DC was measured at eight points chromel-alumel thermocouples of 0.2 mm diameter. Working ends of thermocouple pressed against the surface of the DC by means of four thin rings. According to the testimony of thermocouples determined the average temperature of the surface of the DC. The temperature of neutral particles was measured after the establishment of steady state, as determined by the stability of the input and output parameters. Each mode maintains a constant pressure in the DC, the gas flow rate, current and voltage discharge. Measurements were carried out in the same mode is 5 - 10 times. Thermo-emf thermocouple measured the compensation method potentiometer type PP-63 accuracy class of 0.05. We connected thermocouple circuit using the bipolar switch to one meter. Thermocouples have a common cold junction.

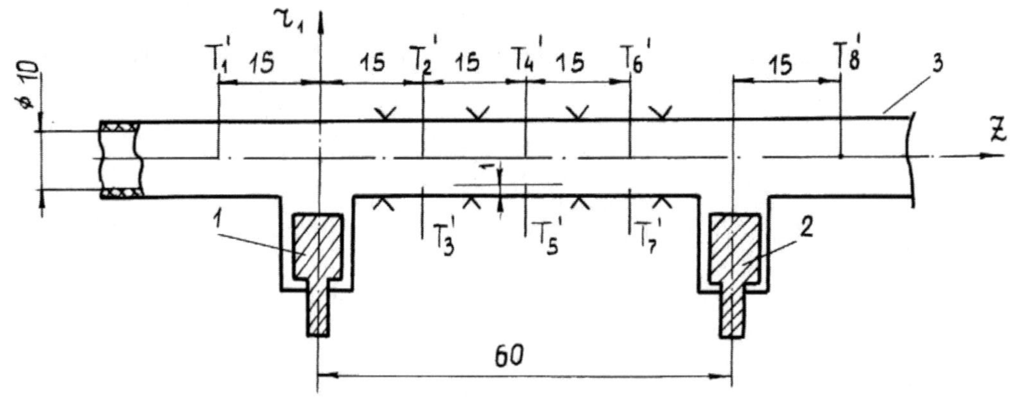

Figure 1. Scheme the DC for studying the temperature field of neutral particles in a longitudinal glow discharge.

3. Results

In this paper experimentally obtained that the air temperature in the longitudinal glow discharge increases linearly with increasing current, discharge capacity and pressure. The temperature increases with increasing discharge power occurs approximately linearly with speed 6,5K / W on axis and 4K / W near the wall of the DC. Temperature field of neutral particles essentially depends on the form of luminous discharge. The two electrodes, between which there is a glow discharge glow on photo (Figure 2), can be seen at the bottom. Also see the six processes, through which the inside of the discharge chamber introduced vitrified thermocouple. The Fig.2 corresponds to a small air flow rate G = 0,007 g / s, when the glow fills the entire cross-section of the discharge tube. Next photo (Figure 3) corresponds to a high air flow rate G = 0,035 g / s, in which the discharge is contracted to the bottom of the discharge tube, and can be clearly seen that the upper ends of the thermocouples are in the non-luminous discharge. The first thermocouple measured the temperature of neutral particles at the entrance to the DC. Its values varied from 292 K to 303 K, depending on the discharge power, the gas dynamic parameters of the flow and the air temperature in the receiver. Thermocouples numbered 2-7 measured the air temperature already directly in the combustion zone of the discharge. The ends of thermocouples with even numbers were located in the center of the DC. The ends of thermocouples with odd numbers were located at a distance of 2 mm from the inner surface of the DC. This arrangement of thermocouples made it possible to measure the gas temperature distributions along and across the DC. The distribution of air temperature and its values are closely related to the geometry of the glowing region of the discharge.

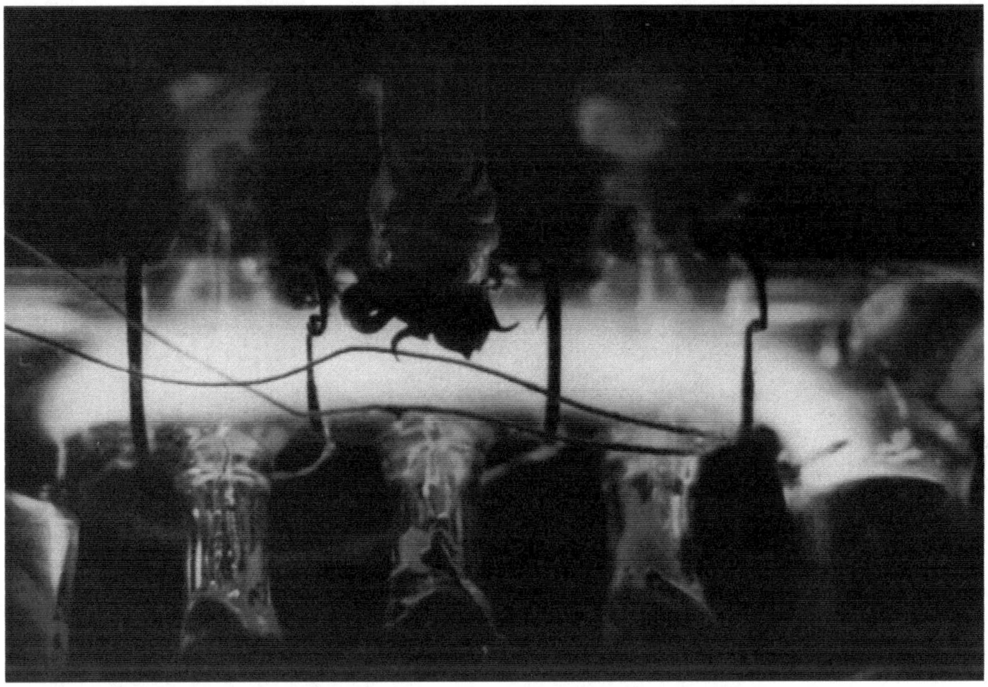

Figure 2. Diffuse form of the discharge

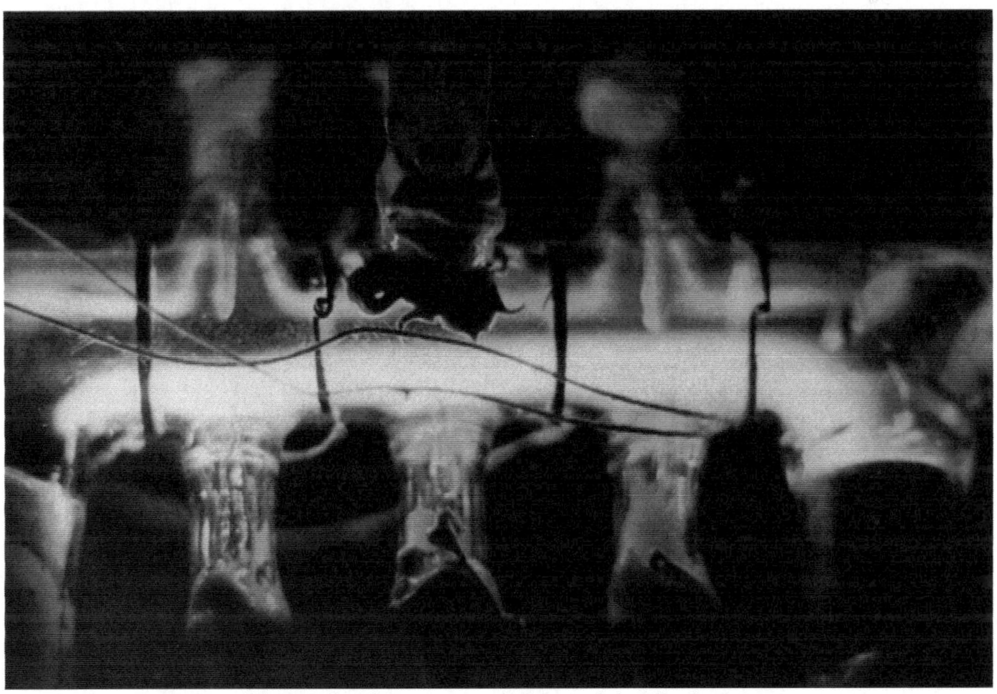

Figure 3. The contracted form of the discharge.

The results of measurements of the gas temperature of the fourth and fifth thermocouples are shown in Figure 4.

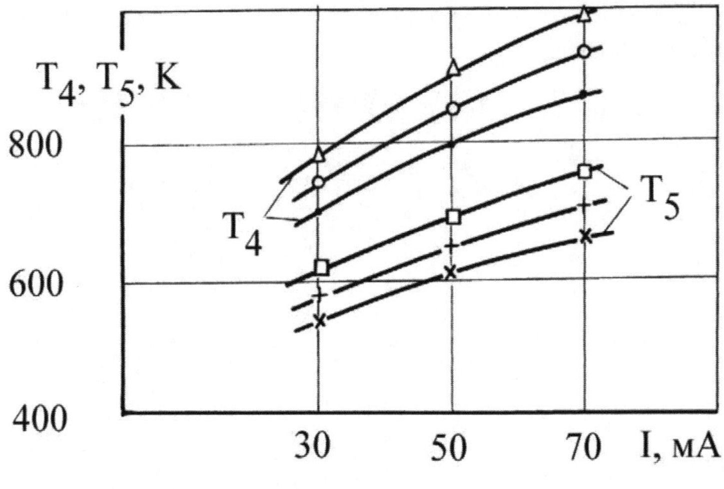

Figure 4. Dependence of air temperature on discharge current: G = 0,007 g / s, $\bullet,\times \rightarrow P = 2,5kPa$; $\circ,+ \rightarrow P = 3,9kPa$; $\Delta,\square \rightarrow P=5,5kPa$

Figure 4 corresponds to the diffuse form of the discharge (Figure 2). Analogous dependences of the air temperature on the discharge current and its power are also obtained for thermocouples with numbers 2 and 3, 6 and 7. The air temperature inside the discharge chamber increases with increasing current, discharge power and pressure. The effect of air flow on the gas temperature is shown in Figure 5.

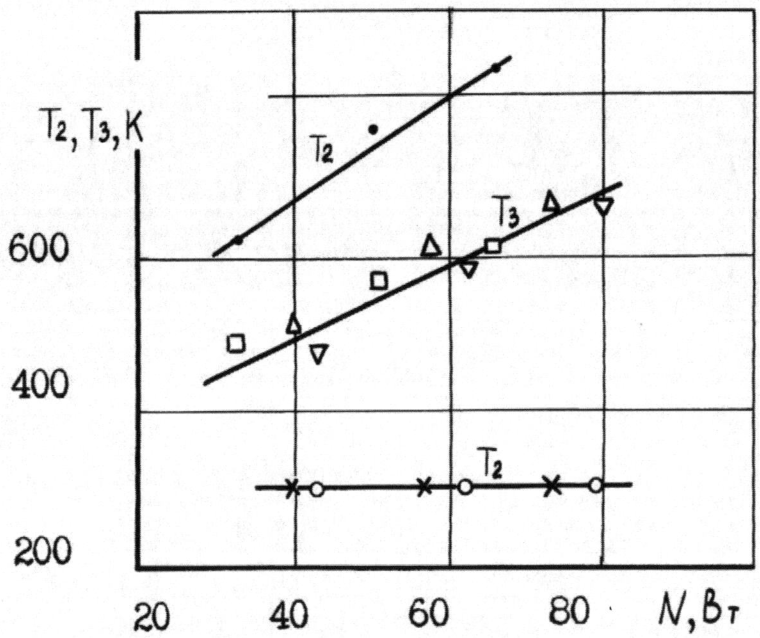

Figure 5. Dependence of air temperature on discharge power: P = 2, 5 kPa: $\bullet$, $\square$ - G = 0,007 g/s; Δ, $\times$ - G = 0,017 g/s; ∇, $\circ$ - G = 0,035 g/s

At low flow rates, the discharge fills the entire space between the electrodes, and the temperature on the discharge axis is maximal. With an increase in the air flow from G = 0.007 g /s to G = 0.017 g/ s, the discharge deviates from the axis under the action of augmented gas flow (Figure 3). This is associated with a sharp decrease in the gas temperature on the axis of the DC with a change in the air flow in the range G = (0.007-0.017) g/ s (Figure 5). When changing the air flow rate in the range G = (0,017 - 0,035) g/s on the discharge axis the temperature decreases nearly to room temperature.

4. Conclusions

In this paper we have experimentally obtained that the air temperature in a longitudinal glow discharge increases linearly with increasing current, discharge power and pressure. Air temperature distributions along the length of the discharge chamber are also obtained, and a significant temperature dependence of the form of the sacred region of the discharge is revealed. The dependence of the temperature of the discharge current and gas flow rate is found. The temperature increase with increasing discharge power occurs approximately linearly with speed 6,5K / W on axis and 4K / W near the wall of the DC. It has been found: the glow discharge is compressed to the boundary of the discharge chamber, when the air flow rate reaches a certain value equal to G = 0,017 g / s.

References

[1] Raizer Y P 1992 *Gas Discharge Physics* (Moscow: "Science") p 536
[2] Yunusov R F 1982 *Journal of Engineering Physics and Thermophysics* **43** 1100-03
[3] Yunusov R F 1983 *Inzhenerno-Fizicheskii Zhurnal* **43** 1100-03
[4] Yunusov R F 1985 *Journal of Engineering Physics and Thermophysics* **48** 591-592
[5] Yunusov R F 1985 *Journal of Engineering Physics and Thermophysics* **48** 214-219
[6] Yunusov R F 1985 Soviet Aeronautics **28** 78-82
[7] Yunusov R F 1988 *Journal of Engineering Physics and Thermophysics* **54** 76 - 80
[8] Yunusov R F 1990 *Journal of Engineering Physics and Thermophysics* **59** 990-994
[9] Yunusov R F 1991 *Inzhenerno-Fizicheskii Zhurnal* **59** 990-994
[10] Israfilov Z K and Kashapov N F 1991 *Journal of Engineering Physics* **60** 364-368
[11] Fayrushin I, Kashapov N and Dautov I 2014 *J. Phys.: Conf. Ser.* **567** 012009
[12] Saifutdinov A I, Fayrushin I I and Kashapov N F 2016 *JETP Lett.* **104** 180-185
[13] Yunusov R F 2017 *Journal of Physics: Conf. Series* **789** 012069
[14] Yunusov F S, Khisamutdinov R M and Yunusov R F 2008 *Russian Engineering Research* **28** 965-973
[15] Yunusov F S and Yunusov R F 2011 *Russian Engineering Research* **31** 15-21
[16] Yunusov F S and Yunusov R F 2011 *Russian Engineering Research* **31** 660-665
[17] Yunusov F S and Yunusov R F 2011 *Russian Engineering Research* **31** 951-959
[18] Denisov D.G, Kashapov N F and Kashapov R N 2015 *IOP Conferernce Series: Materials Science and Engineering* **86** 012005
[19] Zaripov R G, Kashapov N F, Tkachenko L A and Shaydullin L R 2016 *J. Phys.: Conf. Ser.* **669** 012053

Experimental study of voltage fluctuations in a glow discharge

R F Yunusov, M M Garipov

Kazan National Research Technical University named by A.N. Tupolev,420111, K. Marx str., 10, Kazan, Russian Federation

E-mail: optanir@mail.ru

Abstract The article is devoted to the experimental study of voltage pulsations in a glow discharge. The longitudinal discharge was ignited in a stream of air moving in a cylindrical discharge chamber (DC). Copper hollow electrodes were located at a distance of a = (2-8) cm from each other along the axis of the DC. The gas pressure P = (4.7 - 26) kPa and its flow rate G = (0-0.06) g /s were monitored. The discharge current and voltage varied accordingly in the ranges: I = (20-100) mA, U = (1, 5 - 4, 2) kV. The current-voltage characteristics of the discharge were measured. The voltage decreased with increasing current, which is typical for a given type of discharge. The oscillations of the discharge voltage were registered by an oscilloscope; the current-voltage characteristic of the discharge was recorded by a two-coordinate recorder. Three regions of the current-voltage characteristic were detected. The luminous positive discharge column (PC) occupied the entire interelectrode space in the current region I = (100-70) mA. In the current range I = (70-40) mA there is a smooth transition to the discharge without visible glow of the PC with a significant increase in the discharge voltage. With further reduction of the current intensity (the third region), the discharge exists almost at a constant voltage until its quenching. Low-frequency and high-frequency oscillations of the discharge voltage were observed at the boundary of the second and third regions of the current-voltage characteristic.

1. Introduction

A plasma glow discharge is widely used in gas-discharge light sources, as the active medium of gas lasers [1-10], modern technologies for production of various coatings, nanotechnology methods for obtaining unconventional materials [11-14]. Therefore, the study and research of glow discharge plasma parameters is of great scientific and practical interest. Glow discharge in longitudinal and transverse gas flows is actively investigated in modern conditions [15 – 18]. This is due to the needs of practice, since for specific technological tasks various DC and ways of organizing the flow of matter in them are required. Obtaining a stable discharge with the parameters necessary for practice is one of the problems in this area. The glow discharges, as well as the arc discharge, are generally no stationary [19-20]. The purpose of this article is an experimental study of voltage fluctuations in a longitudinal glow discharge.

2. Experiment

The scheme of the experimental setup is shown in Fig. 1. The air through the needle cone **1** and the flow meter **3** (rheometer RDS-10) was supplied to the discharge chamber DC **4**. The DC is a quartz tube with an internal diameter of 10 mm (Fig.1.b). The discharge was ignited in the longitudinal flow between the copper hollow cylindrical electrodes: the cathode and the anode.

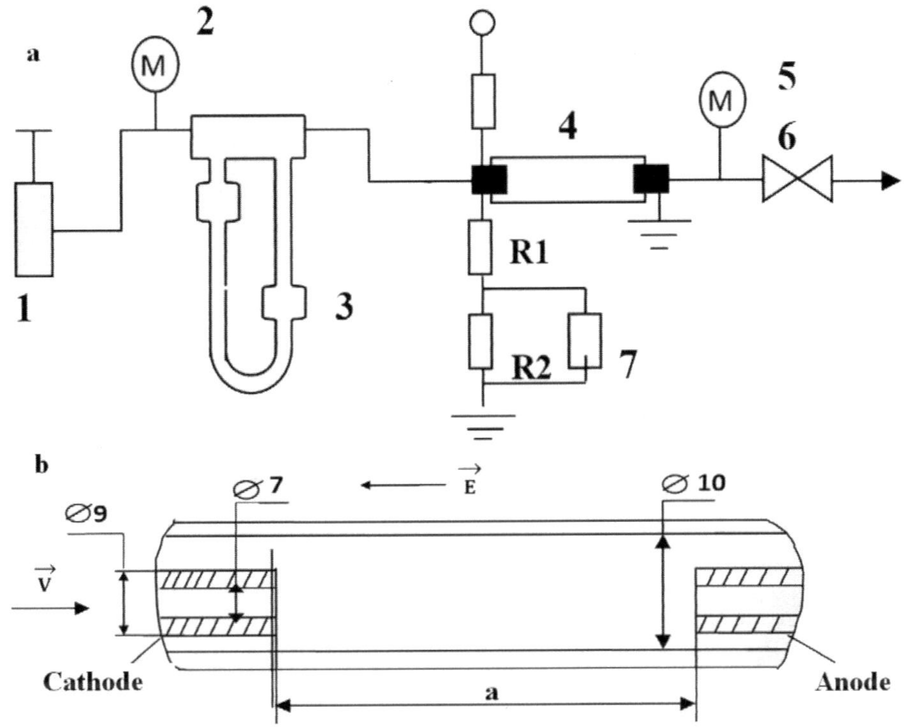

Figure 1. The scheme of the experimental setup: a - gas-dynamic path; b-discharge chamber.

The gas moved in a direction from the cathode to the anode. The distance between the electrodes could be changed in the range from 2cm to 8cm. The spent medium from the DC entered the capacity and was removed into the atmosphere by a vacuum pump. The static pressure in the RC was measured by U-shaped mercury manometers **2** and **5**. The flow rate and the pressure value were controlled by the valve **6**. The oscillations of the discharge voltage were registered by an oscilloscope **7**. The current-voltage characteristic of the discharge was recorded by a two-coordinate recorder of the type H-306.

3. Results

Figure 2 shows the current-voltage characteristics of an electrical discharge in an air stream. The experiment showed three characteristic regions of current-voltage characteristics. After the ignition of the discharge, the starting point falls on the first region I of the current-voltage characteristics. Therefore, for the full description, the inverse branch of the U-I characteristic is considered. Figure 2 shows two volt-ampere characteristics: 1- for a discharge with an air flow, 2 - for a discharge in still air. For the first curve, with a decrease in current strength from 100 to 70 mA (curve 1, region I), the value of the discharge voltage slowly increases. The positive discharge column reaches almost to the cathode. In the current range I = 70 ... 40 mA there is a smooth transition (region II) of the luminous discharge to the discharge without visible glow. With a further decrease in the current intensity (region

III), the discharge exists almost at a constant voltage up to quenching. The glow in this case is observed only from the ends of the hollow electrodes. In the quiescent gas (curve 2), there are also all three regions of the current-voltage characteristic. However, region I shifts towards lower currents, and regions II and III are respectively narrowed.

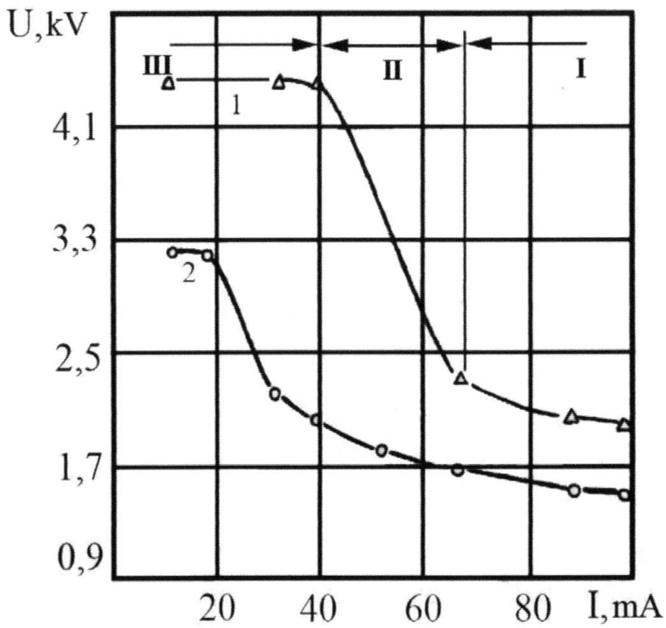

Figure 2. Current-voltage characteristic of an electric discharge in the longitudinal air flow: a = 0.04 m, P = 26 kPa; 1 – G = 0.04 g/s; 2 – G =0.

The current-voltage characteristics practically did not change when the polarity of the electrodes changed.

Photos of the discharge are shown in Figure 3. As can be seen from the photographs (Fig. 3a), the discharge consists of a brightly shining region extending up to the anode and a dark space near the cathode. Directly at the cathode there is the luminous layer. It was found that as the current increases, the length of the dark space decreases, and the surface of the cathode, covered with a luminous layer, increases. In a discharge with a gas flow (Fig. 3b), the positive column of the discharge contracts and loses its axial symmetry.

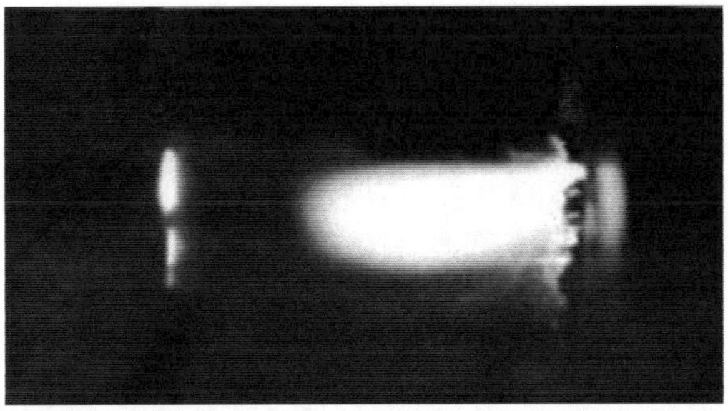

a

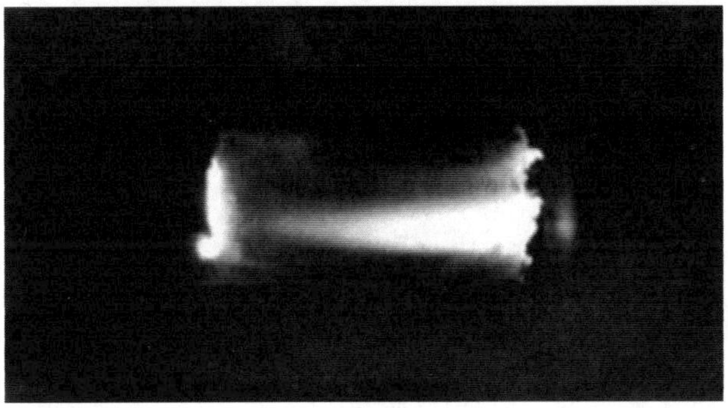

b

Figure 3. Photos of the discharge without a gas flow (a) and with a gas flow (b), P= 3, 6 kPa, a = 0, 03 m, I = 10 mA: a – G=0, b – G = 0, 02 g/s.

With increasing air flow, the discharge voltage first rises, reaches a maximum, then decreases and remains constant (Fig. 4).

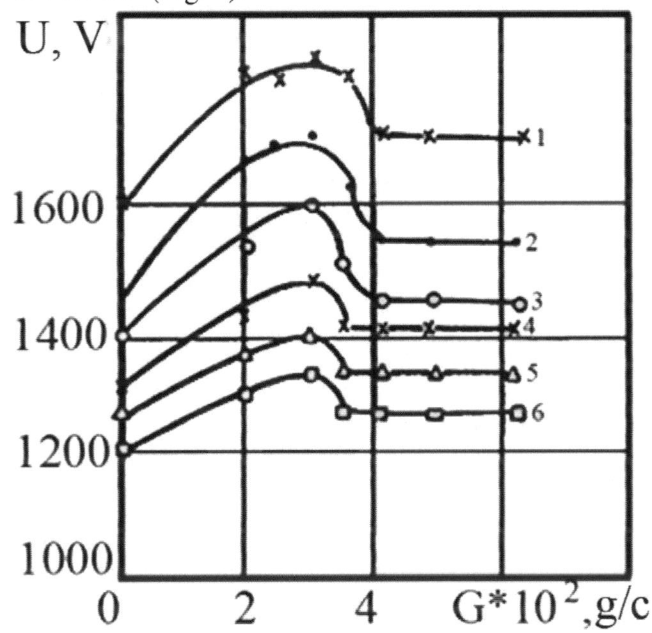

Figure 4. Dependence of the discharge voltage on the flow at different current strengths: 1- I=30 mA; 2- I= 40 mA; 3- I= 50 mA; 4- I= 60 mA; 5- I=70 mA; 6- I= 80 mA; P = 4 kPa.

With increasing pressure and decreasing current strength, discharge stability deteriorates, and voltage pulsations appear (Fig. 5). Low-frequency periodic and aperiodic damped voltage fluctuations occur in a discharge without an air flow. These voltage pulsations occur due to low frequency oscillations of the power supply voltage. The frequency of the variable component is 300 Hz. These low-frequency ripple voltages correspond to the discharge without visible glow of the PC. With further reduction of the current intensity, the discharge is extinguished. In the air flow, the nature of the oscillations varies. High-frequency voltage oscillations (Fig. 5) with a frequency f = (5 – 100) kHz appear in a discharge with an air flow. This feature is observed throughout the investigated range of currents. High-frequency voltage pulsations have the shape of a sinusoid, and their frequency decreases with increasing airflow. The region in which high-frequency voltage oscillations occurred was determined by analyzing the oscillograms. High-frequency voltages oscillations are absent at currents greater than 20 mA.

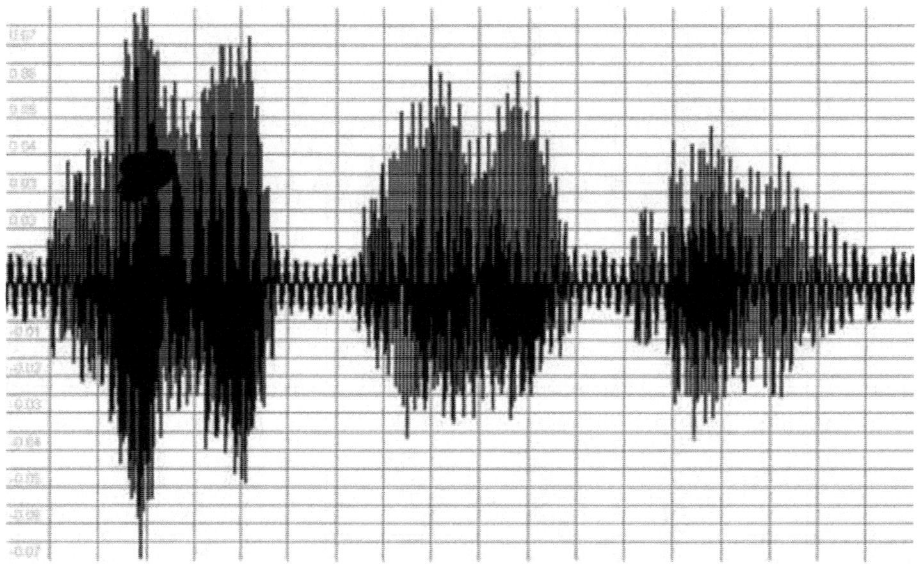

Figure 5. The typical oscillogram of discharge voltage pulsations: G = 0.03 g / s, I=8 mA, U=1380 B

4. Conclusions

A longitudinal glow discharge in the air flow was studied in this paper. The effect of air flow on the current-voltage characteristics of the discharge was obtained. An increase in the air flow in the interval G = (0-0.03) g / s causes an increase in the discharge voltage. With a further increase in the flow rate to the value G = 0.06 g / s, the discharge voltage practically does not change. Simultaneously with this increase in the discharge voltage, its structure changed: the discharge lost its axial symmetry and contracted. With a decrease in the discharge current in the interval I = (70-40) mA, the luminous region of the discharge disappears, and its voltage increases sharply. In this region of transition to the dark form of the discharge, high-frequency and low-frequency oscillations of the discharge voltage are observed. An increase in the air flow rate leads to an increase in the discharge current, at which high-frequency voltage pulsations occur.

References

[1] Raizer Y P 1992 *Gas Discharge Physics* (Moscow: "Science") p 536
[2] Yunusov R F 1982 *Journal of Engineering Physics and Thermophysics* **43** 1100-03
[3] Yunusov R F 1983 *Inzhenerno-Fizicheskii Zhurnal* **43** 1100-03
[4] Yunusov R F 1985 *Journal of Engineering Physics and Thermophysics* **48** 591-592
[5] Yunusov R F 1985 *Journal of Engineering Physics and Thermophysics* **48** 214-219
[6] Yunusov R F 1985 Soviet Aeronautics **28** 78-82
[7] Yunusov R F 1988 *Journal of Engineering Physics and Thermophysics* **54** 76 - 80
[8] Yunusov R F 1990 *Journal of Engineering Physics and Thermophysics* **59** 990-994
[9] Yunusov R F 1991 *Inzhenerno-Fizicheskii Zhurnal* **59** 990-994
[10] Fayrushin I, Kashapov N and Dautov I 2014 *J. Phys.: Conf. Ser.* 567 012009
[11] Yunusov F S, Khisamutdinov R M and Yunusov R F 2008 *Russian Engineering Research* **28** 965-973
[12] Yunusov F S and Yunusov R F 2011 *Russian Engineering Research* **31** 15-21
[13] Yunusov F S and Yunusov R F 2011 *Russian Engineering Research* **31** 660-665
[14] Yunusov F S and Yunusov R F 2011 *Russian Engineering Research* **31** 951-959
[15] Timerkaev B A, Ahmetov M M, Zalyaliev B R, Petrova O A and Israfilov D I 2014 *J. Phys.: Conf. Ser.* **567** 012036
[16] Timerkaev B A and Zalyaliev B R 2014 *High Temperature* **52** 471-474
[17] Saifutdinov A I, Timerkaev B A and Zalyaliev B R 2016 *High Temperature* **54** 669-675
[18] Yunusov R F 2017 *Journal of Physics: Conf. Series* **789** 012069
[19] Thoukhvatoulline R, Dautov G and Feldmann G 2004 *J. Phys. D* **37** 1058-1064
[20] Dautov G Y, Khairetdinova R R and Dautov I G 2016 *J. Phys.: Conf. Series* **669** 012005

IOP Conf. Series: Journal of Physics: Conf. Series **927** (2017) 012078 doi:10.1088/1742-6596/927/1/012078

Generation of quasi-perpendicular Collisionless Shocks by Laser-Produced Plasma to simulate the effects of super-compression of the Earth's magnetosphere*

Yu P Zakharov[1], A G Ponomarenko[1], V A Terekhin[2], A I Golubev[2], V M Antonov[1], E L Boyarintsev[1], K V Vchivkov[1], A V Melekhov[1], V G Posukh[1], P A Prokopov[1]

[1]Institute of Laser Physics (ILP), Siberian Branch of Russian Academy of Sciences (SB RAS), Novosibirsk, Av. Lavrentyeva 13/3, 630090, Russia.
[2]All-Russian Research Institute of Experimental Physics (VNIIEF), Sarov, Av. Mira 37, 607188, Russia.

E-mail: ki1z@mail.ru

Abstract. Giant plasma releases of so called Coronal Mass Ejections (CME, with kinetic energy up to $E_k \sim 10^{36}$ эрг) from the surface of the Sun and their potential catastrophical impact onto Earth's magnetosphere, with the probable opportunity to compress it in 2, 3 or more times [1, 2], represent one of the most important problem in the geophysical and historical bio-evolutional investigations of the past and present of the Earth. It was supposed that the re-connection of dipole magnetic field at magnetopause could play [2] an exclusive role in its inward shift, but from the more general point of view (to perform laboratory simulation [3-6] of magnetopause dynamics), a more important and common features of CME propagation in Solar Wind plasma are the formation of collisionless Quasi-Perpendicular Shocks (Q-PS) ahead of Super-Alfvenic CME. It is provided by the almost radial (along to **R**) motion of CME from the Sun, while the Interplanetary Magnetic Field $\mathbf{B_0}$ has an angle $\theta \sim 45^0$ (relative to **R**) near the Earth orbit. Up to date, in spite of intensive development of laser energetics and energy of Laser-produced Plasma (LP), such Q-PS never were generated by LP [7], excluding recent experiment [8] at KI-1 facility of ILP. In the given work, a first results of our study were presented together with relevant calculations by hybrid code and the basic physics of VNIIEF-model [9] for the collisionless Magnetic Laminar Mechanism (MLM) of interaction between a spherical LP and magnetized Background Plasma (BP). A special analysis was done on the conditions and data of the formation of whistler precursor in front of revealed oblique (Q-PS) shocks as well as on the first data of dipole's field compression by BP and Shocks.

1. Introduction.

Recent numerical MHD-simulations of NASA [1, 2] show that due to the action (in ancient times) of CME (with the Super-High effective energy up to $E_0 \sim 10^{37}$ ergs), a very strong compression of the Earth' magnetopause could occur, from its usual size $R_{mp} \approx 10R_E$ (Earth' radius) up to extreme value $R_{mp}^* \leq 1{,}5R_E$ (!). Therefore a real opportunity could appear for direct contact [2] of interplanetary and Solar energetic plasmas with the ionosphere' and even upper atmosphere' media. As a result, a lot of various chemical reactions and even possible origin of prebiotic life chemistry [1] could take place there, as a basic phenomenon of new scenario for the origin of life at the Earth. Some parts of these effects we could successfully studied in our recent experiment HERMEX [5] at KI-1 facility of ILP, with a blobs of Laser Plasma (LP) and flow of Background Plasma (BP, near impulsive magnetic dipole μ), to simulate an effect of two-fold compression of laboratory model of MagnetoPause (MP). But this simulation was done in a rather simple way – without essential interaction effects between LP and BP, since both of them were directed to dipole along to small uniform magnetic field $\mathbf{B_0}$ = 20 Gs (it was ⊥ μ). Here we present a first results of new simulative experiment **MPS** (MagnetoPause & Shock), in which we try realize for the first time an important part of CME phenomena – a strong forward disturbance of BP in a form of Collisionless Shock Wave (CSW), namely Quasi-Perpendicular Shock (Q-PS), most typical one for the CME case near the Earth. Q-P case means angle $\geq 45^0$ between shock normal **k** and $\mathbf{B_0}$ (| | **Z**, horizontal central axis of facility chamber). At Figure 1 and Table, a general scheme and the parameters (with a relevant *similarity criteria*) of simulative experiment MPS at laser facility KI-1 (*Kosmicheskie Issledovaniya-1*) are presented.

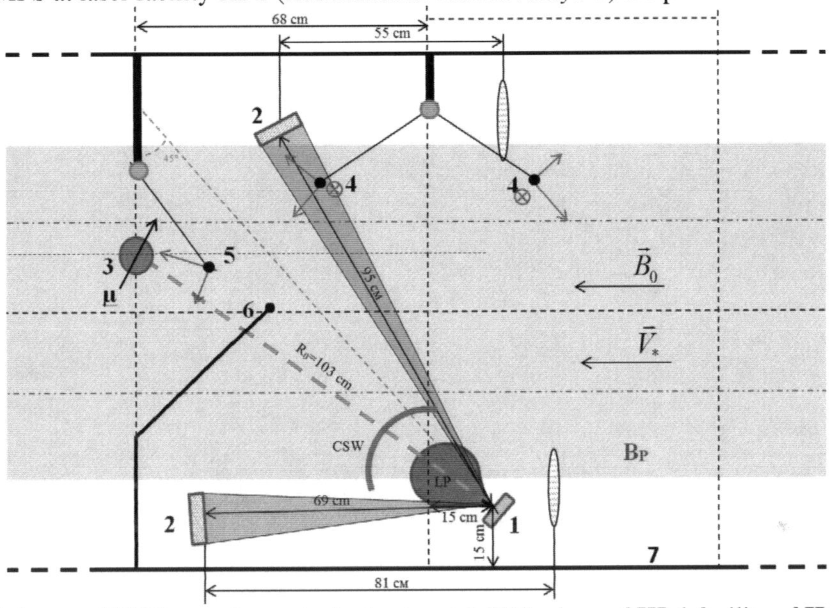

Figure 1. Scheme of **MPS** experiment in the horizontal (ZY) plane of **KI-1** facility of **ILP**:
1– Laser polyethylene target (spot ∅~2,5 cm); **2** – focusing mirrors for CO_2 -laser (up to **500 J**);
3 – dipole; **4, 5, 6** – sets of magnetic and Langmuir probes; **7** – wall of high-vacuum chamber (5 m long and ∅1.2 m, pressure~$3*10^{-6}$ Torr); **5** (is P_1 probe), **6** (P_0). **LP** (Laser Plasma) - blob near target (**1**); **CSW**- Shock, generated near- central part of **BP** (Background Plasma, flowing from the right), with maximal density up to $5*10^{13}$ cm^{-3} (H^+) in ∅~70÷80 cm, μ is a dipole up to 10^6 $G*cm^3$.

2. MLM-processes for Collisionless Shock generation by Laser Plasma Piston.

A basic physical background to generate CSW by LP at KI-1, was a so a called Magnetic Laminar Mechanism (MLM), studied for the first time theoretically at VNIIEF [9] and experimentally, for the first time at KI-1 [10]. Today only unique experimental features of KI-1 facility give a real opportunity to generate both PerPendicular (P-P) Shock [11] and Q-P one (Figure 2a) in laboratory, first of all, due to unique method [12] of producing Laser Plasma Blob with effective kinetic energy up to E_0 ~ 1000 J. Namely such large values of E_0 and effective number electrons $N_{e0} \geq 10^{19}$ in LP are need to supply effective MLM-interaction of LP with BP and generate CSW in it at the need angle up to $\theta \approx 45^0$ due to the enough large value of MLM-parameter $\delta_0 \approx 1{,}5$ [9], determined as $\delta_0 = R_*^2 / R_L R_L^*$. Here $R_* = (3N_{e0}/4\pi n_*)^{1/3}$ is maximal size of diamagnetic cavity of LP in BP [9] and R_L, R_L^* are Larmor radii of both LP' and BP' ions (both determined by V_0 and B_0). So, the effective MLM-interaction for CSW generation could be [9] at angles $\theta_m \geq 90^0 - \arcsin(2\delta_0/3)$, that means essential effect not only near "equator" (where $\theta = 90^0$) of expanding LP-sphere (where initial B_0-field directed from its pole to pole). Since for our case of high Alfven-Mach number ($M_{a0} \geq 4$, see Table), only MLM could supply enough [13] transfer LP energy E_k to kinetic energy E_* of BP (as $E_* \approx \delta_0 * E_k$) for the CSW generation, we need to realize $\delta_0 \geq 1$. For this case also the angle θ_m could reach a need value ~ 45^0 to generate a Q-P Shock. But in the MPS we have used (Figure 1) not a spherical LP [10], but an expanding (within volume angle $\Delta\Omega$ ~ 1 sr) LP-blob [6, 12], directed at the angle θ ~ 45^0 relative to B_0. For this and others cases we can use the relation for effective energy of LP-blob, as $E_0 \approx 4\pi(E_k/\Delta\Omega)$~$10E_k$ up to 1 kJ.

Parameters / *Criteria*	**Notations**	**Values**
Density of Background Plasma - **BP** (its diameter ∅)	$\mathbf{n_*}$ (∅)	$(1\text{-}2)\cdot 10^{13}$ $см^{-3}$(∅~70-80 cm)
Parallel (to $\mathbf{B_0}$) velocity of **BP** (and its ions)	$\mathbf{V_*}$ (H^+)	50-60 km/s (H^+)
Size of **M**agneto**P**ause (**MP**) around dipole in the **BP**	$\mathbf{R_{mp}} = (\mu^2/4\pi n_* m_* V_*^2)^{1/6}$	~ 25 cm
Magnetic moment of dipole and radius $\mathbf{R_d}$	μ	up to $\mathbf{10^6}$ G*cm^3, $\mathbf{R_d}$ =3,5cm
Uniform magnetic field in chamber (along to $\mathbf{V_*}$ of **BP**)	$\mathbf{B_0}$	80 G
Kinetic energy of Laser Plasma-**LP** (Effective Energy)	$\mathbf{E_k}$ ($\mathbf{E_0}$)	~ 60 J (700 J)
Expansion velocity of **LP** (and its solid angle <Ω>)	$\mathbf{V_0}$ (of front)	Up to 200 km/s (<Ω>~1Sr)
Distance between a dipole' center and point of **LP** (or **C**oronal **M**ass **E**jection - **CME**) explosion	$\mathbf{R_0}$	103 cm
Size of **MP** in the stream of **LP** (in vacuum)	$\mathbf{R_m}^* \approx 0{,}75 R_0/K^{1/6}$	~ 15 cm
*Hall parameter **D** for **MP** in plasmas of **BP** and **LP***	$D = R_m/(C/\omega_{pi})$	$D_* \approx 3$ *and* $D \approx 1$
*Alfven-Mach number for **LP** (in **BP** with Alfven C_{A*})*	$M_{a0} = V_0/C_{A*}$	*~ **5-6** ~ 10*
*Energetic criterion **LP** (H^+/C^{+3} ions, <m/z>≈2.5a.e.m.)*	$K = 3E_0R_0^3/\mu^2$	~ $2*\mathbf{10^4}$ >> 10
*Ratio of velocities of **LP** to **BP***	V_0/V_*	*~ **4** (~10)*
*Ratio of the initial **MP**-size to its compressed one*	$\xi = R_{mp}/R_m^*$	*~ **2** (**HERMEX, 2013-without Shock!**)*
For Giant CME($E_0 \geq 10^{34}$эрг) impact onto Earth' MP: M_{ao}~50 ≥10, V_0/V_* ~7(~10), K~10^{22} >>10, D_* ~D~1000 >> 10 and ξ~2 (while even up to ξ~6 for Super CME with E_0~10^{37}erg)		

For the main goal of whole experiment – maximal compression [5] of Magnetosphere Model (MM), we could expect that kJ-LP from its distance R_0 ~ 100 cm (to dipole) with the energetic criterion $K = 3E_0R_0^3/\mu^2$ ~ 10^4 should compress MM up to new size $R_m^* \approx 0{,}75R_0/K^{1/6} \approx 15$ cm (from the initial $R_m \approx 25$ cm). Such near 2-fold compression will be comparable with the data of previous HERMEX, but due to presence of Q-P Shock we could expect some additional effects [1, 2]. Note that up to now non of both shocks (P-P or Q-P) had not reliably observed in LAPD experiments of California University [14].

3. Data on Q-P Shock generation in MPS and its numerical simulations by Hybrid model.

At Figure 2 a main data on the Q-P Shock (**a**) generation are presented with the important feature of such shock [15], as a whistler precursor (curves 2, 3 ∝ $\partial B_x/\partial t$ – signal of probes with area 1 cm^2), ahead front (curve 1) of density jump.

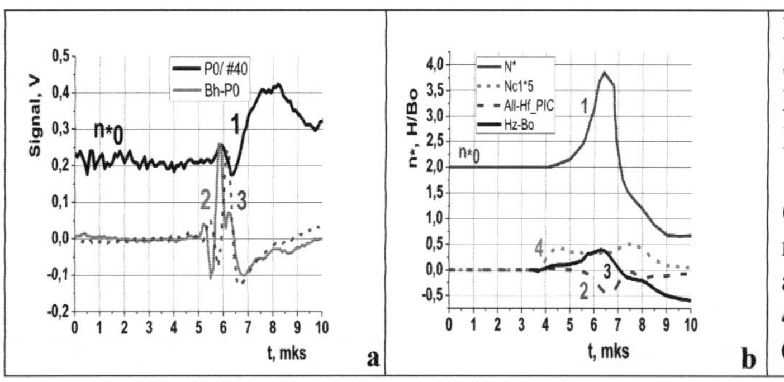

Figure 2. **a** – Data of MPS on Q-P Shock generation by LP at distance R ≈ 77 cm: **1**–density; **2**– magnetic B_x-field;**3**–B_{zy}-field (vector summ). **b** – Data of 2D-PIC simulation (spherical LP expansion into direction 45^0) for close conditions and in the same notations (**1**,**2**,**3**). **4**– density (zoom in 5 times) of LP C^{+3}-ions (H^+-ions stay at 65-70 cm).

The data of Figure 3 present another important feature, a joint and coincidental R-t diagram of MPS-experiment and its simulation by the model «Clouds03» (of Dudnikova and Vshivkov [16]), that early also was applied successfully for the analysis of data KI-1 on perpendicular Shock generation [12].

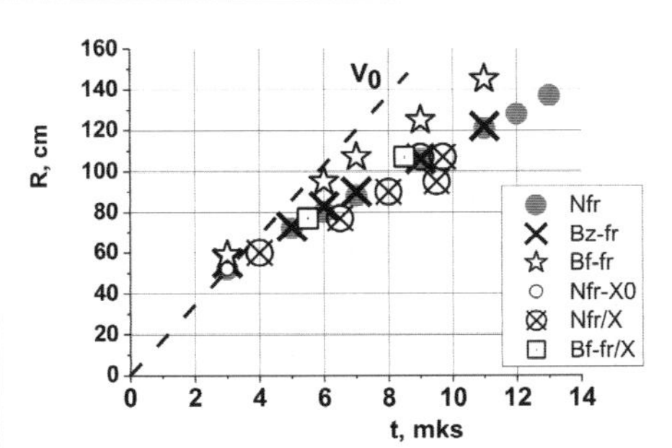

Figure 3. Joint R-t diagrams of MPS experiment (X) and its PIC-simulation, by notations for the case of **MPS**: **Nfr-X0**–front LP-density in vacuum (without B_0), **Nfr/X** – front density n_* in Background Plasma (BP), **Bf-fr/X**– front of magnetic field B_x (or B_φ) component. For the case of **PIC**-simulation (for $n_{*0}=2*10^{13}cm^{-3}$): **Nfr** – front density n_* in magnetized BP, **Bz-fr** – front of the main B_Z-field (#3 at Figure 2b), **Bf-fr** – front of magnetic field B_x (or B_φ) component. **MPS**-data were obtained at $n_{*0} \approx (1\div2)*10^{13}cm^{-3}$.

In spite of rather similar general MHD-features of colisionalless interaction between LP and BP, in MPS and its PIC-simulation (Figure 3), the details of magnetic structures in front of BP-density jump are very different. At Figure 2a, such MPS-structure reveals a typical for whistlers [17] phase shifted two oscillating B-components (2 and 3), really ahead of BP-jump (1), while a simulative data (Figure 2b) have a monotone character (2,3), very probably correalated with a dynamics of heavy LP-ions (4). The reason of absence a whistler precursor (Figure 2a) in MPS-simulation (Figure 2b), may be a zero electron mass [15] in the used «light» version of hybrid code [16]. In the control runs without heavy LP-ions (and similar R-t diagram), a whistler-type precursor also was absent at super-Alfvenic fronts.

4. Preliminary data of MPS experiment on the compressions of dipole magnetic field B_d.

At the Figure 4 a joint data of Magnetic Probes (MP-3,4,5) and Langmuir one (1,2) near the dipole in a flow of BP (1) with the Q-P Shock (2) are presented. In the case of the compression-type both MP-signals, a cumulative effect of BP with QPS leads to the two-fold increasing of the total B_d-field jump (up to $\Delta B_d \approx 180$ G, in comparison with the BP-alone). This ΔB_d-value is not so much in a sense of probable inward shift of MP (around ~30 % only), but it allows us to do a comparison with some MP-model predictions and to plan the development of MPS-experiment. In particular, to test a most simple and general MP-model for ΔB_d-jump, based on the well known solution for the dipole field inside of super-conductive sphere [18], such as additional magnetic field B_s parallel to dipole μ. For the radius R_s of sphere and radius R_d of dipole (its super-conductive sheath) solution has a form: $B_s(r) = \mu*(1/r^3 + K/R^3)/(1-R_d^3/R_s^3)$, where K is a geometric factor: K = 2 for the whole sphere and K ≈ 1 for its one half, as a model of MP front end. So for the given MPS-case (Figure 4), from the maximal value $\Delta B_d \approx 180$ G ~ B_s, one could estimates a probable MP-size as $R_s \approx 17\div18$ cm.

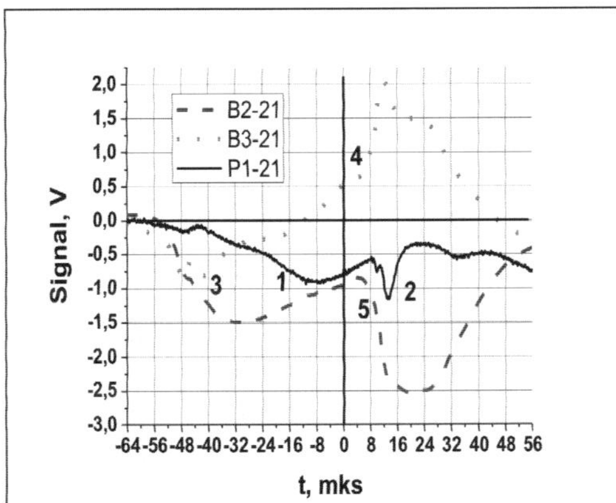

Figure 4. Joint data of Langmuir (1, 2) and magnetic (3-5) probes [19] near dipole (point 5 at Figure 1) in MPS1 experiment: **1** – initial part of Langmuir data is a flow of Background Plasma (BP); **2** – Shock-type disturbance of BP caused by Laser Plasma (LP) injected at the time t = 0; **3** – initial part of Magnetic Probe (MP) data caused by BP; **4** and **5** are disturbances of dipole magnetic field B_d, registered by two orthogonal coils of MP (**B2** and **B3**) at the distances more close to dipole surface (than Langmuir one). Sensitivity of Langmuir probe (**P1**) for its 1 V, gives BP-density (H^+) ~ $5*10^{13}$ cm^{-3}, while 1 V of B2-jump corresponds to 75 G and B3-jump to 50 G.

5. Conclusions.

In the given paper a first data of new type, combined and complicated simulative experiment MPS with a laser-produced plasma, magnetized background and compact impulsive magnetic dipole, at KI-1 facility of ILP are presented. Two most important and fundamental results were obtained for the first time. Due to very high effective energy of laser plasma (up to $E_0 \sim$ kJ) and density of background (up to ~ 3÷ 5*10^{13} cm^{-3}) at scale ~ 1 m, a collisionless Quasi-Perpendicular Shock was generated by laser plasma piston for the first time, with a typical whistler precursor (Figure 2a). At presence of such shock in a flow of background, a strong additional compression of magnetic field near dipole was revealed for the first time (Figure 4), as an important effect for the laboratory simulation of extreme super-compression of the Earth' magnetosphere [1, 2]. Future development of MPS experiment could allow to simulate also a collisionless Bow Shock [20, 21] in front of region of compressed dipole magnetic field and its probable interaction with a generated Quasi-Perpendicular Shocks.

* This work was supported by the ILP Project 0307-2016-0002 of SB RAS on Fundamental Researches and by ILP-Projects 2.3 and 3.3 of Program "Extreme laser radiation" Presidium RAS and SB RAS (Complex Program).

References

[1] Airapetian V et al. 2016 *Nature Geoscience* **9** 452
[2] Airapetian V et al. 2015 *Proc. 18th Cambridge Workshop on Cool Stars, Stellar Systems and Sun* No 320 409
[3] Zakharov Yu P, Antonov V M, Boyarintsev E L et al. 2008 *J. Phys.: Conf. Ser.* **112** 042011
[4] Zakharov Yu P, Ponomarenko A G, Horton W, et al. 2009 *Astrophys. Space Sc.* **322** 151
[5] Zakharov Yu P, Ponomarenko A G, Antonov V M, et al. 2016 *J. Phys.: Conf. Ser.* **688** 012129
[6] Ponomarenko A G, Zakharov Yu P, Antonov V M, et al. 2007 *IEEE Trans. Plasma Sci.* **35** 813
[7] Schaeffer D B, et al. 2017 *Phys. Rev. Let.* **119** 025001
[8] Zakharov Yu P, Ponomarenko A G, Terekhin V A et al. 2016 *Proc. VII Intern. Symp. "Modern Problems of Laser Physics" (22-28 August 2016, Novosibirsk, ILP)* 254.
[9] Bashurin V P, Golubev A I, Terekhin V A 1983 *J. Appl. Mech. Techn. Phys.* **24** 614
[10] Antonov V M, Bashurin V P, Zakharov Yu P, et al. 1986 *J. Appl. Mech. Techn. Phys.* **26** 757
[11] Zakharov Yu P, Ponomarenko A G, Terekhin V A, et al. 2013 *Proc. VI Intern. Symp. "Modern Problems of Laser Physics" (25-31 August 2013, Novosibirsk, ILP)* 193.
[12] Zakharov Yu P, Ponomarenko A G, Terekhin V A, et al. 2016 *Quantum Electronics* **46** 399.
[13] Zakharov Yu P 2003 *IEEE Transactions on Plasma Science* **31** 1243
[14] Bondarenko A S, Everson E T, S. E. Clark S E et al. 2017 *Nature Physics/Letters* 20 Febr.
[15] Balough A, Treumann R A 2013 *Physics of Collisionless Shock* (Springer: New-York)
[16] Dudnikova G I, Vshivkov V A , Zakharov Yu P, et al., 1990 *Proc. Plasma Astrophysics* (Telavi, USSR, ESA: SP-311) 191.
[17] Prokopov P A, Zakharov Yu P, Tishchenko V N, et al. 2016 *J. Phys.: Conf. Ser.* **769** 012086
[18] Goldstein B E et al. 1981 *J. Geophys. Res.* **86A** 5485
[19] Zakharov Yu P, Antonov V M, Shaikhislamov I F, et al. 2011 *Contrib. Plasma Phys.* **51** 182
[20] Ponomarenko AG, Zakharov Yu P, Antonov V M, et al. 2008 *Plasma Phys. Contr. Fus.* **50** 074015
[21] Cruz F, Alves E P, Bamford R A, et al. 2017 *Physics of Plasmas* **24** 022901

The possibilities of control of the characteristics of a glow discharge by using the organization of supersonic gas flow

B R Zalyaliev, B A Timerkaev, I I Amirzyanov

Department of General Physics, Faculty of Physics and Mathematics, Kazan National Research Technical University named after A. N. Tupolev – KAI, 10 Karl Marx Str., Kazan, 420111, Russian Federation

E-mail: bzalyaliev@gmail.com, btimerkaev@gmail.com

Abstract. The possibilities of glow discharges in a supersonic flow which organized in a limited region of discharge are presented.

1. Introduction

The control of the glow discharge characteristics can allow to remove restrictions on its using in the plasma technology under certain conditions. Some results of the novel methods to control parameters of the glow discharge are presented in [1-4]. In [1] a method of enhancing of the glow discharge stability is proposed for a chamber with cathode sections coated by a discontinuous dielectric coating. In this case the average density of electric current in the cathode region decreases and the conditions for the occurrence of instabilities decreasing. The control possibilities of the glow discharge parameters with using magnetic field are described in [2,3]. The method of synthesis of nanomaterials based on ideas [2,3] is presented in [4].

It is known that the glow discharge has self-organized layered structure of near-electrode zones. The zone boundaries correlate with the increasing of electron energy in electrical field. When the gas pressure is changed the zone boundaries are set according to laws of similarity. For example, the law of similarity $pd = const$, where p - pd = constpressure and the d - thickness of the cathode layer, determines the thickness of the cathode layer. But specially organized supersonic flow can break regularities of self-organization of the near-cathode zones and cathode processes in the glow discharge.

Indeed [5,6], although the supersonic gas flow was organized over all interelectrode space, but the concentration of neutral particles could be different in its different regions. This was demonstrated in articles [7-10] where supersonic flow organized in a limited region of the interelectrode space. Because the particle concentration determines the mean free path, the thicknesses of the near-electrode zones will change in the same way.

The next two limiting cases may provide effective control over the distribution of internal parameters of the glow discharge.

The first case is the case of extremely low pressures. In this case the supersonic flow organization can provide necessary conditions for ignition and maintaining the normal glow discharge near the anode. It is worth noting that at excessively low pressures self-sustaining discharge condition is not performed without the organization of supersonic flow.

The second case is the case of medium pressures. In this case, supersonic flow may concentrate discharge in a region where the flow crosses discharge.

2. Experiments

Experiment was conducted in a vacuum chamber where supersonic gas flow was organized perpendicular to the electric field. The flow occupied approximately 10% of the interelectrode space. The air was used as a working gas. Supersonic gas flow was created in the Laval nozzle with a critical section an area of 0.25 mm^2. From the chamber, the gas was pumped out through the confuser, which was located opposite the Laval nozzle. The pump power provided the stationarity of the gas dynamic characteristics in the vacuum chamber. One of the electrodes had the disk form and the second was made in the form of a cylinder. Different electrode polarity variants were investigated.

2.1. Pushing the discharge out of the flow area

Figures 1 and 2 show the photos of the glow discharge at middle pressure.

Figure 1 shows the glow discharge in a transverse supersonic airflow. The pressure in the chamber was generated by air at a pressure 18 torr. The distance between the electrodes was 5 sm. The flow velocity was estimated as 506 m/s (M=1,5). We used a mean-square velocity formula for air molecules at 298 K. The photograph shows the flow in the positive column. At the intersection of the flow and discharge, visible light is not radiated, it seems as if the supersonic flow cuts the positive column.

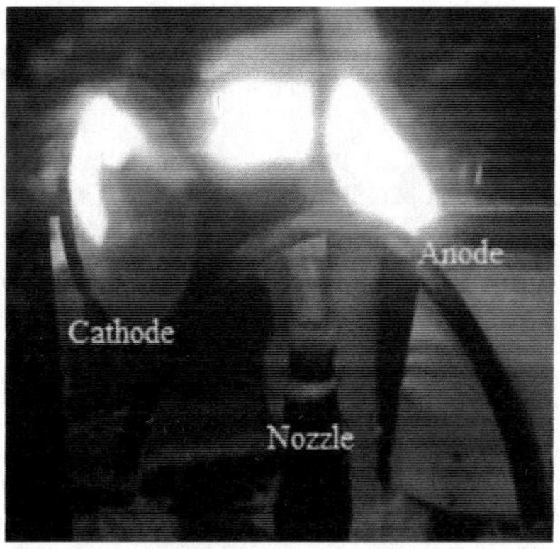

Figure 1. The photo of glow of the glow discharge plasma at the pressure p = 18 Torr in supersonic flow. Interelectrode space is 5 cm.

A similar effect is shown in Figure 2 (b). Just as in the figure 1, visible light is not radiated in the supersonic flow region.

Thus, it becomes possible to create a zone, where the mean free path of electrons differs from the mean free path of electrons in positive column without supersonic flow. In this zone, electrons do not gain energy necessary for excitation of atoms.

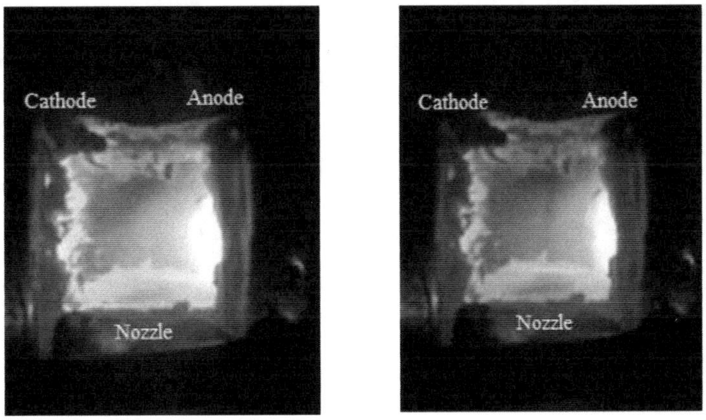

(a) (b)

Figure 2. The photo of glow of the glow discharge plasma at the pressure p = 20 Torr in stationary gas (a) and in supersonic flow (b). Interelectrode space is 5 cm.

2.2. The discharge is concentrated in the flow region

Figures 3 (a) and 4 (a) show the photos of glow discharge at pressure of 2.3 Torr. The entire interelectrode space consists of cathode regions without the positive column. Figures 3 (b) and 4 (b) show photographs of glow discharge in transverse supersonic airflow. We see a brightly luminous volume of plasma, where the flow acts on the discharge. On either side of the luminous volume of plasma visible light is not radiated. In this way, with the help of supersonic flow, the region can be created, where active excitation of neutral particles occurs.

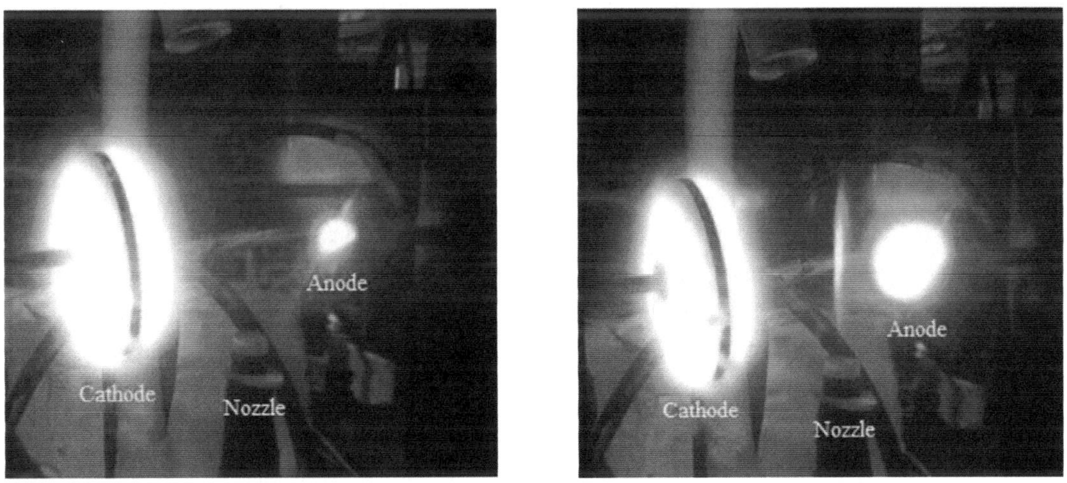

(a) (b)

Figure 3. The photo of the glow discharge plasma at the pressure p = 2,3 Torr in stationary gas (a) and in supersonic flow (b). Interelectrode space is 5 cm.

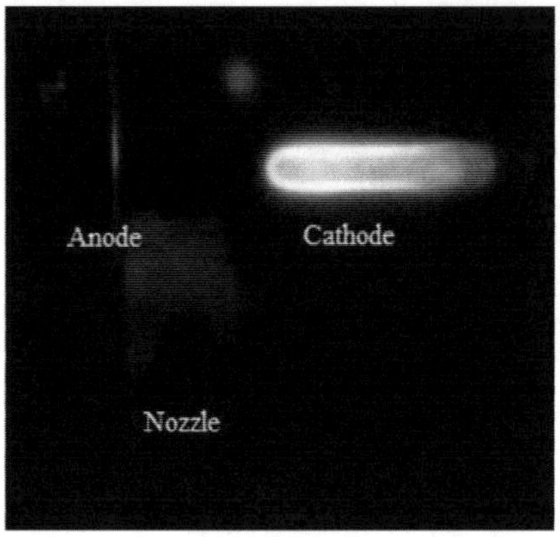

(a)

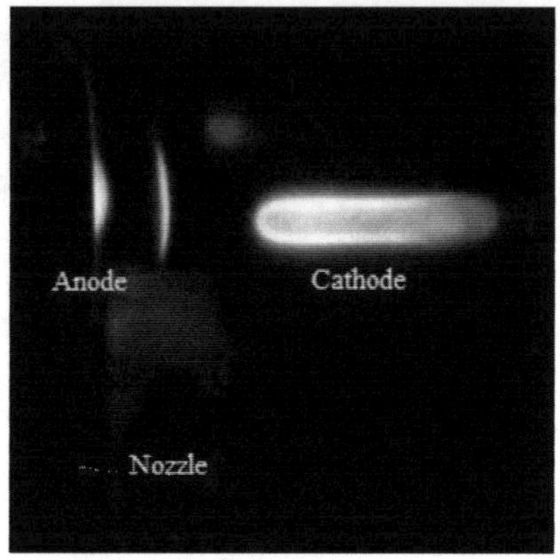

(b)

Figure 4. The photo of glow of the glow discharge plasma at the pressure p = 2,3 Torr in stationary gas (a) and in supersonic flow (b). Interelectrode spacing is 5 cm.

3. Theory

Figure 5 shows the results of a numerical modeling of the glow discharge in the supersonic gas flow. One-dimensional hybrid glow discharge model was based on the balance equations for the concentrations of electrons, as well as the positive and negative ions, on the equation of the electron thermal balance where we account for not only the volume processes but also the thermal conductive spatial transfer, and on the Poisson equation [10]. Simulation of supersonic flow was carried out by tenfold increasing of concentration of neutral atoms over an interval of 2 to 3 cm.

Figures 5 (a) and 5 (b) show distribution of electron concentration of electrons, positive and negative ions. Figure 5 (a) shows cathode discharge zones, the cathode sheath (CS), the negative glow (NG) zone, and the Faraday dark space (FDS). Then the anode sheath (AS) follows, whereas the positive column (PC) is absent. After increasing the concentration of neutral atoms, the structure of discharge changes. Figure 5 (b) shows the reduction of cathode regions to a sharp border, which coincides with the flow boundary. In addition, the PC occurs; the first half is located within the section with the supersonic flow and the second one comes after this section. We obtained similar results in the laboratory experiments [7].

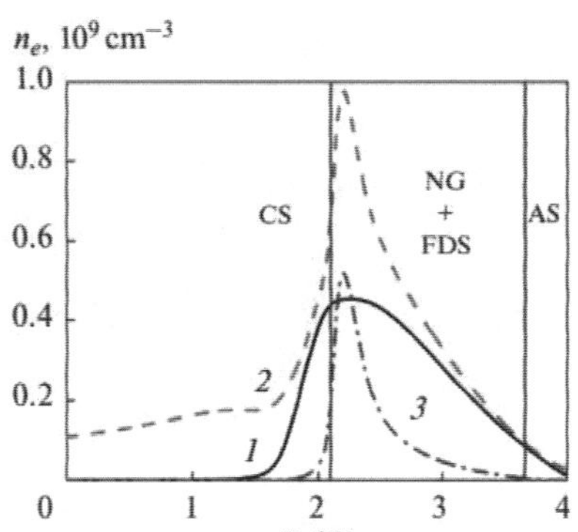

(a)

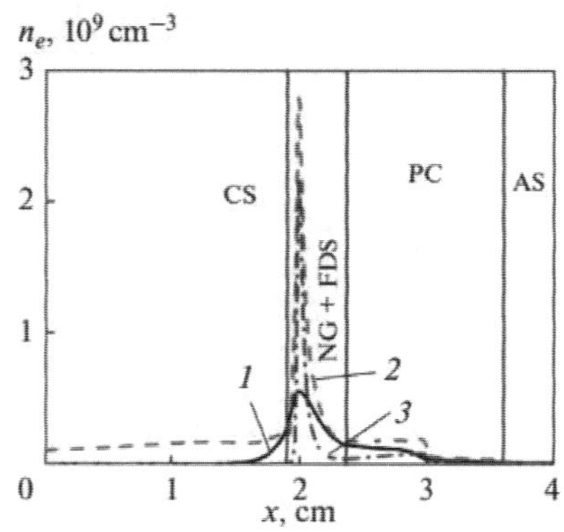

(b)

Figure 5. Parameter distribution of the glow discharge plasma at the pressure p = 0.15 Torr with the interelectrode spacing 4 cm without (a) and with (b) supersonic air flow: concentrations of electrons - 1, positive - 2 and negative - 3 ions [10].

4. Conclusions

In the discharge gap the non-uniform distribution of neutral particles directly affects the distribution of electrical discharge parameters and the distribution of charged particles in the interelectrode space. It is shown that with the help of supersonic gas flow we can control the localization of cathode zones. Supersonic flow in a limited region of the interelectrode space solves the problem of controlling parameters in a discharge chamber and opens up new possibilities for using of a glow discharge.

References

[1] Galeev I G and Asadullin T Ya 2016 *Journal of Physics: Conference Series* **669** 012015.
[2] Galeev I G and Asadullin T Ya 2017 *Journal of Physics: Conference Series* **789** 012003.
[3] Timerkaev B, Andreeva A, Sofronitskiy A 2017 *Journal of Physics: Conference Series* **789** 012063.
[4] Timerkaev B, Sofronitskiy A and Andreeva A 2016 *Journal of Physics: Conference Series* **669** 012062.
[5] Galeev I, Goncharov V, Timerkaev B 1990 *High Temperature* **28** 843.
[6] Galeev I, Goncharov V, Timerkaev B 1992 *High Temperature* **30** 439.
[7] Timerkaev B, Zalyaliev B 2014 *High Temperature* **52** 489.
[8] Timerkaev B, Zalyaliev B, Saifutdinov A 2014 *Journal of Physics: Conference Series* **567** 012032.
[9] Saifutdinov A, Timerkaev B, Zalyaliev B 2014 *Journal of Physics: Conference Series* **567** 0120313.
[10] Saifutdinov A, Timerkaev B, Zalyaliev B 2016 *High Temperature* **54** 669.

Acknowledgments

The work was fulfilled with the assistance of a grant of the Ministry of Education and Science of the Russian Federation №3.6564.2017/8.9.

The authors are grateful to Senior Lecturer of the Department for Foreign Languages of the Kazan National Research Technical University named after A. N. Tupolev – KAI Lidiya Urmanova for valuable comments of the translation of this paper.

The role of the ambipolar field in the formation of the EDF and the criteria of the local approximation

Chengxun Yuan[1], K M Rabadanov[2], E A Bogdanov[2], A A Kudryavtsev[1,2], Zhongxiang Zhou[1]

[1]Department of Physics, Harbin Institute of Technology, Harbin 150001, China

[2]Saint Petersburg State University, St. Petersburg, 199034, Russia

E-mail: akud@ak2138.spb.edu

Abstract. It is shown that in the situation when the ambipolar field exceeds the heating field the local approximation (LA) for determining the EDF is not applicable even if the energy relaxation length of the electron is relatively small. Therefore, accurate results can be obtained only solving the kinetic equation in energy and spatial variables.

1. Introduction

The electron distribution function is usually calculated using the local approximation. In this case the solution of the Boltzmann kinetic equation has a simplified form. The equation depends only on one variable (velocity v or kinetic energy $w = mv^2/2$) and can be written as

$$-\frac{1}{\sqrt{w}}\frac{\partial}{\partial w}\left(\sqrt{w}(eE)^2 D\frac{\partial f_0}{\partial w}\right) = St(f_0) + St^*(f_0) \tag{1}$$

In eq.(1) E – is the electric field at a given point in space, $D = v\lambda/3$ - is the electron diffusion coefficient, $St(f_0)$ and $St^*(f_0)$ - the integrals of elastic and inelastic collisions.

Criteria of applicability of local approximation in the literature are usually produced in a standard way using a condition of smallness of spatial gradients in comparison with energy derivatives in the kinetic equation. The criteria is governed by the relation between the characteristic plasma dimension R and the energy relaxation length λ_ε [1]

$$\lambda_\varepsilon = \sqrt{2D_r\tau_\varepsilon}, \tag{2}$$

where $D_r = V\lambda/3$ is the electron diffusion coefficient and the energy relaxation time has the following form

$$\tau_\varepsilon^{-1} = \delta\nu + \nu^* \tag{3}$$

and determined by the energy losses due to elastic and inelastic collisions (the corresponding frequencies ν and ν^*), $\delta << 1$ is the energy transfer factor for elastic (and quasi-elastic) collisions.

It is suggested that that if $R > \lambda_\varepsilon$, then we can use the local approximation (1) in the calculations. As shown by the results [2-4], the most disputable is the question of the electric field, which enters the initial and the local kinetic equations. The electron reacts to the total field that exists at a given point in space, the kinetic equation includes the total electric field at a given location. As a rule, for numerical simulations by different computational codes [5, 6], the kinetic equation is substituted for the total field, which is found from the solution of the Poisson equation.

In contrast, in [7] it was noted the ambipolar field is also determining by spatial gradients. And when we use the local approximation in the local kinetic equation (1), the ambipolar electric field should be neglected. In other words, the ambipolar field should be removed (subtracted) from the total field when it is substituted into (1).

The question of which field should be substituted into the local kinetic equation (complete or minus ambipolar) is controversial, to which there is still no unequivocal opinion in the literature.

In this paper we present a continuation of the studies [2-4]. It is shown that at the periphery of the plasma, where the ambipolar field is dominant, the local approximation for determining the EDF and other characteristics of the electron gas is inapplicable. In condition when the $R > \lambda_\varepsilon$ reliable results can be obtained only when solving the kinetic equation in both energy and spatial variables.

2. Results and discussion

The object of the research was a positive column (PC) of a DC discharge in argon in a tube of radius R = 1 cm in a wide range of pressures.

For carrying out the numerical experiments, a computational model was developed that includes a number of related modules in COMSOL Multiphysics [2-4]: a) the kinetic module for solving the 2D Boltzmann equation in coordinate variables and kinetic energy *(r, w)*; b) a diffusion-drift module that includes equations for finding the densities of all neutral and charged particles; c) Poisson equation for calculating the field and potential profiles. A set of plasmachemical reactions in argon includes direct ionization, excitation from the ground state, radiative excitation, stepwise ionization from the metastable state, and Penning ionization.

For example, fig. 1 shows normalized EDFs at different points of the radius for the average pressures of 12 Torr, when, by criterion (2.3), there must certainly be a local regime for the EDF ($\lambda_\varepsilon > R$).

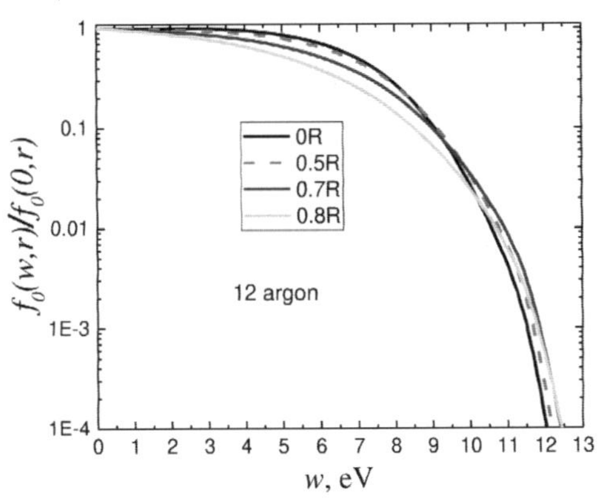

Figure 1. Comparison of EDF at different values of the radius for argon *p = 12 Torr. I = 3 mA*

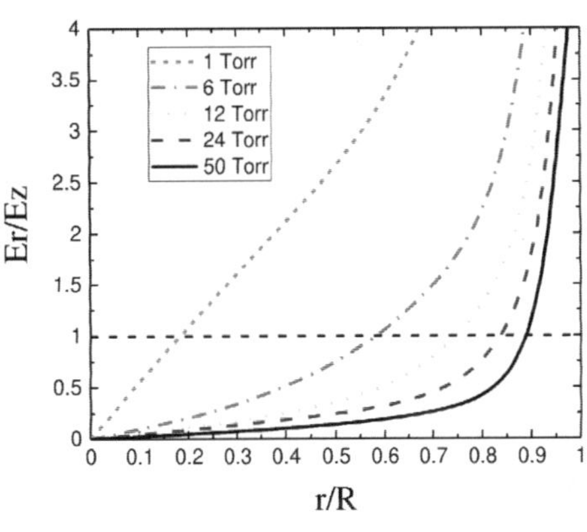

Figure 2. Axial (heating) and transverse (ambipolar) fields in argon at different pressures

It can be seen that the EDF at different points of the radius do not coincide, i.e. depend not only on energy, but also on the spatial variable. It means that the EDF is nonlocal (meaning that when calculating the definition it is necessary to use the complete kinetic equation in the variables both the energies and the coordinates).

As preliminary studies have shown [2-4], the main reason for the appearance of the radial dependence of the EDF is the contribution of the ambipolar field, which, as is known [7], increases from the center to the periphery of the discharge.

Figure 2 shows the values of the ratio of ambipolar and heating fields at different pressures. It is seen that with increasing pressure (already from several Torr and higher), i.e. when the energy relaxation length of the electron (2) is small and $R > \lambda_{\varepsilon}$, the ambipolar field (sooner or later) begins to exceed the longitudinal field, at the periphery of the PC.

In order to determine which field should be substituted into the local kinetic equation, total or only heating field (as recommended in [1, 7]), were performed calculations of the local Boltzmann equation (1) with total and with only heating field (see Fig. 3.4).

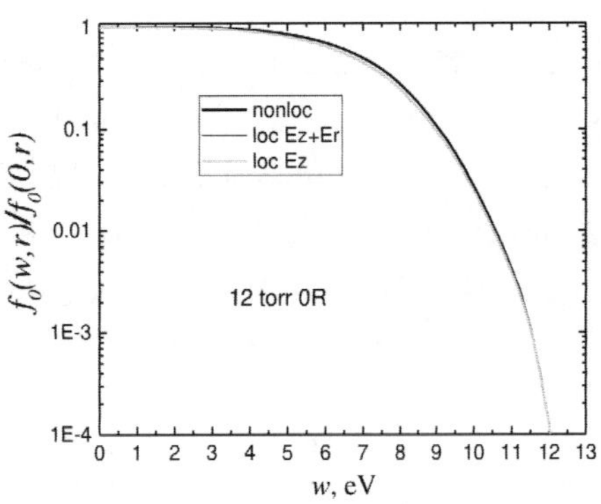

a

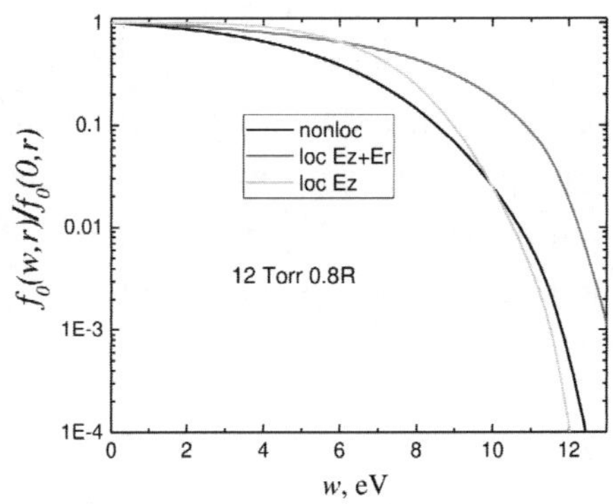

b

Figure 3 (a, b). Comparison of the EDF with the nonlocal and local approximation for argon at p = 12 Torr, I = 3 mA. **a** - on the tube axis r = 0. **b** - r = 0.8R

It can be seen that on the axis (in the near-axis regions), where the ambipolar field E_a is small in comparison with the longitudinal E_z (see Fig. 3), the curve of the local EDF calculation are in good agreement with the complete solution of the kinetic equation. At the same time, local calculations with the total field $E^2 = E_a^2 + E_z^2$ practically coincide with the solution of the complete kinetic equation at the periphery of the discharge, where the ambipolar field dominates. In this case, large discrepancies occur when substituting in the local kinetic equation (1) only one longitudinal field E_z^2 (as recommended in [1, 7]).

For the argon, the elastic collision cross sections are growing with energy. In this case, as the energy increases, the friction ("resistance") of the electrons increases with their energy diffusion in the field. Therefore, it is of interest to consider the case in which the elastic scattering cross sections are incident. Figures 4 (a, b) show corresponding results, similar to fig. 3 but using the model cross section of elastic collisions of the type $\sigma = 1/w$.

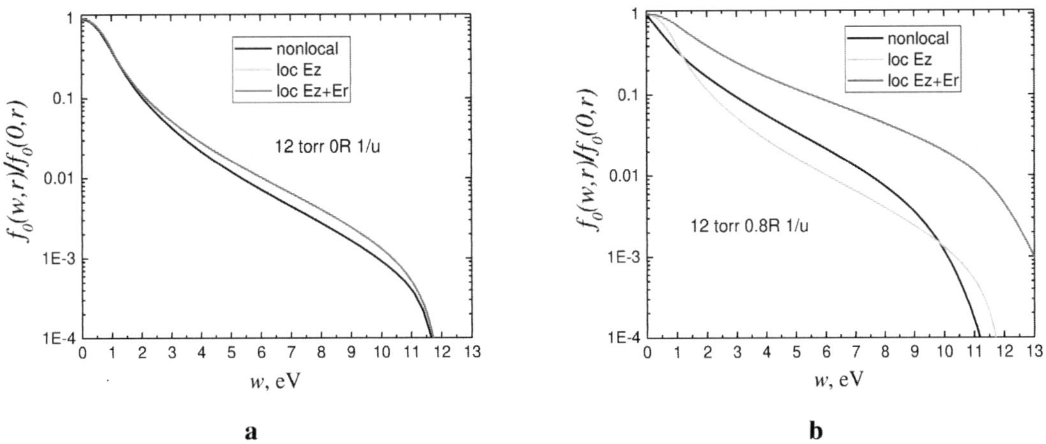

Figure 4 (a, b). A comparison of the EDF with a nonlocal and local approximation for a cross section of the type at 12 Torr (0r). The current is 3 mA. a) - on the tube axis r = 0, b) - r = 0.8R

From this comparison, it is seen that the opposite situation is observed in fig. 3: the calculations of local EDF in accordance with (1) and with the substitution of the longitudinal field (as recommended in [1, 7]) correspond to the results of modeling, while with a full field give large differences.

3. Conclusions

Consequently on the periphery of the plasma, where the ambipolar electric field exceeds the heating electric field, even if the condition $R > \lambda_\varepsilon$ is satisfied, one can't use the local approximation (1) to determine the EDF with the substitution of both the total field at a given point in space and only the heating field (minus the ambipolar field). Moreover, depending on the behavior of the cross section for the elastic scattering of electrons from energy (increasing or falling), errors occur both in the ambipolar field in the local kinetic equation (1) and in the use of the total field there. Therefore, reliable results can be obtained only when solving the kinetic equation in both energy and spatial variables.

Acknowledgements
This work was supported in the frame of the SPbGU 11.37.212.2016.

References

[1] U. Kortshagen, C. Busch, and L. D. Tsendin, Plasma Sources Sci.Technol. 5, 1 (1996).
[2] Kirill D. Kapustin, Anatoly A. Kudryavtsev. M.V. Krasilnikov. J. Phys. of Plasmas, 2014. Phys. Plasmas 21, 120701(1-4), (2014);
[3] Kapustin, M. B. Krasil'nikov, A. A. Kudryavtsev. Technical Physics Letters, 2015, Vol. 41, No. 1, pp. 43–45.
[4] E. A.Bogdanov, A.A. Kudryavtsev. Technical Physics, 2009, Vol. 54, No. 6, pp. 810–819.
[5] https://www.comsol.com/
[6] https://www.esi-group.com/
[7] M. Lieberman and A. Lichtenberg, Principles of Plasma Discharges and Materials Processing (Wiley, NY, 2005).

IOP Conf. Series: Journal of Physics: Conf. Series **927** (2017) 012081 doi:10.1088/1742-6596/927/1/012081

Photoionization of hydrogen molecular ion by ultrashort photo-pulse in a wide range of field magnitudes

D N Arkhipov, S A Astashkevich, A A Mityureva and V V Smirnov*

Faculty of Physics, St. Petersburg State University, 3 Ulianovskaya Street, Petrodvorets, St. Petersburg, 198504, Russia

*E-mail: valery_smirnov@mail.ru

Abstract. The probabilities of photoionization of hydrogen molecular ion by ultrashort laser field were evaluated on base of earlier developed trajectory method. The results are obtained in a wide range of field magnitudes up to superatomic fields approaching relativistic intensities. In a weak field gained results are in good agreement with the data obtained by other methods. In a strong field the photoionization probability behaviour resembles some features of the so called stabilization effect in atoms.

1. Introduction

The process of photoionization of atoms and molecules is of great importance in various plasma media, for example, in astrophysical plasma, where hydrogen is of especial interest ([1]; also see bibliography in [2]). It has been studied for a long time, however the interest to this process resumed in connection with development of ultrafast strong field radiation sources. There are several theoretical approaches to address this problem [3]. In this respect the trajectory based method for calculating transition probabilities in a quantum system developed in our works [4–9] has many advantages. It can be treated as the variant of semiclassical approximation in Weyl representation. The method is nonperturbative and takes into account all orders of multiphoton and recollision processes. It takes into account electron's correlations for multielectron systems. The photoionization probabilities can be calculated in a wide range of field frequencies and intensities. It makes possible to evaluate photoionization probability directly for entire ionization continuum without integration over probabilities for each state in continuum. Computation of each continuum state (especially for multielectron system) may be rather complicated problem. At the same time total ionization probability is often required in practice. The possession of totality of these capabilities is unique compared to the other methods.

2. Method

The transition probability w is calculated as follows (atomic units are used throughout the paper if not stated otherwise)

$$w(t',t)=\frac{1}{V}\int W_f(\zeta_l)\,\left|exp(i\,S(\zeta))-K(t',t)\right|^2 W_i(\zeta_0)\,d\zeta_0\,, \quad (1)$$

$$K(t',t)=\frac{1}{V}\int exp(i\,S(\zeta))W_i(\zeta_0)\,d\zeta_0\,, \quad (2)$$

where $\nu = (2\pi)^n$ is the quantum unit volume of the phase space, $2n$ is the phase space dimension, t' and t are the initial and final moments in time, respectively, W_i and W_f are the Weyl's symbols of initial and final states, respectively, $S(\zeta) = -\int_{t'}^{t} V\,dt$ is the integral of the interaction V (the phase function) taken along stationary phase trajectory $\zeta = (q, p)$ from the initial point $\zeta_0 = (q_0, p_0)$ to the final one $\zeta_l = (q_l, p_l)$. These trajectories are solutions of the Hamilton's equations with initial phase space coordinate ζ_0. The parameter $K(t', t)$ (2) excludes contribution of the system evolution without interaction.

In this paper we apply this method in calculating the probabilities of hydrogen molecular ion photoionization under the exposure of ultrashort laser radiation. The results are represented for equilibrium internuclear distance ($R = 2$).

The Weyl symbol W_i of the ground electronic state in hydrogen molecular ion was evaluated using the wave functions taken from the works [10, 11].

We consider transitions into all inelastic channels and into entire continuum. Final states can be represented by means of the projectors on corresponding groups of states. Note that the projector on the whole set of all eigenstates of Hamiltonian $\hat{H}_0$ for the system without interaction is equal to the identity operator due to the completeness of the set. The Weyl's symbol of identity operator is equal to one

$$W_f(\zeta) = 1.$$

The usage of this symbol in (1) yields the total probability of all inelastic transitions since the transition into initial state is excluded by means of parameter (2). The projector into entire continuum can be represented by subtracting from the identity operator the projector to all states of the discrete spectrum. So it is reasonable to introduce the following model of the final distribution excluding the energy gap containing discrete levels which in boolean notation reads as

$$W_f(\zeta) = H_0(\zeta) < \varepsilon_0 \vee \varepsilon_1 < H_0(\zeta),$$

where $\varepsilon_0 = -0.67$, $\varepsilon_1 = 0$ are the energies of the first excited state of hydrogen molecular ion and its ion respectively. Such model distribution was tested in our works [4–6] and had given reliable results.

Interaction of a system (of electrons) with an electromagnetic field in velocity gauge can be written up to the second relativistic order as

$$V(\boldsymbol{r}, \boldsymbol{p}, t) = \frac{e}{mc}\boldsymbol{p} \cdot \boldsymbol{A}(\boldsymbol{r}, t) - \frac{e^2}{2mc^2} A^2(\boldsymbol{r}, t),$$

where $\boldsymbol{A}$ is a vector potential and $\boldsymbol{r}$ and $\boldsymbol{p}$ are coordinate and kinetic momentum of electron. We consider a plane-wave photo-pulse with linear polarization and a Gaussian modulation.

3. Results

The result in a weak field is usually represented in a form of cross section. Gained hydrogen molecular ion photoionization cross section averaged over molecular orientations is shown in figure 1.

The calculations are presented for the photon-pulse duration of $\tau = 50$ (~ 1.2 fs). There are several works with calculations of ionization cross section for a stationary radiation with agreement of results. For example, in figure 1 our results are compared with the data of [12, 13]. Our calculations are in good agreement with the data obtained by other methods especially taking into account the simplicity of our method.

Figure 2 shows photoionization probability dependence on field magnitude. The magnitude of the field is represented by the dimensionless Faisal parameter $\chi = \frac{eA}{p_0 c}$ (the ratio of the addition to electron's kinetic momentum in a field $\frac{eA}{c}$ to the atomic scale of electron momentum $p_0 = \frac{\hbar}{a_0}$).

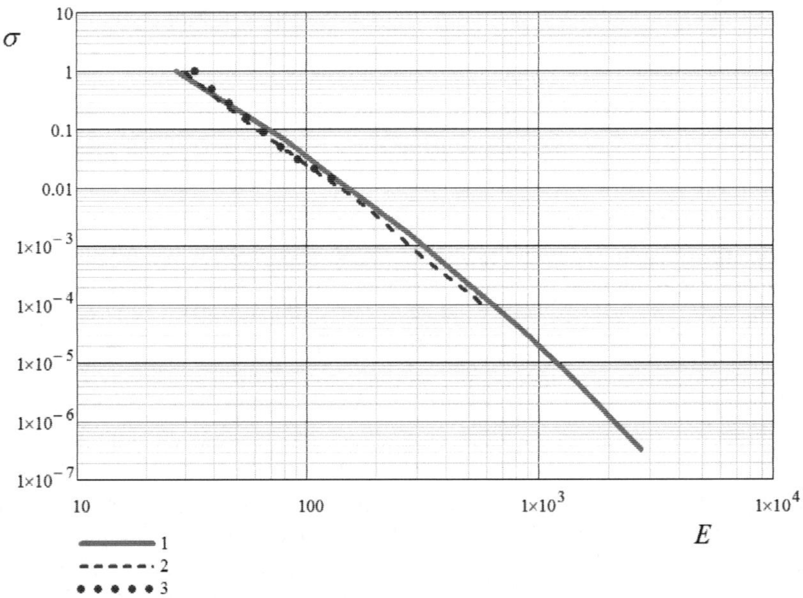

Figure 1. H_2^+ photoionization cross section σ(Mb) dependence on photon energy E (eV). Line 1 – our result, line 2 – [12], line 3 – [13].

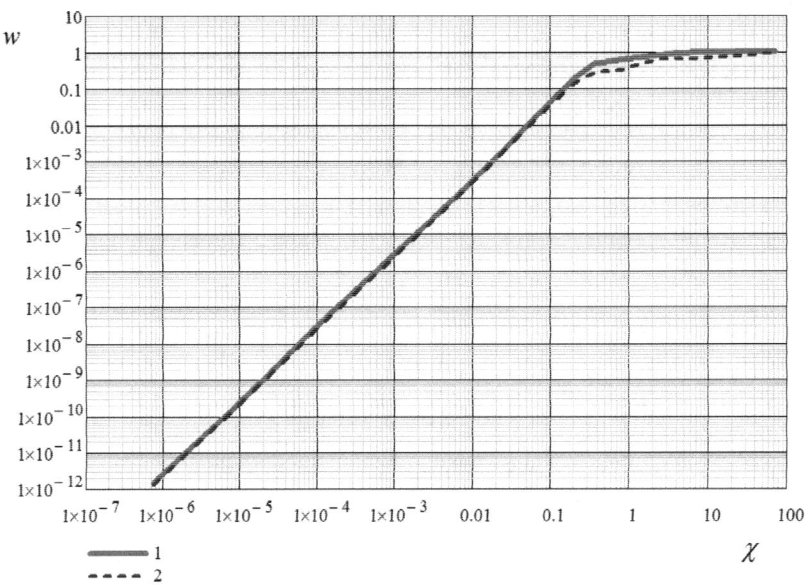

Figure 2. H_2^+ transition probabilities w dependence on Faisal parameter χ: into all inelastic channels (line 1) and into entire continuum (line 2).

If $\omega\tau >> 1$ then $\chi \sim 1/\gamma$, where γ is the Keldysh adiabatic parameter. The rang of Faisal parameters $\chi << 1$ corresponds to a weak field, $\chi \sim 1$ corresponds to an intermediate field, $1 < \chi < c$ corresponds to a strong superatomic field, but with lower than relativistic intensities, and $c < \chi$ – to a field with relativistic intensity (c = 137).

The calculations are presented for the frequency of the radiation field of $\omega = 1.1$ (~ 30 eV) at ionization threshold, the photon-pulse duration of $\tau = 50$ (~ 1.2 fs) and the wave propagating along molecular axis.

In figure 2 we see how the linear dependence of the probability on the intensity χ^2 in a weak field turns into a nonlinear one with a Faisal parameter greater than 0.1. In the field magnitude region $0.3 < \chi < 10$ the photoionization probability behaviour resembles some features of the so called stabilization effect in atoms [14].

4. Conclusions

The probabilities of photoionization of hydrogen molecular ion by ultrashort laser field are obtained in a wide range of field frequencies and magnitudes up to superatomic fields approaching relativistic intensities. It has been found that in a weak field gained results are in good agreement with the data obtained by other methods and in a strong field its behaviour resembles some features of the so called stabilization effect in atoms.

In addition to the studies carried out, it is planned to investigate the dynamics of photoionization.

References

[1] Pankaj Jain 2016 *An Introduction to Astronomy and Astrophysics* (CRC Press)
[2] Astashkevich S A and Lavrov B P 2015 *J. Phys. Chem. Ref. Data* **44** 023105
[3] Popruzhenko S V 2014 *J. Phys. B: At. Mol. Opt. Phys.* **47** 204001
[4] Bichkov A B, Kozhina A S and Smirnov V V 2016 *Phys. Lett. A* **380** 1263
[5] Bichkov A B and Smirnov V V 2015 *Europ. Phys. J. D* **69** 190
[6] Bichkov A B and Smirnov V V 2013 *Laser Phys.* **23** 055302
[7] Bichkov A B and Smirnov V V 2013 *J. Phys. A: Math. and Theor.* **46** 015303
[8] Bichkov A B, Mityureva A A and Smirnov V V 2011 *J. Phys. B: At. Mol. Opt. Phys.* **44** 135601
[9] Kozhina A S and Smirnov V V 2017 *Russian J. Phys. Chem. B* **11** 12
[10] Coulson C A 1941 *Proc. Cambridge Philos. Soc.* **37** 55
[11] Wind H 1965 *J. Chem. Phys.* **42** 2371
[12] Richards J A and Larkins F P 1986 *J. Phys. B: At. Mol. Phys.* **19** 1945
[13] Colgan J, Pindzola M S and Robicheaux F 2003 *Phys. Rev. A.* **68** 063413
[14] Emelin M Yu and Ryabikin M Yu 2014 *Phys. Rev. A* **89** 013418

IOP Conf. Series: Journal of Physics: Conf. Series **927** (2017) 012082 doi:10.1088/1742-6596/927/1/012082

Formation of high intensity ion beams with ballistic focusing

Koval T.V.[1)], Ryabchikov A.I.[2)], Shevelev A.E.[3)], Tran My Kim An[4)] , Tarakanov V. P.[5)]

1 Professor, the Department of Software Engineering,the Institute of Cybernetics, National Research Tomsk Polytechnic University,Tomsk, Russia
[2]Head of Laboratory, the Institute of Physics and Technology,National Research Tomsk Polytechnic University,Tomsk, Russia
[3]Graduate student, the Institute of Physics and Technology, National Research Tomsk Polytechnic University,Tomsk, Russia
[4]Graduate student, the Institute of Cybernetics, National Research Tomsk Polytechnic University,Tomsk, Russia
[5]Lead researcher, the Joint Institute of High Temperatures of RAS, Moscow, Russia

E-mail: [1]tvkoval@mail.ru, [2]ralex@tpu.ru

Abstract. This investigation presents the results of experimental investigation and theoretical simulations of the influence of plasma and negative bias parameters on formation, transportation and focusing of high intensity ion beams of titanium and nitrogen (with an ion current density up to 1 A/cm^2 and pulsed power density up to 2.6 kW/cm^2). It was shown that the conditions of space charge neutralization of the focusing beam have a significant influence on the distribution and magnitude of the ion current at the collector.

1. Introduction

Ionimplantation provides an important tool for the surface modification of a wide range of materials for a broad array of applications in fundamental research and applied technology. Typically, ion implantation involves bombarding the sample surface with beams of metal or gaseous ions with energy in the 1 – 100 keV range and at current density in the 1 – 100 μA/cm2 range to achieve modification of the ion-bombarded material surface to a depth of up to about 1 μm. Traditionally the high ion energy beam is provided by an ion source with high extraction voltage or with post-acceleration of the extracted beam. In recent years an alternative approach called plasma immersion ion implantation (piii or pi3) has evolved [1,2] in which the ions are accelerated across the high voltage plasma sheath generated by a repetitively-pulsed or DC negative bias applied to the substrate immersed in the plasma.

In many ion implantation applications there are aneed for high implantation dose, in turn calling for high current density of metal and gaseous ions. Metal ion beams can be generated by vacuum arc ion sources. The vacuum arc discharge, however, as well as providing a copious flux of metal plasma also generates a significant component of cathode debris – solid, initially molten, particulates called "macroparticles" – that can lead to contamination of the implanted substrate. Here we describe a novel approach to the formation of high intensity beams of gaseous and metal ions with extremely high current density with the use of gas discharge plasma source with hot cathode, as well as metal plasma source based on DC vacuum arc discharge accompanied with the originally developed system, which excludes accumulation of macroparticles in the area of ion beam interaction with a target. The method of the formation of high intensity ion beam, investigated in Ref. [3,4] and based on plasma immersion extraction and acceleration of ions and their subsequent ballistic focusing. The system is the combination of the grid electrode, made in the form of hemisphere, and a cylindrical section, providing a field-free drift space for ballistic focusing and transportation of the ion beam formed by the grid. The hemispherical grid, in the certain conditions, can provide ballistic focusing of the beam, however, the efficiency of focusing and transportation is substantially depends on neutralization of the beam space charge in the drift space.

This work is devoted on experimental investigations and theoretical simulations of the influence of plasma and negative bias parameters on formation, transportation and focusing of high intensity ion beams of titanium and nitrogen.

2. Experimental results

A DC vacuum arc discharge is used to form dense metal plasma. We used a water-cooled titanium cathode located within a modest axial magnetic field (about 60 G); the arc discharge current was 160 A. A hemispherical metal grid is positioned on-axis some 40 cm from the cathode. Nitrogen plasma was generated with the use of arc discharge source with hot cathode PINK [5], the discharge current was 20 A. The hemisphere grid with radius of 7.5 cm is mounted on and electrically attached to cylindrical section, providing equipotential drift space for ballistic focusing and transportation of the ion beam formed by the grid. A current collector plate is positioned at the geometric focus of the system and is connected to a repetitively pulsed negative bias generator with amplitude varied from 1.2 kV to 2.6 kV; pulse duration from 1 to 8 μs, with the fixed pulse repetition rate of 10^5p.p.s. A Rogowski coil is used to monitor the collector current. Macroparticles in the plasma flux incident upon the hemispherical grid are blocked from viewing the sample holder by a metal plate mounted on-axis; we used a disk of radius 2 cm.

Typical oscillograms of the ion current of titanium and nitrogen beams with the amplitude of negative bias of - 2.6 kV, pulse durations of 2, 4, 6 and 8 μs are presented in Fig. 1a,b. The oscillograms of titanium ion beam current (Fig. 1a) shows a significant influence of pulse duration on the formation of ion beam. With the pulse duration of 2 and 4 μs, an ion beam current reaches up to 0.8 A. It should be noted that the amplitude of the spike at the beginning of a pulse is conditioned by inductive and capacitive characteristics of Rogowski coil and has no relations with the measured ion current. The delay of the ion current pulse, in comparison with a bias pulse, is related to the time of sheath formation near the grid electrode and ion time-of-flight through the drift space. The increase in bias pulse duration up to 6 μs leads both to the manifold decrease of an ion current amplitude at the beginning of the pulse and to the delay in the ion beam formation (Fig. 1a).

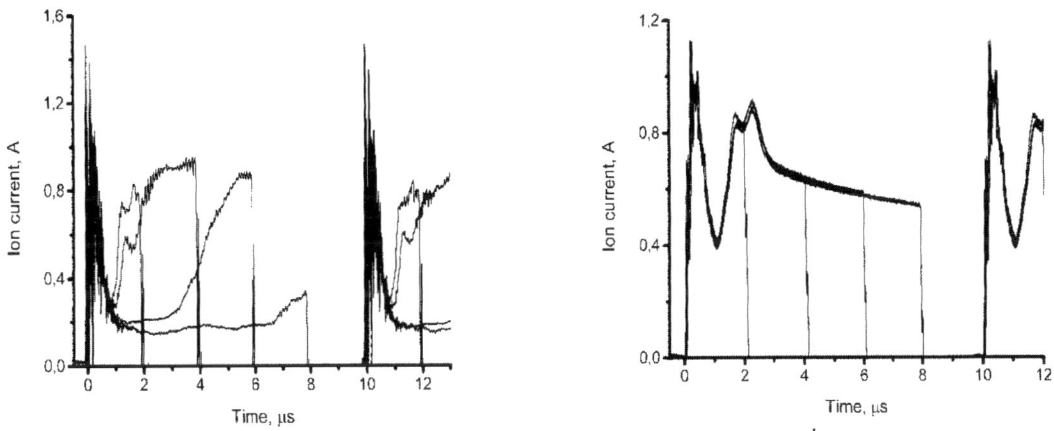

a b

Figure 1. Typical oscillograms of titanium (a) and nitrogen (b) ion beam currents

The distribution of titanium ion current at bias pulse duration of 4 μs and bias pulse frequency of 100 kHz was defined with the use of thermographic camera by analyzing the temperature changes of rapidly heated thin tungsten foil and presented in Fig. 2. The obtained distribution reveals the substantial influence of the beam space charge on the efficiency of beam focusing.

In case of a plasma flow, formed by the PINK source, the high stability of the nitrogen ion current pulses and the similarities in their shape at different negative pulse durations were observed (Fig. 1b). With the increase of chamber gas pressure up to 0.4 Pa, the ion beam is forming and focusing even at pulse duration of 8 μs.

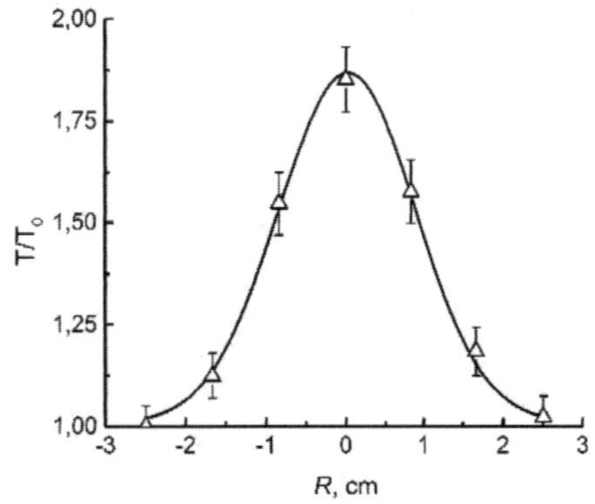

Figure 2. The relative distribution of the temperature of tungsten target after its irradiation by titanium ions

3. Simulation results

If the pulse repetition rate is large, the off-bias time can be small enough to fill equipotential drift space with plasma. The propagation velocity of vacuum arc plasma usually amounts to approximately $1.5 \cdot 10^6$ cm/s and with the off-bias time 2 μs, the plasma flux can penetrate through drift space only 3 cm from the grid electrode. The distribution of plasma in drift space is defined by its velocity and diffusion process tightened with the concentration gradient.

The simulations of the ion beam formation were performed using PIC code KARAT [6].

Near the entrance of the drift space plasma concentration is almost one order of magnitude larger than ion beam concentration; however, in the spherical geometry the density of fully neutralized ion beam is increasing as $1/r^2$ with the distance from the grid electrode to the current collector plate. Thus, the focusing of ion beam is accompanied with the formation of inhomogeneous potential well, which defines the movement of charged particles of plasma (ions and electrons). Fig. 3 demonstrates the

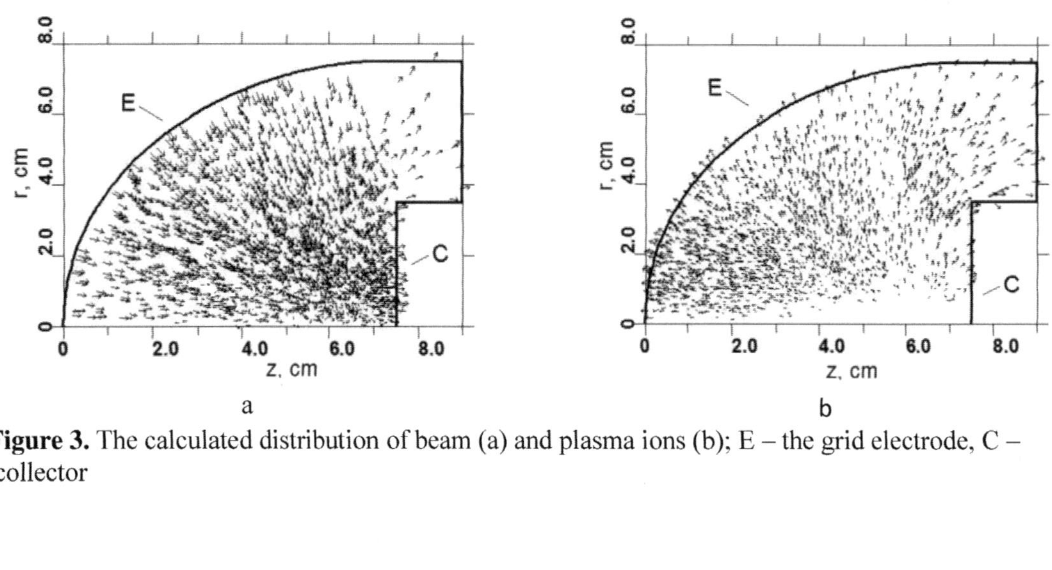

Figure 3. The calculated distribution of beam (a) and plasma ions (b); E – the grid electrode, C – collector

calculated distributions of ions in beam (a) and plasma (b), where the arrows designate the direction of ion velocity. The ion energy at the entrance of the drift space corresponds to the accelerating voltage of 2.6 kV, the plasma density $n = 2{\cdot}10^{10}$ cm^{-3}. With the off-bias time t > 4 μs, focused ion beam is forming. Fig. 4 shows simulation results on the current distribution of such beam.

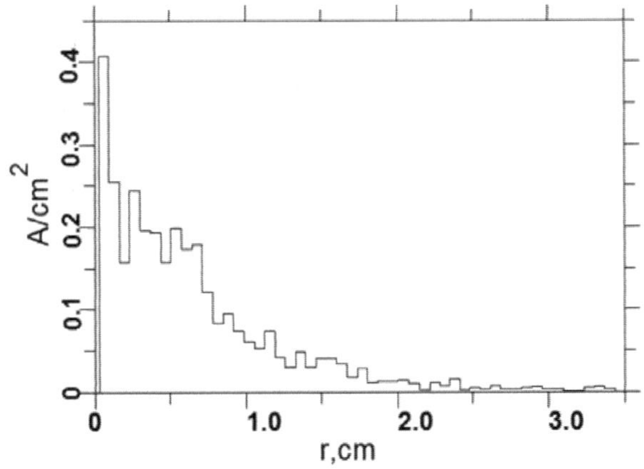

Figure 4. The ion beam current density distribution on the collector

With $t < 4$ μs, the number of plasma electrons in the drift space during the formation of the beam, may be insufficient to neutralize its charge. In this case, the beam formation is accompanied by redistribution of the nonuniform plasma density and the formation of a virtual anode. Fig. 5 shows the dynamics of the number of plasma particles and ions beam (increased 2-fold) in time. Fig. 6 shows the configuration portraits of the beam at different moments, plasma concentration $n = 10^{10}$ cm^{-3}. The part of the reflected ions from the potential well minimum falls on the grid, increasing its current. The other part passes into the sheath, reducing the current density in the ion layer, which leads to fluctuations in the emission plasma boundary and beam current. As the calculations show, the lower the plasma density, the greater the time of the emission of plasma ions, and a correspondingly longer time to form the beam and the less current, that is observed in experiment (Fig. 1).

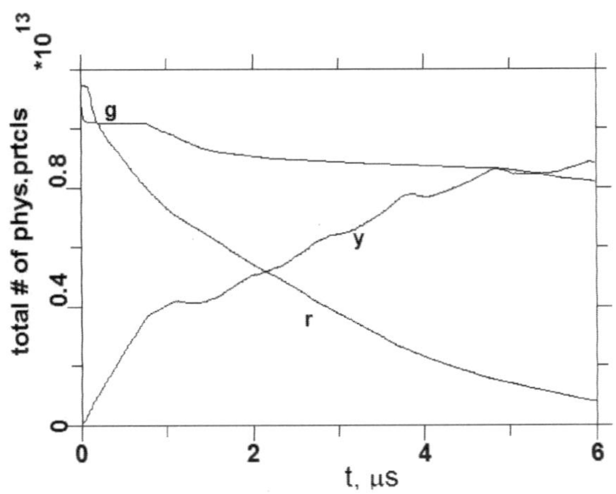

Figure 5. The dynamics of the number of charged particles in the drift space: g and r – plasma electrons and ions, y – beam ions

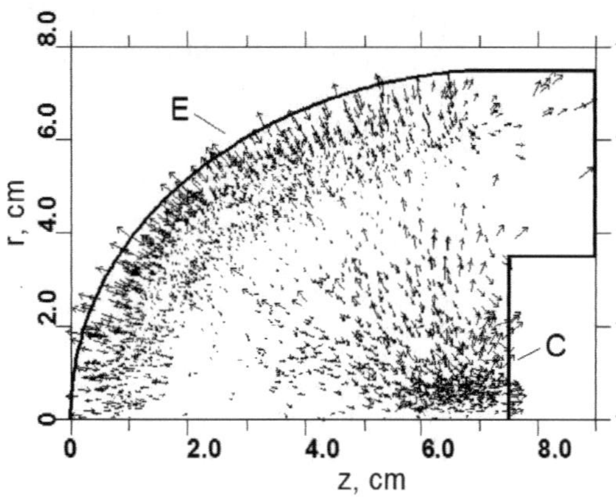

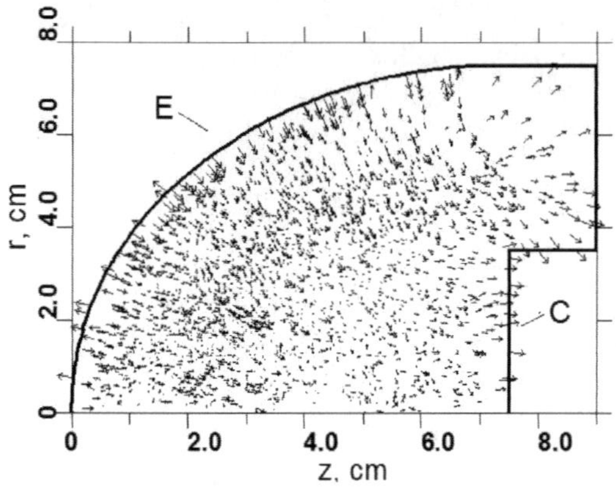

Figure 6. The configuration portraits of the beam at different moments of time: 1.5 μs (a) and 4.5 μs (b); E – the grid electrode, C – collector

In addition, the formation of the beam and the dynamics of the collector current are influenced by the ionization of residual gas by ions, plasma and the secondary electrons captured in the potential well. Secondary electrons formed as a result of ion-electron emission from the collector surface, also compensate the space charge of the transported ion beam.

4. Conclusions

The experimental and theoretical studies have shown that the formation of ion beam of a very high density, and the ion current at the collector are significantly depends on conditions of space charge neutralization of the focusing beam. The beam formation may be accompanied by the appearance of a virtual anode, if the drift space doesn't fill with plasma or residual gas.

The work is performed with financial support of Russian Science Foundation, Grant №17-19-01169.

References

[1] Conrad J.R., Radtke J.L., Dodd R.A. et al. // J. Appl. Phys. 1987. V. 62. P. 4591.
[2] Arbuzov N.M., Vaulin V.A., Isaev G.P. et al. // USSR Patent № 1412517. 1990. №33.
[3] A.I. Ryabchikov, P.S. Ananin, S.V. Dektyarev et al. // Vacuum 2017. V.143. P. 447-453
[4] A.I. Ryabchikov, P.S. Ananin, S.V. Dektyarev et al // Technical Physics Letters, 2017 (in press)
[5] Goncharenko I.I., Grigoriev S.V., Lopatin I.V., Koval N.N. et al. // Surface and Coatings Technology 2003. V. 169-170. P. 419-423.
[6] Tarakanov, V. P. User's manual for code KARAT. Springfield, VA: Berkley Research. 1992.

Generation of acoustic disturbances by the moving charged gas-suspension

D A Tukmakov[1, 2] and N A Tukmakova[2]

[1] Kazan national research technical university named after A N Tupolev-KAI, 10 K. Marx St, Kazan, 420111, Russia
[2] Institute of Mechanics and Engineering, Kazan Science Center, Russian Academy of Sciences (IME KazSC RAS), 2/31 Lobachevsky St, Kazan, 420111, Russia

e-mail: tukmakovda@imm.knc.ru

Abstract. In the work dynamics of the charged gas-suspension in external electric field is considered. Process of generation of acoustic waves in the carrying medium at instantaneous inclusion of the external electric field setting a disperse phase in motion is described. Influence of the particles size and the volume content of a disperse phase on intensity of acoustic disturbance is considered.

1. Introduction

As a rule, the movement of a disperse phase is considered for the second time concerning the movement of the carrying medium and happens due to interphase power interaction in dynamics of multiphase mediums. At the same time, in some cases dynamic processes in the carrying medium are initiated by the disperse phase which is starting moving under the influence of gravitational, electric or magnetic forces. Mutual hydrodynamic interaction of phases is observed, for example, in processes of powder coloring, and also in technology of receiving the composite metal-polymeric powders consisting of the metal particles covered with a layer of polymer [1]. Such powders are used for receiving composite materials with a metal disperse matrix. At the same time disperse powders are delivered in the field of formation of composite mixture by a dusting in external electric field. The mathematical model considering not only the movement of particles in aerodynamic and electric fields, but also the return impact of moving particles on the carrying medium is necessary for a numerical research of similar processes. In this paper on the basis of the dynamics model of the charged gas-suspension in electric field generation of acoustic waves in the carrying medium owing to the movement of a disperse phase caused by electric field is numerically investigated.

2. The equations of the movement of the charged gas-suspension and method of their decision

Medium represents electrically charged gas-suspension which particles are under the influence of aerodynamic resistance force, buoyancy force, force of the attached masses and force operating from electric field which is created by the distributed gas-suspension charge. The system of the equations of dynamics of a two-speed two-temperature gas-suspension with high-speed and temperature sliding of phases [2, 3] is applied to the description of her movement. The system includes the equations of the movement of the carrying medium and a disperse phase [4, 5, 6].

3. Results of calculations and discussions

Let the gas-suspension with the set average density and volume contents be in the right half of the channel at L/2 in initial time point.

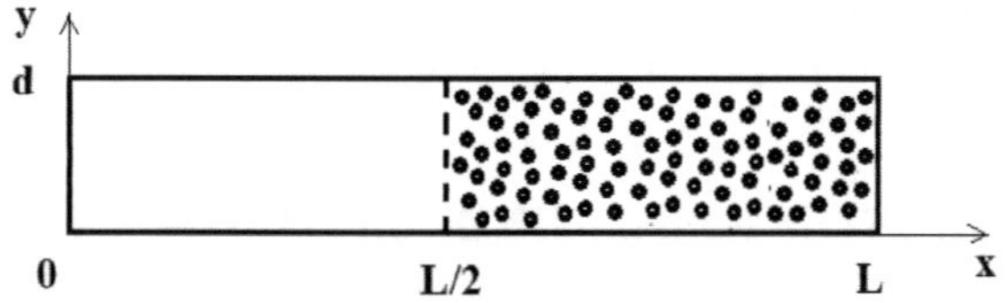

Figure 1. Distribution of particles in initial time point.

Physical density of particles material is ρ = 2700 kg/m^3, specific electric charge of a disperse phase comes to value q = 0.001 C/kg. Charged gas-suspension forms self-coordinated with distribution of density electric field under the influence of which there is a scattering of particles and the subsequent transition to an equilibrium state to uniform distribution of a gas-suspension in the channel. The movement of a disperse phase causes the movement of the carrying medium which nature depends on the size of particles and the volume content of a solid phase.

In figure 2 process of the wave front origin in gas owing to the initial shift of a disperse phase under the influence of the electromagnetic field is shown. In the carrying medium the compression wave propagating in the direction of the movement of a solid phase and the discharge wave moving in the opposite side is formed (figure 2, a). In the course of solid electrically charged particles transfer cocurrent flow of gas is formed (figure 2, b). At the same time the velocity of the movement of the carrying medium is less than the velocity of the movement of a disperse phase (figure 2, c). In the course of the gas-suspension movement there is a redistribution of average density of the disperse phase (figure 2, d).

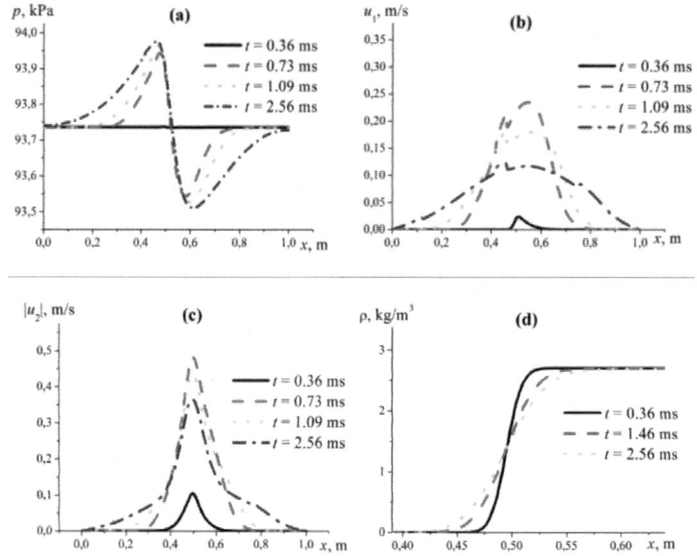

Figure 2. Formation of the acoustic field under the influence of a moving gas-suspension: (a) - distribution of pressure; (b) - distribution of a longitudinal component of gas velocity; (c) - distribution of the longitudinal velocity component of a disperse phase in various time points; (d) - distribution of average density of a disperse phase.

In figure 3, a, b - influence of the particles size on the acoustic disturbance intensity produced by the movement of a gas-suspension is shown. The volume content of a disperse phase was α = 0.001 in the calculations given below. Distributions received for the time point t = 2.56 ms of the carrying medium pressure show that the disperse phase consisting of larger particles forms acoustic disturbance of smaller intensity (figure 3, a). At that the form of small intensity pressure wave in gas (figure 3, b) is similar to a form of final intensity pressure wave (figure 3, a). The revealed regularity is explained by the fact that at the set volume content of a disperse phase, reduction of the size of particles leads to increase in the area of interphase interaction, to growth of integrated force of interphase friction and to fuller transfer of impulse of the disperse phase to the carrying medium. As a result intensity of pressure waves grows and the gas flow velocity increases.

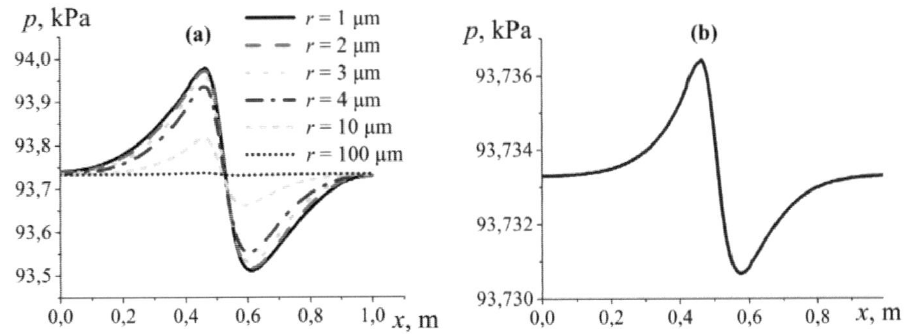

Figure 3. The intensity of the pressure wave produced in the carrying medium by the movement of a disperse phase at its initial volume contents $\alpha = 0.001$ in time point $t = 2.56$ ms: a - at various sizes of the charged gas-suspension particles; b - the distribution of pressure of the carrying medium produced by movement of a gas-suspension with the particles size of 100 microns.

In the figure 4, a gas flows velocities for the charged gas-suspensions with various sizes of particles are presented. From the distribution of velocities follows that the movement of the disperse fractions consisting of larger particles creates gas streams with a smaller velocity. So, particles with a radius of 100 microns form the flow with velocity of 0.006 m/s (figure 4, b) whereas at the same volume content and radii of particles of 1 micron, the velocity of the flow of gas increases up to 0.3 m/s. At the same time, velocities of particles with growth of radius increase (fig 4, c). Such regularity is explained by the fact that force of aerodynamic resistance of gas depends on the contact area of disperse inclusion and the carrying medium, and the Coulomb force setting particles in motion depends on the charge proportional to the volume of particles.

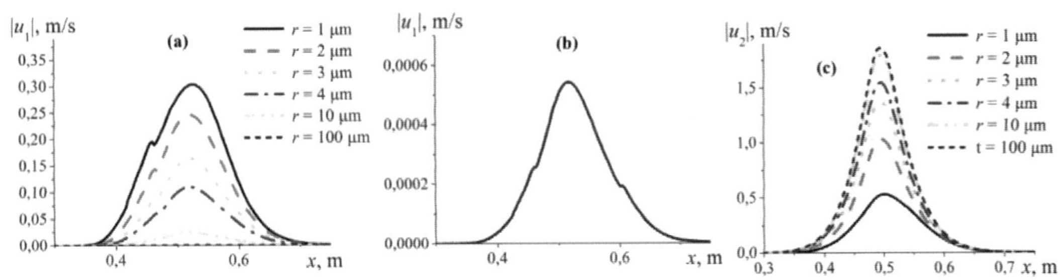

Figure 4. Velocity waves of the carrying and disperse phase: a - distributions of longitudinal velocity of gas in the waves created by the movement of gas-suspensions with various dispersion of particles; b - distribution of longitudinal velocity of gas in the wave created by moving gas-suspension with particles radius of 100 microns; c - distributions of longitudinal velocities of the disperse phase with various dispersion.

Thus, the analysis of the movement of the gas-suspension caused by movement in the self-coordinated electric field of the charged disperse phase shows that at the identical volume content of a solid phase, reduction of the size of particles increases intensity of acoustic disturbance, pressure differential in waves of compression and depression, and also the velocity of a cocurrent flow of gas.

At the same time the velocity of the charged disperse phase moving under the influence of the self-coordinated electric field with growth of the size of particles increases in the carrying medium that is connected with the fact that Coulomb force is volumetric in the assumption of particle charge proportionality to its weight and force of aerodynamic resistance is superficial.

Acknowledgments
The research was supported by RFBR and Tatarstan Republic government (Project No. 17-48-160359).

References

[1] Bogomolova O Yu and Danilaev M P 2016 Parameters of multiphase gas flows for encapsulation of submicron particles with polymer *The Scientific and Technical Volga region Bulletin* **3** 25-7
[2] Tukmakov A L 2014 Model of the movement and sedimentation of the loaded gas-suspension in electric field *Journal of Engineering Physics and Thermophysics* **87** 35-44
[3] Tukmakov A L 2015 Numerical model of the electro-gas-dynamics of a gas–particle system based on the equations of motion of a two-velocity two-temperature gas–particle mixture *Journal of Applied Mechanics and Technical Physics* **56** 112-20
[4] Tukmakov A L and Tukmakov D A 2017 Dynamics of the charged gas-suspension with initial spatially nonuniform distribution of average density of a disperse phase during the transition to an equilibrium state *High Temperature* **55** 509-12
[5] Nigmatullin R I 1978 *Fundamentals of the mechanics of heterogeneous media* Izdatel'stvo Nauka Moscow 336 pages
[6] Kutushev A G 2003 *Mathematical modeling of Wave Processes in Aerodisperse and Powder Media* Nedra St. Petersburg 283 pages

Interface layer formed in a low-temperature plasma of hydrocarbons on iron

A M Lyakhovich[1] and A A Shakov[2]

[1]Kazan Federal University, ul. Kremlevskaya, 18, Kazan, 420008, Russia,
[2]Physicotechnical Institute, Ural Branch, Russian Academy of Sciences, ul. Kirova 132, Izhevsk, 426001 Russia

Abstract. The method of uncovering the interphase layer in the systems "iron-nanosized polymer film" obtained in a low-temperature plasma on iron from hydrocarbons was used. A method of interface layer uncovering based on electrochemical dissolution of the iron substrate was proposed, which allowed one to achieve a high rate of metal. Ammonium sulfate solution was used as an electrolyte. A method of interface layer uncovering based on electrochemical dissolution of the iron substrate was proposed in the work. This allowed one to achieve a high rate of metal dissolution and make it possible to control the composition of the dissolution product. Salt solution was used as an electrolyte. Interface layer studies were carried out using the AFM and XPS methods, which allowed establishing the relief and chemical composition of the interphase boundary, as well as the location on the interphase surface of the metal-containing phase. The application of a double-layer iron substrate consisting of a plate of ^{56}Fe nonresonance isotope with a ^{57}Fe ultrafine layer deposited on its surface allowed one to use the Mössbauer spectroscopy for investigation of the chemical state of iron atoms in the interfacial region of "iron-nanoscale polymer film" systems.

1. Introduction.

The formation of nanoscale protective organic and polymer layers on metals to protect them from corrosion is an urgent problem. It is known that the protective effect of polymer coating is caused by both the isolating properties of the coating itself and the energy state of the interphase boundary of the "coating-substrate" system, which is especially important for nanoscale coatings. One method of coating thin protective organic layers is the plasma-chemical method. Earlier, we found [1, 2] that in the low-temperature plasma forming polymer coatings from hydrocarbons on iron, the chemical structure of the coating is due to the radical formed during the conversion of the hydrocarbon in the plasma. In this connection, the study of the chemical constitution of the interphase boundary and processes proceeding during its formation is quite important for understanding the action of the protective mechanism of such coatings.

The interfacial layer obtained by depositing a polymer coating from heptane on iron in a low-temperature low-pressure plasma was investigated.

2. Experimental

A plasma polymerization reactor had a working volume of 200 cm^3 and the external arrangement of the electrodes. Low-temperature plasma was excited by a radiofrequency (40.68 MHz) field. Polymer films were deposited in the dynamic mode onto an insulated metal (iron) mounted at the center of the reactor at equal distances from the electrodes. The heptane vapor pressure was 12 Pa, the discharge power was 40 W and plasma processing time 300 s.

The methods of investigations were used: X-ray photoelectron spectroscopy (XPS), atomic-force microscopy (AFM), and Mössbauer spectroscopy . XPS investigations were performed on an ES2401 spectrometer with a magnesium anode. The vacuum in analyzer chamber was 10^{-5} Pa. The spectrometer was calibrated with respect to the $Au4f_{7/2}$ peak at bond energy E_b equal 84.0 eV. Investigations by AFM methods were carried out on a Solver P47 microscope in a tapping mode. The Mössbauer investigations were performed at room temperature on a YaGRS-4M spectrometer operating in a regime of constant accelerations with a source of γ-radiation of ^{57}Co in a Cr matrix. The Mössbauer spectra were mathematical treated in a discrete representation with the use of a Levenberg–Marquardt algorithm by the damped least-squares method. Samples on the side of plasma-polymerized film were isolated by epoxy adhesive.As was shown by X-ray photoelectron spectra of the film surface, the residual air in the reactor had no effect on the chemical composition of the films at these parameters. The substrates from both ARMCO-iron and double-layer substrate from ^{56}Fe (^{57}Fe content is not more than 0.01%) and ^{57}Fe with isotopic purity of 95% served as samples for development of production techniques for investigation of the interfacial region in "iron–nanoscale polymer film" systems. In all cases, the iron content in the substrates was 99.9%. Before coating of film, the ARMCO- iron surface was subjected to electrochemical polishing with immediate washing by water, acetone, and petroleum ether. The double layer substrates were washed in the same way after ^{57}Fe deposition [3]. The substrates prepared in this way were immediately transferred into a reactor for plasma-treatment. The substrates were oxidized upon transfer into the reactor: according to XPS data, there were 5 at % of oxygen in the surface layer of iron substrates.

A method of interface layer uncovering based on electrochemical dissolution of the iron substrate was proposed, which allowed one to achieve a high rate of metal dissolution and make it possible to control the composition of the dissolution product. Ammonium sulfate solution was used as an electrolyte. To preserve the interfacial layer and properly nanoscale film, the sample on the side of film was sealed with epoxy resin. We gave up on application of an electrochemical cell in which the cathode and anode spaces are not separated. The natural mixing of anode and cathode space content leads to the formation of iron (II) hydroxide, the suspension of which fills the whole volume of the cell and settles on the sample surface.

Application of a wedge-shaped substrate (Figure 1) allowed one to completely dissolve iron and, thus, prepare the "metal-plasmapolymerized film" interface layer for tests. Hereafter, complete dissolution was achieved with the use of substrate with a square shape by progressive dipping of it into electrolyte during dissolution.

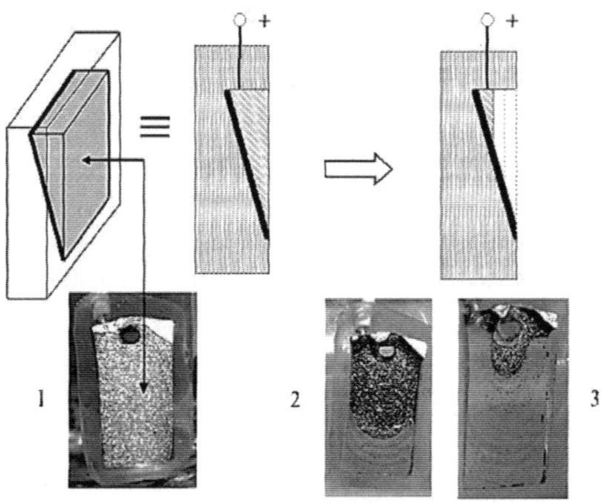

Figure 1. Dissolution scheme of wedge-shaped iron substrate and photographs of sample at different time of dissolution τ , h: (1) 0.5, (2) 4, and (3) 8.

3. Results and discussion

The investigation of the sample after the iron-substrate dissolution by the X-ray photoelectron spectroscopy showed the absence of iron atoms on the external side of sample (epoxy resin surface) (Figure 2a) that testifies the qualitative washing of sample after the substrate dissolution.

At the same time, the investigation of the interface layer that remained on film revealed the signal of Fe atoms (Figure 2b) forming part of the interface layer. The low intensity of the Fe2p-electron spectrum at a rather high level of noise did not allow one to decompose it correctly into components; however, the range of the bond energy maximum of 707...712 eV allows one to assume the presence of metallic iron or its carbide and iron in higher degrees of oxidation [4] in the analyzed layer, which is confirmed by the intense O1s- electron line. There are a low-intensity line with Eb = 283.5 eV (carbides) and line with Eb = 289.0 eV characteristic of metal carboxylates in the C1s- spectrum along with the main line with Eb = 285.0 eV [4].

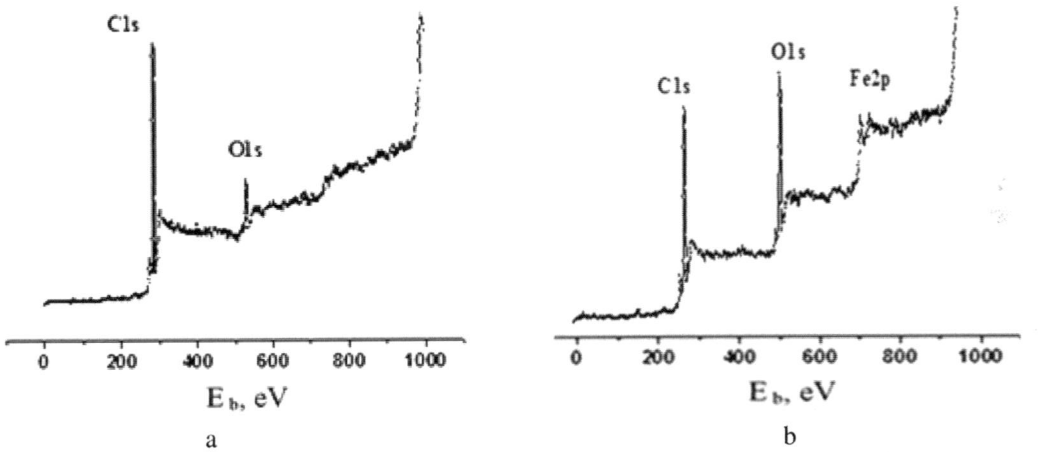

Figure 2. X-ray photoelectron spectra: (a) free side of auxiliary film; (b) interface layer.

Investigation of interface layer by atomic-force microscopy showed that the distribution of Fe-containing components is uniform, which is confirmed by a phase-contrast map (Figure 3).

To investigate the interface layer by the "volume" method of Mössbauer spectroscopy, double-layer substrates consisting of substrate of ^{56}Fe nonresonance isotope with a ^{57}Fe ultrafine layer where the polymer coating was deposited obtained on its surface [3]. This allowed one to increase the signal fraction of ^{57}Fe atoms of the interfacial layer. In this case, the chemical compositions of the interface layer obtained with the use of substrates from both ARMCO- iron and ^{57}Fe are identical.

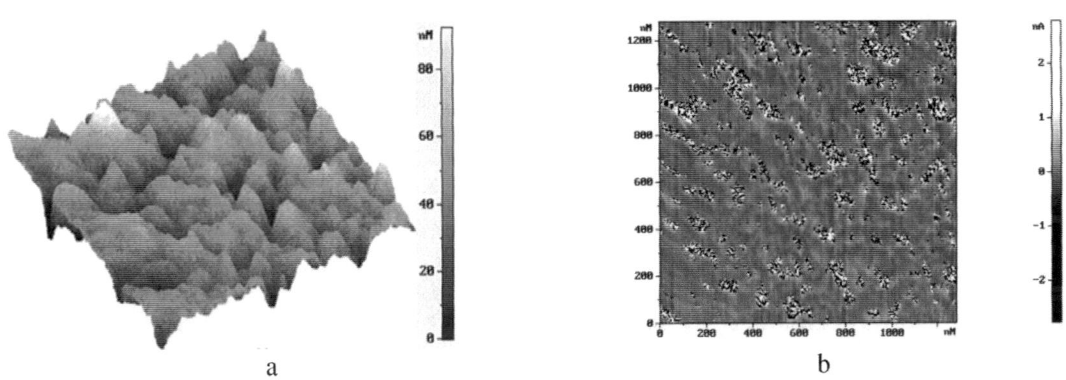

Figure 3. AFM images of interfacial region obtained after dissolution of iron substrate: (a) relief; (b) map of phase contrast.

Before the deposition of ^{57}Fe, the substrate of $^{56}Fe0$ representing a foil of 30 μm thickness was annealed in vacuum of 100 Pa at 700°C for 1 h. ^{57}Fe was deposited by an electrolytic method from $^{57}FeSO_4{\cdot}(NH_4)_2SO_4$ solution. $^{57}FeSO_4{\cdot}(NH_4)_2SO_4$ solution was obtained at anode dissolution of ^{57}Fe plate in $(NH_4)_2SO_4$ at room temperature. The electrolytic deposition of ^{57}Fe on the substrate was performed under conditions of natural aeration (pH 2.5–3.5) at room temperature.

Mathematical treatment of the Mössbauer spectra of the samples under investigation showed that the spectra consist of two sextets (Figure 4) corresponding to local atomic configuration of Fe with the parameters characteristic for the α-Fe phase (intensity of ultrafine magnetic field H = 330 kOe) and Fe carbide phase (H = 202 kOe) [5] with a phase ratio of 0.86 : 0.14. The signals of α-FeOOH, Fe_3O_4, γ-Fe_2O_3, and α-Fe_2O_3 iron compounds with H parameters equal to 356, 454, 503, and 511 kOe, respectively [6], can be also revealed in the Mössbauer spectra, but the intensities of these lines are negligibly small.

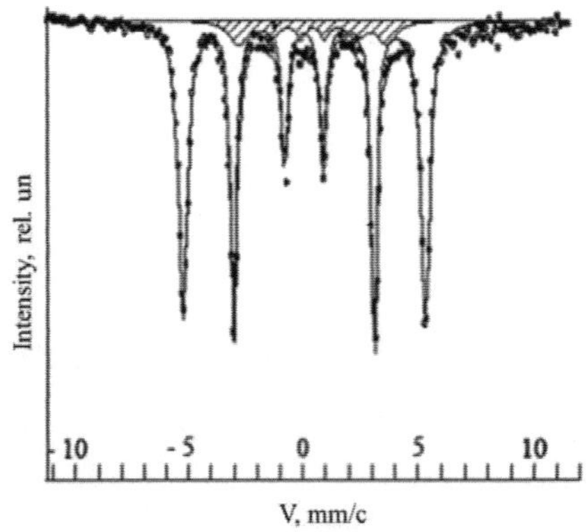

Figure 4. Mössbauer spectrum of iron sample with ^{57}Fe layer and film deposited from heptane in plasma. Sextet of lines with greater intensity is referred to iron and that with low intensity to iron carbide.

4. Conclusions

The investigation of an "iron-nanoscale polymer film" system obtained in low-temperature plasma of hydrocarbons showed that the interfacel layer consists of iron carbides and oxygen- and carbon-containing compounds of iron. The efficiency of the combination of the techniques proposed for investigation of the interface layer providing more complete information on the layer chemical composition was shown.

Acknowledgements

The results were obtained as part of the state task of the Ministry of Education and Science of Russia 11.1300.2017/4.6 and of the Federal Agency of scientific organizations of Russia No AAAA-A16-116021010085-9.

References

[1] Lyakhovich A M, Lyalina N V and Reshetnikov S M 2010 *Protection of Metals and Physical Chemistry of Surfaces* **46** (6) 666
[2] Lyakhovich A M 2014 *High Energy Chemistry* **48** (1) 54
[3] Lyakhovich A M, Shakov A A, Lyalina N V and Elsukov E P 2014 *Protection of Metals and Physical Chemistry of Surfaces* **50** (7) 884
[4] Nefedov V I *X-Ray Electron Spectroscopy of Chemical Compounds* (Moscow: Khimiya) 1984 p 160
[5] Maksimov Yu V, Suzdalev I P and Arents R A 1972 *Solid state physics* **14** 3344
[6] Sej J O, Cook D C and Townsend H E 1998 *Hyperfine Interact.* **112** (1) 59

IOP Conf. Series: Journal of Physics: Conf. Series **927** (2017) 012085 doi:10.1088/1742-6596/927/1/012085

Investigations of the growth of the vapor-air shell of a gas discharge with a liquid electrolytic cathode of sodium hydroxide solution

R N Kashapov[1,2], L N Kashapov[1] and N F Kashapov[1]

[1] Engineering Institute, Kazan Federal University, Kazan, 420008, Russia
[2] Laboratory of Radiation Physics, Kazan Physical-Technical Institute of the Kazan Scientific Center of the Russian Academy of Sciences, Kazan, 420029, Russia

E-mail: kashramiln@gmail.com

Abstract. Gas discharges with liquid electrodes are widely used in cleaning, polishing, hardening, forming a surface microrelief with specified parameters. Burning gas discharge in a number of cases occurs in the vapor-air shell, with film boiling of the electrolyte. The properties of the vapor-air shell strongly influence on the physical processes of the discharge. Investigation of the growth mechanism of the vapor-air shell and the determination of the main factors of influence are relevant for the optimization of applied processes. The structure of the discharge in the vapor-air shell is determined, and the presence of the discharge leader structure is established. Anomalous growth of the vapor-air shell of a gas discharge with a liquid electrolytic cathode was detected.

1. Introduction

The low-temperature plasma produced in gas discharges with liquid electrodes is increasingly used in the field of thermal and chemical treatment of alloys [1], the formation of surface microrelief [2], the formation of functional oxide coatings, the production of metallic micro- and nanoparticles [3], the activation of chemical reactions in polymer solutions [4], generation of shock waves [5], etc. Under the terms of the initiation of discharges in a liquid, it decided to allocate the following varieties: initiating discharge inside the liquid electrode, the discharge in the gas phase with the liquid electrode, the discharge inside of gas bubbles in the liquid [6]. This process is generally called plasma electrolysis, and depending on the polarity of the metal electrode, add the "anodic" or "cathodic" effect to the name, and in the case of an immersed electrode, call the contact glow discharge electrolysis (CGDE). However, we can't call this glow discharge, since the nature and mechanism of charge transfer from the electrolyte to the plasma are not good studied.

From the point of view of application, the discharge in the electrolyte is interesting by the possibility of generating active radicals ($^{\cdot}OH$, H_2O_2, etc.) used to destroy phenols [7] and organic dyes [8]. In this case, the discharge, depending on the shape of the applied voltage, burns either in the vapor-air shell formed around the metal electrode or in the form of separate microdischarges on the surface of the electrode. Researchers Sengupta and Singh [9] found that the discharge ignition voltage in the aqueous solutions of electrolytes on the anode is 420 V, and at the cathode 160 V. At the same time, the voltage value does not depend on the composition, concentration, temperature and surface tension. This controversial statement, which requires additional research. It is considered to be the reason for the formation of the vapor-air shell overheating of the metal electrode and the appearance of film boiling. The structure of the vapor-air shell remains unexplained, just as the discharge in it is burning. When the discharge burns between a metal anode and an electrolytic cathode, a vapor-air shell 2-3 mm thick is formed. It is of interest to investigate the mechanism of the formation of a vapor-air shell under conditions different from the standard thermal mechanism, when film boiling is observed. The aim of the work was to study the mechanism of the appearance of a vapor-air shell around the metal anode in a solution of sodium hydroxide.

2. Main part

Figure 1 shows the functional diagram of the experimental setup, which consists of an electric power supply system 1, an electrolytic bath - 2, an electrode system - 3, an oscilloscope - 4, an additional resistance - 5, a voltmeter - 6, an ammeter - 7, a thermocouple - 8. Using the electrode system, the distance between the anode and the electrolyte solution was monitored. With the help of an oscilloscope 4, the shape of the applied voltage and current was monitored, the voltmeter and ammeter were used to measure the voltage and discharge current. Two power sources were used in the work: the first is a DC power source with continuously variable voltage, provides regulated rectified voltage of various shapes, consists of a diode bridge (diodes SD 246) and a laboratory autotransformer of adjustment type 1M with a voltage range from 1 to 300V depending on the experimental conditions, a smoothing capacitive filter is connected (C = 1560 μF)). Measurement of voltage and discharge current was carried out using two digital universal measuring devices APPA 305 and APPA 109N, the relative error of measurement is 0.8%.

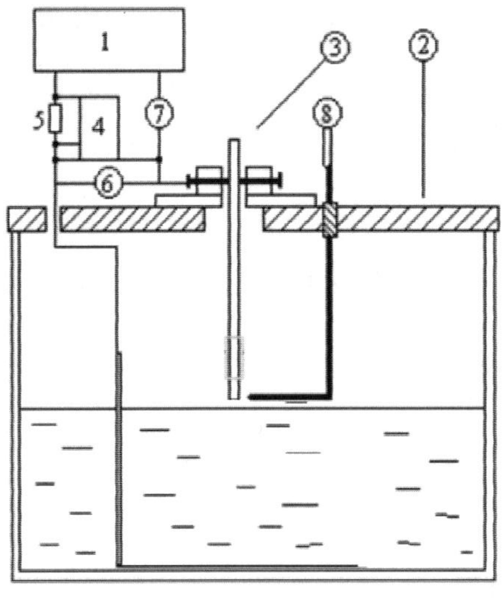

Figure 1. The scheme of the experimental setup.

The combustion of a gas discharge occurs between a metal electrode made of copper and a liquid electrolyte. As an electrolytic cathode, aqueous solutions of sodium hydroxide with a concentration of 1 to 5% by weight were used. In the experiments, a copper anode and a cathode of 0.65 cm^2 and 34 cm^2, respectively, were used. The electrode system 3 allows, by rotating the side screw, to vertically and horizontally move the anode electrode to the surface of the electrolyte or the wall of the discharge chamber. In the experiments, two positions were used, when the anode was immersed in the electrolyte and was above it. Figure 2 shows images of discharge burning on a metal anode in a vapor-air shell. Figure A shows a uniform luminescence of the entire surface of the electrode, the intensity of the glow decreases from the bottom up. At the initial moment of formation of the vapor-air shell, the intensity of the glow is minimal and increases with increasing applied voltage. With the formation of a steam-air shell, there is a sharp drop in current from 8 A to 1 A and an increase in voltage by 15-20 V. When the electrode was moved horizontally to the surface of the observation window, the presence of a lead structure of the discharge was detected (Figure 2B). There are many discharge channels starting from the tip of the metal anode and propagating along the surface of the electrolytic cathode. In this case, intensive vaporization of the electrolyte is observed. This allows us to say that the discharge in the vapor-air shell burns between the tip of the metal electrode and the electrolyte of the vapor-gas shell. The discharge channels run parallel to the surface of the metal electrode over the surface of the liquid cathode, and not between the surfaces of the electrode and the electrolyte.

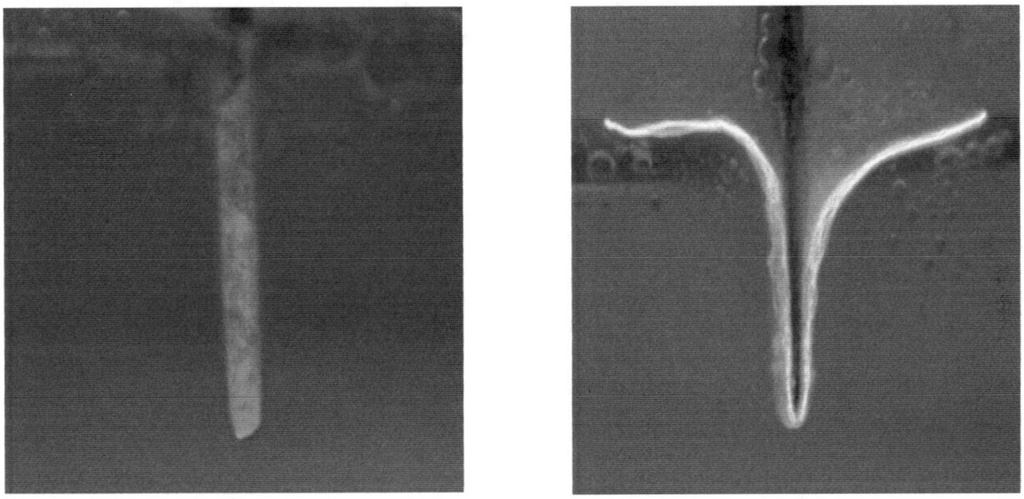

A B

Figure 2. The image of the steam-air shell in the electrolyte. A – the electrode is surrounded by electrolyte. B – section of vapor-air shell.

In the second case, when the metal anode was vertically immersed in the electrolyte, the formation of the discharge leader structure was observed at the initial moment of contact (Figure 3A). At a voltage of 80-100 V, electrolyte accumulation can occur on the surface of the electrode above the combustion zone of the discharge. The electrolyte is accumulated by spattering the electrolytic cathode with a discharge. At voltages above 150 V, the electrolyte does not become hot, since the evaporation process occurs due to the anode heating. Figure 3B shows the increase in the size of the accumulated electrolyte above the discharge. Further immersion of the metal anode results in a displacement behind the tip of the electrode of the discharge leader structure. In this case, intensive foaming of the electrolyte over the discharge zone is observed (Figure 3C).

Continued immersion of the electrode leads to a situation where discharge channels propagate from the tip of the anode between the main electrolyte layer and the expanded electrolyte located above the discharge zone. In this case, one can observe a clear glow boundary separating the foamed vapor-air shell. Gradually, the foamed shell grows, its rotation around the metal anode is observed, but the center remains stationary and attached to the tip of the anode. Figure 4 shows the growth pattern of the shell. It consists of the following stages: A - splashing and formation of an electrolyte droplet, B - further immersion of the anode, foaming and dripping of the drop to the tip of the electrode into the discharge zone, C - formation of a foamed shell with the center of the discharge burning at the tip of the anode, D- growth of the foamed-vapor-air shell.

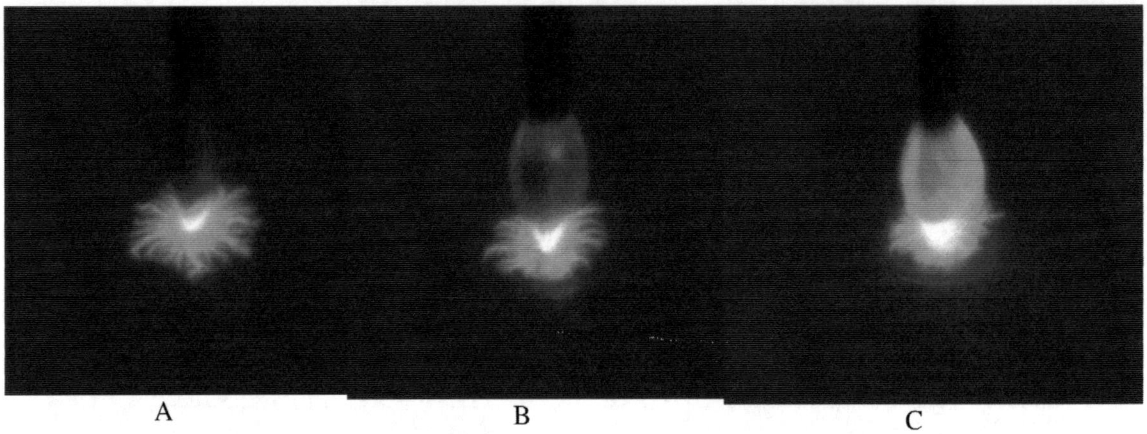

Figure 3. Formation of the vapor-air shell during smooth immersion of the electrode.

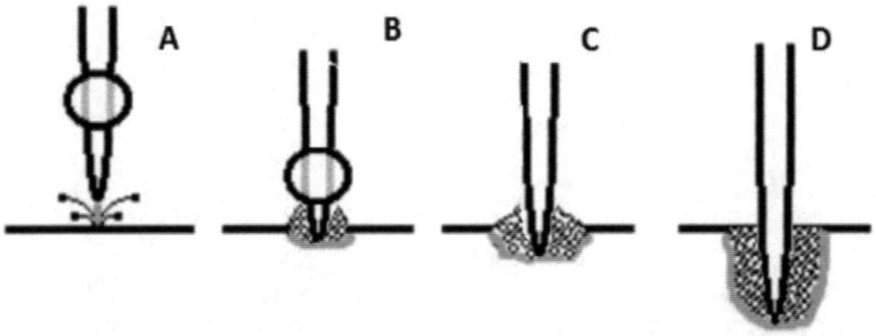

Figure 4. Schematic diagram of anomalous growth of a vapor-air shell in NaOH solution.

It is established that for the plasma electrolyte process in NaOH solutions with a smoothed voltage obtained after full-wave rectification and using a 450 μF capacitive filter, a vapor-air envelope is observed: for 1% solution at 97 V, 3% solution for 93 V, 5% solution - 81 V. The "foamed" shell gradually grows and fills the entire space between the metal electrodes. The decisive factor is the applied voltage.

3. Conclusions

The structure of a gas discharge with an electrolytic cathode inside a vapor-air shell is determined. The discharge has a multitude of channels emanating from the tip of the anode and propagating along the surface of the electrolyte.

The leader structure of the discharge is determined by slow immersion of the anode in the electrolyte. The conditions for the anomalous growth of the steam-air shell were determined: for a 1% solution at 97 V, a 3% solution for 93 V, and a 5% solution for 81 V. The discharge was initiated when the metal anode was in contact with the electrolyte surface.

Acknowledgments

This work was financed by grant MK-1861.2017.8 allocated by the Council on grants to the President of the Russian Federation for state support of young Russian scientists and for state support of the leading scientific schools of the Russian Federation.

This work was funded by the subsidy allocated to Kazan Federal University for the state assignment in the sphere of scientific activities (11.1300.2017/4.6).

This work was financed within the framework of state support of leading universities in order to increase their competitiveness among the world's leading scientific and educational centers.

References

[1] Kashapov L., Kashapov N. and Kashapov R. 2013 *Journal of Physics : Conference Series* **479** 012011
[2] Kashapov L., Kashapov N. and Kashapov R. 2013 *Journal of Physics : Conference Series* **479** 012003
[3] Kashapov R., Kashapov L. and Kashapov N. 2017 *IOP Conference Series: Materials Science and Engineering* **240** 012071
[4] Chen Q., Saito K., Takemura Y. and Shirai H. 2008 *Thin Solid Films* **516** 6688-6693
[5] Denisov D., Kashapov N. and Kashapov R. 2015 *IOP Conference Series: Materials Science and Engineering* **86** 012005
[6] Bruggeman P. and Leys C. 2009 *Journal of Physics D: Applied physics* **42** 053001
[7] Tezuka M. and Iwasaki M. 1998 *Thin solid films* **316** 123
[8] Wang L. and Hazard J. 2009 *Mater.* **171** 577
[9] Sengupta S., Singh O. 1991 *J. Electroanal. Chem.* **301** 189

Research of plasma-electrolyte discharge in the processes of obtaining metallic powders

R N Kashapov[1,2], L N Kashapov[1] and N F Kashapov[1]

[1] Engineering Institute, Kazan Federal University, Kazan, 420008, Russia
[2] Laboratory of Radiation Physics, Kazan Physical-Technical Institute of the Kazan Scientific Center of the Russian Academy of Sciences, Kazan, 420029, Russia

E-mail: kashramiln@gmail.com

Abstract. The use of the plasma electrolyte process has never been considered as a simple, cheap and fast method of obtaining powders used in selective laser melting processes. Therefore, the adaptation of the plasma-electrolyte process to the production of metal powders used in additive production is an urgent task. The paper presents the results of studies of gas discharge parameters between a metal and liquid electrode in the processes of obtaining metallic iron powders. The discharge combustion conditions necessary for the formation of metal powders of micron size are determined. A possible mechanism for the formation of powder particles in a discharge plasma is proposed.

1. Introduction

The phenomenon of electric discharge between the metal and liquid electrodes has been of interest to researchers since the end of the XIX century. Interest was of a fundamental and applied nature. From a fundamental point of view, the case of burning a discharge between a metal anode and an electrolytic cathode is interesting from the point of view of studying charge transfer from an electrolyte solution to plasma. Even more interesting is the study of the mechanism for converting the electrical energy of a discharge into mechanical energy with the appearance of shock waves. From a practical point of view, it is possible to identify many areas: the production of oxide coatings, the purification of sewage, the micro-dimension processing of non-conducting objects, the formation of the microrelief of the surface of metals, the cleaning and modification of the surface (cyanidation, nitration), the heat treatment and hardening with increasing wear resistance, the preparation of nanostructures and particles, creation of plasma torches and new types of plasma chemical reactors.

Studies of plasma-electrolyte discharge on the thermal treatment of metal objects and on the production of metal nanoparticles show the possibility of obtaining metallic powders. The search for new methods for obtaining metallic powders used in selective laser melting (SLM) is an urgent task. Selective laser melting refers to one of the types of additive technology, when metal powder is applied layer by layer on the platform and its selective melting is performed with a laser [1]. This technology allows you to obtain complex geometric products, which previously couldn't be produced by standard production methods. To the powders used in SLM, a number of requirements are established: to have a certain dispersed composition, the particles must be spherical, there must be no gas pores in the powder. [2]. Unfortunately, modern technologies for producing powders, namely gas, water, plasma with a rotating electrode, plasma in crossed flows and HF discharge of atomization, have a number of disadvantages:

1) High scatter of particles from 1 to 200 μm. This requires an additional sieving procedure.
2) The difficulty of transferring the plant to the production of another type of material.
3) High energy consumption.
4) Use of expensive equipment.

In the case of gas atomization, the sputtering of molten metal occurs in a supersonic jet of an inert gas. Spraying is carried out in a liquid stream during water atomization. In the third case, an arc discharge is ignited between the electrodes, and the electrode is rotated around its axis, which leads to the spattering of the metal. In the fourth case, the metal rod is sprayed in crossed plasma streams. In the fifth case, the pre-ground metal particles are passed through the installation with a high-frequency induction or capacitive discharge, and as a result they are shaped into spheroids.

The idea of additive production is the rapid production of prototypes of products from the required materials [3]. Therefore, the development of a new cheap method for obtaining metallic powders, which is devoid of all the above-mentioned drawbacks and is capable of rapid replacement of metallic material, is an urgent task. One solution is to use a gas discharge with liquid electrodes [4,5,6], this method is simple and does not require expensive equipment.

To create a technology, it is necessary to investigate the parameters of a gas discharge during the formation of metallic particles and to determine how the parameters of the gas discharge plasma affect the physical phenomena of the formation of a part. Therefore, the aim of the work was to study the plasma-electrolyte discharge and the mechanism of formation of metallic particles in plasma.

2. Main part

Burning discharge can occur both on the metal anode and the cathode, while the electrode can be immersed in the liquid, and may be above it. The combination of all these conditions with the variation in the shape of the current and the magnitude of the applied voltage gives a variety of options for searching for micron-size powder production regimes. Our starting point for the studies was a survey paper [7], which describes the main methods for obtaining metal nanoparticles and their oxides. Based on this review, a list of possible research options was developed, the description of which is given in Table 1. It is evident from the table that for the type of experiment No. 1 the active metal electrode was the anode that was immersed in the electrolyte, and the voltage obtained after the smoothing capacitive filter. In total eight types of experiments were singled out.

Table 1. Explanation of the types of experiment.

Experiment Type Number	The electrode is immersed in the electrolyte	Electrode over electrolyte	Active electrode ANOD	Active electrode CATOD	Voltage form: smoothened	Form of voltage: pulsating
1	+		+		+	
2	+			+	+	
3	+		+			+
4	+			+		+
5		+	+		+	
6		+	+			+
7		+		+	+	
8		+		+		+

Figure 1 shows the functional diagram of the experimental setup, which consists of an electric power supply system 1, an electrolytic bath 2, an electrode system - 3, an oscilloscope - 4, an additional resistance - 5, a voltmeter - 6, an ammeter - 7, a thermocouple - 8.Using the electrode system, the distance between the anode and the electrolyte solution was monitored. With the help of an oscilloscope 4, the shape of the applied voltage and current was monitored, the voltmeter and ammeter were used to measure the voltage and discharge current. Two power sources were used in the work: the first one is a DC power source with continuously variable voltage, provides regulated rectified voltage of various shapes, consists of a diode bridge (diodes SD 246) and a laboratory autotransformer of regulation type 1M with a voltage range from 1 to 250V (to the power supply, depending on the experimental conditions, a smoothing capacitive filter is connected (C = 1560 μF)), the second is a high-voltage direct current source with a smooth adjustment of the output voltage in the range from 0.4 to 3 kV and current 0 to 10 A. Measurement of voltage and discharge current was carried out using two digital universal measuring devices APPA 305 and APPA 109N, the relative error of measurement is 0.8%.

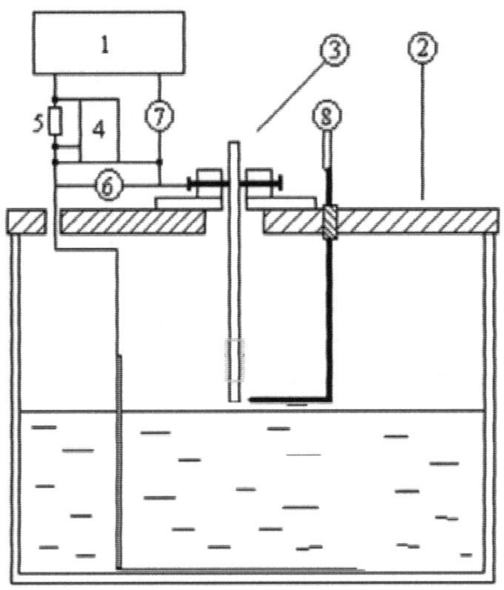

Figure 1. Diagram of the experimental setup.

Burning gas discharge occurs between a metal electrode made of steel grade 07Х16Н4Д4Б-Ш or from copper and liquid electrolyte. As the electrolytic cathode, aqueous solutions of NaCl and K_2CO_3 were used with a concentration of 0.1-1% by weight. The resulting powder was washed with deionized water and dried in a drying oven. The morphology of the powder was studied using a scanning electron microscope Carl Zeiss EVO 50. The dispersion composition was determined by the sieve method with a set of sieves from 10 to 300 μm. The average sieving time was 30 minutes. Analysis of the chemical composition was determined with the aid of a scanning electron microscope attachment of an energy-dispersive electron-probe spectrometer Oxford Instruments Inca X-act.

Consider the group of experiments under the numbers 1,2,3,4. It is established that in the case of using an aqueous solution of NaCl under anodic polarization and a pulsating form of stress, the electrochemical dissolution process predominates without forming a metallic powder. In experiment No. 1, the formation of a vapor-air shell is observed, the process proceeds stationary, but the power input is insufficient to melt the electrode surface and form a powder. In cases Nos. 2 and 4, the formation of an ultradisperse metallic powder was observed. The formation of a metal powder is observed during anodic and cathodic polarization of the electrode in experiments 5, 6, 7, 8. However, the use of a pulsating voltage does not allow maintaining discharge combustion at inter-electrode distances greater than 0.5 mm. In the case of using a metal cathode, intense heat generation occurs, which leads to uncontrolled melting of the electrode without forming a powder. Thus, the most appropriate options for the production of metallic powders are experiments No. 2 and No. 5.

The current-voltage characteristic of experiment No. 2 is shown in Figure 2a, the active electrode was a copper cathode. At the initial moment, the process of classical electrolysis proceeds and a linear increase in the current intensity occurs with increasing voltage. The formation of a vapor-air shell occurs at voltages of 60-70 V, which surrounds the metal electrode. With increasing voltage, the area of the discharge burns on the surface of the electrode, and then the intensity of the glow. The formation of a copper metallic powder is observed at voltages above 130 V.

Studies of the combustion of a gas discharge between a metal electrode 07Х16Н4Д4Б-Ш and liquid electrodes have shown the possibility of obtaining a spherical powder with the dispersion of particles from 0.5 to 40 μm. In the voltage range from 500 to 800 V, particles smaller than 40 μm in size with a productivity of 1 • 10-2 g / s occur. Above this range, the electrode melts and the process for obtaining the powder ceases. When certain values of current and voltage are reached, the process of spraying the metal anode is observed, most of the powder enters the electrolytic cathode and crystallizes. In parallel, there is a process of evaporation of the liquid electrode.

The main parameters of plasma-electrolyte production of steel powder are voltage, current, discharge power, used for heat generation, physical and chemical properties of the metal anode. The power put into the discharge is determined by the current-voltage characteristic, on the basis of which it is possible to estimate the energy contribution. Figure 2b shows the current-voltage characteristic of the plasma-electrolyte process for three interelectrode distances.

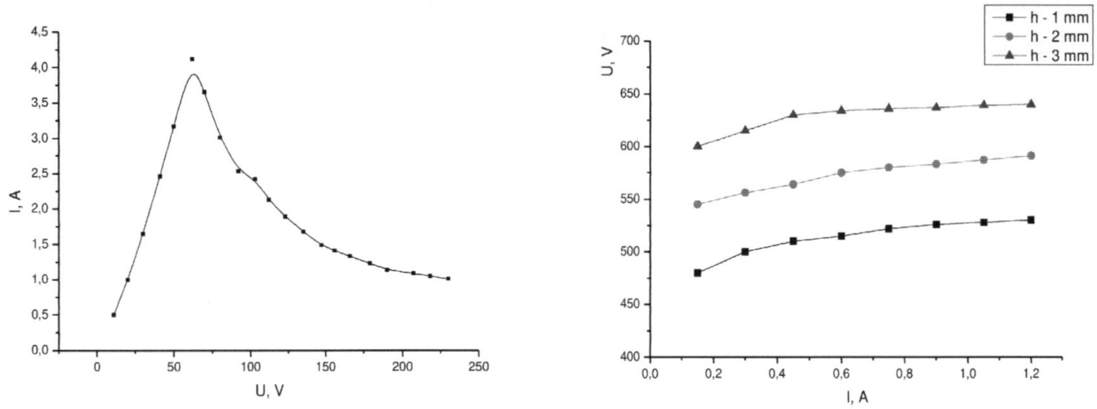

a b

Figure 2. IV characteristics of the plasma-electrolyte process.

The significant effect on the discharge combustion process is exerted by the electrolyte concentration, it is established that the electric field strength decreases with increasing concentration. Therefore, for the stability of the results, solutions with a concentration of less than 1% were used.

The formation of the powder occurs when the surface temperature of the anode reaches the corresponding solidus temperature. When the liquidus temperature is reached, the electrode is melted and large drops of metal are formed. Proceeding from this, the powder can be produced in the temperature range of the anode - liquidus and solidus. An increase in temperature will lead to an increase in the productivity of the powder, but with an increase in the particle size of the powder. The productivity of the powder increases nonlinearly with increasing temperature. A linear increase in the average particle size with increasing anode temperature is established. The largest amount of powder is obtained by a size smaller than 40 μm. Figure 3 shows SEM images of the resulting powders. The powder has a spherical shape, the smallest particle size is 0.5 μm. The resulting powder is suitable for use in selective laser melting installations. However, further research is needed to increase the productivity of the process by increasing the anode area and determining the dynamic characteristics of the powder.

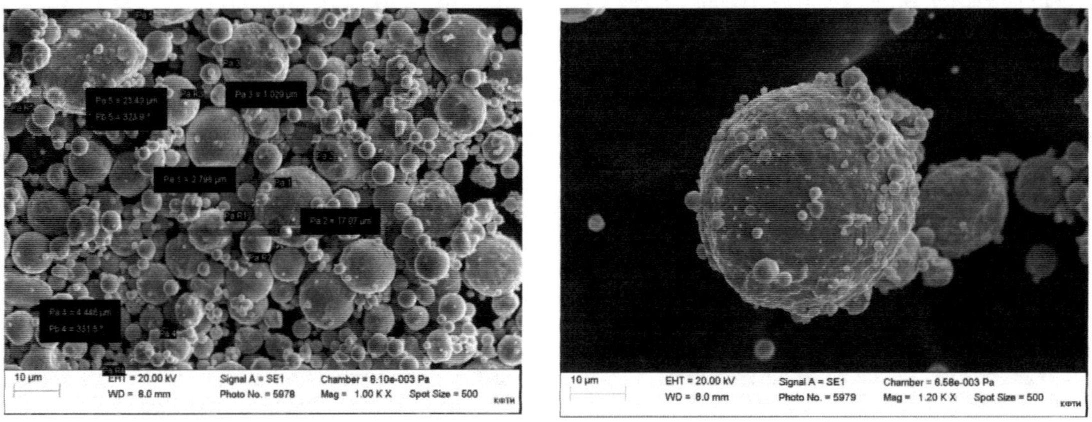

a б

Figure 3. SEM images of steel powder obtained in the plasma-electrolyte process.

The formation of a metal powder occurs as a result of the fusion of its surface with plasma of gas discharge. The main factors influencing the production of powder are the power of the gas, the amount of thermal energy supplied to the surface of the electrode, and the energy of the ions bombarding the surface of the cathode, the amount of heat withdrawn from the system through the electrode. By changing all of the above factors, it is possible to control the rate of fusion of the electrode surface and the size of the resulting metallic particles.

3. Results

Studies of the combustion of a gas discharge between a submerged metallic copper cathode and a liquid anode have shown the possibility of obtaining an ultradisperse powder with a particle dispersion of 0.5 to 2 μm. It is established that with increasing time of burning of the discharge an increase in the dimensions of the powder particles is observed.

Studies of the combustion of a gas discharge between the metal anode 07Х16Н4Д4Б-Ш and a liquid cathode have shown the possibility of obtaining a spherical powder with the dispersion of particles from 0.5 to 40 μm. The analysis of the obtained SEM photographs shows similarity with the powder obtained by the gas atomization method. The influence of the gas discharge parameters on the productivity of the process and the size of the particles obtained is determined. In the voltage range from 500 to 800 V, particles smaller than 40 μm in size with a productivity of $1 \cdot 10^{-2}$ g / s occur. Above this range, the electrode melts and the process for obtaining the powder ceases. The location of the metal electrode affects on the mechanism of particle formation and on the physicochemical processes occurring in the discharge.

The parameters of the plasma-electrolyte discharge at which the powder particles are formed are determined.

Acknowledgments

This work was financed by grant MK-1861.2017.8 allocated by the Council on grants to the President of the Russian Federation for state support of young Russian scientists and for state support of the leading scientific schools of the Russian Federation.

This work was funded by the subsidy allocated to Kazan Federal University for the state assignment in the sphere of scientific activities (11.1300.2017/4.6).

This work was financed within the framework of state support of leading universities in order to increase their competitiveness among the world's leading scientific and educational centers.

References

[1] 2017 *Journal of Alloys and Compounds* **691** 316-322
[2] 2017 *International Journal of Heat and Mass Transfer* **104** 665-674
[3] 2017 *Scripta Materialia* **126** 41-44
[4] Kashapov L., Kashapov N. and Kashapov R. 2013 *Journal of Physics : Conference Series* **479** 012011
[5] Denisov D., Kashapov N. and Kashapov R. 2015 *IOP Conference Series: Materials Science and Engineering* **86** 012005
[6] Kashapov L., Kashapov N., Kashapov R. and Denisov D. 2016 *Journal of Physics: Conference Series.* **669** 012029
[7] Kareem T. Abdul and Kaliami A.Anu 2012 *Ionics* **18** 315-327